Preface

In 1973, there existed substantial material in EMC literature covering the subject of electromagnetic shielding, but many of these documents appeared confusing to an individual who was only recently introduced to EMI/EMC. Other sources were simply not organized for use by a design engineer. To fill this void, we then published our first book on the subject, *Electromagnetic Shielding: Materials and Performances.*

Twelve years later, the literature was replete with data on the shielding effectiveness (SE) of metals, conductive paints and related test data. Yet there existed little information on how to orchestrate all this information to obtain a given method of building a given attenuation for a box. So, along with the making of the first comnputer-aided design (CAD) software of this kind (ICT Program #5500), we issued a book describing an orderly approach to define and build a shielded equipment or enclosure, *Shielding Design Methodology and Procedures.*

With the constant evolution of materials, technologies, test specifications and digital and analog processing speeds, there now is a need to reconcile shielding materials information and packaging methodology into a single handbook. This new handbook, gathering much of the contents of the two previously mentioned ones plus much more, is designed to lead the prospective designer, in a step-by-step process, from the understanding of elementary field and shielding theory, through the properties of all available materials and hardware, to the final completion of an equipment or system housing which meets the required EMC performance criteria.

The first part (Chapters 1 through 3) is devoted to shielding theory and material shielding performances. Regarding the impedance of E-fields, H-fields or plane waves, literature and manufacturers' data are often quite misleading. For example, since the wave and circuit impedance which produced the field are interlocked, and since a circuit impedance is not infinite, E-field SE data are generally optimistic. Conversely, H-field SE data are pessimistic (too low) since a magnetic source circuit impedance is not zero. Chapter 1 clarifies and quantifies these points.

Another example of possibly misleading information is the use of MIL-STD-285 to measure and report the SE of test items to E- and H-fields—misleading because the reference test distance per MIL-STD-285 is 1 ft (0.3 m). This issue is addressed in Chapter 3, and several testing alternatives are described.

The discussions and design data on SE in this portion are not restricted to homogeneous metals. In fact, no real-life and useful shielded compartment, box, cabinet or room is homogeneous since, many penetrations of a six-sided shield configuration are usually necessary. Shielding performance of nonhomogeneous metals are also discussed in this handbook. Some examples are shields made from metal deposition and flame-spray processes. Shields made of small-aperture metals, screens, wire meshes, cable braids and metalized textiles are discussed herein with design data.

The second part of this book explains how to orchestrate the various material properties, the characteristics of the field to be attenuated and the packaging constraints to obtain an equipment with satisfactory EMC performance. This second part explains the general procedure, its building blocks and how they are put together to determine how the SE objectives, the various iterations and the hardening are controlled. Then a rigorous technique for developing the required SE ("How much shielding is needed?") is explained, based on defining the EMI ambient threats and the victim circuit's response.

This goal having been set, the next chapters (7 through 10) explain how to select the optimum material for the box skin, whether it is metal, metalized plastic, loaded composite or whatever. A very important part of this relates to the prediction and treatment of all aperture leakages (displays, cooling slots, seams, etc.). A seldom mentioned aspect, viz. shielding against emission sources inside the box, is also given wide coverage, including a prospective of future ultra high-speed processing with clock rates exceeding 100 MHz.

A Handbook Series on
Electromagnetic Interference and
Compatibility

Volume 3

Electromagnetic Shielding

Donald R.J. White
and
Michel Mardiguian

Interference Control Technologies, Inc.
Gainesville, Virginia

Interference Control Technologies, Inc.
Route 625, Gainesville, VA 22065
TEL: (703) 347-0030 FAX: (703) 347-5813

Published 1988
Printed in the United States of America
95 94 93 92 91 90 89 88 5 4 3 2 1

Library of Congress Catalog Card Number: 88-80529
ISBN: 0-944916-03-1

The equipment design part of this book is followed by an important aspect, **cable shielding**, which is covered in Chapter 11. It is vital to remember that a cable SE cannot be dissociated from the entire system shielding performance.

Finally, abundant data on shielded rooms, building SE and vehicle SE are provided in Chapters 12 and 13. Appendices A through H contain many SE design graphs for several metals whose thicknesses range from 0.0001 mil (2.54 nm) to 1 inch (2.56 cm). For both near-field and far-field calculations and associated frequencies, the design graphs cover source-to-metal distances ranging from 10 cm to 10 km. Frequency coverage is from 10 Hz to 30 GHz.

The methodology and procedures described in this handbook have been incorporated into the shielding portion of Interference Control Technologies' intensive training course titled "Grounding and Shielding." They have also been implemented in CAD Program #5500, which is available for IBM and IBM-compatible computers. We invite and welcome the users of this book to submit comments, questions and critiques of the contents.

Donald R.J. White
Interference Control
Technologies, Inc.
Gainesville, Virginia

Michel Mardiguian
Interference Control
Technologies, Inc.
Paris, France

Other Books in the 12-Volume Series

VOLUME 1
Fundamentals of Electromagnetic Compatibility

VOLUME 2
Grounding and Bonding

VOLUME 3
Electromagnetic Shielding

VOLUME 4
Filters and Power Conditioning

VOLUME 5
EMC in Components and Devices

VOLUME 6
EMI Test Methodology and Procedures

VOLUME 7
EMC in Telecommunications

VOLUME 8
EMI Control Methodology and Procedures

VOLUME 9
United States Commercial Standards

VOLUME 10
European and International Commercial Standards

VOLUME 11
Military EMC Standards

VOLUME 12
Supporting Military EMC Standards

Contents

Common Terms and Abbreviations in EMC Literature

Prefixes for Decimal Multiples

10^{12}	tera	T
10^{9}	giga	G
10^{6}	mega	M
10^{3}	kilo	k
10^{2}	hecto	h
10	deka	da
10^{-1}	deci	d
10^{-2}	centi	c
10^{-3}	milli	m
10^{-6}	micro	μ
10^{-9}	nano	n
10^{-12}	pico	p

Technical Terms

absolute	abs
alternating current	ac
American wire gage	AWG
ampere	A
ampere per meter	A/m
ampere-hour	Ah
amplitude modulation	AM
amplitude probability distribution	APD
analog to digital	A/D
analog-to-digital converter	ADC or A/D converter
anti-jamming	AJ
arithmetic logic unit	ALU
audio frequency	AF
automatic data processing	ADP
automatic frequency control	AFC
automatic gain control	AGC
average	avg
bandwidth	BW
binary coded decimal	BCD
bit	b
bit-error rate	BER
bits per second	bps
British thermal unit	Btu
broadband	BB
byte	B
bytes per second	Bps
centimeter-gram-second	cgs
central processing unit	CPU
characters per second	cps
common-mode coupling	CMC
common-mode rejection ratio	CMRR
complementary metal-oxide semiconductor	CMOS
continuous wave	CW
coulomb	C
cubic centimeter	cm^3
decibel	dB
decibel above 1 milliwatt	dBm
decibel above 1 volt	dBV
decibel above 1 watt	dBW
degree Celsius	°C
degree Fahrenheit	°F
degree Kelvin	°K
diameter	dia
differential-mode coupling	DMC
digital multimeter	DMM
digital to analog	D/A
digital voltmeter	DVM
digital-to-analog converter	DAC or D/A conv.

diode-transistor logic .... DTL
direct current .... dc
double pole double throw . DPDT
double sideband .... DSB
double sideband suppressed carrier .... DSB-SC
dual in-line package .... DIP
electric field .... E-field
electromagnetic compatibility .... EMC
electromagnetic interference .... EMI
electromagnetic pulse .... EMP
electromotive force .... EMF
electron volt .... eV
electronic countermeasures ECM
electrostatic discharge .... ESD
emitter-coupled logic .... ECL
extremely high frequency . EHF
extremely low frequency... ELF
farad .... F
fast Fourier transform .... FFT
field intensity .... FI
field intensity meter .... FIM
field-effect transistor .... FET
foot .... ft or ’
frequency .... freq
frequency division multiplex FDM
frequency modulation .... FM
frequency shift keying .... FSK
gauss .... G
gram .... g
ground .... gnd
ground loop coupling .... GLC
ground support equipment GSE
hazards of electromagnetic radiation to ordnance .... HERO
henry .... H
hertz (cycles per second)... Hz
high frequency .... HF
high-power transistor-to-transistor logic .... HTTL
high-speed complementary metal-oxide semiconductor .... HCMOS
high-threshold logic .... HTL
hour .... hr
inch .... in or ”
inch per second .... ips
industrial, scientific and medical .... ISM
infrared .... IR
input/output .... I/O
inside dimension .... ID
instantaneous automatic gain control .... IAGC
insulated-gate field-effect transistor .... IGFET
integrated circuit IC
interference-to-noise ratio . I/N
intermediate frequency .... IF
joule .... J
junction field-effect transistor .... JFET
kilogram .... kg
kilohertz .... kH
kilovolt .... kV
kilowatt .... kW
kilowatt-hour .... kWh
lambert L
large-scale integration .... LSI
least significant bit .... LSB
length l
length (of cable) .... l_c
line impedance stabilization network .... LISN
line of sight .... LOS
liter l
local oscillator .... LO
low frequency .... LF
lower sideband .... LSB
lumen .... lm
lux lx
magnetic field .... H-field
master oscillator power amplifier .... MOPA
maximum .... max
maxwell .... Mx
mean time between failure MTBF
mean time to failure .... MTTF
mean time to repair .... MTTR
medium frequency (300 kHz to 3 MHz) .... MF
metal-oxide semiconductor MOS
metal-oxide semiconductor field-effect transistor .... MOSFET
metal-oxide varistor .... MOV
meter .... m
microfarad .... μF
microhenry .... μH
micron (10^{-6} meter) .... μ
micro-ohm .... $\mu\Omega$
microwave .... MW
mile .... mi
military specification .... MIL-SPEC
military standard .... MIL-STD
milliamp .... mA

million instructions per second	MIPS
millisecond	ms
millivolt	mV
milliwatt	mW
minimum	min
minimum discernable signal	MDS
minute	min
modulator-demodulator	modem
most significant bit	MSB
multilayer board	MLB
multiplex, multiplexer	mux
nanofarad	nF
nanohenry	nH
nanosecond	ns
narrowband	NB
negative	neg
negative-positive-negative (transistor)	npn
negative-to-positive (junction)	n-p
newton	N
noise equivalent power	NEP or P_n
non-return to zero	NRZ
N-type metal-oxide semiconductor	NMOS
nuclear electromagnetic pulse	NEMP
oersted	Oe
ohm	Ω
ohm-centimeter	Ωcm
ohms per square	Ω/sq
ounce	oz
outside dimension	OD
peak	pk
peak-to-peak	p-p
phase lock loop	PLL
phase modulation	PM
positive	pos
positive-negative-positive (transistor)	pnp
positive-to-negative (junction)	p-n
pound (sterling)	£
pound per square centimeter	p/cm^2
pound per square inch	psi
power factor	PF
printed circuit board	PCB
private branch exchange	PBX
P-type metal-oxide semiconductor	PMOS
pulse per second	pps
pulse position modulation	PPM
pulse repetition frequency	PRF
pulse-amplitude modulation	PAM
pulse-code modulation	PCM
pulse-duration modulation	PDM
pulse-width modulation	PWM
quasipeak	QP
radiation hazard	RADHAZ
radio frequency	RF
radio interference and field intensity	RI-FI
radio-frequency interference	RFI
random access memory	RAM
receiver	RX
reference	ref
relative humidity	RH
resistance-inductance-capacitance	RLC
return to zero	RTZ
revolutions per minute	rpm
roentgen	R
root-mean-square	rms
second	s
sensitivity time control	STC
shielding effectiveness	SE
sideband	SB
siemens	S
signal-to-interference (ratio)	S/I
signal-to-noise (ratio)	S/N
silicon controlled rectifier	SCR
single sideband	SSB
square meter	m^2
standing-wave ratio	SWR
super high frequency	SHF
super low frequency	SLF
surface acoustic wave	SAW
surface-mount technology	SMT
surface-mounted component	SMC
surface-mounted device	SMD
television	TV
temperature coefficient	TC
tesla	T
time division multiplex	TDM
transistor-to-transistor logic	TTL
ultra high frequency (360 MHz to 3 GHz)	UHF
ultraviolet	UV
very high frequency (30 MHz to 300 MHz)	VHF
very high-speed integrated circuit	VHSIC
very large-scale integration	VLSI
very low frequency (3 kHz to 30 kHz)	VLF

volt V
volt meter .... VM
voltage standing wave ratio VSWR
voltage-to-frequency converter .... VFC
voltampere .... VA
volt-ohm meter .... VOM
watt .... W
waveguide beyond cuttoff . WGBCO
weber............................... Wb
words per minute .... wpm
yard .... yd

Mathematical Functions and Operators

absolute value .... abs
approximately equal .... ≈
argument .... arg
cosine.............................. cos
cosine (hyperbolic) .... cosh
cotangent .... cot
cotangent (hyperbolic)....... coth
determinant...................... det
dimension .... dim
exponential...................... exp
imaginary........................ im
inferior............................ inf
limit................................ lim
logarithm, common ($base_{10}$) log
logarithm, Napierian ($base_e$).......... ln
sine................................. sin
tangent............................ tan
tangent (hyperbolic).......... tanh

Common Variables in EMC Equations

attenuation constant, absorption factor .... α
Boltzmann's constant .... K
capacitance (in farads) C
charge .... Q
coefficient of self-inductance L
conductance in mho .... G
conductivity, propagation constant, leakage coefficient, deviation...... σ
current I
dielectric constant, permittivity .... ϵ
frequency (in Hz) .... f
impedance........................ Z
induced voltage .... E
inductance (in henrys)....... L
infinity .... ∞
length (coil turn, ground loop, etc.) l
length in millimeters .... l_{mm}
magnetic susceptibility ..... χ
magnetizing force .... H
parasitic capacitance......... C_p
permeability of free space μ_o
permeability of medium relative to μ_o .... μ_r
phase constant .... β
radius............................... r
relative permittivity.......... ϵ_r
resistance (in ohms).......... R
rise time .... τ_r
shield thickness............... d
time .... t
time constant, transmission factor .... τ
velocity, volume............... V
wavelength .... λ

Acknowledgement

The authors extend thanks to Jeff Eckert and Gil Fitzpatrick for their assistance in the editing and production of this book. Special gratitude is extended to Martin Green, who provided invaluable inputs to the cable shield section and Monique Lutin who typed all of the manuscript.

Chapter 1

Shielding Theory

The first part of this handbook consists of design data regarding shielding effectiveness (SE) relative to several variables, including metal type, metal properties, thickness, distance, frequency, etc. The basics of shielding are presented in Chapter 1. Chapter 2 covers information on how to use the design graphs, including constraints and illustrative examples. As an introduction, Chapter 1 presents tutorial information on shielding theory and materials. This includes field theory, near and far-field definitions, wave impedance, metal barrier impedance, absorption loss, reflection loss and SE of an ideal metal barrier.

1.1 Field Theory

The purpose of this section is to present some relationships regarding magnetic, electric and electromagnetic (EM) fields as pertinent background to understanding and applying field theory. Since EMC literature is replete with discussions of Maxwell's equations and field theory, only a few aspects are presented here. The electric (E_θ, E_r) and magnetic (H_ϕ) fields existing around the oscillating doublet of Fig 1.1 (tiny dipole in which its length $D \ll \lambda$) are obtained from applying Maxwell's equations:

$$E_\theta = \frac{Z_0 I D \pi \sin\theta}{\lambda^2} \left[-\left(\frac{\lambda}{2\pi r}\right)^3 \cos\psi - \left(\frac{\lambda}{2\pi r}\right)^2 \sin\psi + \left(\frac{\lambda}{2\pi r}\right) \cos\psi \right] \tag{1.1}$$

$$E_r = \frac{2 Z_0 I D \pi \cos\theta}{\lambda^2} \left[\left(\frac{\lambda}{2\pi r}\right)^3 \cos\psi + \left(\frac{\lambda}{2\pi r}\right)^2 \sin\psi \right] \tag{1.2}$$

$$H_\theta = \frac{I D \pi \sin\theta}{\lambda^2} \left[\left(\frac{\lambda}{2\pi r}\right)^2 \sin\psi + \left(\frac{\lambda}{2\pi r}\right) \cos\psi \right] \tag{1.3}$$

where,

Z_0 = free-space impedance (for $r >> \lambda/2\pi$ = $\sqrt{\mu/\epsilon}$ = 120π = 377 Ω

I = current in short wire (doublet)

D = length of short wire (doublet) in which $D << \lambda$

θ = zenith angle to radial distance r

λ = wavelength corresponding to frequency, $f = c/\lambda$

r = distance from short wire (doublet) to measuring or observation point

$\psi = (2\pi r/\lambda) - \omega t$

ω = radial frequency = $2\pi f$

t = time = 1/f

$c = 1/\sqrt{\mu\epsilon} = 3 \times 10^8$ m/s

Several observations can be made about the near and far fields in Eqs. (1.1) and (1.3):

1. When the multiplier, $\lambda/2\pi r = 1$ in the electric- and magnetic-field terms, all coefficients of either sine or cosine terms are unity and equal. Thus, when $r = \lambda/2\pi$ (about one-sixth wavelength), this corresponds to the transition-field region or boundary between near field (first term of equations) and far field (last term).

2. When $r >> \lambda/2\pi$ (far-field conditions), only the last term of Eqs. (1.1) and (1.3) is significant.* For this condition the wave impedance $Z_0 = E_\theta/H_\phi = 377\ \Omega$. This is called the radiation field (plane waves) and both E_θ and H_ϕ are in time phase ($\cos\psi$, common to both terms) although they are in directional quadrature.
3. When $r << \lambda/2\pi$ (near-field conditions), only the first term of each equation is significant.* For this condition the wave impedance $E_\theta/H_\phi = Z_0\ \lambda/2\pi r$. Note that the wave impedance is now $>> Z_0$ since $\lambda/2\pi r >> 1$. This is sometimes called an electric field or a **high-impedance field,** i.e., high relative to a plane-wave impedance. it is also the induction field, and E_θ and H_ϕ are in time and spatial quadrature.
4. Had the oscillating source not been a small, straight wire or doublet exhibiting high circuit impedance, but rather a small wire loop of Area A, exhibiting low circuit impedance, the first term appearing in Eqs. (1.1) and (1.2) would vanish, and a similar (cubed) first term would appear in Eq. (1.3). For this condition, the wave impedance in the near field $E_\theta/H_\phi = Z_0\ 2\pi r/\lambda$. Note that the wave impedance is now $<< Z_0$ for $\lambda/2\pi r >> 1$. This is sometimes called a magnetic field or a **low-impedance field,** i.e., low relative to Z_0, the plane wave (radiation impedance).

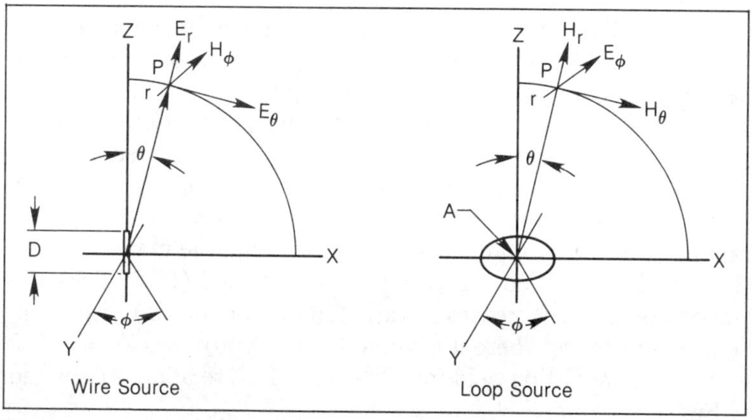

Figure 1.1—Spherical Coordinates Used in Eqs. (1.1-1.3)

*The term E_r is dropped from further discussion because it represents a non-propagating component of the E-field.

Figure 1.2 illustrates the first three of the above observations for the amplitude of each of the electric-field terms in Eq. (1.1). Note that the quasi-stationary field is the largest term in the far field and the induction is next largest; whereas, the radiation term is the largest in the far field.

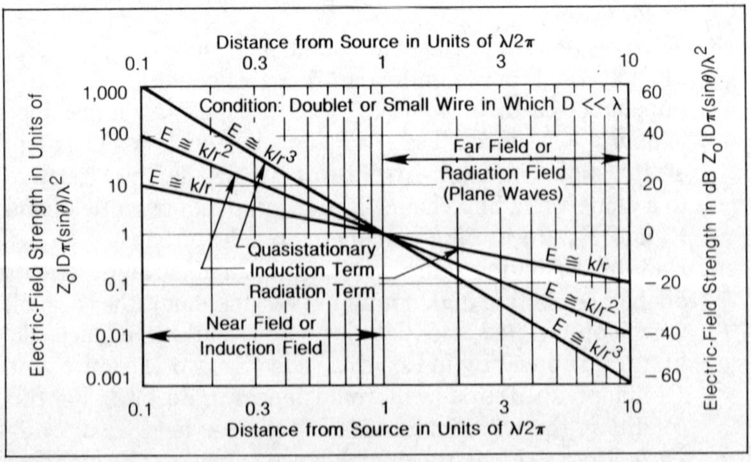

Figure 1.2—Electric-Field Strength vs. Source Distance

For $r = \lambda/2\pi$, the coefficients of all three terms are equal. Figure 1.3 illustrates the preceding third and fourth observations in the near or induction field. Figure 1.3a shows a short monopole or straight wire in which the circuit current is low.* Consequently, the source or circuit impedance = V/I is a high impedance. As previously shown, the wave impedance close to the source is also high, relative to 377 Ω. The associated electric field attenuates more rapidly ($1/r^3$) with an increase in distance than the magnetic field ($1/r^2$) in the induction region [cf. Eqs. (1.1) and (1.3)]. Thus the wave impedance decreases linearly (20 dB per decade) with an increase in distance where it asymptotically approaches $Z_0 = 377\ \Omega$ in the far or radiation field for $r >> \lambda/2\pi$, This relation is depicted in Fig. 1.4.

The converse to the above applies for Fig. 1.3b, wherein a low-impedance radiation source creates a low-impedance wave, i.e.,

*The conduction current in the wire is the same as the displacement current flowing in the capacitance from wire to ground or from one end of the wire to the lower return side.

a magnetic field. This wave impedance increases linearly (20 dB per decade) with distance where it asymptotically approaches 377 Ω in the far field. Figure 1.4 also illustrates this wave impedance as a function of distance, r.

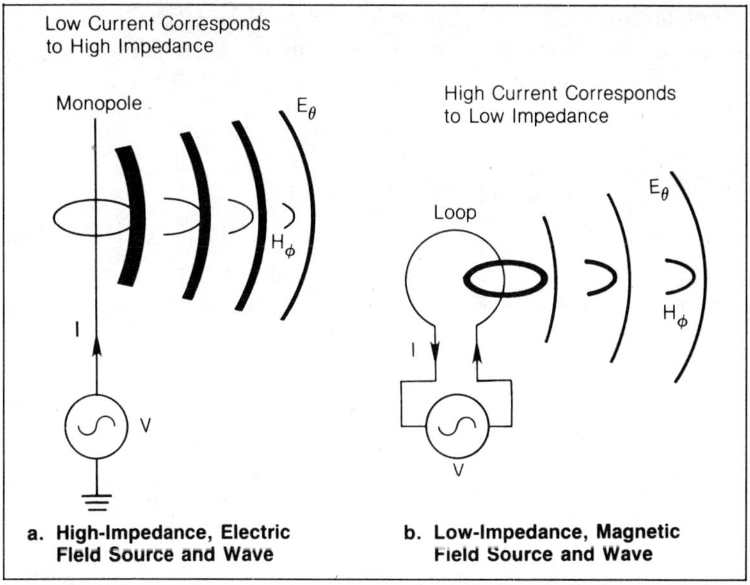

Figure 1.3—Conceptual Illustration of Field Strengths vs. Source Type and Distance

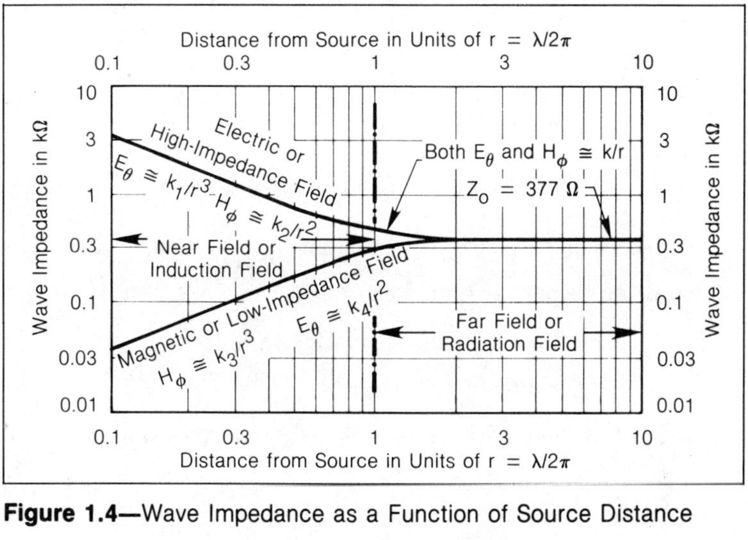

Figure 1.4—Wave Impedance as a Function of Source Distance

When the dimension D of an emission source (e.g., a wire or an antenna) becomes an appreciable fraction of a wavelength, the expressions presented in Eqs. (1.1) to (1.3) no longer apply. Different relations must then be used since the curvature of the arriving wave front no longer permits the incremental elements of an antenna or radiating source to be in phase. This situation is described in Vol. 6 of this handbook series.

1.2 Wave Impedance

The previous section showed that the electromagnetic wave impedance is a function of the impedance of the source circuit generating the wave (the distance r from source-to-metal barrier measuring points) and the frequency. Summing up, the wave impedance, Z_w, is:

$$Z_w = E/H = kZ_0 = k\sqrt{\mu/\epsilon} = 120\pi k = k\ 377\ \Omega \qquad (1.4)$$

provided that the length D of the source emission-generating circuit element is $<< \lambda$.

where,

$$k \approx \frac{\lambda}{2\pi r} \quad \text{for current source (very high impedance) and } r \leqslant \lambda/2\pi \qquad (1.5)$$

$$\approx \frac{2\pi r}{\lambda} \quad \text{for voltage source (very low impedance) and } r \leqslant \lambda/2\pi \qquad (1.6)$$

$$\approx 1 \text{ for any network impedance and } r \geqslant \lambda/2\pi \qquad (1.7)$$

$\mu = \mu_0\mu_r$ = permeability of medium

μ_0 = absolute permeability of free space = $4\pi \times 10^{-7}$ H/m

μ_r = permeability of medium relative to air

$\epsilon = \epsilon_0\epsilon_r$ = permittivity of medium

ϵ_0 = absolute permittivity of free space = $(1/36\pi) \times 10^{-9}$ F/m

ϵ_r = permittivity of medium relative to air

In the far field, therefore, the plane-wave impedance, 377 Ω, is independent of the circuit source impedance. In the near field, wave impedance depends on the source impedance but can never exceed 377 × λ/2πr, nor be less than 377 × 2πr/λ.

The above raises the question about just how the source-generating circuit and wave impedances are related in the near field. This relation is important since it will be shown in Section 1.5 that the SE in the near field is developed in terms of a high-impedance source (electric fields) and low-impedance source (magnetic fields). When designing shields, the use of electric-field sources may give rise to optimistic results (relatively high SE), and the use of magnetic-field sources may give rise to pessimistic results (relatively low SE). The former could result in an underdesign condition, and the latter could result in overdesign of any shielded circuit, box, equipment, cabinet, etc.

The development of a discrete relation between circuit, Z_c, and wave, Z_w, impedance in the near field is beyond the scope of this handbook. However, the following mathematical relations are suggested for all conditions in which the circuit dimensions, D << λ:

$$Z_w = \frac{Z_0\lambda}{2\pi r}, \text{ for } Z_c > \frac{Z_0\lambda}{2\pi r} \geqslant Z_0 \tag{1.8}$$

$$\cong Z_c, \text{ for } \frac{Z_0\lambda}{2\pi r} \geqslant Z_c \geqslant Z_0 \tag{1.9}$$

$$\cong Z_0, \text{ for } Z_c = Z_0 \text{ or } \frac{\lambda}{2\pi r} \geqslant 1 \tag{1.10}$$

$$\cong Z_c, \text{ for } Z_0 > Z_c \geqslant \frac{Z_0 2\pi r}{\lambda} \tag{1.11}$$

$$\cong \frac{Z_0 2\pi r}{\lambda}, \text{ for } Z_0 > \frac{Z_0 2\pi r}{\lambda} > Z_c \tag{1.12}$$

Equations (1.8) through (1.12) are plotted in Fig. 1.5 for several values of Z_c, including common transmission-line impedances of 50, 100, 300, and 600 Ω. To the extent that these conditions exist, the common transmission line impedances, then, never permit either a very high or very low wave impedance condition to exist when r << $\lambda/2\pi$.

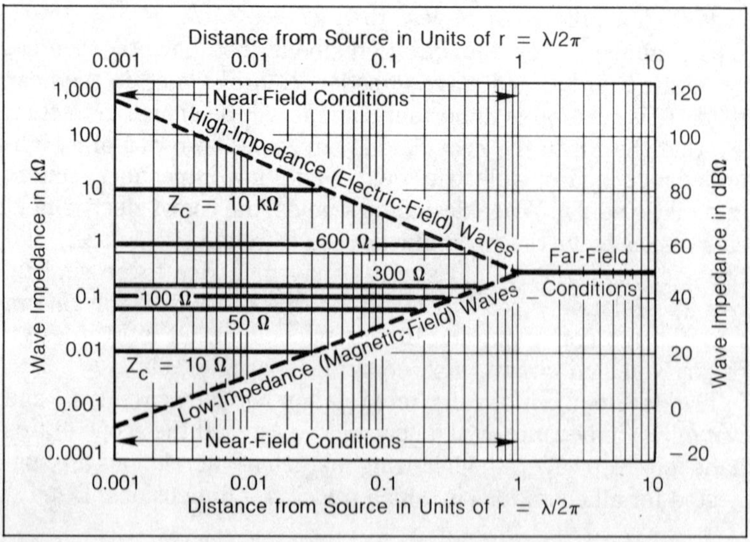

Figure 1.5—Wave Impedance for Several Circuit Impedances

As discussed later, sometimes it is more useful to present Eqs. (1.4) through (1.7) in terms of frequency, f, rather than wavelength. Since, for free space:

$$f \lambda_m = c = 1/\sqrt{\mu_0 \epsilon_0} = 3 \times 10^8 \text{ m/s} \tag{1.13}$$

or,

$$f_{MHz} = 300/\lambda_m = \text{frequency in MHz} \tag{1.14}$$

where,

f = frequency in Hz

f_{MHz} = frequency in MHz

λ_m = wavelength in meters

c = velocity of light in free space

$$= 0.066/\sqrt{f_{MHz}} \text{ mm for copper } (\mu_r = \sigma_r = 1) \quad (1.29)$$

$$= 2.6/\sqrt{f_{MHz}} \text{ mils for copper } (\mu_r = \sigma_r = 1) \quad (1.30)$$

$$= 0.066/\sqrt{\mu_r \sigma_r f_{MHz}} \text{ mm for any metal} \quad (1.31)$$

$$= 2.6/\sqrt{\mu_r \sigma_r f_{MHz}} \text{ mils for any metal} \quad (1.32)$$

Equations (1.31) and (1.32) are plotted in Fig. 1.6 for various materials.

1.3.2 Barrier Impedance of Metals ($t < 3\delta$)

In the previous section, the development and applications of Eqs. (1.24) and (1.25) are based upon the thickness, t, of the metal barrier being many skin depths, i.e., $t >> \delta$. In order to establish the applicability of the equations, a somewhat arbitrary situation is selected in which $t > 3\delta$. When $t = 3\delta$, 95 percent of the current flows between the surface and a depth of 3δ, and 5 percent of the current flows beyond a depth of 3δ. Thus, when $t = 3\delta$, the barrier impedance must actually be 5 percent greater than indicated in Eqs. (1.24) and (1.25).

To accommodate any metal thickness, a generalization of the previous equation is needed, that is:

$$Z_B = \frac{Z_m}{(1 - e^{-t/\delta})}, \text{ for } t/\delta \text{ any value} \quad (1.33)$$

$$= Z_m \text{ for } t/\delta >> 1 \quad (1.34)$$

$$\cong \frac{Z_m}{[1 - (1 - t/\delta)]}, \text{ for } t/\delta << 1 \quad (1.35)$$

$$\cong \delta Z_m/t, \text{ for } t/\delta << 1 \quad (1.36)$$

The error in Z_B in Eq. (1.36), which may develop from applying Z_m under conditions for which t/δ is not << 1, is:

$$\Delta Z_{dB} = -20 \log (1 - e^{-t/\delta}) \text{ dB} \quad (1.37)$$

Equation (1.37) is plotted in Fig. 1.7. The error is 4 dB when $t = \delta$ and becomes proportional to t/δ (20 dB/decade slope) when $t/\delta < 1$. Substituting Eq. (1.33) in Eq. (1.25) yields the more general situation for Z_B of any thickness:

$$Z_B = 369 \sqrt{\mu_r f_{MHz}/\sigma_r} \, (1-e^{-t/\delta})^{-1} \; \mu\Omega/\text{sq} \tag{1.38}$$

Equation (1.38) is plotted in Fig. 1.8 for copper and iron corresponding to several values of t/δ. The $t/\delta > 3$ parameter line for copper in Fig. 1.8 is the same as the copper line in Fig. 1.6.

Sometimes it is useful to know the barrier impedance corresponding to very thin deposits of metal on glass or clear plastic substrates like those used for electrically conductive transparent EMI-shielded windows. The conductively coated windows may have deposits of much less than a micron in thickness in which $t/\delta << 1$. For this situation, Eq. (1.36) is directly obtained from Eq. (1.26):

$$Z_B = \delta Z_m/t = \frac{\sqrt{2}}{\sigma t} \; \Omega/\text{sq for } t/\delta << 1 \tag{1.39}$$

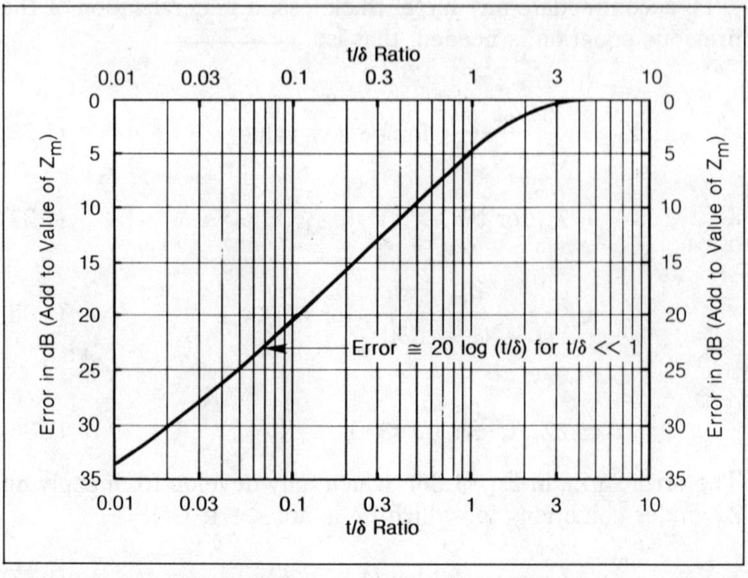

Figure 1.7—Barrier Metal Impedance Error in Z_m Expression by Assuming $t/\delta >> 1$

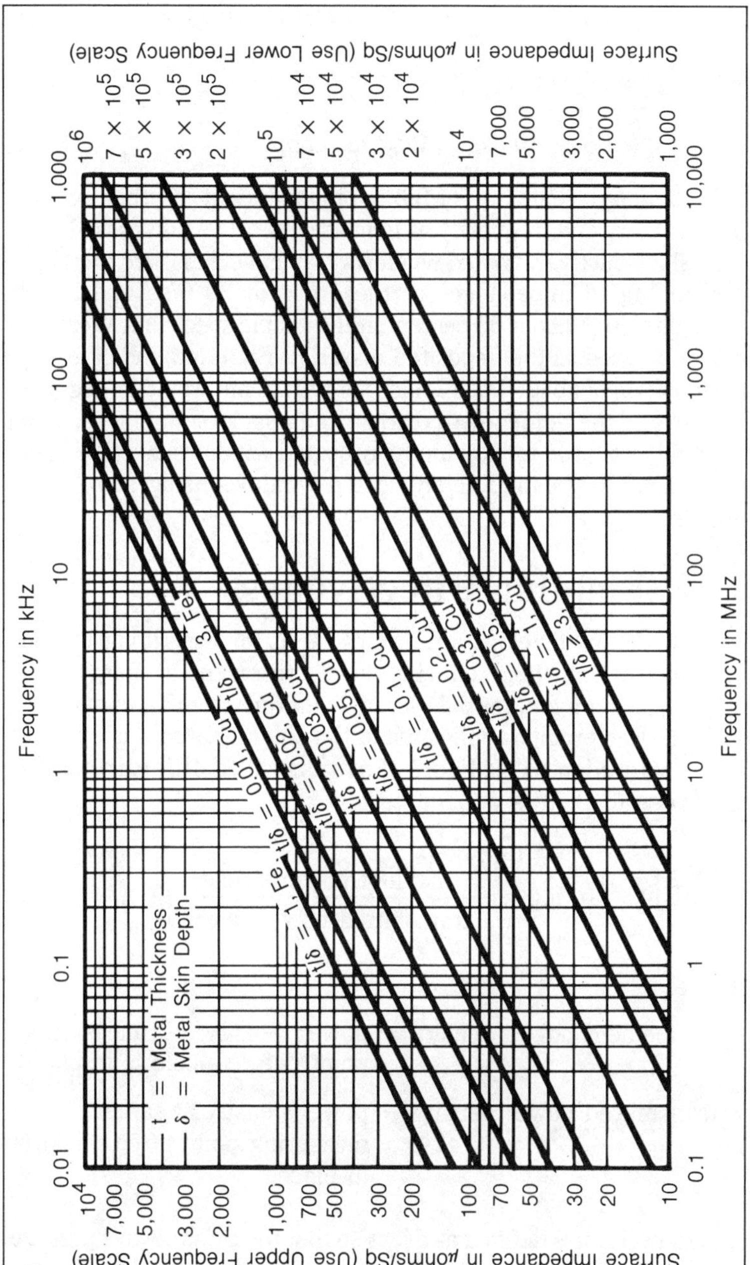

Figure 1.8—Surface Impedance of Copper and Iron vs. Frequency and Skin Depth in Units of t/δ Ratio

$$= 0.244 \times 10^{-4}/t_{mm}\ \Omega/\text{sq for copper} \quad (1.40)$$

$$= 0.0244/t_{microns}\ \Omega/\text{sq for copper} \quad (1.41)$$

$$= 96 \times 10^{-5}/t_{mils}\ \Omega/\text{sq for copper} \quad (1.42)$$

It is noted that Eq. (1.39) is independent of the material permeability and frequency as long as $t/\delta << 1$. As discussed later, typical conductively coated windows may have deposited metal corresponding to impedances of the order of 20 Ω/sq and a light transmission of about 65 percent in the visible spectrum. Were copper to be used as the deposited material, Eq. (1.40) indicates that a deposition of about 0.0012 microns or 1.22 nm would be required. Because of the tendency to oxidize and a lack of stability, copper is usually not used for conductively coated windows. Section 10.4.2 discusses SE of conductively coated windows using more stable metals.

1.4 Shielding Effectiveness

This section develops a figure of merit for describing the performance of a shield in reducing the electromagnetic energy either impinging upon a potential victim or exiting from an undesirable source. (Reciprocity applies when the shield material is operated below saturation.) This figure of merit is called **shielding effectiveness** (SE). It is defined as:

$$SE_{dB} = 10 \log \left(\frac{\text{incident power density}}{\text{transmitted power density}} \right) \quad (1.43)$$

where,

incident power density = power density at a measuring point before shield is in place

transmitted power density = power density at same measuring point after shield is in place.

Equation (1.43) is defined as a loss so that the SE is always positive.

Since power density in W/m^2 can also be expressed as the product of E(V/m) multiplied by H(A/m), SE can be written as:

$$SE_{dB} = 10 \log \frac{E_b H_b}{E_a H_a}$$

where,

E_b, H_b = Value of E- and H-Fields before shielding

E_a, H_a = Value of E- and H-Field after shielding

If the wave impedance $Z_w = E/H$ is the same before and after i.e., if the attenuation of the barrier is identical for the E and the H terms, or if the emerging field is measured at a far-field distance from the shield in stable field conditions:

$$H_b = \frac{E_b}{Z_w} \text{ and } H_a = \frac{E_a}{Z_w}$$

$$\text{so, } SE_{dB} = 10 \log \frac{E_b^2/Z_w}{E_a^2/Z_w} = 10 \log \frac{E_b^2}{E_a^2} = 20 \log \frac{E_b}{E_a} \tag{1.44}$$

In stabilized field conditions, therefore the measure of SE in power density terms or in field terms gives the same results.

In terms of the field strength, Eq. (1.43) may also be defined as follows as long as the two fields are measured in the same medium having the same wave impedance:

$$SE_{dB} = 20 \log \left(\frac{E_b}{E_a}\right) \text{ for electric fields} \tag{1.45}$$

$$= 20 \log \left(\frac{H_b}{H_a}\right) \text{ for magnetic fields} \tag{1.46}$$

where,

E_b = electric-field strength **before** shield is installed

E_a = electric-field strength **after** shield is installed

H_b = magnetic-field strength **before** shield is installed

H_a = magnetic-field strength **after** shield is installed

Equations (1.45) and (1.46) give identical results in the far field ($r >> \lambda/2\pi$) but do not result in the same SE_{dB} in the near field since the wave impedances are different.

Figure 1.9 shows the conceptual mechanism for determining SE. The details are shown in Fig. 1.10. The explanation for this mechanism is that the incident field strength (an electric field is illustrated here) is considered as unity relative to itself. The

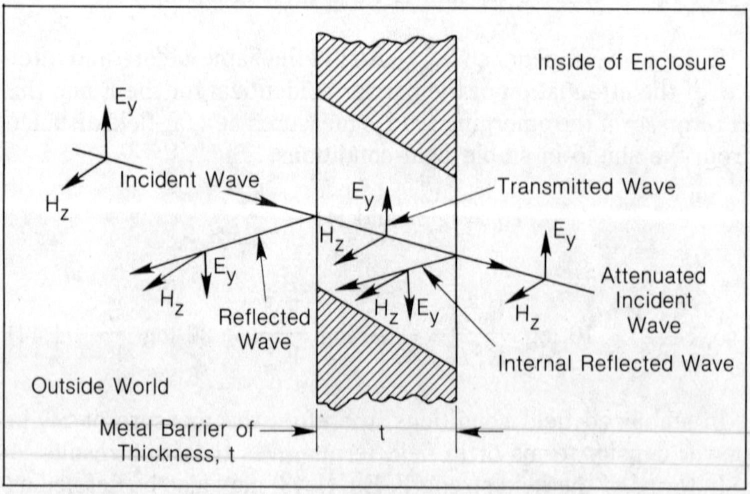

Figure 1.9—Representation of Shielding Phenomena for Plane Waves

reflected field is: (equal to the reflection coefficient ρ, since field is unity) is:

$$\rho = \frac{1 - K}{1 + K} = -1 \text{ for } K >> 1 \tag{1.47}$$

$$= 0 \text{ for } K = 1 \tag{1.48}$$

$$= +1 \text{ for } 0 \leqslant K << 1 \tag{1.49}$$

where,

$K = Z_w/Z_m$

Z_w = wave impedance defined in Section 1.2

Z_m = metal impedance defined in Section 1.3

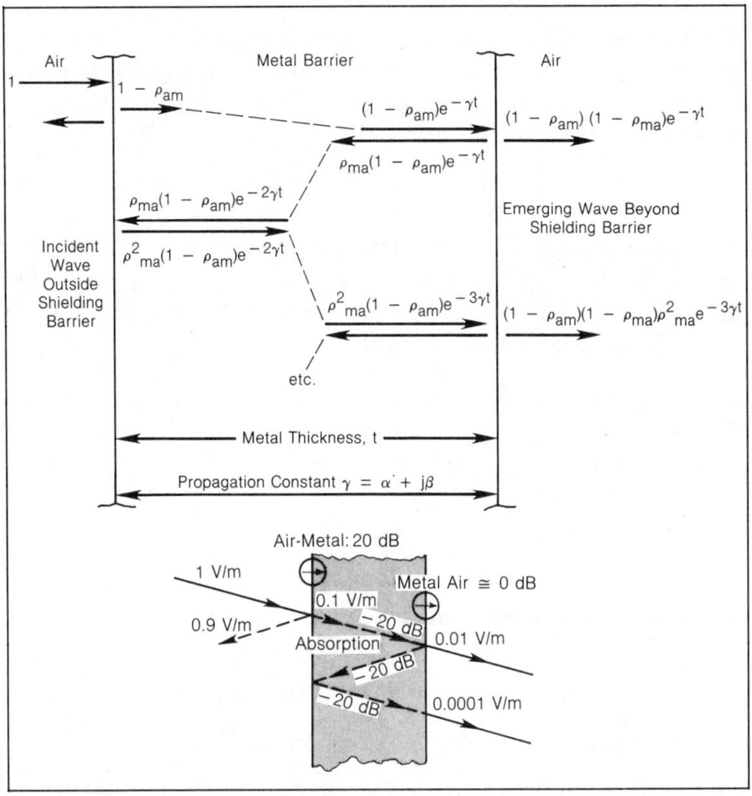

Figure 1.10—Geometry of Metal Barrier Used in Explaining Shielding Effectiveness

and, $$\rho_{am} = \frac{1 - K}{1 + K} \qquad (1.50)$$

= reflection coefficient at the air-to metal interface, i.e., wave propagation from air medium into metal

$$\rho_{ma} = \frac{K - 1}{K + 1} \qquad (1.51)$$

= reflection coefficient at the metal-to-air interface, i.e., wave propagation from metal medium into air, in which case the roles of Z_w and Z_m are interchanged, whereby $K_{ma} = 1/K_{am}$ in Eqs. (1.50) and Eq. (1.51)

The K term may be thought of as the voltage standing-wave ratio (VSWR) for $Z_w/Z_m \geqslant 1$ in Eq. (1.50). When $Z_w/Z_m \leqslant 1$, the VSWR concept still applies if Eq. (1.50) is defined as Eq. (1.51) in which $Z_m/Z_w \geqslant 1$. The relative transmitted field, Γ_{am}, just inside the left edge of the metal-to-air barrier is:

$$\Gamma_{am} = 1 - \rho_{am} \tag{1.52}$$

This field undergoes an attenuation in traversing the thickness of the metal barrier, which turns the associated loss into heat. The arriving field Γ_{ar} results in a lower field strength impinging on the right, inside edge of the barrier:

$$\Gamma_{ar} = \Gamma_{am}e^{-\gamma t} = e^{-(\alpha + j\beta)t}\Gamma_{am} = e^{(-\alpha + j\beta)t}(1 - \rho_{am}) \tag{1.53}$$

where,

γ = propagation constant = $\alpha + j\beta$

α = attenuation constant

β = phase constant = $2\pi/\lambda$

t = metal thickness

The rereflected relative field strength Γ_{RR}, at the inside right, edge of the metal-to-air barrier of Fig. 1.10 is:

$$\Gamma_{RR} = \rho_{ma}\Gamma_{AR} = e^{-\gamma\tau}\rho_{ma}(1 - \rho_{am}) \tag{1.54}$$

where, ρ_{ma} = metal-to-air reflection coefficient Eq. (1.51).

The relative transmitted field, Γ_{RT}, to the right, outside the metal barrier is:

$$\Gamma_{RT} = 1 - \rho_{ma}\Gamma_{AR} = e^{-\gamma t}(1 - \rho_{am})(1 - \rho_{ma}) \tag{1.55}$$

Equation (1.55) is the SE expressed as a gain (a negative number) when $\gamma >> 1$.

When the propagation constant is not significant, one or more round-trip rereflections must be considered. For example, the rereflected field of Eq. (1.54) undergoes a second attenuation and phase shift in propagating back to the inside edge of the left metal-to-air interface of Fig. 1.10:

$$\Gamma_{LR} = \Gamma_{RR}\,e^{-\gamma t} = e^{-2\gamma t}\rho_{ma}(1 - \rho_{am}) \tag{1.56}$$

The rereflected field strength, Γ_{LL}, from the left edge inside barrier is:

$$\Gamma_{LL} = \rho_{ma}\Gamma_{LR} = e^{-2\gamma t}\rho^2_{ma}(1 - \rho_{am}) \tag{1.57}$$

Undergoing a third attenuation and phase shift of this rereflected field in arriving back at the inside face of the right edge, the relative field strength, Γ_{AR}, becomes:

$$\Gamma'_{AR} = \Gamma_{LL}\, e^{-\gamma t} = e^{-3\gamma t}\rho^2_{ma}\,(1 - \rho_{am}) \qquad (1.58)$$

Finally, the transmitted component of Γ'_{AR} to the right, outside the metal barrier is:

$$\Gamma'_{RT} = (1 - \rho_{ma})\,\Gamma'_{AR} = e^{-3\gamma t}\rho^2_{ma}\,(1 - \rho_{am})\,(1 - \rho_{ma}) \qquad (1.59)$$

Since the rereflected field component of Eq. (1.59) is coherent with the direct, transmitted field of Eq. (1.55), they are coherently added. Therefore, the **total** transmitted field, equal to transmission coefficient Γ_T, is:

$$\Gamma_T = e^{-\gamma t}(1 - \rho_{am})(1 - \rho_{ma})[1 + \underbrace{e^{-2\gamma t}\rho^2_{ma}}_{\text{First Round-Trip Rereflections}} + \underbrace{e^{-4\gamma t}\rho^4_{ma} + \ldots}_{\text{Multiple Round-Trip Rereflections}}] \qquad 60)$$

The terms in the bracket constitute an infinite series, i.e., an infinity of rereflections. The bracket expression can be simplified by writing this series in terms of its reciprocal. Thus, Eq. (1.60) becomes:

$$\Gamma_T = e^{-\gamma t}(1 - \rho_{am})(1 - \rho_{ma})(1 - \rho^2_{ma}\, e^{-2\gamma t})^{-1} \qquad (1.61)$$

Equation (1.61) may be expressed in terms of the impedance ratio of the air-metal and metal-air interfaces by substituting Eqs. (1.50) and (1.51) therein:

$$\Gamma_T = e^{-\alpha t}\left(\frac{2K}{1 + K}\right)\left(\frac{2}{1 + K}\right)\left[1 - \left(\frac{K - 1}{K + 1}\right)^2 e^{-2\gamma t}\right]^{-1} \qquad (1.62)$$

$$= \underbrace{e^{-\alpha t}}_{\text{Absorption Term (A)}}\ \underbrace{\frac{4K}{1 + K^2}}_{\text{Reflection Term (R)}}\ \underbrace{\left[1 - \left(\frac{K - 1}{K + 1}\right)^2 e^{-2\gamma t}\right]^{-1}}_{\text{Rereflection Correction Term (B)}} \qquad (1.63)$$

Expressing Eq. (1.63) as a loss (i.e., SE) rather than a gain, and converting it to decibels, results in:

$$SE_{dB} = 20 \log (1/\Gamma_T) \quad (1.64)$$

$$= 20 \log \left\{ e^{\alpha t} \frac{(1 + K)^2}{4K} \left[1 - \left(\frac{K - 1}{K + 1} \right)^2 e^{-2\gamma t} \right] \right\}$$

where,

$$\text{Absorption Loss, } A_{dB} = 8.686\alpha t \quad (1.65)$$

$$\text{Reflection Loss, } R_{dB} = 20 \log (1 + K)^2/4K \quad (1.66)$$

$$\cong 20 \log \frac{Zw}{4\, Z_m} \quad \text{if } K \geqslant 10$$

$$\text{Rereflection Correction, } B_{dB} = 20 \log |1 - [(K - 1)^2/(K + 1)^2] e^{-2\gamma t}| \quad (1.67)$$

Equation (1.64) is plotted in graphs in the appendices in this handbook for several metals, metal properties, metal thicknesses, distances and frequencies. This relationship will now be examined in further detail in terms of its three loss components: (1) absorption loss, (2) reflection loss and (3) rereflection loss correction.

1.4.1 Absorption Loss

Equation (1.65) may be expanded:

$$A_{dB} = 8.686\alpha t = 8.686 t \sqrt{\pi f \mu \sigma}$$

where,

$$\gamma = \alpha + j\beta = \sqrt{j\omega\mu(\sigma + j\omega\epsilon)} \quad (1.68)$$

$$= \sqrt{j\omega\mu\sigma} \quad \text{since } \sigma >> \omega\epsilon \text{ for metals} \quad (1.69)$$

$$= (1 + j) \sqrt{\pi f \mu \sigma} \quad (1.70)$$

or, $\alpha = \beta = \sqrt{\pi f \mu \sigma}$ for metals (1.71)

If Eq. (1.68) is defined in terms of t in mils (thousandths of an inch), and f in MHz and t in mm, it becomes for both the English and metric system of units:

$$A_{dB} = 3.338\, t_{mils} \sqrt{f_{MHz}\, \mu_r\, \sigma_r} \text{ dB, English units} \quad (1.72)$$

$$= 131.4\, t_{mm} \sqrt{f_{MHz}\, \mu_r\, \sigma_r} \text{ dB, metric units} \quad (1.73)$$

where, μ_r and σ_r are permeability and conductivity, respectively, relative to copper.

Equations (1.72) and (1.73) are plotted in Appendix A for various metals: copper, aluminum, brass, beryllium, monel, iron, stainless steel and the exotic high-permeability metals. For magnetic materials ($\mu_r > 1$), the graphs are accurate only for the stipulated μ_r condition which varies with both magnetic-field strength (or flux density) and frequency, especially above several kilohertz (see Section 2.1.4). Above about 1 MHz, the relative permeability of most magnetic materials has already become or approaches unity. Thus, the superior performance of magnetic materials only shows up for low-frequency shielding of magnetic fields. Beyond this, their performance is equal to or less than that of nonmagnetic, high-conductivity metals.

1.4.2 Reflection Loss

The reflection loss relations are based upon an impedance mismatch at the metal-barrier interfaces. Thus, it is useful to replace the impedances of Eq. (1.49) by their equivalents from Eq. (1.4) for Z_w and Eq. (1.23) for Z_m:

$$K = \frac{Z_W}{Z_m} = \frac{k \sqrt{\mu_o/\epsilon_o}}{(1 + j) \sqrt{\pi f \mu/\sigma}} \quad (1.74)$$

where,

$$k = \lambda/2\pi r = 1/\pi r f \sqrt{\mu_o \epsilon_o} \text{ for high-impedance E-fields} \quad (1.75)$$

$$k = 2\pi r/\lambda = 2\pi r f \sqrt{\mu_o \epsilon_o} \text{ for low-impedance H-fields} \quad (1.76)$$

$$= 1 \text{ for far fields, } r \geqslant \lambda/2\pi \quad (1.77)$$

Combining Eq. (1.74) through (1.77) yields:

$$K = 1/2\ \pi r f \epsilon_o \sqrt{2\pi f \mu/\sigma} \text{ for high impedance fields} \quad (1.78)$$

$$= r\sqrt{2\pi f \sigma \mu_o/\mu_r} \text{ for low impedance fields} \quad (1.79)$$

$$= 1/\sqrt{2\pi f \mu_r \epsilon_o/\sigma} \text{ for plane waves} \quad (1.80)$$

The reflection loss term of Eq. (1.66) may be expressed in terms of the wave and metal barrier impedances:

$$R_{dB} = 20 \log (1 + K)^2/4K \quad (1.81)$$

$$= 20 \log (K/4 = 20 \log (Z_W/4Z_m) \text{ for } K >> 1 \quad (1.82)$$

Equations (1.78) through (1.80) may be substituted into Eq. (1.82) to yield:

Electric (High-Impedance) Fields:

$$R_{dB} \cong 152.0 - 10 \log (\mu_r f^3_{MHz} r^2_{ft}/\sigma_r) \text{ dB, English units} \quad (1.83)$$

$$\cong 141.7 - 10 \log (\mu_r f^3_{MHz} r^2_m/\sigma_r) \text{ dB, metric units} \quad (1.84)$$

where, r_{ft} = source-to-barrier distance in feet
r_m = source-to-barrier distance in meters

Magnetic (Low-Impedance) Fields:

$$R_{dB} = 64.3 - 10 \log (\mu_r/f_{MHz}\sigma_r r^2_{ft}) \text{ dB, English units} \quad (1.85)$$

$$\cong 74.6 - 10 \log (\mu_r/f_{MHz}\sigma_r r^2_m) \text{ dB, metric units} \quad (1.86)$$

Plane Waves (Far-Field Conditions):

$$R_{dB} = 108.1 - 10 \log (\mu_r f_{MHz}/\sigma_r) \text{ dB, both units} \quad (1.87)$$

(independent of r as long as $r >> \lambda/2\pi = 47.8/f_{MHz}$)

When the concept of circuit impedance is introduced rather than simply high- or low-impedance waves [cf. Eqs. (1.8) through (1.20)], the SE in the near-field becomes:

$$R_{dB} \cong 20 \log \left(\frac{Z_c}{4 \times 369 \times 10^{-6} \sqrt{\mu_r f_{MHz}/\sigma_r}} \right) \quad (1.88)$$

$$\left.\begin{array}{ll} & = 56.6 - 10 \log (\mu_r f_{MHz}/\sigma_r Z^2{}_c) \\ \text{or} & = \text{Eqs. (1.83) and (1.84)} \end{array}\right\} \begin{array}{l}\text{whichever} \\ \text{is less for} \\ \text{E-fields}\end{array} \quad (1.89)$$

$$\left.\begin{array}{ll} & = 56.6 - 10 \log (\mu_r f_{MHz}/\sigma_r Z^2{}_c) \\ \text{or} & = \text{Eqs. (1.85) and (1.86)} \end{array}\right\} \begin{array}{l}\text{whichever} \\ \text{is less for} \\ \text{H-fields}\end{array} \quad (1.90)$$

$$= \text{Eq. (1.87) for plane waves}$$

For circuits whose impedance $Z_c = 377\ \Omega$, the reflection loss appears to be that of far-field conditions, even if r is in the near field.

1.4.3 Rereflection Correction

Equation (1.67) may be expanded in terms of the wave and metal-barrier impedance (see Eq. (1.74):

$$B_{dB} = 20 \log \left| 1 - \left((K - 1)^2/(K + 1)^2 \right) e^{-2\gamma t} \right| \quad (1.91)$$

$$= 20 \log \left| 1 - \left(\frac{K - 1}{K + 1} \right)^2 10^{-.1A_{dB}} \left(\cos 0.23A_{dB} - j\sin 0.23A_{dB} \right) \right| \quad (1.92)$$

$$= 20 \log \left(1 - e^{-2(\alpha + j\beta)t} \right), \text{ for } K >> 1 \quad (1.93)$$

$$= 20 \log \left(1 - e^{-2t\sqrt{\pi f \mu \sigma}} e^{-j2t\sqrt{\pi f \mu \sigma}} \right) \quad (1.94)$$

where, A_{dB} is given in Eqs. (1.72) and (1.73). Equation (1.92) is plotted in Fig. 1.11.

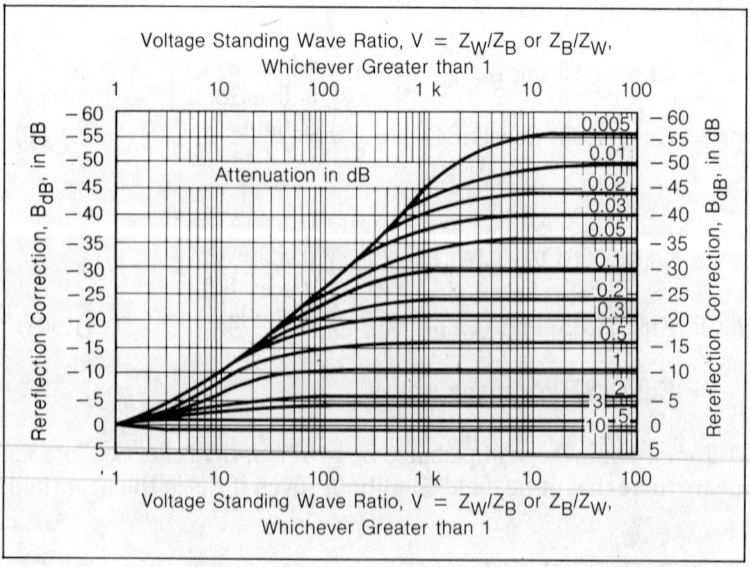

Figure 1.11—Re-Reflection Correction vs. VSWR and Material Absorption Loss

In most shielding situations, the rereflection mechanism plays a negligible role since the thickness is generally enough to cause a significant attenuation across the two-way trip of the rereflected wave. For instance, in lower part of Fig 1.10 see the miniscule contribution of the first rereflected wave emerging on the right of an arbitrary metal barrier having 20 dB of reflection loss (R) and 20 dB of absorption loss (A): This second emerging field is two orders of magnitude below that of the first emerging one and is negligible ¡in comparison. However, in situations where t is smaller than δ, the absorption loss is small and the rereflection term is no longer negligible. This causes the second emerging field to combine constructively or destructively with the first one (such would be the case of thin platings used to coat plastic or glass).

1.4.4 Total Losses for K >> 1

It may be a bit misleading to think of a rereflection loss and rereflection correction as separate terms. After all, a reflection loss should be the entire loss including rereflection. For certain conditions, Eq. (1.64) may be greatly simplified:

where,

$$SE_{dB} = 20 \log \left\{ e^{t/\delta} \frac{(K+1)^2}{4K} \left[1 - \left(\frac{K-1}{K+1} \right)^2 e^{-2\gamma t} \right] \right\} \tag{1.95}$$

where, $1/\delta = \gamma/\sqrt{2} = \alpha = \sqrt{\pi f \mu \sigma}$

For conditions in which a substantial metal barrier impedance and wave impedance mismatch exists, i.e., for K >> 1, Eq. (1.95) becomes:

$$SE_{dB} \cong 20 \log \left[e^{t/\delta} \frac{K}{4} \left(1 - e^{-2\gamma t} \right) \right] \quad \text{for } K >> 1 \tag{1.96}$$

$$= 20 \log \left[\underbrace{e^{t/\delta}}_{\text{Absorption Loss}} \underbrace{\frac{K}{4} \left(1 - e^{-2t/\delta} e^{-j2t/\delta} \right)}_{\text{Total Reflection Loss}} \right] \quad \text{for } K >> 1 \tag{1.97}$$

When γt (or t/δ) is very small (i.e., $\gamma t << 1$), the exponential term in Eq. (1.96) may be expanded as a power series and only the first two terms retained. This results in:

$$SE_{dB} \cong 20 \log \left[e^{t/\delta} \left| \frac{K}{4} (2\sqrt{2}\, t/\delta) \right| \right] \tag{1.98}$$

$$\cong 20 \log \left[(1 + t/\delta) \frac{K\sqrt{2}\, t/\delta}{2} \right] \tag{1.99}$$

$$\cong 20 \log (0.707Kt/\delta), \text{ for } t/\delta << 1 \text{ and } K >> 1 \tag{1.100}$$

The t/δ term of Eq. (1.100) may be compared with Eq. (1.35) which involves the metal impedances for $t/\delta << 1$. Equation (1.100) applies only when the metal thickness is a small fraction of its skin depth at any frequency. On the other hand, when t/δ is >> 1, Eq. (1.97) becomes:

$$SE_{dB} = 20 \log (Ke^{t/\delta}/4), \text{ for } t/\delta >> 1 \text{ and } K >> 1 \tag{1.101}$$

Using the relation $\delta = \sqrt{2}/\sigma Z_m$ defined in Eq. (1.26), replacing K by k $377/Z_m$ (as defined with Eq. 1.4) and inserting it into Eq. (1.100) yields:

$$SE = 20 \log 0.707\ k \frac{377}{Z_m}\ t \frac{\sigma Z_m}{\sqrt{2}}$$

$$= 20 \log (188.5\ k\sigma t) \tag{1.102}$$

Assuming $\sigma = \sigma_{copper} \times \sigma_r$ and $\sigma_{copper} \cong 10^6/1.8$ mho/cm,

$$SE = 160.8 + 20 \log k\sigma_r\ t_{cm} \tag{1.103}$$

$$\text{for } t/\delta < 0.74 \text{ and } K \geqslant 10$$

While many sources (Refs. 1 through 4) give the calculation routine for applying the rereflection correction term B in the case of thin shields, the advantage of expressing the total (A + R + B) losses in one equation is evident. Equation (1.103) also shows that SE for extremely thin shields (or conductive coatings) becomes independent of frequency in the far field (k = 1) as long as t is smaller than δ. This is due to the fact that when frequency increases, the rereflection term decreases while the surface resistance increases. This constant value of SE will go on until the film thickness approaches one skin depth, where shielding efficiency will start to increase exponentially. This mechanism is characteristic of the conductive coatings described in Chapter 2, Section 2.2.

Equations (1.100) and (1.101) become equal when $2\sqrt{2}\ t/\delta = e^{t/\delta}$, i.e. when $t/\delta = 0.74$. This may be considered to be a **transition frequency**. Susbstituting Eq. (1.27) into Eq. (1.101) yields:

$$SE_{dB} = 20 \log \left(66.6\ ke^{t/\delta}/R_s \right) \tag{1.104}$$

$$= 8.68t\sqrt{\pi f \mu \sigma} + 20 \log (66.6k/\sqrt{\pi f \mu/\sigma}) \text{ for } t/\delta > 0.74 \text{ and } K \geqslant 10 \tag{1.105}$$

$$= 108 + 1{,}314t_{cm}\sqrt{f_{MHz}\mu_r\sigma_r} + 20 \log \left(k/\sqrt{f_{MHz}\mu_r/\sigma_r} \right) \tag{1.106}$$

where, $k = 1$ for far fields ($r \geqslant /2\pi$)

$= \lambda/2\pi r$ for near E-fields

$= 2\pi r/\lambda$ for near H-fields

R_s = surface resistivity in Ω/sq

$$= Z_m/\ (1 + j) = 1/\sigma\delta \tag{1.107}$$

To select which equation to use, $t/\delta = 0.74$ may be determined from the graph of δ of Fig. 1.6 at any frequency for the metals used. Equation (1.106) is plotted in Fig. 1.12 for values of R_s ranging from 10 mΩ to 30 Ω, k from emission-source-to-metal-barrier distance values of $10^{-5}\ \lambda/2\pi$ to $100\ \lambda/2\pi$ and $t/\delta = 0.74$.

The SE will ordinarily be greater than that shown in Fig. 1.12 (because of absorption loss) provided to $t/\delta > 0.74$. Note that for all values of k, S_{dB} is basically a reflection loss, viz, an impedance mismatch between the wave impedance ana the metal impedance. Thus, the objective is to economically achieve a low surface resistance using foils, conductive paints, flame-spray metals, meshes, screens, etc., as explained in Chapter 2.

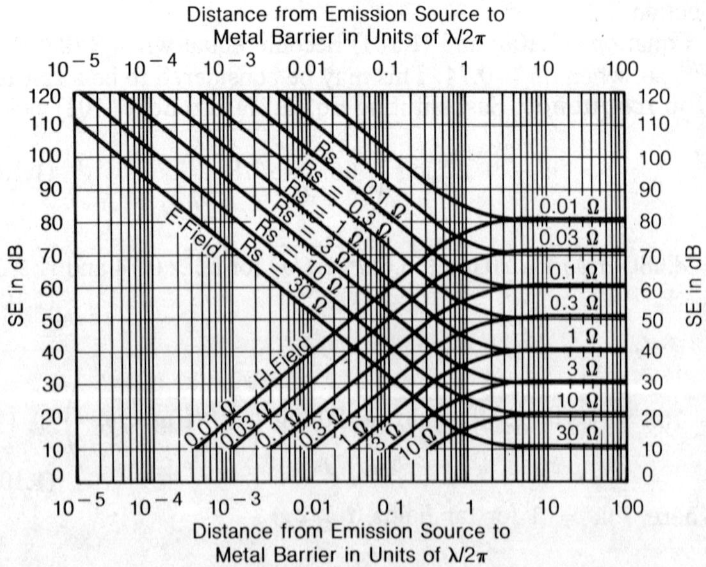

SdB = 20 log $(66.6ke^{t/\delta}/Rs)$
Rs = Surface Resistance in Ω/sq.
t/δ = 0.74, $K > 10$

Figure 1.12—Shielding Effectiveness vs. Metal-to-Emission Distance and Surface Resistances

1.4.5 Low-Frequency Magnetic SE

Equation (1.64) presented the overall SE expression applicable for far fields and for electric and magnetic fields in the near field for any thickness of any metal. The following equation is a substantial simplification of Eq. (1.64) results in and below very low frequency for magnetic materials when t/δ, K, and $2\gamma t \ll 1$:

$$SE_{dB} \cong 20 \log \left| \frac{1}{4K} [1 - (1 - 4K)(1 - 2\gamma t)] \right| \qquad (1.108)$$

$$\cong 20 \log \left| \frac{1}{4K}[1 - (1 - 4K - 2\gamma t)] \right| \tag{1.109}$$

for $8K\gamma t << 4K$ and $2\gamma t$

$$\cong 20 \log (1 + \gamma t/2K) \tag{1.110}$$

where, from Eqs. 1.68 and 1.79:

$$\gamma = (1 + j)\alpha = \sqrt{2\pi f\mu\sigma}$$

$$K = Z_w/Z_m = r\sqrt{2\pi f\sigma\mu_o/\mu_r} \quad \text{for magnetic fields}$$

Replacing γ and K by their values yields:

$$SE_{dB} = 20 \log \left(1 + \frac{t\sqrt{2\pi f\mu_o\mu_r\sigma}}{2r\sqrt{2\pi f\sigma\mu_o/\mu_r}} \right) \tag{1.111}$$

$$= 20 \log \left(1 + \frac{\mu_r t}{2r} \right), \text{ for } t/\delta, \text{ K and } 2\gamma t << 1 \tag{1.112}$$

Equation (1.112) is a useful and simple expression for very low frequency and dc SE with magnetic fields. It is equally useful to determine the SE of a wire pair, coaxial cable or harness having a magnetic shield of thickness, t, and radius, r. If a wire or harness shield is used, be sure that the **effective** relative permeability, μ_e, is used for braided shield since much of the shield may be air or other material. Also, the designer must check the μ_r he assumes for magnetic material against the EMI frequency of concern (see Chapter 2, Fig. 2.7). Equation (1.112) is plotted in Fig. 1.13.

Illustrative Example 1.1

Determine the low-frequency magnetic SE of a tin-plated, copper-clad steel braid in which tin = 3 percent, copper = 37 percent, and steel = 60 percent. The braid is 80 percent material and 20

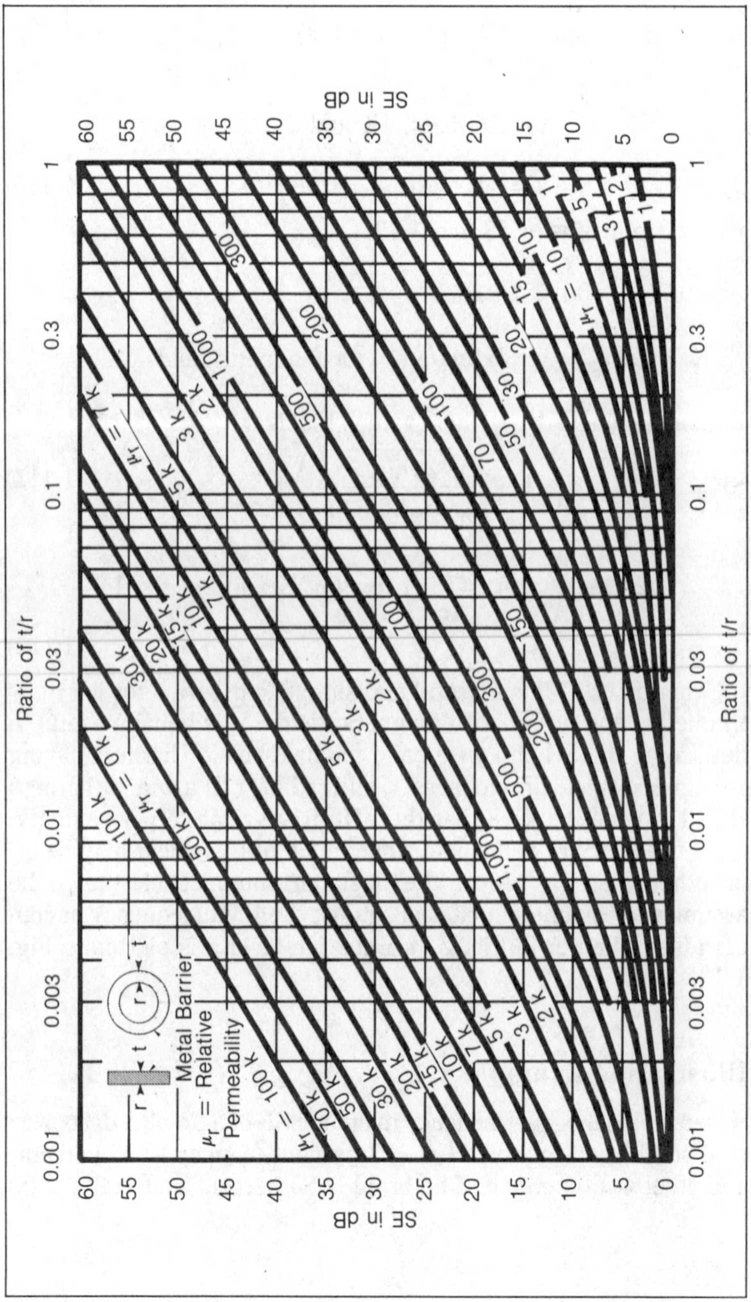

Figure 1.13—Low Frequency, Shielding Effectiveness to Magnetic Fields

percent air by volume and has an average thickness of t = 0.020" (5 mm).

The braided shield, when slipped over a harness has an average radius of 0.50" (12.7 mm):

$$\mu_e = M_{de}\, MM_{de}\, \mu_r + NMM_{de} + A_{de} \tag{1.113}$$

$$= M_{de} MM_{de}\, \mu_r, \text{ since } NMM_{de} << MM_{de}\, \mu_r \tag{1.114}$$

where,

M_{de} = decimal equivalent of total material by volume

MM_{de} = decimal equivalent of magnetic material by volume referenced to total material

μ_r = relative permeability of magnetic material ≅ 1,000 for steel

NMM_{de} = decimal equivalent of nonmagnetic material by volume referenced to total material

A_{de} = decimal equivalent of air by volume = $1 - M_{de}$

Eq. (1.114) for the above example becomes:

$$\mu_e = 0.80 \times 0.60 \times 1000 = 480$$

Thus, the low-frequency magnetic SE is:

$$SE_{dB} = 20 \log \left(1 + \frac{480 \times 0.020}{2 \times 0.50} \right) = 20.5 \text{ dB}$$

1.4.6 Performance Degradation

All preceding discussions assume that the shielding material is both homogeneous and large in planar dimensions such that neither leakages nor edge effects are significant. This would otherwise result in a shield whose performance is compromised. The SE expressed by Eq. (1.64) may then be conceptually redefined:

$$SE'_{dB} = A_{dB} + R_{dB} + B_{dB} - \text{leakage effects} - \text{standing wave effects} \qquad (1.116)$$

where, A_{dB} = absorption loss, Eq. (1.65) or (1.72)

R_{dB} = reflection loss, Eq. (1.66) or (1.82)

B_{dB} = rereflection correction, Eq. (1.67) or (1.91)

Leakage effects in Eq. (1.116) may be identified as due to one or more of the following situations which exist in any practical, real-life shield:

1. Seams:
   Mating members (screwed, riveted, welded etc.)
2. Holes and apertures for:
   Connectors, components, cable entries, cooling, displays etc.
3. Doors and access covers
4. Vents:
   Ventilation, heating etc.
5. Nonhomogenous areas:
   Screens, meshes etc.

A complete review of these leakages, their calculation and their reduction is presented in Chapters 7, 8 and 10.

Degradation due to standing waves in Eq. (1.116) involves resonance effects at higher frequencies where enclosures act as resonant cavities. This results in areas or regions within a shielded enclosure which exhibit poorer performance, i.e., lower SE.

Equation (1.116) is expressed in conceptual rather than explicit form. The method of combining effects due to aperture leakages (L_{dB}) and the base metal SE (without these effects), can be

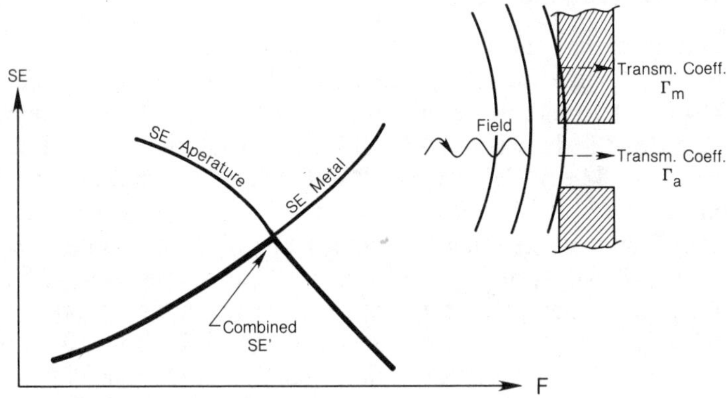

$$SE_{metal} = 1/\Gamma_m \qquad SE' = \frac{1}{\Gamma_m + \Gamma_a}$$

$$SE_{aperture} = 1/\Gamma_a$$

Example: if $SE_m = 80 \text{ dB} \rightarrow \Gamma_m = 0.0001$

$SE_a = 60 \text{ dB} \rightarrow \Gamma_m = 0.001$

$SE' = -20 \text{ Log } (10^{-4} + 10^{-3}) \cong 59 \text{ dB}$

Figure 1.14—Conceptual View of the Combination of Metal Attenuation with Aperture Attenuation

visualized as the addition of two transmission coefficients (Fig. 1.14). The combined effect is:

$$SE'_{dB} = -20 \log \left[\log^{-1}\left(-SE_{dB}/20 \right) + \sum_{i=1}^{n} \log^{-1}\left(-L_{idB}/20 \right) \right] \tag{1.117}$$

Note that Eq. (1.117) coherently adds all leakage effects together and coherently adds these results with SE_{dB}. This is a worst case since the phases may range from nearly all in phase (coherent at low frequencies) to approaching random phase (incoherent at high frequencies). However, usually one or a few leakages dominate so that coherent worst-case combination is justified. If coherent addition did not take place (i.e. effects combine randomly or incoherently), the three "20s" in Eq. (1.117) would be replaced by three "10s."

Illustrative Example 1.2

Suppose that the metal of an enclosure results in an SE of 110 dB at some frequency, but the leakages separately yield an SE of: (1) filter/connector panel = 101 dB, (2) vent leakage = 92 dB, (3) door leakage = 88 dB and (4) shield panel seam leakage = 83 dB. Calculate the total leakage attenuation and the combined SE of the enclosure.

Application of the second half of Eq. (1.117) yields a total leakage SE of 76.8 dB. Applying all of Eq. (1.117) gives SE'_{dB} = 76.6 dB. Thus, in this example, it does not particularly matter what the basic enclosure material performance is as long as it is about 10 dB better than the lowest, or least, leakage point. This may be illustrated by applying Eq. (1.117) for several different values of SE_{dB} in the above example having a total leakage SE of 76.8 dB.

Basic Enclosure Metal SE_{dB}	Combined Attenuation, All Leakages L_{dB}	Combined Shielding SE'_{dB}
60 dB	76.8 dB	58.8 dB
70 dB	76.8 dB	66.7 dB
80 dB	76.8 dB	72.2 dB
90 dB	76.8 dB	75.1 dB
100 dB	76.8 dB	76.2 dB
110 dB	76.8 dB	76.6 dB
120 dB	76.8 dB	76.7 dB
130 dB	76.8 dB	76.8 dB
⩾150 dB	76.8 dB	76.8 dB

The above example demonstrates the importance of controlling leakages in shielded enclosures. To state it in simple terms: **The SE of a finished box cannot be better than that of its worst leakage.** Usually, but not always, a leakage results in a poorer performance than that of the base metal. One exception is for low-frequency magnetic fields where the phase shift through the leakage can combine out of phase with the penetration of the base shield metal. The topics of leakages and the techniques for controlling them in shielded housing designs are detailed in Chapters 5 through 11 of this book.

As it develops then, the SE of a useful six-sided homogeneous box may be regarded as an upper limit which approaches the analytical relations and design graphs of this handbook. Where overall SE is relatively low, say less than about 40 dB, there usually exists a good correlation between theory and practice, except possibly for low-frequency magnetic-field sources. When SE_{dB} is between 40 and 100 dB, for example, the correlation between theory and practice may range from good to poor depending upon (1) the degree of departure from an homogeneous enclosure, (2) the frequency and (3) the degree to which the leakage points or areas have been shored up. Where correlation is not good, the measured values are almost always below theoretical values. For theoretical SE_{dB} values in excess of about 100 dB, correlation between theory and practice is generally poor since there are many leakage joints in practical enclosures.

1.4.6.1 More about Leakages

Metals which, across all their surface, do not exhibit nearly identical conductivity or surface impedance and the same thickness throughout, may be regarded as nonhomogeneous barriers (see Chapter 2). Examples of these include:

(1) The situations listed in the previous section

(2) Any practical and useful six-sided enclosure, whether at the circuit, equipment, rack, console or room-size level

(3) Conductive paints, flame-spray deposits, very thin depositions, metalized textiles, screens, wire mesh and the like

From the above list, it would appear that no useful shielded enclosure qualifies as a homogeneous metal for which the relations in this handbook can be used. In a strict sense, this is true since at the very least any enclosure, to be useful, must have wires (or fiber optics) entering or leaving the enclosure.

Not only do these wires constitute either a radiating or pickup **antenna-like** source (except fiber optics), all require holes in the enclosure in order to be admitted. Again, Chapters 5 through 11 of this handbook discuss these situations at length.

For the relations of Section 1.4 to apply, the metal should be homogeneous and infinite in the planar dimensions. As a practical matter, as long as either (1) the metal sheet planar dimensions are much greater than the distance between an emission source and the shield or (2) a six-sided enclosed box or parallelpiped is used, then the SE relations become an upper limit of performance.

Lack of enclosure homogeneity gives rise to low-frequency resonance effects which result from unequal phase shifts in wave transmission to a receiving point via two or more paths: one via the base shielding material and another or others via the seam-leakage paths. The resonant frequencies correspond to those situations for which the path lengths through the metal* are odd multiples of a half wavelength greater than through the leakages (usually air gaps). Thus, from Eq. (1.71), for the path length difference to be $\beta t = \pi$ (i.e., 180° path delay), the lowest resonant frequency is obtained:

$$e^{j\beta t} = e^{j\pi} = e^{jt\sqrt{\pi f \mu \sigma}} \tag{1.118}$$

or,

$$f_r = \frac{\pi}{t^2 \mu \sigma} = \frac{6.68 \times 10^7}{t^2_{mils} \mu_r \sigma_r} \text{ English units} \tag{1.119}$$

$$= \frac{4.31 \times 10^4}{t^2_{mm} \mu_r \sigma_r} \text{ metric units} \tag{1.120}$$

Illustrative Example 1.3

Calculate the first magnetic-field resonant frequencies of a shielding box fabricated from 1/32" aluminum and 1/32" steel. Here, t_{mils} = 31.25 mils (1/32" = 0.794 mm) with $\sigma_r = 0.63$ and $\mu_r = 1$ for aluminum, and $\sigma_r = 0.17$ and $\mu_r = 1{,}000$ for steel. From Eq. (1.119), the resonant frequencies are 109 kHz for aluminum and 402 Hz for steel. Notice that subsequent resonances will exist at all next multiples for $\beta t = n \times \pi$.

Below this resonance, the SE of a leaky box is fairly well defined by the SE relations of the base material. At and very near low-frequency resonance, the SE is greater; and above resonance, SE_{dB} is less than that defined by the performance of the base material. A further discussion of these topics is beyond the scope of this handbook because the results are seam-type and geometry related. For futher information, the reader is referred to the work

*Path length differences outside the metal result in negligible phase differences in air at low frequencies and are neglected.

of Schultz, Cooley[5] and others mentioned in the references in Chapter 4 of this handbook and again in Chapters 5 to 11.

1.4.6.2 Standing Waves

As a result of either penetrations or leakages, a six-sided shielded box or cabinet will also exhibit internal reflections. It can act as a cavity with standing waves at higher frequencies. Resonant frequencies, f_{MHz}, of such an enclosure exist at:

$$f_{MHz} = 150\sqrt{\left(\frac{k}{L}\right)^2 + \left(\frac{m}{h}\right)^2 + \left(\frac{n}{w}\right)^2}\ \text{MHz} \tag{1.121}$$

where, L = length of box or enclosure in meters

h = height of box or enclosure in meters

w = width or depth of box or enclosure in meters

k, m, n = a positive integer 0,1,2,3... but not more than one of k, m or n can be zero at the same time to support any resonant mode

The microwave cavity resonance effect here is not to be confused with the low-frequency resonance discussed in the previous section, Section 1.4.6.1. The two effects are entirely different phenomena.

From Eq. (1.21), if the box or enclosure dimensions are all comparable, i.e., are approximately equal, then the resonant frequencies, f_{MHz}, corresponding to the lowest order TE_{011}, TE_{101} or TE_{110} modes are:

$$f'_{MHz} \cong 212/L \cong 212/h \cong 212/w \tag{1.122}$$

Thus, the SE of boxes or enclosures will be expected to exhibit wide variations on the inside, above those frequencies corresponding to approximately 212/d, where d is the average dimension in meters. Actually, this is not the overall box SE (defined as incident power/inside power) which changes, but its apparent value due to the nonuniform E/H distribution inside. These variations come about as a result of the tangential electric-field intensity becoming the least at the inside of the metal enclosure and rising to greater values within.

Illustrating the above, the lowest resonant-frequency effects associated various size enclosures are listed below:

Enclosure Size and Type	**Frequency**
0.305 m³ (1 ft³) box	696 MHz
1.52 m × 19" wide rack (11" or 0.56 m I.D.), 0.76 m (30") depth	220 MHz
2.44 m (8') × 3.66 m (12') × 6.1 m (20') shielded enclosure	47.8 MHz

The SE relations in this handbook do not account for any standing-wave effects. In general, the standing waves inside a six-sided box or enclosure will result in fields which vary from twice that through the basic enclosure material to very low values at the metal interfaces. Thus, a maximum 6 dB degradation in SE can be expected from enclosure resonances.

A summary of principal SE equations is presented in Fig 1.15 and Table 1.1 provides an overview of the SE of permeable and nonpermeable materials.

1. Near-Field Zone ($r < \lambda/2\pi$) of Magnetic-Loop Antenna (Low-Impedance Field)
1a. Low-Frequency Magnetic Fields, down to DC
2. Near-Field Zone ($r < \lambda/2\pi$) of Electrical Rod or Dipole Antenna (High-Impedance Field)
2a. Low-Frequency Electric Fields, Down to DC
3. Far-Field Zone ($r >> \lambda/2\pi$) of Either Type of Antenna (Wave Impedance 377 Ω)

Cases 1., 1a., High Magnetic Field Components:

$$SE = 131.43\, t\sqrt{f\mu_r\sigma_r} + 74.6 - 10 \log \left(\frac{\mu_r}{\sigma_r \cdot f \cdot r^2} \right) \text{dB} \geqslant 0$$

Cases 2., 2a., High Electrical Field Components:

$$SE = 131.43\, t\sqrt{f\mu_r\sigma_r} + 141.7 - 10 \log \left(\frac{\mu_r f^3 r^2}{\sigma_r} \right) \text{dB} \geqslant 0$$

Case 3., Antenna Far-Field, Plane Waves:

$$SE = 131.43\, t\sqrt{f\mu_r\sigma_r} + 108.1 - 10 \log \left(\frac{\mu_r}{\sigma_r} f \right) \text{dB} \geqslant 0$$

Case 1., Extremely Thin Shield, $t/\delta << 1$:

$$SE = 20 \log (1 + 500\, \mu_r t/r) \text{ dB}$$

Note: All cases have their lower and upper frequency limits for which $SE \geqslant 0$.

r = Source-to-Shield Distance in m
t = Shield Metal Thickness in mm
f = Frequency in MHz
σ = Skin Depth in mm
μ_r = Permeability Relative to Copper
σ_r = Conductivity Relative to Copper

Figure 1.15—Summary of Principal SE Equations Depending on Shield-to-Source Relationship

Table 1.1—Summary of Shielding Effectiveness of Permeable and Nonpermeable Materials

Permeable Materials	Radio Frequency	Absorption Loss A_{dB} All Fields	Reflection Loss, R_{dB}		
			Electric Fields	Magnetic Fields	Plane Waves
Magnetic ($\mu \geqslant 1{,}000$)	Low < 1 kHz	Bad	Excel.	Fail	Good
	Medium 1 - 100 kHz	Good	Good	Bad	Fair
	High > 100 kHz	Excel.	Fair	Poor	Fair
Non-Magnetic ($\mu = 1$)	Low < 1 kHz	Fail	Excel.	Bad	Good
	Medium 1 - 100 kHz	Bad	Excel.	Poor	Good
	High > 100 kHz	Good	Good	Fair	Fair

Assumptions:
Material Thickness: .8 mm (1/32")
Source Distance: 3m (10')
Radio Frequency: as shown

Attenuation Scores:
Excellent: > 150 dB
Good: 100-150 dB
Fair: 50-100 dB
Poor: 30-50 dB
Bad: 10-30 dB
Fail: < 10 dB

1.5 References

1. Bannister, P.R., "New Theoretical Expressions for Predicting Shielding Effectiveness for the Plane Shield Case," *IEEE Transactions on Electromagnetic Compatibility*, Vol. EMC-10, no. 1, March 1968.
2. Bridges, J.E.; Miller, D.A.; and Schulz, R.B., "Comparison of Shielding Calculations," *IEEE Transactions on EMC*, Vol. EMC-10, no. 1, March 1968, p. 175.
3. Ficchi, R.F., *Electrical Interference*, (New York: Hayden, 1964).
4. Schulz, R.B., et. al., "Shielding Theory and Practice," *Proceedings, Ninth Tri-Service Conference on Electromagnetic Compatibility*, Armour Research Foundation, Chicago, Ill., October 1963, pp. 597-636.
5. Cooley, W.W., "Low-Frequency Shielding Effectiveness of Nonuniform Enclosures," *IEEE Transactions on EMC*, Vol. 10, no. 1, March 1968, pp. 34-43.

1.6 Bibliography

Abramowitz, M. and Stegun, I.A., *Handbook of Mathematical Functions* (Boulder, Colorado: National Bureau of Standards, 1964).

Adams, W.S., "Graphical Presentation of Electromagnetic Shielding Theory," *Proceedings of the Tenth Tri-Service Conference on Electromagnetic Compatibility* (Chicago, Ill.: Armour Research Foundation, November 1964), pp. 421-499.

AF/BSD Exhibit 62-87, "Electro-Interference Control Requirements for Minuteman (WS 133B)," 6 December 1962.

Albin, A.L., "Optimum Shielding of Equipment Enclosures," *Electronic Design,* Vol. 8, no. 3, February 3, 1960.

Bannister, P.R., "The Quasi-near Fields of Dipole Antennas," *IEEE Transactions on Antennas and Propagation*, Vol. AP-15, September 1967, pp. 618-626.

Blewett, J.P., "Magnetic Field Configuration Due to Air Core Coils," *John Hopkins Applied Physics*, Vol. 18, p. 968, November 1947.

Boeing Aircraft Corporation D2-2-2444, "Electro-Interference Control Requirements (Equipment)," 5 March 1959.

Bozorth, R.M., *Ferromagnetism*, (Princeton, New Jersey: Van Nostrand, 1951).

Bridges, J.E.; Huenemann, R.G.; and Hegner, H.R., "Electric Field Shielding and Measurement," *IEEE EMC Symposium Record* (New York: IEEE), Vol. 27C80, Washington, D.C., July 18-20, 1967, pp. 173-177.

Bronwell, A.B., and Bean, R.E., *Theory and Application of Microwaves* (New York: McGraw-Hill, 1945), pp. 337-339.

Brush, D.R.; Schulz, R.B.; and Jorgensen, L., "Initial Magnetic Permeability Measurements of RF Shielding Materials," *8th IEEE EMC Symposium Record* (New York: IEEE), July 1966.

Brush, E.R.; Schulz, R.B.; and Jorgensen, L., "Low-Frequency Electrical Characteristics of RF Shielding Materials," *IEEE Transactions on Electromagnetic Compatibility,* Vol. 10, no. 1, March 1968, pp. 67-72.

Cole, N.H., "A Comparison of RF Shielding Materials," *Electronic Design*, Vol. 10, no. 20, September 27, 1962.

Cowdell, R.B., "New Dimensions in Shielding," *IEEE Transactions on EMC,* Vol. 10, no. 1, March 1968, pp. 158-167.

Cowdell, R.B., "Simplified Shielding," *1967 IEEE EMC Symposium Record* (New York: IEEE), Vol. 27C80, Washington, D.C., July 18-20, 1967.

Elight, H.B., *Tables of Integrals* (New York: MacMillan, 1947).

Erdelyi, A., Ed., *Tables of Integral Transforms*, Vol. 2 (New York: McGraw-Hill, 1954).

Filtron Company, Inc., "Interference Reduction Guide for Design Engineers," Prepared for U.S. Army Electronics Labs., Fort Monmouth, New Jersey, Vols. 1 and 2, Accession AD 619666 and AD 619667.

Foster, J.; Buegal, K.; and Sowa, C., "Electromagnetic Compatibility Lunar Orbiter," NASA, Document no. NAS 1-3800, January 29, 1965.

Goubau, G., "Waves on Interfaces," *IRE Transactions on Antennas and Propagation*, Vol. AP-7, December 1959, pp. S140-S147.

Haber, F., "Generation of Standard Fields in Shielded Enclosures," *Proceedings of IRE*, Vol. 42, November 1954.

Haber, F., "Study of GSFC Radio Frequency Interference (RFI) Design Guideline for Aerospace Communication Systems," Moore School of EE, University of Pennsylvania, Report no. 66027 for NASA, April 30, 1966.

Harrison, C.W., Jr., "Transient Electromagnetic Field Propagation through Infinity Sheets into Spherical Sheets and into Hollow Cylinders," *IEEE Transactions Antennas and Propagation*, Vol. AP-12, May 1964, pp. 319-334.

Harrison, C.W., Jr., and Papas, C.H., "On the Attenuation of Transient Fields by Imperfectly Conducting Spherical Shells," *IEEE Transactions on Antennas and Propagation*, Vol. AP-13, November 1965, pp. 960-966.

Hays, J.B., "Protecting Communications Systems from EMP Effects of Nuclear Explosions," *IEEE Spectrum*, Vol. 1, May 1964, pp. 115-122.

Hollway, D.L., "Screen Rooms and Enclosures," *Proceedings IREA*, Vol. 21, no. 10, October 1960, pp. 660-668.

Jackson, J.D., *Classical Electrodynamics* (New York: Wiley and Sons, 1962).

Jahnke, E., and Emde, F., *Tables of Functions with Formulas and Curves* (New York: Dover, 1945).

Jordan, E.C., *Electromagnetic Waves and Radiating Systems* (Englewood Cliffs, New Jersey: Prentice-Hall, 1950).

Kaden, H., *Wirbelstrome und Schirmung in der Nachrichtentechnik* (Berlin: Springer, 1959), pp. 71-85, 93-107.

Karakash, J.J., *Transmission Lines and Filter Networks* (New York: MacMillan, 1950).

King, L.V., "Electromagnetic Shielding at Radio Frequencies," *Phil. Mag. J. Sci.*, Vol. 15, February 1933, pp. 201-223.

Kozakoff, D.J.; Bolt, A.T.; and Howard, F.D., "Shielding Effectiveness of Various Shaped Geometrical Enclosures in Terms of Normalized Parameters," *1970 IEEE Regional Electromagnetic Compatibility Symposium Record*, Vol. 70C64-REGEMC, San Antonio, October 6-8, 1970, pp. V-A-1 to V-A-11.

Larmor, J., "Electromagnetic Induction in Conducting Sheets and Solid Bodies," London, Edinburgh, and Dublin Phil. Mag. and J. Science, Ser. 5, Vol. 17, January 1884.

Lien, R.H., "Radiation from a Horizontal Dipole in a Semi-infinite Dissipative Medium," *John Hopkins Applied Physics.*

Lyons, W., "Experiments on Electromagnetic Shielding at Frequencies Between One and Thirty Kilocycles," *Proceedings of IRE,* Vol. 21, April 1933, pp. 574-590.

Maxwell, J.C., "On the Induction of Electric Currents in an Infinite Plane Sheet of Uniform Conductivity," London, Edinburgh, and Doublin Phil. Mag. and J. Science (Proc. Royal Soc.), Ser. 4, Vol. 43, no. 289.

Mei, K., and Van Bladel, "Low-Frequency Scattering by Rectangular Cylinder," *IEEE Transactions Antennas and Propagation*, Vol. AP-11, January 1963, pp. 52-56.

Meindl, J.D., "The Calculation of Field External to Shielded Transmission Lines," Ph.D. dissertation, Carnegie Institute of Technology, Cleveland, Ohio, 1958.

Miedzinski, J., "Electromagnetic Screening Theory and Practice," British Electrical and Allied Research Assn., Technical Report N/T 135, 1959.

Miller, D.A., and Bridges, J.E., "Geometrical Effects on Shielding Effectiveness at Low Frequencies," *IEEE Transactions on Electromagnetic Compatibility*, Vol. EMC-8, December 1966, pp. 174-187.

Miller, D.A., and Bridges, J.E., "Review of Circuit Approach to Calculate Shielding Effectiveness," *IEEE Transactions on Electromagnetic Compatibility*, Vol. 10, no. 1, March 1968, pp. 52-62.

Morecroft, J.H., and Turner, A., "The Shielding of Electric and Magnetic Fields," *Proceedings of the IRE*, Vol. 13, August 1925, pp. 477-505.

Moser, J.R., "Low-Frequency Shielding of a Circular Loop Electromagnetic Field Source," *IEEE Transactions on EMC*, Vol.

EMC-9, March 1967, pp.6-18.

Moser, J.R., "An Empirical Study of ELF and VLF Shield Cans," *IEEE Transactions on EMC,"* Vol. 10, no. 1, March 1968, pp. 112-125.

Plantz, V.C.; Schulz, R.B.; and Goldman, R., "Temperature Stress and Nuclear Radiation Effects on Electromagnetic Shielding" (Abstract), *IEEE Transactions on EMC*, Vol. 10, no. 1, March 1968, p. 129.

"Proceedings of the Magnetics Workshop," NASA Technical Memo 33-216, 1965.

Ramo, S., Whinnery, J.R., *Fields and Waves in Modern Radio*, 2nd Edition (New York: John Wiley and Sons, 1953), p. 244.

Ramo, S.; Whinnery, J.R.; and Van Duzer, *Fields and Waves in Communication Electronics* (New York: John Wiley and Sons, 1965), p. 294.

Ryan, C.M., "Computer Expression for Predicting Shielding Effectiveness for the Low-Frequency Plane Shield Case," *IEEE Transactions on EMC*, Vol. EMC-10, September 1967, pp. 83-94.

Salati, O.M., "Recent Developments in Interference," *IRE Transactions on Radio Frequency Interference*, Vol. RFI-4, no. 2, May 1962, pp. 24-33.

Sanford, R.L., "Basic Magnetic Quantities and the Measurement of the Magnetic Properties of Materials," NBS Monograph 47, 1962.

Schelkunoff, S.A., "Transmission Theory of Plane Electromagnetic Waves," *Proceedings of the IRE*, Vol. 25, November 1937, pp. 1457-1492.

Schelkunoff, S.A., *Electromagnetic Waves* (Princeton, New Jersey: Van Nostrand, 1943), pp. 223-225, 303-306.

Schulz, R.B.; Plantz, V. C.; and Brush, D. R., "Low-Frequency Shielding Resonance," *IEEE Transactions on EMC*, Vol. 10, no. 1, March 1968, pp. 7-15.

Schwartz, R.F., "Bibliography on Radio-Frequency Shielding," Moore School of Electrical Engineering, University of Pennsylvania, Philadelphia, 1954.

Shenfeld, S., "Shielding of Cylindrical Tubes," *IEEE Transactions on EMC*, Vol. 10, no. 1, March 1968, pp. 29-34.

Sommerfeld, A., "On the Propagation of Waves in Wireless Telegraphy," (in German), *Ann. der Phys*., Vol. 81, December 1926, pp. 1135-1153.

Stirrat, W.A., "Antenna Model for RFI Work," *Proceedings, Ninth Tri-Service Conference on Electromagnetic Compatibility*, Chicago, Ill., 1963.

Stirrat, W.A., "USA ECOM Contribution to Shielding Theory," *IEEE Transactions on EMC*, Vol. EMC-10, no. 1, March 1968, pp. 63-66.

Stratton, J.A., *Electromagnetic Theory* (New York: McGraw-Hill, 1941).

Weber, E., *Electromagnetic Fields* (New York: John Wiley and Sons, 1950).

CHAPTER 2

Shielding Materials and Performances

Among other things, the previous chapter indicated that for the equations of Section 1.4 to apply, the shielding material characteristics must be homogeneous and independent of frequency, field strength or other saturating effects (i.e., they are lincar). These equations apply, with restrictions, to pseudohomogeneous metals such as conductive paints, coatings, depositions, flame-spray processes and the like. Except at lower frequencies, they do not apply for small aperture materials such as screens, meshes and metalized textiles, which follow an entirely different law. These and other related topics on shielding materials are presented in this chapter.

2.1 Shielding Materials

This section summarizes the properties of homogeneous metals, pseudo-homogeneous metals and small-aperture metals.

2.1.1 Homogeneous Metals

Many of the metals and their associated conductivities and permeabilities which the designer might expect to encounter are listed in Table 2.1 The fourth column, entitled Product $\sqrt{\sigma_r \mu_r}$, is a ranking of the latent absorption loss of metals relative to copper. For nonmagnetic materials, except silver, all σ_r μ_r values relative to copper are less than 1.0. Thus nonmagnetic materials provide less absorption loss than copper.

Table 2.1—Relative Conductivity and Permeability of Metals

(1) Metal	(2) Relative Conduct. σ_r	(3) Relative Permbly. μ_r @ ≤10kHz	(4) Product $\sqrt{\sigma_r\mu_r}$ $A = k_1\sqrt{\sigma_r\mu_r}$	(5) Quotient $\sqrt{\sigma_r/\mu_r}$ $R = k\sqrt{\sigma_r\mu_r}$	(6) Relative Reflection R_{dB}
1. Silver	1.064	1	1.03	1.3	+0.3 dB
2. Copper (Solid)	1	1	1	1	0
3. Copper (Flame Spray)	0.10	1	0.32	0.32	−10.0
4. Gold	0.70	1	0.88	0.88	−1.1
5. Chromium	0.664	1	0.81	0.81	−1.8
6. Aluminum (Soft)	0.63	1	0.78	0.78	−2.1
7. Aluminum (Tempered)	0.40	1	0.63	0.63	−4.0
8. Aluminum (Household Foil, 1 mil)	0.53	1	0.73	0.73	−2.8
9. Aluminum (Household Foil, 1 mil)	0.61	1	0.78	0.78	−2.1
10. Aluminum (Flame Spray)	0.036	1	0.19	0.19	−14.4
11. Brass (91% Cu, 9% Zn)	0.47	1	0.69	0.69	−3.3
12. Brass (66% Cu, 34% Zn)	0.35	1	0.52	0.52	−5.7
13. Magnesium	0.38	1	0.61	0.61	−4.3
14. Zinc	0.305	1	0.57	0.57	−4.9
15. Tungsten	0.314	1	0.56	0.56	−5.0
16. Beryllium	0.33	1	0.53	0.53	−5.5
17. Cadmium	0.232	1	0.48	0.48	−6.3
18. Platinum	0.17	1	0.42	0.42	−7.6
19. Tin	0.151	1	0.39	0.39	−8.2
20. Tantalum	0.12	1	0.33	0.33	−9.6
21. Lead	0.079	1	0.28	0.28	−11.0
22. Monel (67 Ni, 30 Cu, 2 Fe, 1 Mn)	0.041	1	0.20	0.20	−13.9
23. Manganese	0.039	1	0.20	0.20	−14.1
24. Titanium	0.036	1	0.19	0.19	−14.4
25. Mercury (Liquid)	0.018	1	0.13	0.134	−17.4
26. Nichrome (65 Ni, 12 Cr, 23 Fe)	0.0012	1	0.035	0.035	−29.2
27. Supermalloy	0.023	100,000	53.7	0.0005	−65.4
28. 78 Permalloy	0.108	8,000	29.4	0.0037	−48.7
29. Purified Iron	0.17	5,000	29.2	0.0058	−44.7
30. Conetic AA	0.031	20,000	28.7	0.0011	−58.8
31. 4-79 Permalloy	0.0314	20,000	25.1	0.0013	−58.0
32. Mumetal	0.0289	20,000	24.0	0.0012	−58.4
33. Permendur (50 Cu, 1-2V, & Fe)	0.247	800	14.1	0.018	−35.1
34. Hypernick	0.0345	4,500	12.5	0.0028	−51.1
35. 45 Permalloy (1200 Anneal)	0.0384	4,000	12.4	0.0031	−50.2
36. 45 Permalloy (1050 Anneal)	0.0384	2,500	9.80	0.0039	−48.1
37. Hot-Rolled Silicon Steel	0.0384	1,500	7.59	0.0051	−45.9
38. Sinimax	0.0192	3,000	7.59	0.0025	−51.9
39. 4% Silicon Iron (Grain Oriented)	0.037	1,500	7.45	0.0050	−46.1
40. 4% Silicon Iron	0.029	500	3.81	0.0076	−42.4
41. 16 Alfenol	0.0113	4,500	7.13	0.0016	−56.0
42. Hiperco	0.069	650	6.70	0.010	−39.7
43. Monimax	0.0216	2,000	6.57	0.0033	−49.7
44. 50% Nickel Iron	0.0384	1,000	6.20	0.0062	−44.2
45. 45-25 Perminvar	0.091	400	6.03	0.015	−36.4
46. Commercial Iron (0.2% Impure)	0.17	200	5.83	0.29	−30.7
47. Cold-Rolled Steel	0.17	180	5.53	0.031	−30.2
48. Nickel	0.23	100	4.70	0.047	−26.6
49. Stainless Steel(1Cu,18Cr,8Ni,&Fe)	0.02	200	2	0.010	−40.0
50. Rhometal (36 Ni)	0.019	1,000	4.36	0.0044	−47.2
51. Netic 53-6	0.172	300	7.18	0.024	−32.4

All magnetic materials, on the other hand, have relative absorption loss values exceeding two and are relatively good absorbers of energy at low frequencies compared to nonmagnetic materials. On the other hand, since the relative permeability degrades with frequency, magnetic materials offer absorption losses less than nonmagnetic materials at higher frequencies (above approximately 100 kHz).

For thin-metal thicknesses (thicknesses for which $t/\delta << 1$, i.e., the metal thickness is a small percentage of the skin depth) the absorption loss is negligible (see Eqs. (1.98) through (1.100)) and the choice of metal, whether magnetic or nonmagnetic, for shielding effectiveness (SE) based on absorption loss alone is unimportant. However, when the metal is not thin (when t/δ is not $<< 1$), the absorption loss becomes significant. Apart from cost and other considerations, columns 2 and 3 are useful in choosing a metal for shielding or simply in extracting the applicable σ_r and μ_r values for elsewhere in this handbook.

The fifth column of Table 2.1, entitled **Quotient** $\sqrt{\sigma_r/\mu_r}$, is a ranking of the latent reflection loss of metals relative to copper. Expressed in dB form in column six, the nonmagnetic materials outperform nearly all magnetic materials for reflection-loss performance. For thin-metal thickness (thicknesses for which $t/\delta << 1$), the reflection loss is the only important term for overall SE. For this condition then, the designer would select a nonmagnetic material of higher conductivity.

For conductive coatings on glass or plastic substrates used for RF shields in which optical viewing is required, where t is measured in nanometers (a tiny fraction of δ), the more stable (lower on the electrochemical series), higher-conductivity metals would be selected, such as gold. The SE for gold corresponding to $t/\delta << 1$ is plotted in Figs. 2.1 and 2.2.

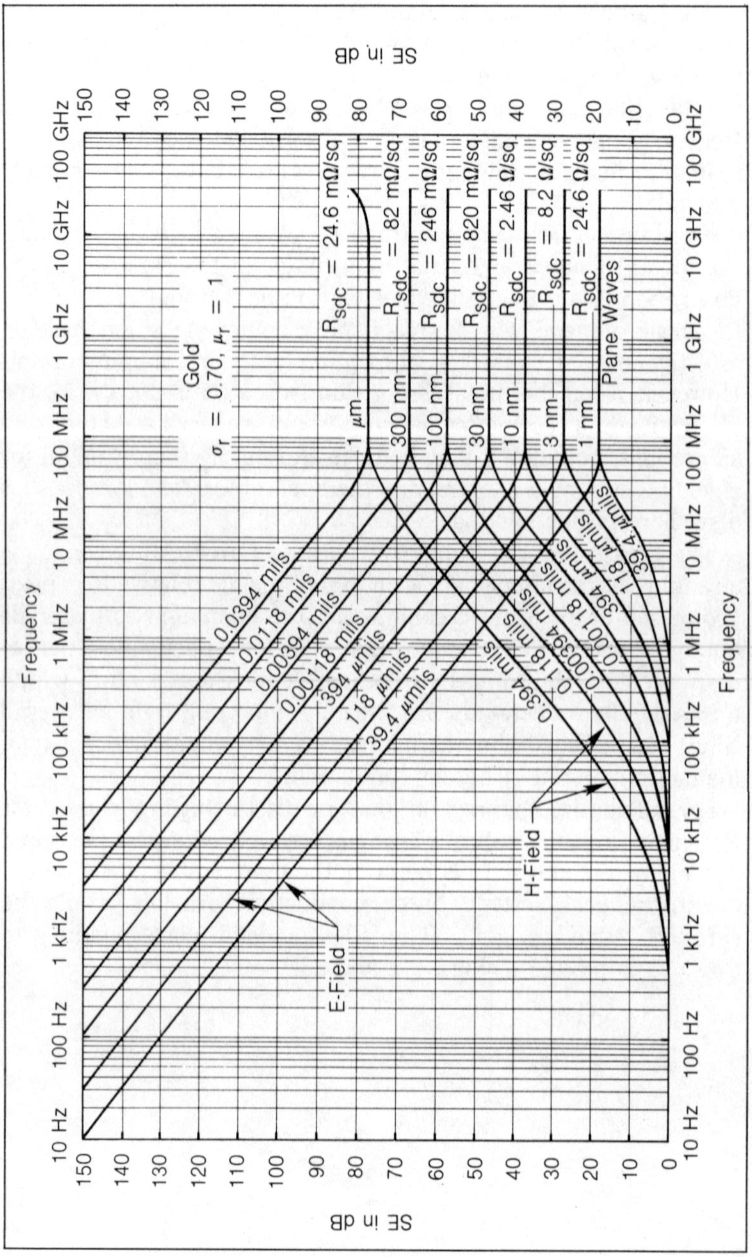

Figure 2.1—Shielding Effectiveness of Gold vs. Frequency and Thickness for Source-to-Metal Distance of 1 m

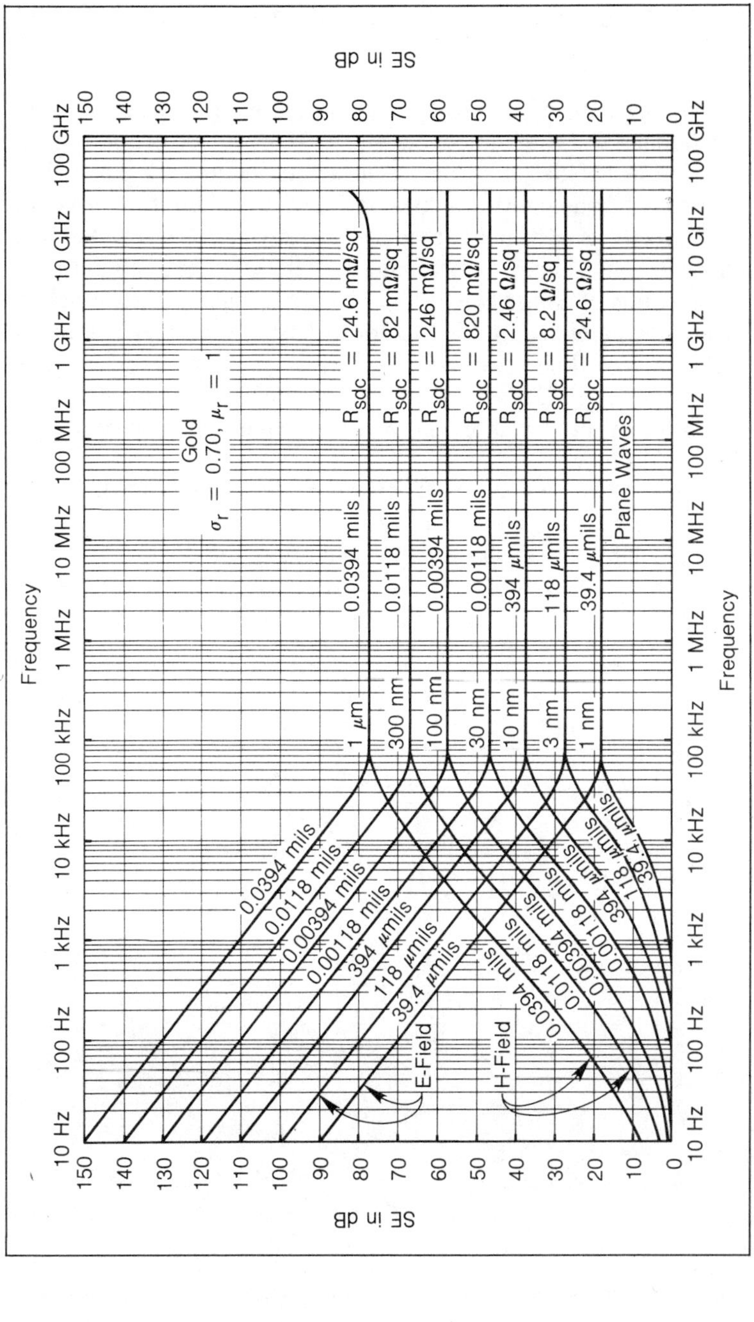

Figure 2.2—Shielding Effectiveness of Gold vs. Frequency and Thickness for Source-to-Metal Distance of 1 km

2.1.1.1 Numerical Value of Relative Permeability

When dealing with magnetic shielding materials, the question arises regarding the numerical value of relative permeability to be used in the calculation of SE discussed in Section 1.4. The values of relative permeability, μ_r, given in the various literature are often used directly as a fixed, material parameter which is unvarying with external conditions. Unfortunately, materials that fall into the classification of **magnetic**, i.e., $\mu_r > 1$, are sensitive to external environmental effects, not the least of which is the ambient magnetic environment. The evaluation of SE_{dB} in a realistic manner depends on the SE of the correct value of μ_r. Factors needing consideration in the development of accurate permeability values are discussed below.

Figure 2.3 shows the magnetization curve and saturation-to-saturation hysteresis curve of a **typical** material. In this figure various characteristics of the material are defined. H_c is the coercive force or the value of applied reverse magnetic field strength required to drive the magnetic induction (flux density) to zero.

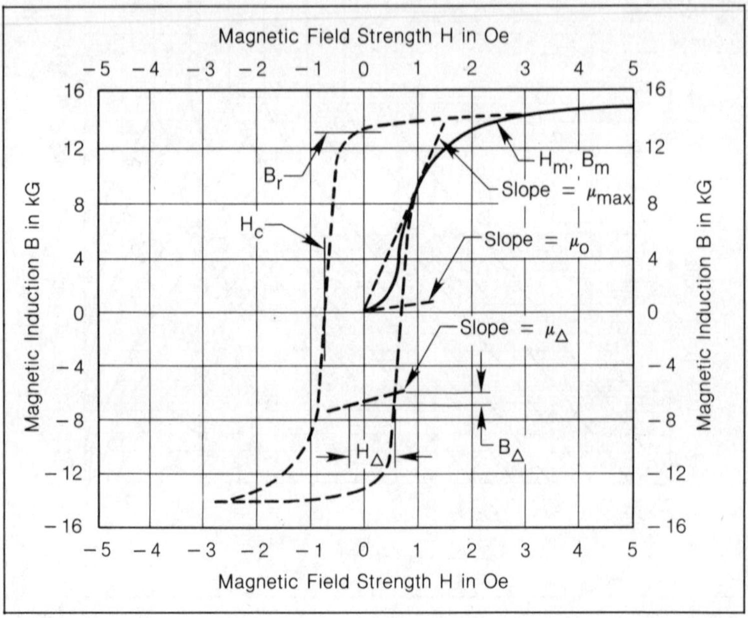

Figure 2.3—Magnetization Curve (Solid) and Hysteresis Loop (Dotted)

B_r is the residual magnetism or value of magnetic induction retained after a saturation value of induction, B_m, has been achieved by application of a magnetic field strength, H_m, which is then reduced to zero.

Relative permeability, μ_r, is defined as the ratio of magnetic induction, B, to applied magnetic field strength, H. Examination of Fig. 2.3 shows the definition of three different relative permeabilities, μ_r: μ_0, the initial permeability; μ_{max}, the maximum permeability; and μ_Δ, the incremental permeability. These three values of permeability can vary by more than an order of magnitude (or more than 10 dB of SE_H for reflection loss and by any amount for absorption loss). Since magnetic SE_{dB} values are usually low at low frequencies, this can result in a significant error.

The incremental permeability, μ_Δ, is of particular interest in this discussion. As shown in the figure as a minor loop on the hysteresis curve, it shows the effect of a low-level signal applied in addition to the hysteresis loop generation signal, at the single point shown, in opposition to the signal. This represents the situation when a magnetic field strikes a permeable shield.

Figure 2.4 shows the relative permeability μ_r as a function of applied magnetic field strength and induction. This clearly portrays the dependence of permeability upon the magnetic environment, a factor which must be considered in evaluating for the computation of SE_{dB}.

Removing the hysteresis curve while retaining the presence of the externally applied magnetic signal results in the situation shown

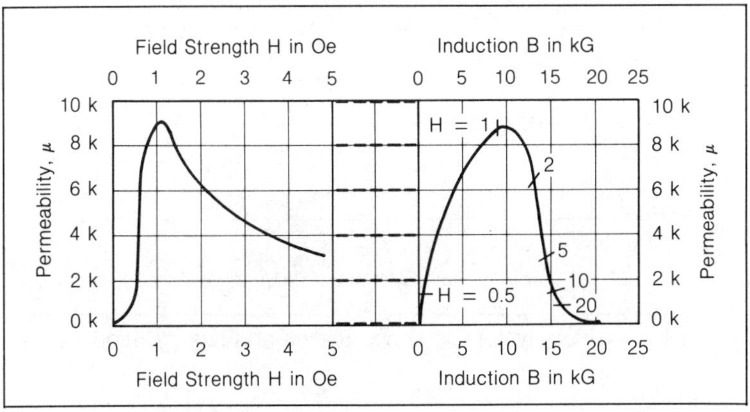

Figure 2.4—Permeability Curves of Iron, with μ Plotted Against H and B.

in Fig. 2.5, where the small loops generated by applied fields are shown on the magnetization curve. Note that μ_Δ will vary with applied field, in that the B/H slope varies, which defines μ_Δ.

The situation experienced by a shield material in real life is, of course, even more complex. Shields are only used in the presence of alternating fields. Thus, one typical situation would be a shield

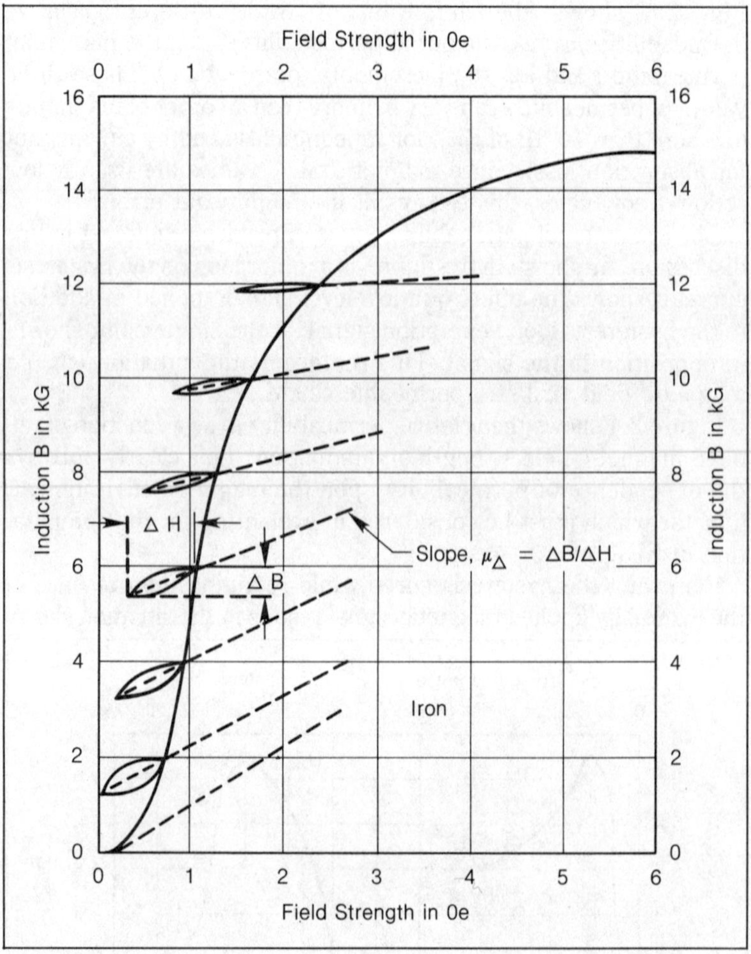

Figure 2.5—Minor Hysteresis Loops Shown on Magnetization Curve

existing in the earth's magnetic field ($\approx$ 0.5 Oersted)* of dc bias with a 60 Hz H-field superimposed. This causes a hysteresis effect and (at a minimum) the signal to be shielded. As shown in Fig. 2.6, this situation allows the estimation of an effective permeability, μ_e, to be used for computation of SE_{dB}:

$$\mu_e \cong \mu_\Delta = 2\ \mu_0 = 0.1\ \mu_m$$

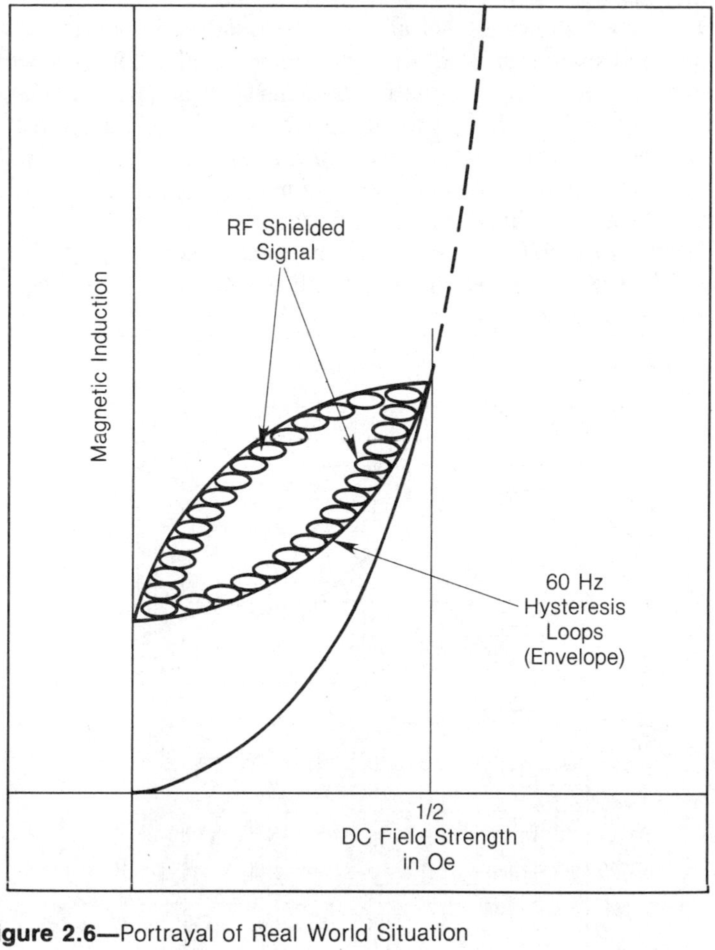

Figure 2.6—Portrayal of Real World Situation

*One oersted = $10^3/4\pi$ = 79.6 ampere-turns/meter.

In the more general situation, several effects take place including eddy current losses which are a function of (1) magnetic field strength, (2) frequency, (3) distance to the metal barrier, (4) metal thickness and (5) metal permeability and resistivity. Thus, it is concluded that unless all conditions are known, it is impossible to precisely define an equivalent μ_r. Consequently, the use of the μ_r values in Table 2.1 and the equations in Section 1.4 for magnetic materials bears some degree of approximation.

Figure 2.7 shows $\mu_r \rightarrow 1$ above a few hundred kilohertz for an undefined magnetic field strength; whereas Fig. 2.8 shows the behavior of μ_r versus magnetic flux density at 60 Hz. (1 gauss = 10^{-4} tesla = 160 dBpT). About all one might conclude from this is that below saturation and at very low frequencies, the permeability can increase up to about an order of magnitude above its 10 kHz value listed in Table 2.1. At and above medium frequencies (300 kHz to 3 MHz), $\mu_r \rightarrow 1$ for all materials except ferrites. Thus, the SE curves presented in the appendices apply only for the stated values of μ_r used.

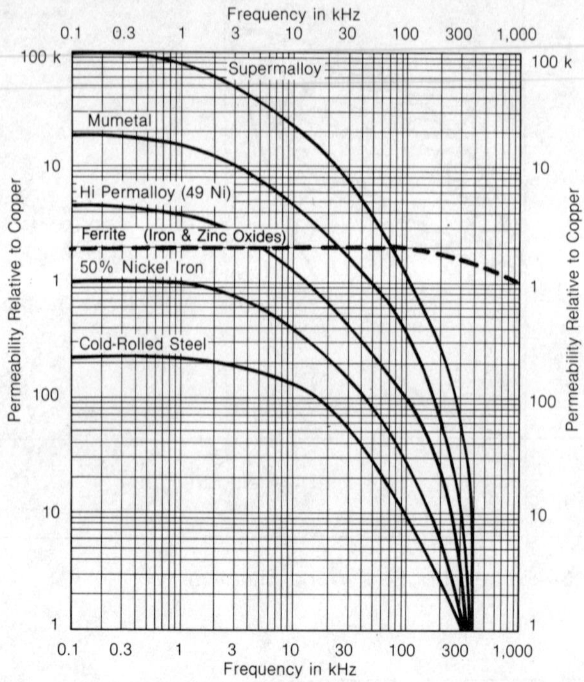

Figure 2.7—Relative Permeability vs. Frequency

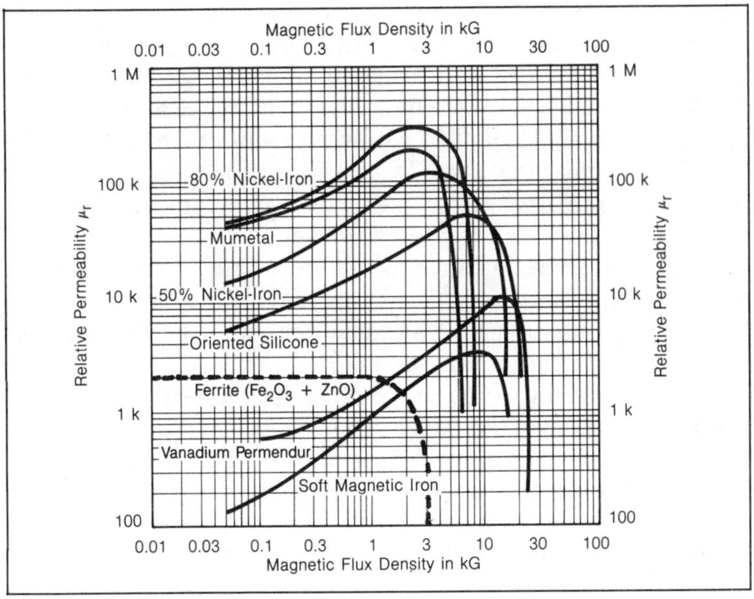

Figure 2.8—Relative Permeability vs. Magnetic-Flux Density at 60 Hz (courtesy of Magnetic Metals Co.)

When the maximum magnetic flux density may be exceeded, a double shield containing either a nonmagnetic metal or a higher saturation magnetic metal should face the more hostile magnetic field source. For hostile emissions coming from outside a box or cable shield, this means that this first layer of protective metal should face the outside, whereas if the emissions originate within, the protective metal should face the inside.

2.1.1.2 Availability and Applications

Several of the metals in Table 2.1 are available off-the-shelf in sheet stock form from thicknesses of about 0.4 mm (0.0156") or less to about 3.2 mm (0.125") or more. Metals having thicknesses less than 0.4 mm (0.0156") are sometimes regarded as foils. Many of the high-permeability metals come in foil thicknesses ranging from about 0.025 to 0.254 mm (1 to 10 mil). They are usually available in both sheet and tape form. The foil stock is also available in the form of adhesive-backed foil in roll lengths typically up to 30.5 m (100').

The thinner nonmagnetic foils, whose thickness is of the order of a few mils, are widely used for radio frequency (RF) shielding. While there are many stories spread around about how kitchen-type aluminum foil saved the day, the appendices clearly show that one mil of household aluminum foil can produce SE to plane waves and electric fields in excess of 80 dB below 1 GHz (see Section 2.2). Its performance against low-frequency magnetic fields, however, is very poor, i.e., it is nearly transparent.

Metal foils are used in a number of ways. One rather interesting application involves metal-foil wallpaper (MFWP) for converting an entire room into a limited shielded enclosure. It exhibits considerably less SE than purchased shielded rooms, but also at a much lower cost. These materials have been used on occasion to construct shielded chambers of thousands of square feet (hundreds of square meters) in surface area down to small equipment enclosures.

The MFWP must be used with other materials such a pressure-sensitive metal-foil tape, conductive adhesives, conductive epoxy or caulking compounds. They also require taking measures to reinstate lost shielding integrity at doors, windows and the like.

MFWP usually comes in thicknesses of 2 or 3 mils (51 to 76 microns) and is made of either aluminum or copper foil or special stainless steel foil of relatively high conductivity and high permeability.

Representative SE is 25 to 40 dB for magnetic fields at 200 kHz, 80 to 100 dB for electric fields from 200 kHz to 10 MHz, and 60 to 80 dB for plane waves above 400 MHz (see Section 2.2) when measured in accordance with MIL-STD-285 (see Chapter 3). Sometimes thin foils of the order of 25.4 μm (1 mil) are bonded to (metalized on) a plastic base such as 125 to 150 μm (5 to 10 mil) mylar. They can then be used as an air-inflatable structure which performs as a shielded enclosure. Some of the magnetic sheet metal and foil manufacturers who make the basic stock supplied to the electromagnetic interference/electromagnetic compatibility (EMI/EMC) and related communities are listed in Appendix I of this book.

2.1.2 Metallic and Other Conductive Coatings

Metals which lack homogeneity but which do not **intentionally** have holes, slits, or other apertures (whether small or large) are called pseudohomogeneous metals (PHMs) in this handbook. Examples of PHMs include conductive paints and coatings and the arc or flame-spray process of metalizing an insulator. A PHM may have thin areas because lack of adequate quality control of the process does not lend itself to homogeneity. Consequently, the theoretical SE of a PHM may or may not correspond to measured results.

The electrical properties of pseudohomogeneous, conductive coatings are usually measured either in terms of surface resistivity per unit thickness or in units of volume resistivity or absolute conductivity. Occasionally, it is measured in conductivity relative to copper. The surface resistance, R_s, is measured in ohms per square* (Ω/sq) and is related to volume resistivity, ρ:**

$$R_{sdc} = 100\ \rho_{\Omega m}/t_{cm} \qquad \Omega\text{, metric system} \tag{2.1}$$

$$= 3{,}970\ \rho_{\Omega m}/t_{mil} \qquad \Omega\text{, English system} \tag{2.2}$$

where, t = thickness of surface coating in indicated units

The volume resistivity is related to conductivity, σ, and surface resistance in the following manner at dc:

$$\sigma_{mhos/m} = \frac{1}{\rho_{\Omega m}} = \frac{100}{R_{sdc} \times t_{cm}} \qquad \text{metric} \tag{2.3}$$

$$= \frac{39{,}370}{R_{sdc} \times t_{mil}} \qquad \text{English} \tag{2.4}$$

Finally, the conductivity relative to copper (σ_r for copper = 1, σ_{cu} = 5.8 x 10^7 mhos/m) is:

$$\sigma_r = \frac{\sigma_{metal}}{\sigma_{cu}} = 1.72 \times 10^{-8}/\rho_{metal} \tag{2.5}$$

$$= 1.72 \times 10^{-6}/R_s t_{cm} \qquad \text{metric} \tag{2.6}$$

$$= 6.79 \times 10^{-4}/R_s t_{mil} \qquad \text{English} \tag{2.7}$$

Equations (2.1) and (2.2) are plotted in Fig. 2.9 to yield surface resistance versus volume resistivity with metalized coating thickness as a parameter. Equations (2.6) and (2.7) are plotted in Fig. 2.10 to present coating conductivity relative to copper versus surface resistance with thickness as a parameter. This figure is especially

*The term **ohms per square** refers to the dc resistance of a square cross section

**(The author has observed a more or less equal split in the literature between those reporting resistivity in units of ohm-centimeters (Ωcm) and ohm-meters (Ωm)

useful since it allows developing an equivalent relative conductivity, σ_r, which can be used in the many expressions of SE throughout this handbook.

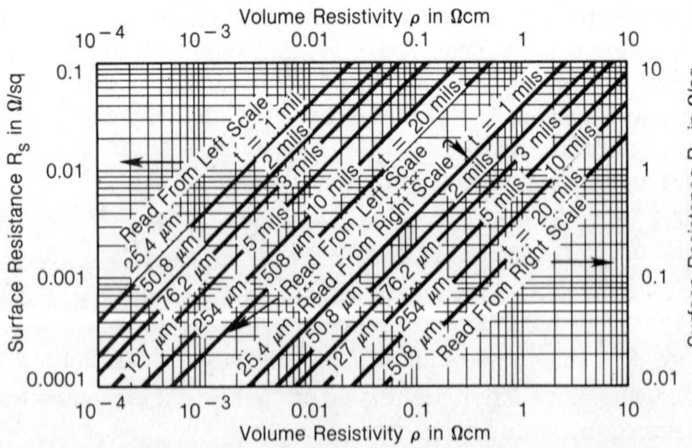

Figure 2.9—Surface Resistance of Copper vs. Volume Resistivity for Various Metal Thicknesses

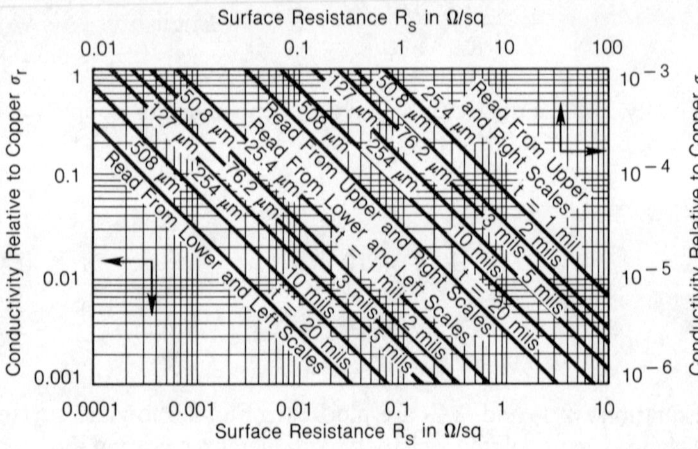

$$\sigma_r = 1.72 \times 10^{-6}/R_s\, t_{cm}$$
$$= 6.8 \times 10^{-4}/R_s,\, t_{mil}$$

Figure 2.10—Conductivity Relative to Copper vs. Surface Resistances

The principal techniques used for creating a conductive layer on top of an insulating (or mediocre conductor) material are the following:

1. Conductive paints
2. Plating (electrolytic)
3. Electroless plating
4. Flame spray
5. Arc spray
6. Ion (plasma torch) spray
7. Vacuum deposition

These techniques and their main features will be described next. While the sections of this chapter discuss the basic SE of these "pseudometals," their application in creating or restoring shielding integrity to a plastic box, nonconductive coating, display panels, etc., will be covered in Chapters 6 and 7.

2.1.2.1 Conductive Paints

Conductive paints (also called conductive coatings) are available from many manufacturers (see references). They are usually a lacquer, elastomeric, silicone resin, vinyl, acrylic or latex base and require careful surface preparation. The most frequently used conductive additives (see also Table 2.2) are:

Silver (resistivity as low as a few milliohms per square)

Copper (resistivity as low as 100 mΩ/sq)

Nickel (approximately 0.5 to 1 Ω/sq)

Graphite (greater than 2 Ω/sq).

Some conductive paints require an overcoat for protection. To assure good electrical and mechanical reliability, surface preparation of plastics, woods, ceramics and other base materials requires removing all greases, waxes, oils, dirt, mold releases and foreign matter until the surface is water-break free. Most conductive paints may be applied by one or more of the following processes: dipping, spraying, silk screening, roll coating or brushing. Several conductive paints are also available in aerosol spray cans.

Table 2.2—Typical Properties of Modern Conductive Paints (Source: ACC/Coates Electrographics, Midsomer, Norton, Bath BA3 4BQ UK)

ACC/Coates Code	Pigment	Binder	Reducer	Color	% Solids by weight (BASE)	%Pigment by weight (BASE)	Weight per litre (kgs)	Touch Dry Time at 25°C	Coverage per mixed Litre at 0.050 mm DFT*	EMI/RFI Attenuation dBs (E-field)	ESD Performance	Time to Attain Lowest Resistance at 25°C	Resistance ohm/sq at 0.050 mm DFT*
6-B-17	Graphite	Acrylic	TH27	Black	23	12	0.93	15 min	2.4 sq m	30-50	Excellent	48 hr	< 50
49-B-5	Graphite	Acrylic	TH27	Black	26	18	0.95	15 min	2.5 sq m	40-60	Excellent	48 hr	< 5
—	Graphite	Polyurethane	TH27	Blue-Grey	45	33	1.28	30 min	5 sq. m	20-40	Excellent	72 hr	$10^5 - 10^7$
49 × 4	Nickel	Acrylic	CN-203-A	Dark Grey	68	58	1.94	10 min	5 sq m	60+	Excellent	72 hr	0.5
49-B-3	Nickel	Acrylic	CN-203-A	Black	67	57	1.89	10 min	5 sq. m	60+	Excellent	72 hr	0.5
6-1X-1	Nickel	Polyurethane	CN-207-P	Grey	63	55	1.83	30 min.	5 sq. m	60+	Excellent	72 hr	0.5
6-1B-1	Nickel	Polyurethane	CN-207-P	Black	63	56	1.78	30 min	5 sq m	60+	Excellent	72 hr	0.5
—	Nickel	Epoxy	CN-210-E	Black	85	70	2.55	30 min	5 sq. m	50-60	Excellent	24 hr	2.5
60-S-1	Silver	Vinyl	TH27	Silver	57	51	1.65	15 min	4.3 sq m	70+	Good	24 hr	0.010
60-1S-1	Silver	Polyurethane	TH27	Silver	60	55	1.78	30 min	5 sq m	70+	Good	24 hr	0.025
1-1S-20	Silver	Epoxy	TH27	Silver	65	61	1.91	30 min	5 sq m	70+	Good	24 hr	0.025

*DFT = Dry Film Thickness

Silver lacquer paint is highly conductive, and a film will stretch over 100 percent. Thus, it is particularly useful where the substrate to which it is applied is subject to flexing or stretching. Air drying typically results in a surface resistance of about 50 mΩ/sq for a 25 μm (1 mil) coating thickness. This corresponds to a $\sigma_r = 0.014$. Oven curing results in a surface resistance of about 1 mΩ/sq or a $\sigma_r = 0.68$. It provides a resistance to salt spray and has an operating temperature range of −54 to 121°C (−65 to 250°F). Extended use at 121°C is excellent.

The silver-based silicone surface coating exhibits both high conductivity and high operational temperature properties. The surface resistance is 40 mΩ/sq for a 25 μm (1 mil) coating thickness (σ_r = 0.39). It can be used continuously at a temperature of 315°C (600°F). When applied to a substrate which will withstand even higher temperatures, the surface coating can be fused to the substrate. Moisture and chemical resistance are good.

Paints consisting of a pure silver filler in a vinyl copolymer result in a conformal coating having surface resistance of about 10 mΩ/sq for a 25 μm (1 mil) deposit. It will stretch, compress and deflect with its substrate while maintaining basic properties. The operating temperature range is −54 to 121°C (−65 to 250°F).

The conductive acrylic coatings developed for the plastic industry (e.g., Tecknit 72-00025) exhibit a surface resistance of about 50 mΩ/sq for a 1 mil deposit. The variation depends upon the percentage of solids used, the base material, the deposition depth, etc. In preparation for spraying, the surface is typically cleaned with either alcohol or naphtha. Should environmental, decorative or dielectric insulation protection be required on conductive acrylic paint, the overcoat must be a nonsolvent paint. Typical overcoats exhibit dielectric strengths of 300 V/mil (12 kV/mm) and have temperature ranges which match the conductive coating: −54 to 93°C (−65 to 200°F).

Next below silver paint are the copper-loaded paints. Copper's excellent conductivity and low cost make this coating an attractive solution which is intermediary between the expensive silver and the cheap but poorly performing graphite. This good conductivity also allows a lower percentage of metal by weight, such that a 50 micron (2 mil) thick coat will weigh about 60 g/m^2. Typical surface resistance of copper acrylic ranges from 0.3 to 1 Ω/sq for a 50 micron thickness, providing about 40 dB of plane wave SE. After extended exposure to humidity (e.g., MIL-STD-810 method), there is a typical loss of 4 to 10 dB of SE. Passivation processes are now available which render the copper film non-oxidizing.

Close to copper and favored by industry is nickel paint. Available in waterborne or standard solvent mix, the nickel paint is typically heavier and requires thicker applications than silver or copper. Surface resistances are in the 0.8 to 2 Ω/sq range. The strong advantage of nickel is its unmatched resistance to oxidation.

Down on the scale are the graphite coatings, which range from a few ohms per square for the least resistive to several kilohms per square for the more resistive. In the 10 Ω/sq range, these low-cost coatings can still provide about 20 dB of plane wave SE, which may be enough in many consumer or industrial applications. More resistive coatings can function as static drains or E-field-only shields in less-critical applications.

A number of different conductive surface coatings, called **Electrodag**, are made by Acheson Colloids. They range in surface resistance from 15 mΩ/sq/mil for more expensive highly reflective applications to less-expensive 200 Ω/sq/mil for low-SE use.

Figures 2.11a and 2.11b illustrate typical applications. Because of their inhomogeneous nature, these metal-loaded paints require very careful stirring before spraying. Prices for the above conductive coatings vary over a wide range depending upon the quantity, method of application and type of conductive paint selected. In average quantities, prices may range from about \$1 per m^2 per 1 mil (25 microns) thickness for higher surface resistances to \$40 per m^2 per 1 mil thickness for low surface resistances with other special properties. A list of some manufacturers of the above conductive coatings will be found in Appendix I. Extracted from one manufacturer's catalog, Table 2.2 lists typical properties of conductive paints.

A particular coating is selected based on the abrasion resistance; surface hardness; adhesion to plastics, glass or ceramics; chemical and solvent resistance; and service temperature. While conventional paint spraying is the usual application method, brushing or dipping techniques may also be used.

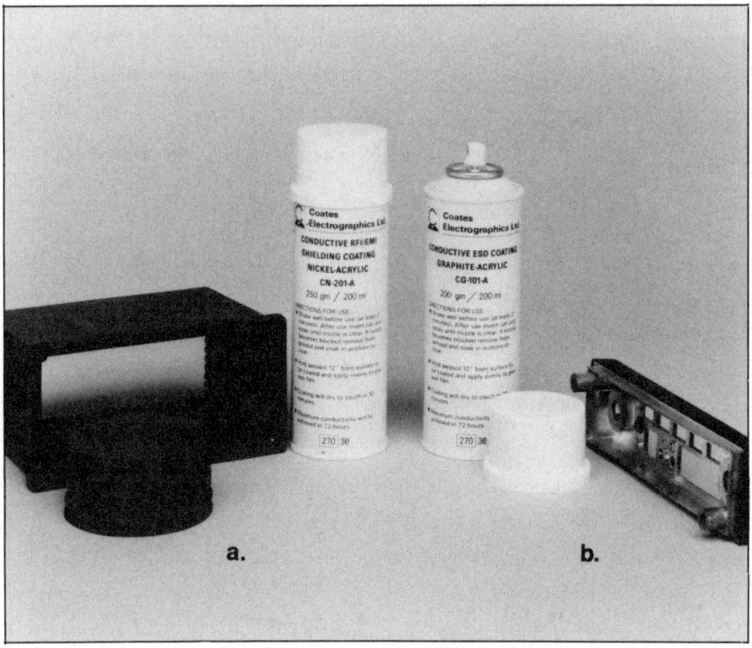

Figure 2.11—Conductive Coatings (courtesy of Acheson Colloids Co.)

2.1.2.2 Flame, Arc and Plasma (Ionizing) Sprays

Hot spraying of metals is a process which applies in this handbook to metalizing a surface by vaporization of the melted basic metal which becomes deposited upon a surface. This process includes flame spraying, thermal spraying and plasma spraying (Fig. 2.12).

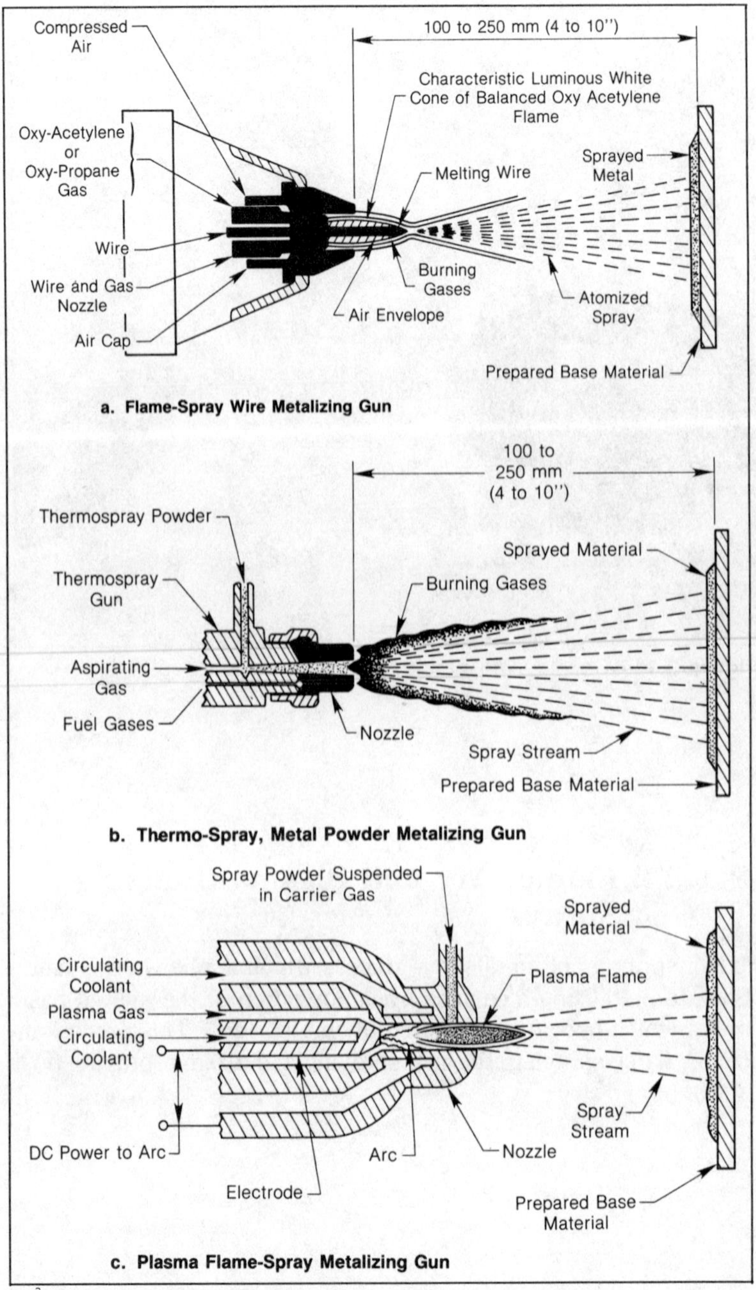

Figure 2.12—Several Methods of Spraying Melted Metal on Substrates (courtesy of METCO, Inc.)

In flame spraying, a desired wire rod of copper, aluminum or other metal is fed into a special metalizing gun. This gun also accepts a special gas such as oxyacetylene or oxypropane and compressed air. One typical configuration is shown in Fig. 2.12a. The wire rod is drawn through the gun and nozzle by powered feed rods. The wire is continually melted in the oxygen-fuel gas flame and atomized by a compressed air blast which carries the metal particles to the previously prepared surface.

Any metal (with the exception of tungsten) obtainable in a uniform wire or rod size can be sprayed using special guns. Flame spraying speeds depend on a number of factors. Using 6.8 kg/cm^2 (15 psi) of acetylene pressure, for example, 5.5 kg (12 lb) of aluminum, 9.1 kg (20 lb) of copper, 8.2 kg (18 lb) of monel or up to 7.3 kg (16 lb) of steel per hour can be deposited. Higher spraying speeds can be achieved using propane gas. Typically, metal wire and rod diameters from 0.9 mm (AWG #18) to 4.76 mm diameter can be accommodated in the special spray guns. Since the flame (about 6,000°F = 3,300°C) is located in the nozzle and does not dwell over any surface area, surface materials such as plastics and fiberglass are not damaged.

However, air pockets in the deposition can develop and, depending on the process control, the metalized surface may range from a sandpaper-type finish to a relatively smooth one. Chemical analysis indicates that oxides of the base metal deposited on the surface are low. The flame-sprayed deposits of copper and aluminum listed in Table 2.1 exhibit conductivities relative to their own base metals of about 10 percent for copper to 6 percent for aluminum. This develops because most of the deposit is trapped air particles or tiny voids in the deposited air surface, i.e., the basic metal density is low. Consequently, it takes about 10 times for copper and 17 times for aluminum of the base metal thickness, to yield the same conductivity. For thin deposits there is also the danger of many small area voids in the flame-spray process.

A variation of the flame-spray process (called **Electrospray** by Wall Colmony Corp.), or more generally **arc spray**, uses an electric arc gun. Used for higher volume metalizing operations, it works faster and costs somewhat less to operate. The 3,900°C (7,000°F) temperature of the arc melts a wire rod faster and deposits particles having higher heat and greater fluidity. There is some danger, however, that plastic substrates could be deformed if the flame spray dwells too long on a surface. The electric-arc spray rate is claimed to be three to five times faster than that of oxyacetylene equipment, and bond strength is greater because welded adhesion is added to the usual mechanical bond.

This latter property, of course, cannot be used with plastic substrates, which means that a minimum speed across a surface must be maintained. Good bonds can be achieved with less than ideal surface preparation. Since hotter particles mean denser coatings, coating oxide content is reduced compared to conventional oxyacetylene methods.

One of the most popular arc spray processes is zinc arc (promoted in the United States by TAFA Corp., among others). Two zinc wires from a wire dispenser are driven by a 5 kVA power supply and form the electrodes with an established arc voltage of approximately 20 V. By using a compressed air throw at 1,000 liters/min, the melted metal is blown off the gun and "crystallizes" on the substrate. The ideal spraying distance is 10 to 20 cm. If properly applied, the plastic substrate keeps a temperature rise around 50°C. The resulting coating is about 95 percent zinc rich, with minimum air bubbles. The coating is generally in the 0.03 to 0.15 mm range to keep flexibility so as to comply with possible substrate bending, torsion, etc. The difficulties arise when the part to coat has a lot of ribs, chamfers and hard-to-reach places, which will result in irregular coverage. The application requires considerable skill and equipment adjustment during spraying; coarse particles may roughen the surface or even detach later inside the equipment.

If the original plastic surface is glossy, it should be cleaned and sandblasted for a good adhesion. Finally, the process is not recommended for types of plastics which produce significant out-gassing. When properly applied, zinc arc is a dependable shielding method, producing surface resistances of less than 0.1 Ω/sq.

A different type of flame-spray process, called Thermospray™ by Metco Inc. (see Fig. 2.12, part b) accepts the metal to be sprayed in powder form. The powdered aluminum, copper or other metal is held in a hopper atop the gun and is gravity fed into the gun where it is picked up by oxyacetylene or hydrogen gas mixtures and carried to the gun nozzle. Here, it is melted almost instantly due to the thermal efficiency of the gun and carried to the surface being sprayed by means of a siphon-jet arrangement at the gun nozzle. Because the burning gases extend well beyond the nozzle, the Thermospray process requires that the distance to plastic base materials be increased over that shown in the figure or that increased surface speed be used to prevent deforming the plastic.

Plasma spraying uses a technique involving the application of electrical energy to ionize an inert gas (see Fig. 2.12c). This creates a higher energy level which generates temperatures as high as 16,650°C (30,000°F). The primary value of such high temperatures is that the melting point of the material being sprayed is reached quickly. Thus, it need not remain in the hot zone as long as that corresponding to lower-temperature oxyacetylene flames. Requiring a shorter time at heat means it can be propelled faster through the plasma arc to reach the substrate with less heat loss and with greater impact. These two advantages make plasma coatings denser and give them higher bond strengths.

The plasma gun generates its high temperatures without combustion, so even the most reactive materials can be melted with little or no oxidation, and without altering the original chemical composition to any significant degree. Also, the spraying can be done within a protective atmosphere chamber, to further protect the sprayed material. The elimination of oxides produces a more cohesive coating which will finish to a better surface and protect itself better in service. Most metals can be sprayed onto almost any material. The substrate needs to be heated to over 150°C (300°F) to eliminate heat distortion problems (most materials are unaffected using a 204°C (400°F) maximum).

2.1.2.3 Plating

In plating processes, a layer of pure metal is deposited on a plastic, fiberglass or other support by either electrolytic or electroless plating. Electrolytic plating is the oldest technique, having been used for many years to plate conventional metals. For plastic, the process is rather expensive. It requires pretreatment to make the plastic surface receptive to the metal and also create a ductile interface between plastic and metal, necessitated by their different thermal expansion coefficients. The pretreatment consists of etching the plastic with an acid solution, then depositing an underlayer of bright-acid copper and finally making the electrolytic deposit of nickel or chromium. Tin has also been used because of its excellent contact properties and lower surface resistance than nickel. Also, tin does not oxidize like copper does. Because of the high cost, energy waste and toxicity of electroplating, it is gradually being superceded by electroless plating. However, it is still applied when one wants to improve the surface conductivity of an ordinary steel frame to provide both better contact quality (for gasket or finger stock interface) and a corrosion-free contact.

In electroless plating, the normally hydrophobic plastic is etched with an oxidizing solution (or sometimes vapor) which creates micro pores, making the plastic surface "wettable".[1] An aqueous solution of copper or nickel is then applied. Once rinsed, a thin layer of metal stays captive over the plastic surface. The metal film is tight and continuous, with a consistent thickness, regardless of the complexity of the shapes and edges. This plating also goes into the cooling slots, creating the "waveguide beyond cutoff" effect described in Chapter 7 of this book. Also, normally, the two sides of the housing or panels are covered, making the overall shielding more efficient. This in turn may be a disadvantage if the aesthetic aspect requires a decorative painting.

Since the metal is not diluted in a binder (as for paints) the metal density is that of a pure metal and less thickness is required. One micron thickness produces about the same effect as 50 microns of copper-filled paint. The metals used most often for electroless plating are nickel and copper (passivated copper is now available which overcomes the problem of corrosion).

2.1.2.4 Vacuum Metalization

This process consists of depositing a metallic film by evaporation in a high vacuum. The metal used most often is aluminum, with coating thicknesses from 3 to 10 μm. A base coat is necessary. The film will stick to complex shapes, with excellent adhesion properties. The coating is free of porosity, is dense and ductile and can withstand temperature shocks of -40 to 100°C (-40 to 212°F).

Seen on an electron beam microscope, the structure resembles a dense "fur" of columnar crystals. The fur is rather flexible, causing no stress on the metal-plastic interface. After vaporization, a 200 Å (0.02 μm) maximum oxide layer forms on the top of the aluminum. The process is expensive, requiring a high capital investment (vacuum chamber, etc.).

2.1.2.5 General Considerations about Conductive Coatings

There are many features that conductive coatings share, which ought to be known when making the decision. First, better results are always obtained if the design of the plastic part "foresees" the conductive coating.[2] Of course, sales-oriented literature tends to publicize the fact that conductive spray or plating are an excellent retrofit solution, allowing an otherwise unprepared housing to become a shielded envelope and relieving the unaware project manager from FCC, VDE or MIL-STD constraints.

But the fact is that SE performance and durability of the coating are optimized if some basic rules are followed in the design of the plastic part. These rules are similar to those commonly applied for economical, defect-free production of plastics, but they must be adhered to more carefully because the coating often accentuates surface defects. These rules are enumerated below and shown in Fig. 2.13.

1. Make wall sections as uniform as possible.
2. If changes in thickness are necessary, they should be progressive rather than abrupt.
3. Give generous radii at angles (1 mm being a safe minimum).
4. Eliminate sharp corners, which create stress concentrations.
5. To support thin edges, bosses, etc., provide ribs with thicknesses of 50 to 60 percent of the wall thickness.
6. Avoid large mass-to-area ratios.
7. Avoid deep and narrow recesses.
8. Holes, especially blind holes, should not exceed a depth-to-diameter ratio of 1:5.
9. For mating edges, do not use V grooves; round-shaped groove and tongue designs are preferred. This will also provide a good seat for a conductive gasket, if needed.

If the coating must be invisible from the outside, make sure that it still extends long enough from the inner side across the butting edges to provide electrical contact. By avoiding localized stresses and cracks, these guidelines can ensure a durable adhesion and integrity of the conductive layer.

The durability of the coating is another criterion to observe. It is generally combined with environmental considerations such as the adhesion, surface resistance, etc. Parameters are measured with a freshly applied coat, then remeasured after an accelerated aging (Fig. 2.14). For instance, the Society of Plastic Industries (SPI) in the United States recommends the following aging cycles before abrasion and adhesion tests be made:

1. **Thermal cycling**: 30 minutes at −40°C, 5 minutes at room temperature, 30 minutes at 70°C; 10 cycles

2. **Thermal conditioning**: 70°C for 3 days, then 14 days, then 30 days
3. **Humidity**: 95 percent relative humidity at 70°C for 3 days, then 14 days, then 30 days

The German DIN 50017 stipulates 5 alternating hot-and-cold cycles of 120 hours.

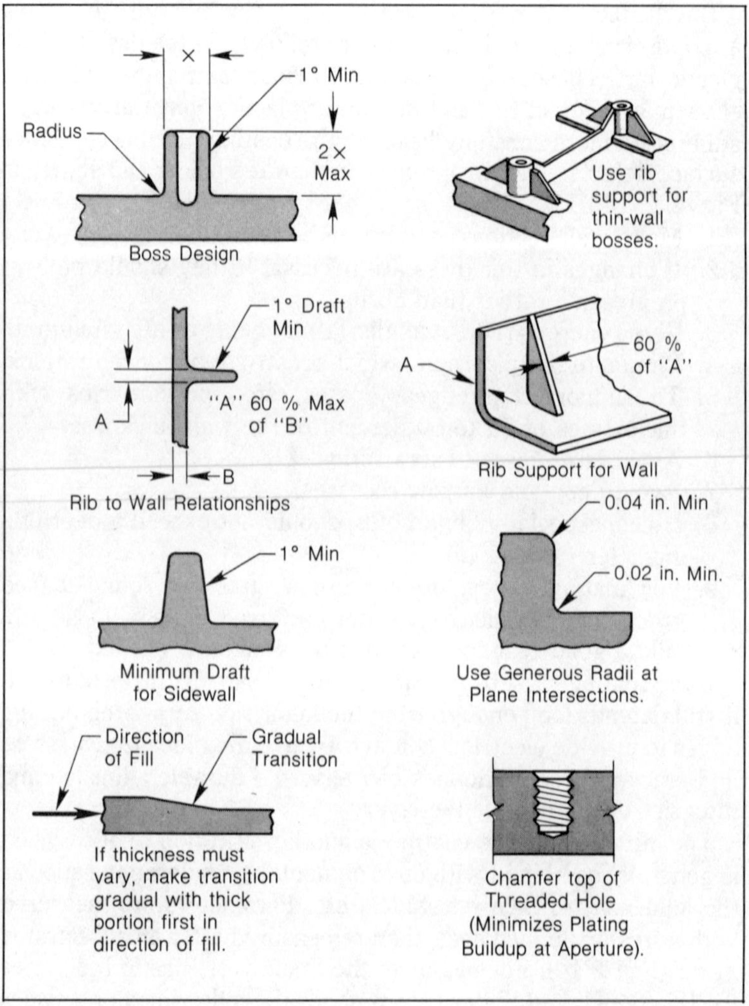

Figure 2.13—A Few Guidelines to Follow when Designing Plastic Parts for Metalization (continued next page)

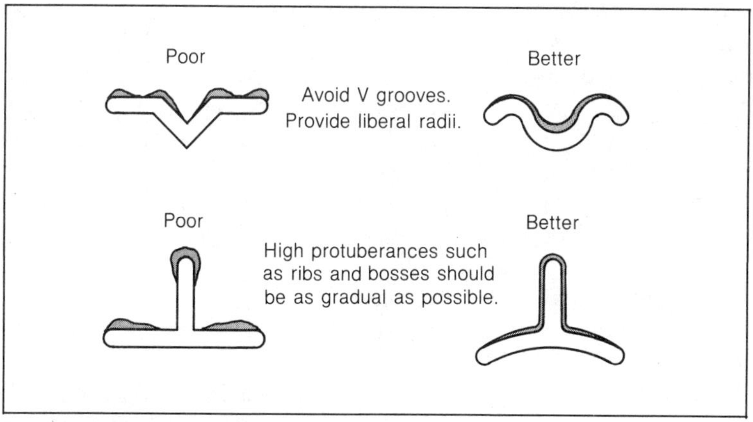

Figure 2.13—(continued)

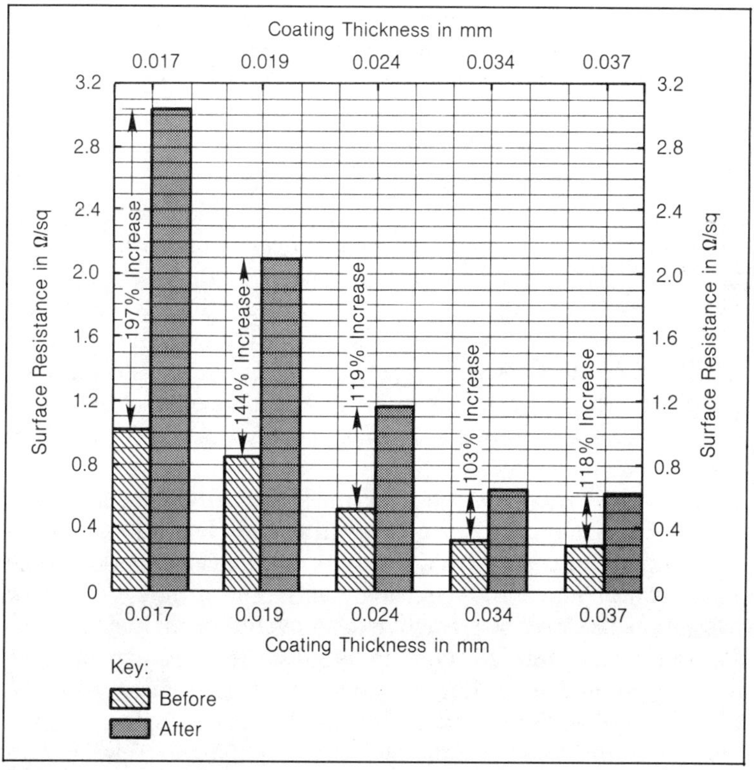

Figure 2.14—Resistance before and after Temperature Cycling, Varying Coating Thicknesses

After these climatic exposures, several tests can be performed, among them an **adhesion test** (DIN 53151 or similar). In this procedure, a round target is selected on the sample surface and a round disk (anvil) glued to it. Then the disk is pulled away with a gradually increasing force until the coating pulls free. The adhesion force is calculated from the disk area and the removal force.

Typical values obtained, for instance, with zinc coating, are:

Zinc on Noryl and polycarbonate: 25 kg/cm^2

Zinc on ABS: 17 kg/cm^2

Zinc on glass-filled nylon: 74 kg/cm^2

The tested part is not destroyed, and the test spot can be recoated.

Another post-environmental exposure test is the abrasion and scratch test (e.g., ASTM D-3359 or UL 746 C). Here, an array of orthogonal scars is made with a cutter, then a specified highly adhesive tape is pressed over this array and pulled out (Fig. 2.15). The sample fails the test if chips of the paint or plating are rubbed off by the test. This also indicates a possible safety issue (UL concern) by flakes falling inside the equipment and causing shorts.

Another key issue in selecting a coating method is the production costs. The base material cost, the surface preparation, the capital investment of the tooling and the application workmanship, the number of coats, the air dry versus oven cure are all financial parameters. Tables 2.2 and 2.3 give some qualitative and quantitative comparison data. An important point (though seldom addressed by conductive coatings vendors) is the way metallic parts like filters, connector shells, transformer shields, cable shields, etc., are bonded to the conductive surface. Abrasion and corrosion evaluation need to be made at these bonding interfaces to avoid later problems.

Application is another aspect to consider when using conductive paints. The same coatings applied differently (variations in thinner, waiting time between consecutive coats, drying method, type of substrate) may produce widely different results. Very low viscosity paints may be attractive to spray, but metal filler will settle, creating a metal-rich layer on the substrate side and poor adhesion. To minimize settling of metal particles, coating with the highest possible viscosity should be used. The best manufacturers produce coatings with high solid contents (50 to 70 percent by volume) along with easy handling. Finer particles mean easier mixing and less settling.

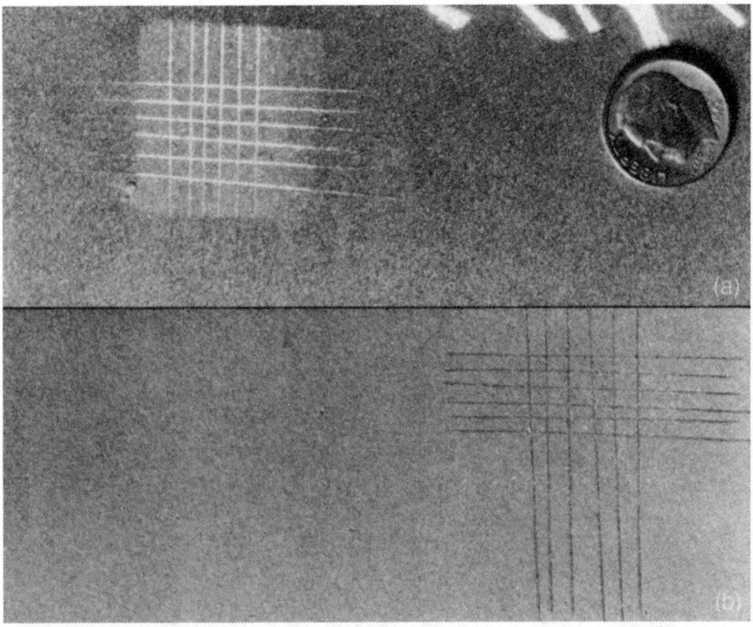

Figure 2.15—Poor Adhesion Due to Dry Spray (Gun Too Far Away from Substrate in a., Correct in b.)

Table 2.3—Metalizing Costs

Process	Cost (1985) Includes Application Cost (for paints/sprays manual operation assumed). Dollars/ft^2	(Dollars/m^2)
Electroplating	$0.25-2.00	($2.5-20.00)
Electroless Plating (Copper & Nickel, Inside and Outside	$1.30	($13.00)
Metal Spray	$1.50-4.00	($15.00-40.00)
Vacuum Deposition	$0.50-2.25	($5.00-22.00)
Carbon Coating (0.5 mil = 12.7 μm)	$0.05-0.50	($0.50-5.00)
Graphite Coating (2 mil = 51 μm)	$0.10-1.10	($1.00-11.00)
Copper Coating (2 mil = 51 μm) Overcoated with Conductive Graphite at 0.2 mil = 5 μm	$0.30-1.50	($3.00-15.00)
Nickel Coating	$0.70	($7.00)
Silver Coating (0.5 = 12.7 μm)	$1.25-3.00	($12.5-30.00)

Figures 2.16 and 2.17 show the effects of improper thinner. Each manufacturer can recommend the appropriate thinner for each substrate.

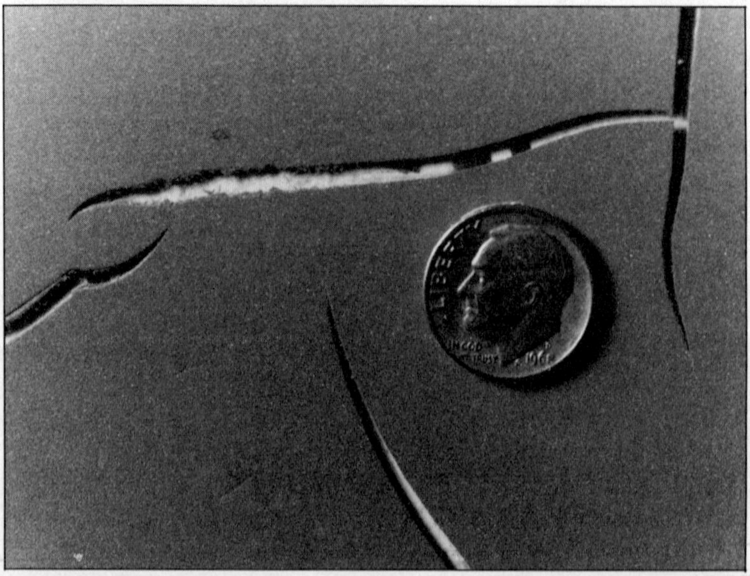

Figure 2.16—Severe Stress Cracking Due to Incorrect Thinner

Figure 2.17—Poor Adhesion Due to Use of Incorrect Thinner

2.1.2.6 Performance of Conductive Coatings

Since practically no conductive coatings in use today are laid with a sufficient thickness to provide several skin-depths below a few hundred MHz, the basic shielding mechanism is reflection. Section 1.4 of this book demonstrated that reflection losses are determined by the factor of K, which is defined as the ratio of the impinging wave impedance Z_W to the conductive barrier impedance Z_m. Therefore, a measure of the coating surface resistance is a valid figure of merit of the SE. This surface resistance is defined from Eq. (2.1) in ohms per square and can be simply measured by applying over the coated sample a set of two 5 cm long electrodes, separated by 5 cm to encompass a square area. The reading of the resistance between the two probes corresponds to the ohms per square value, provided the contact resistance of the probe (a few milliohms) is negligible compared to the sample resistance (generally a few hundred milliohms or more). Otherwise, a zero calibration has to be made first to offset the probes' contact resistances.

Figure 2.18 shows comparative data of SE for various coatings. The most severe situation has been assumed. In other words, for the frequency range where near-field conditions exist, the field is assumed to be predominantly magnetic (low Z-field).

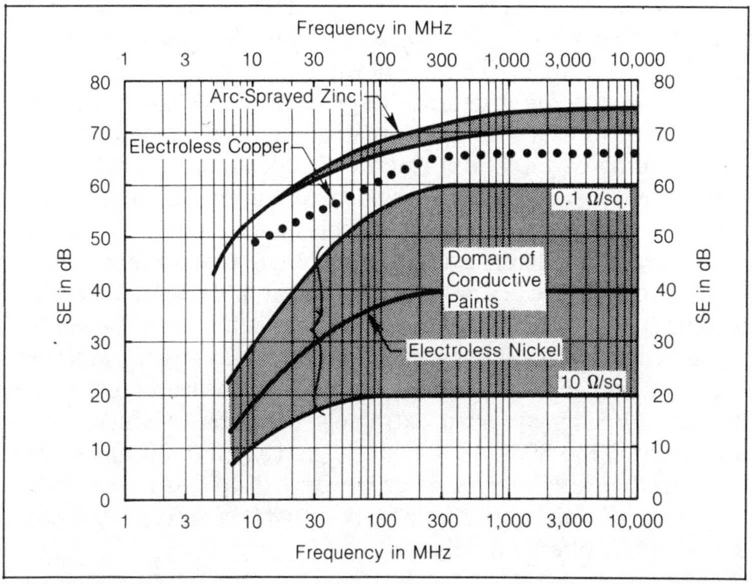

Figure 2.18—Shielding Effectiveness of Conductive Treatments (By Standard 30 cm Distance Test, Near-Field Attenuation Given Against H-Field)

2.1.3 Composite Materials

Composite materials like carbon or boron fibers are increasingly used as replacement for aluminum alloys in aircraft, helicopters, missiles and some other vehicles. By 1990, 40 to 50 percent of the structural areas of a military fighter will be made of composite materials, and the figure will be approximately 20 percent for commercial airliners. These materials pose serious concerns relating to shielding bonding and lightning/electromagnetic pulse (EMP) coupling because of their mediocre conductivities. Structurally sound composite joints tend to be "leaky" in terms of shielding. Composite manufacturing processes stress the production of high structural quality and disregard electrical properties.

The most widely used composite material is carbon fiber composite (CFC), made of carbon fibers in an epoxy resin binder. Resistivity is quite high. The fiber itself has a resistivity of 1 to 2×10^{-3} Ωcm (1 to 2×10^{-5} Ωm). This is about 1,000 times the resistivity of copper and 300 to 500 times that of typical aircraft aluminum alloys. But since the carbon constitutes only 60 to 65 percent of the composite (the rest being a nonconductive resin), overall resistivity is even poorer.

Furthermore, the material is highly anisotropic. For unidirectional CFC, where fibers are laid in one direction only, the resistivity perpendicular to the fibers is several hundred times higher than parallel to the fibers. This is shown in Table 2.4.

Since, usually, the structural designers need mechanical strength in both directions, composites are generally made of several plies with crossing patterns. But electrically speaking, this still makes two perpendicular directions which are privileged, while the others are more resistive. More complex schemes have been developed where, from the reference direction, the composite is a hybrid structure of Kevlar® (the best strength-to-weight modulus) and graphite plies with 0°, −45°, 90°, +45° angles, etc. In any case, the fabrication process strongly influences the composite electrical properties.

The two other widely used composites are based on boron and Kevlar. Boron/epoxy is an extremely mediocre conductor, with resistivity in the 30×10^{-3} Ωcm range, i.e., 15 to 30 times more resistive than CFC. Kevlar is even worse since it is a nonconducting material (dielectric) with a resistivity of 1.6×10^{10} Ωcm. Therefore, it offers no SE at all.

Table 2.4—Comparison of Some Key Data for Composite Materials

Material	Directivity	Vol. Resistivity ρ Ω × cm	Skin Depth at (in mm)		
			1 MHz	10 MHz	100 MHz
Aluminum Alloy (Reference)	Isotropic	2.5×10^{-6}	0.08	0.025	0.008
Carbon	Isotropic	1.5×10^{-3}	2	0.6	0.2
CFC, Unidirectional	∥ to Fibre	2×10^{-3}	2.3	0.7	0.23
65% Vol. Graphite	⊥ to Fibre	⊥	51	16	5.1
CFC Crossed Fibres	0 or 90°	5×10^{-3}	3.6	1.1	0.36
90°	45°	0.1			
Carbon Fibre Filled	∥ to Fibre	0.1 to 1	16-50	5-16	1.6-5
Nylon, 30% Carbon	⊥ to Fibre	1 to 10			
Boron Epoxy	∥ to Fibre	30×10^{-3}	9	2.8	0.9

Table 2.4 shows also the skin depth of various composites for 1, 10 and 100 MHz. It is clear that below 10 MHz, a typical 1 mm thick composite panel does not even represent one skin depth, so no absorption loss can be expected, and all the shielding in the HF range down to VLF frequencies can only occur by reflection. Reflection losses being proportional to Z_W/Z_b, one can predict that, for example, CFC will exhibit 50 to 60 dB less reflection loss than conventional aluminum alloys.

For near E-fields or plane waves, this still leaves a decent reflection shielding, better than 50 dB. But in the case of near magnetic fields, the reflection loss approaches 0 dB below 100 kHz.

Above a few tens of MHz, the handicap of composite materials is less pronounced since they enter into their skin effect regime, and their handicap compared to aluminum is no longer like $\rho_{Al}/\rho_{composite}$ but like $\sqrt{\rho_{Al}/\rho_{composite}}$ instead. Figures 2.19 through 2.22 show the SE values of some common composites.

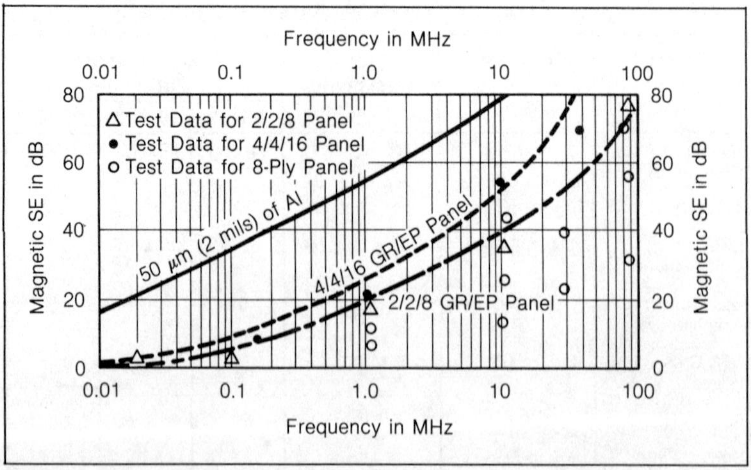

Figure 2.19—Magnetic Shielding Effectiveness for 2/2/8 and 4/4/16, Unprotected Graphite/Epoxy Panel and 2 Mil Aluminum Foil

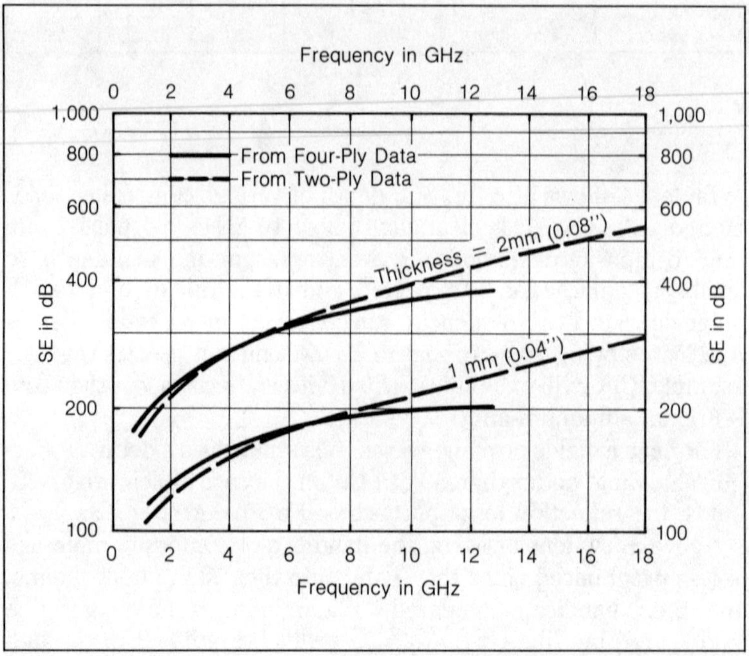

Figure 2.20—Calculated Graphite-Epoxy Plane Wave Shielding (Symmetric Cross-Ply)

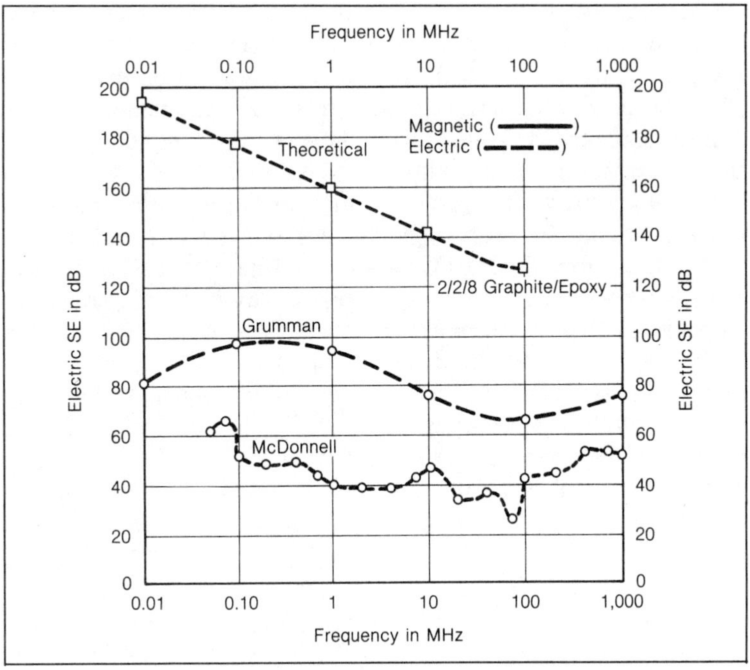

Figure 2.21—Published Graphite/Epoxy Electric Shielding Effectiveness

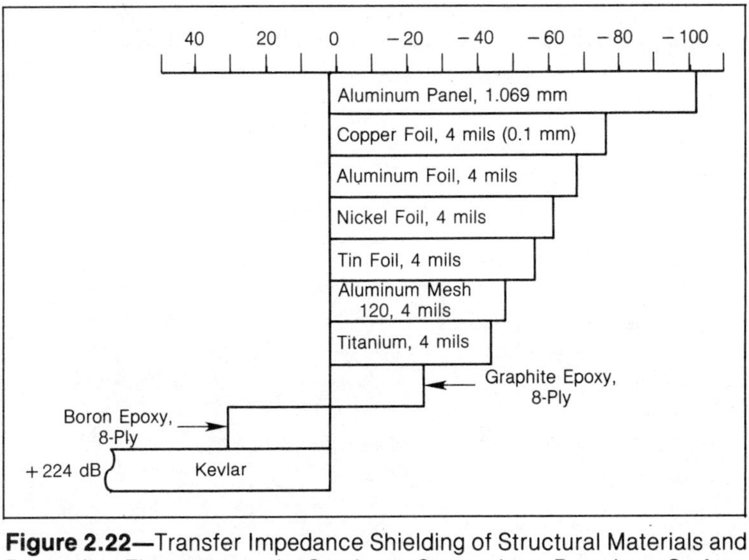

Figure 2.22—Transfer Impedance Shielding of Structural Materials and Protective Electromagnetic Coatings, Comparison Based on Surface Transfer Impedance in the Ohm's Law Region, below approx. 0.1 MHz

EMP and lightning survivability is another critical issue with the shielding properties of composites.[3] The lightning survivability is marginal with CFC and bad with Boron. It depends on the induced current concentration around bolts, rivets and butted joints, especially if these joints (CFC-to-metal or CFC-to-CFC) cause change in the fiber orientation.[4] With lightning currents of 100 kA flowing on the vehicle skin, simple Joule heating can cause delamination and charring around the fastening points. If the current distribution were uniform, a 1 mm thick CFC panel would have to dissipate 5 kW of peak power per square meter, which is already considerable for a material with poor thermal conductivity. But since carbon resistivity decreases when the temperature increases, the current concentrates more and more in the hot fibers in an "avalanche" heating process. Typical service temperature for the material is 200°C (300°C for polyamid matrix), beyond which irreversible damage may occur. To prevent hot spots, the surface conductivity at the joints can be improved by removing the epoxy-rich surface layer. This can be done by drilling a slight countersink under bolt and nut heads, or via a thin milling of mating surfaces.

Much better results are obtained by applying conductive coatings like aluminum by some of the processes described in Section 2.1.2. Figure 2.23 shows the improvements gained by a top conductive coating, as well as the subsequent weight penalties (Figs. 2.24 and 2.25). In the Air Bus A 310 airliner, metalization of the CFC caused

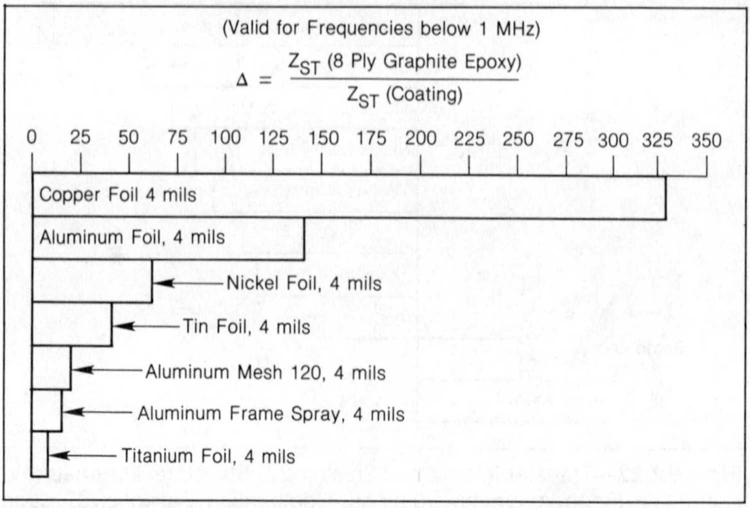

Figure 2.23—Improvement Protective Coatings Provided Relative to 8-Ply Graphite/Epoxy (Based on Surface Transfer Impedances Z_{ST})

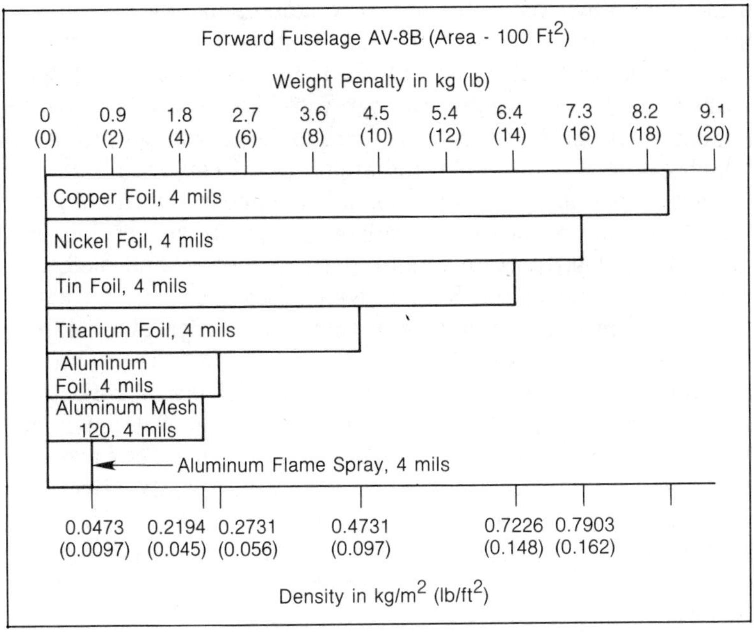

Figure 2.24—Weight Penalty Imposed by Protective Coatings

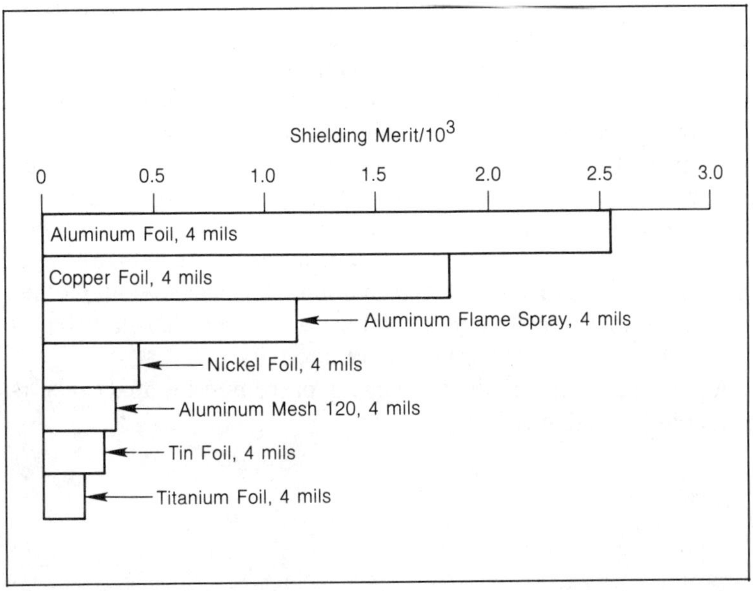

Figure 2.25—Weight/Shielding Figure of Merit of EM Protective Coatings (Protection Beyond 8-Ply Graphite/Epoxy Provided by the Weight of 1 ft^2 (0.1 m^2) of Protective Coating)

a 4.6 percent increase in the rudder weight, though the overall savings over a metal rudder were still a substantial 20 percent.

Finally, leakages at the joints (CFC-to-metal or CFC-to-CFC with a fiber orientation change) are aggravated by the wide resistance mismatch at the interface. This causes large voltage gradients which excite the joint "slot radiation" (see Chapter 7).

Another possible technique to modify composite electrical properties is by internal additives, or doping. High-frequency shielding depends on resistivity, magnetic permeability μ_r, and dielectric constant ϵ_r. Except for Kevlar, which is a dielectric by itself, dielectric constant is dominated by that of the binder, ranging from 3 to 5. Increasing the dielectric permittivity is not a very attractive technique because it will mainly affect electrical reflection, which is not the critical problem of composite materials shielding. Increasing the magnetic permeability by adding magnetic material particles or fibers is feasible but can very rapidly affect the mechanical strength of the composites. So the most widely used techniques, besides top coating, are the insertion of a thin metal mesh within the composite plies or chemical doping of the fiber material.

Thin metal meshes have been used successfully. They pose the problem of a weight penalty greater than metal coatings plus an added complexity of bonding at joints.

Doping the fibers with various impurities is a promising technique. For instance, graphite fibers doped with boron exhibit an improved conductivity by a factor of 5 to 50. A recent process uses the intercalation of graphite fiber with various molecules. Pure graphite is a crystallized form of carbon where atoms are arranged hexagonally into planar layers. The peripheral atoms' attraction is large in the coplanar direction, but interplane bonds are weak and allow for insertion of electron donors (typically alkali metals) or acceptors (typically nonmetals). Fibers doped this way show a 20 to 40 times improvement in conductivity.

A summary of the EMI aspects of composite materials is presented in Table 2.5.

Table 2.5—Summary of Electromagnetic Features of Aircraft Composites

- Boron/Epoxy and Graphite/Epoxy Filaments Used in Surfaces
  - Increase in Strength-to-Weight Ratio
  - Less Waste Material in Surface Fabrication
- Shielding Effectiveness
  - Graphite Better than Boron
  - 20-70 dB Poorer than Aluminum for E-Fields (100 kHz-10 GHz)
  - 15-80 dB Poorer than Aluminum for H-Fields (10 kHz-10 MHz)
  - Can Add Metalized Coating
- Lightning
  - Much More Prone to Damage
  - Can Reduce Effects by Metalized Paint, Wire Mesh & Flame Spray Al
- Electrical Systems
  - No Ground Plane Return
  - Add Metalizing or Dedicated Wire Returns
- Static Electricity
  - More Prone to Static Charging/Discharging Effects
  - Add Metalizing Paints
- Antenna Performance
  - Some Pattern and VSWR Effects Due to Counterpoise Material
- Radar Cross Section (RCS)
  - Early Tests Suggest Small Change in RCS

2.1.4 Particle-Loaded Materials

Particle or filament loading is another way to provide plastics with some shielding properties (see Fig. 2.26). Although it looked very appealing when introduced in the 1970s, this technique still ranks as but a small percentage of all EMI-shielded plastic production. One reason seems to be some shortcomings in its practical application. The process consists of adding a certain percentage of conductive fillers to the plastic before molding. Typical fillers include:

1. Aluminum flakes
2. Stainless steel fibers
3. Nickel fibers
4. Aluminum, nickel or silver-coated graphite filaments
5. Graphite granulates
6. Aluminum, nickel or silver-coated microspheres

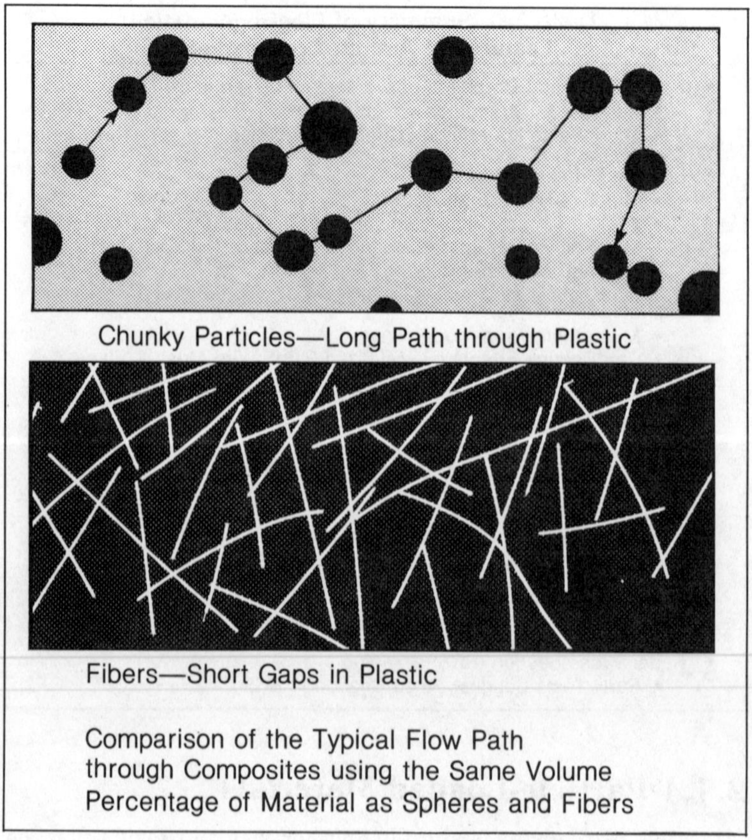

Figure 2.26—Comparison of Chunky Particles vs. Filament Loading of Conductive Plastics

Using surface resistance as a figure of merit for the SE of such conductive plastics is rather meaningless because (1) the top layer is resin-rich, giving erroneous readings, and (2) the whole thickness of the material is conductive and constitutes the shielding barrier. So volume resistivity is a more adequate indicator of the shielding properties. Figure 2.27, using the equations established in Chapter 1, shows the plane wave SE versus volume resistivity in ohm-centimeters. Notice that a "ρ" of 1 Ωcm barely provides 35 dB of SE.

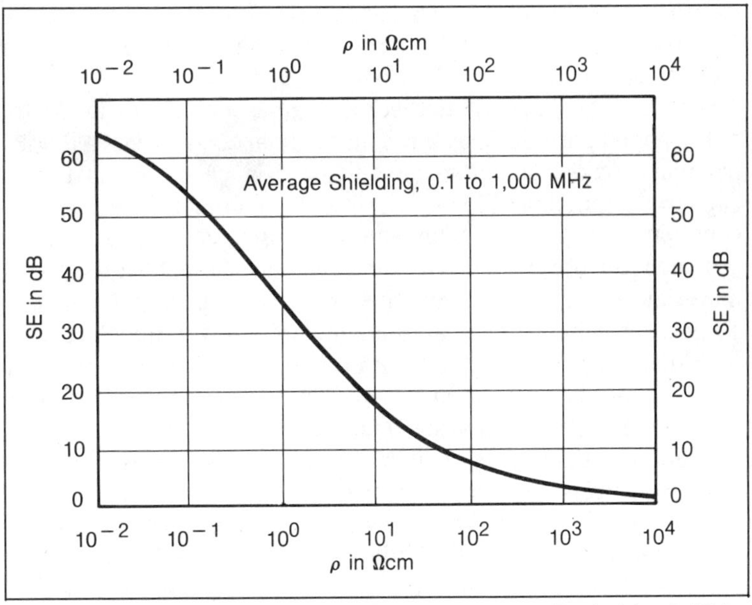

Figure 2.27—Measured Shielding Effectiveness vs. Resistivity 2.27 for Typical Plastic Thickness of 3 to 5 mm

We could also calculate that with $\rho = 1$ Ωcm, the conductivity relative to copper is about 1.8×10^{-6}, corresponding to a skin depth of 48 m at 1 MHz and 4.8 mm at 100 MHz. Therefore, in most frequency ranges of EMI problems, conductive-filled plastics will not reach their skin depth (absorption loss) regime. We will use, as a rule of thumb, this 1 Ωcm ≡ 35 dB as a minimum criterion to discuss conductive-filled plastics. Table 2.6 describes the volume resistivity reached with several types of fillers. Figure 2.28 shows the influence of concentration.

Pure aluminum flakes were the first technique used.[5] It takes more than 40 percent in weight (about 20 percent in volume) of filler to get a volume resistivity less than 1 Ωcm, i.e., the 35 dB minimum performance. This is quite a large volume of aluminum, and it raises questions about whether the housing should still be considered plastic. Stainless steel fibers can achieve the 1 Ωcm, 35 dB figure with only 5 percent weight (1 percent volume) loading. However, the abrasion of molds and extrusion screws is severely increased. Nickel coated graphite fibers (Fig. 2.29) produce the 1 Ωcm figure with about 10 percent volume loading and a lesser wear penalty for the molding hardware.[6]

Besides the simple volume-percentage aspect, the shape factor of the added particles play a role in the final conductivity. Chunky particles, flakes or spheres result only in casual contacts and capacitive coupling. This is why it takes so much to attain a decent conductivity. Long filaments with shape factors of 100 to 300 provide more opportunities for direct contact throughout the plastic material, so they can achieve the conductivity goal with only 5 to 10 percent volume. Typical filaments are a few millimeters long, with diameters in the 10 to 60 μm range. In both cases, the metal adder should be passivated to avoid oxidation and chemical reactions when plastic outgasses over the product life.

A uniform dispersion of fibers or particles within the material is also important. Nonuniform dispersion can lead to samples showing low resistivity while the whole part will have a poor SE because of fiber-lean "pockets" that will leak. So periodic testing of the molded parts is important. The SE can be improved by up to 16 dB if a base polymer with low molecular weight is selected, which also improves the spiral flow properties for molding. Additives used to decrease viscosity can affect SE either way. Better results (up to 8 dB improvements) have been realized by incorporation of 0.75 percent amid wax.

Unfortunately, as in the case of conductive coatings, particle-loaded plastic has not clearly addressed (and even less solved) some practical problems which are crucial to the designer. The means of achieving a good electrical contact for grounding straps, earth wires, filter boxes and shielded connectors over this material is less than obvious. With ordinary metals, wire terminal lugs and connectors or filter flanges are attached by screws. Ideally, the screw applies the mechanical force while the mating surfaces provide the electrical contact. Nothing like this happens with conductive-filled plastics; since the surface conductivity of the resin-rich top layer is very poor, the mounting screw has to be relied on to provide the electrical contact. Not only is the surface of this contact limited to the fillets, but these fillets make only a few random contacts with the conductive chunks or filaments. Therefore, such contact is rather unreliable, especially with aging. This problem, somewhat like the earth rod in a resistive soil, is not insur-

Table 2.6—Conductive—Filled Plastics: Resistivity vs. Filler Type and Percentage

Filler Type	Percentage of Weight	Plastic	Volume Resistivity in Ω cm
Nickel-Coated Carbon Fibres	3 5	Thermosetting Resins and BMC Polyester	1.15 0.45
0.5 μm Ni over 8 μm	8		0.14
	10	Nylon	4.30
	20	Nylon	0.04
	30	Urethane	0.60
	13	ABS	1.00
Carbon Fibers	10	Non-Specified	10^4
	18		10^2
	40		3.00
Stainless Steel Fibres	5-6 (1-1.3% vol)		1.00
	10	(Polycarbonate	0.30
Aluminum Flakes	20 (10% Volume)		100
	45 (22% Volume)		1.00
Silver Particles (Chunks)	36	Bakelite	10^2
	40		0.02

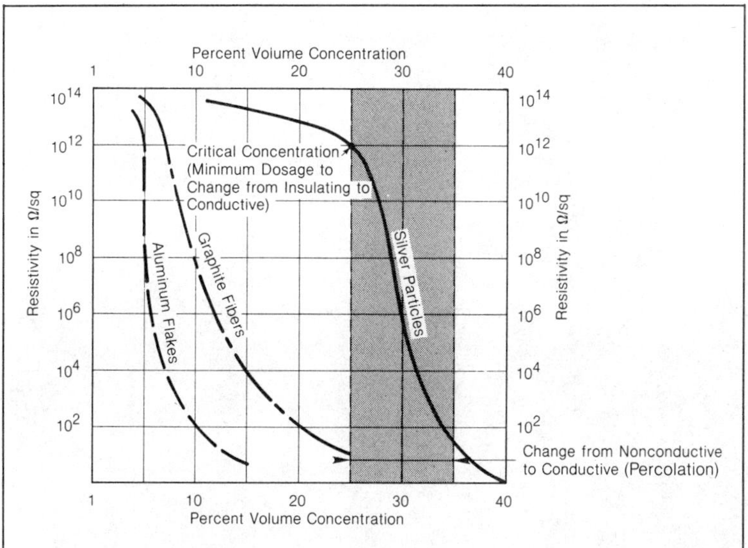

Figure 2.28—Resistivity vs. Volume of Filler for Conductive Filled Plastic

mountable. But it should receive more attention in the manufacturers' application notes.

Another unaddressed issue is the category in which a conductive plastic housing will fall, in terms of International Electrotechnical Commission (IEC) Safety Class I or II. Strictly speaking, an insulating housing allows equipment to be considered as Class II, double insulated, provided certain leakage currents and inaccessibility criteria are met. And such an equipment does not need an earthing wire in its power cord. But changing this housing to a loaded plastic version will bring the product into the Class I realm with all its different constraints.

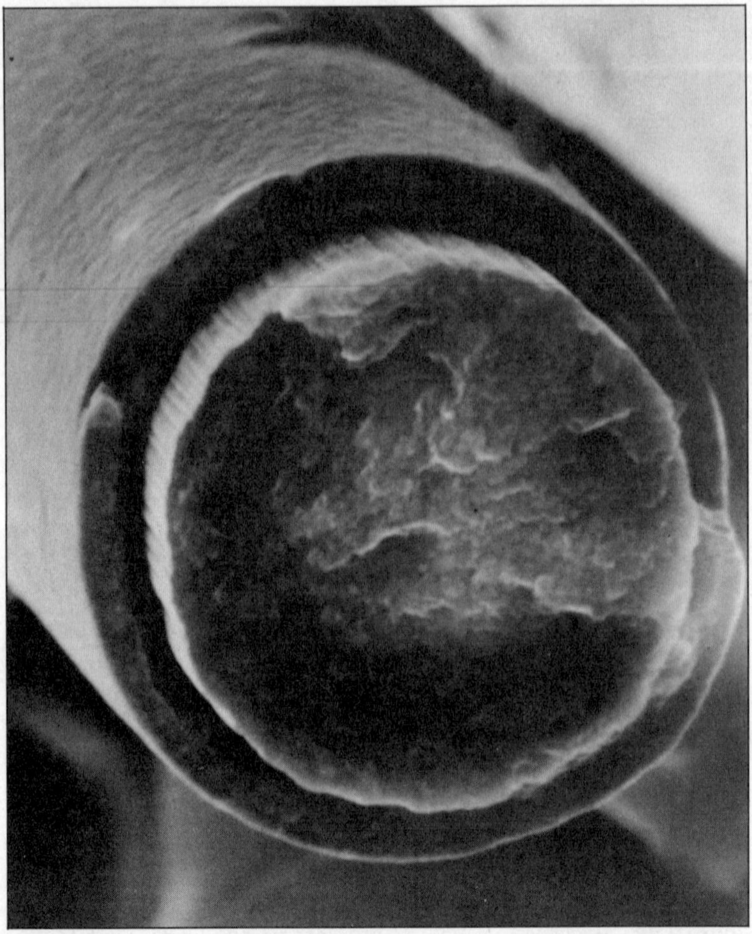

Figure 2.29—Magnified View of a Nickel-Coated Carbon Fiber (courtesy of CYCOM® Div. of Cyanamid Corp.)

All that has been said so far sounds like a deterrent to the use of conductive plastics. However, this would be an unfair conclusion. First, there are many cases where 30 to 40 dB is all that is needed to bring a product within some commercial EMI standards. Also, this distributed conductivity makes it sufficient for drainage of electrostatic discharge (ESD). Unlike metallic coatings, conductive plastics do not require an additional process, so the application (coating or plating) and capital investment costs are saved. For instance, assuming an equipment housing measuring 55 × 50 × 50 cm, the cost comparison in 1986 prices is as follows:

Ordinary plastic + zinc or nickel coating = $50

Filament loaded plastic = $38

The mass conductivity also translates into a significant heat transfer improvement. This can eliminate the need of a fan or reduce the size of the cooling apertures, which ends up being beneficial to overall shielding performance. And finally, conductive plastics have a peerless feature over metals or highly conductive films: their mediocre conductivity makes them behave as low-Q enclosures when EMI frequencies reach the box resonance, while most metallic enclosures will exhibit very sharp tuning effects. So it could be concluded that conductive-filled plastics are appropriate in many applications, but they have not yet fully matured. An extensive use of conductive-filled materials already exists with conductive neoprenes and elastomers used in EMI gaskets (see Chapter 10).

2.1.5 Small-Aperture Metals

Metals such as screens, metal mesh, conductive textiles, which in this handbook are called **small-aperture metals** (SAMs) do not follow the SE relations of Section 1.4. In fact, even if the bonding of the joints of screens and mesh were perfect, the SE in terms of the resistance per square still would follow a different relationship at higher frequencies. At HF and above, there is little evidence of the relationship of the base material conductivity and permeability of SAM to the SE in contrast to both the homogeneous and PHM. Except for some metalized textiles, the SAM family may typically have from 1 to 1,000 times as much air path as basic metal, i.e., the base metal fibers or wires occupy about 0.1 to 50 percent of the optical path. These materials can constitute an enclosure by themselves (e.g., many of the first-generation shielded rooms were made with copper mesh) or the basis of some shielding integrity hardware (shielded glass, shielded ventilation panels, etc.) used to reduce the leakage of an existing hole in a conductive housing. This latter aspect will be addressed in Chapter 7. The present section covers the material, as when a whole enclosure is made of this type of metal.

2.1.5.1 Wire Screens

As suggested above, the SAM family is characterized by metal-thickness-to-air-gap area ratios on the order of 0.05. (They may range from 0.01 for chicken wire to more than 0.1 for some screens, and approximately 0.5 for some metalized textiles.) The air-gap spacings, g, may vary from about 0.25 mm (10 mils) to 5.1 cm (2"). Above low frequency, the SE model which fairly describes this family in the far field (plane waves) is:

$$SE_{dB} \cong 20 \log \left(\frac{\lambda/2}{g} \right) \quad \text{for } g < \lambda/2 \tag{2.8}$$

$$\cong 0 \quad \text{for} \geqslant \lambda/2 \tag{2.9}$$

where, λ = wavelength
g = air gap dimension in same units as λ

Thus, the SAM family may be described as an array of miniature dipoles, with partial cancellation at their adjacent edges. The SE is caused by reflection loss with essentially no loss due to absorption. This array becomes virtually transparent (0 dB) when the length of the elementary dipoles reaches $\lambda/2$.

Equation (2.8) may be expressed in both the English and metric systems for frequency instead of wavelength, i.e.:

English:

$$SE_{dB} = 20 \log (5{,}906/g_{in}f_{MHz}) \text{ dB, for } g_{in} \leqslant 5{,}906/f_{MHz} \tag{2.10}$$

$$= 0 \text{ for } g_{in} \geqslant 5{,}906/f_{MHz} \tag{2.11}$$

Metric:

$$SE_{dB} = 20 \log (15{,}000/g_{cm}f_{MHz}) \text{ dB, for } g_{cm} \leqslant 15{,}000/f_{MHz} \quad (2.12)$$

$$= 0 \text{ for } g_{cm} \geqslant 15{,}000/f_{MHz} \quad (2.13)$$

As the frequency continues to decrease, SE_{dB} does not rise at 20 dB/decade indefinitely for far-field conditions. Measurements indicate that a maximum of about 110 dB (5.5 decades below cutoff = $\lambda/2$) results for copper and aluminum screen an upwards of 140 dB for galvanized steel. To support this theory, it appears that when the screen mesh gap, g, becomes a very tiny fraction of a wavelength (i.e., 10^{-5} to 10^{-7}), the material approaches a homogeneous media from a macroscopic point of view. To the extent that this applies, the relationships of Section 1.4 may once again govern for electric and magnetic fields ($R << \lambda/2\pi$) and plane waves ($R >> \lambda/2\pi$) with the further restriction that g < about $10^{-6}\ \lambda$.

The accuracy of the 110 to 140 dB boundary is probably questionable because it approaches the dynamic range of the best measuring equipment available, knowing that we assume a far-field source with wavelengths as large as 30 km (for 10 kHz). But it checks out quite well with the SE of a thin metal barrier (see Section 1.4.2): evidently, a perforated barrier cannot give better shielding than an homogeneous one, everything else being equal.

Equations (2.8) and (2.9) are plotted in Fig. 2.30. The figure shows that for plane waves the SE of screen becomes relatively small above a few gigahertz whereas it is very significant below 1 MHz. The SE to electric fields ($R << \lambda/2\pi$) would be greater than shown in Fig. 2.30, whereas to magnetic fields ($R << \lambda/2\pi$) it would be less. It is assumed here that the **crossing mesh wires are making good electrical contact** with each other by welds or other bonding processes.

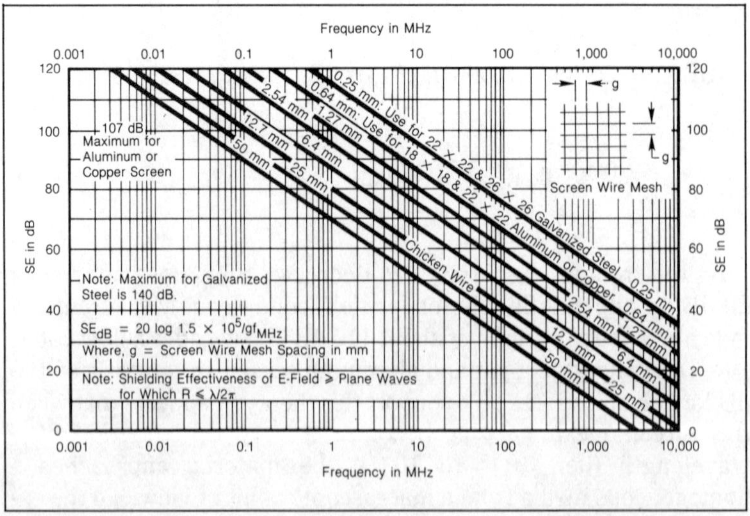

Figure 2.30—Shielding Effectiveness of Screen Wire for Plane Waves

Near-Field Shielding Effectiveness of Screen Meshes

While the plane wave attenuation of metal meshes is rather easy to estimate, based on the reradiation efficiency (or, rather, inefficiency) of the elementary dipoles per Eq. (2.8), the near-field value is more complicated to predict. If we simply consider the case of a low-frequency H-field (at 1 kHz for instance) using Eq. (2.8), a grid with 1 m wire spacing would represent the formidable attenuation of 20 log $(300 \times 10^3)/2 = 104$ dB. Anyone who has tried to shield against LF magnetic fields knows that this figure cannot be true. In fact, the figure is true, but only if the source is beyond the near-to-far-field transition distance. In our example this is $300 \times 10^3/2\pi$, or about 50 km!

A simple way exists to derive low-frequency attenuation of meshes and grids, as shown conceptually in Fig. 2.31. Note that in the figure, the SE values to lower frequencies are bounded by those of a plain metal foil of the same thickness. At mid frequencies and in the near field, the $SE_{(H)}$ is flat since Z_W increases at the same rate that λ decreases.

Given a certain mesh with grid dimension "g," we know from Eq. (2.8) that its plane wave SE is given by $\lambda/2g$. Moving backward towards lower frequencies (i.e., longer wavelengths), we know that we reach a frequency where, for the given source-to-shield distance, we cross the transition border and enter the near-field region. This critical frequency (already encountered in Section 1.1 and other parts of this book) will be called the **near-far field transition frequency, $F_{NF\text{-}FF}$.**

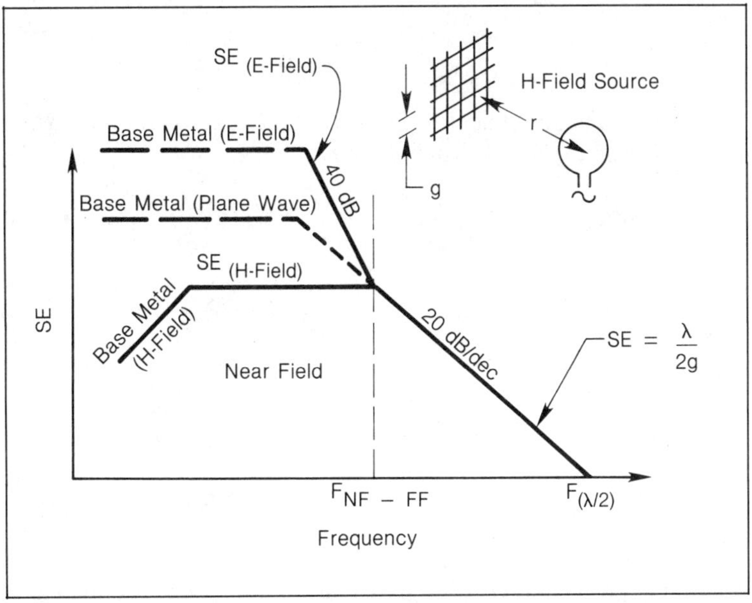

Figure 2.31—Conceptual View of the Evolution of SE for a Metal Grid at Low Frequency

Below that frequency, we can state that the SE of a grid will change its plane wave frequency dependence and evolve differently depending upon the dominantly magnetic or electric nature of the near field. Since there is no "through" attenuation for a mesh and the shielding occurs by reflection, it follows that below $F_{NF\text{-}FF}$, the SE for H- and E-fields will depart from the plane wave figure proportionally to their $Z_W/377\ \Omega$ (H or E) ratio. For a magnetic near field, we know that the wave impedance **decreases** with decreasing frequency like $2\pi r/\lambda$. For an electric near field, the wave impedance **increases** with decreasing frequency like $\lambda/2\pi r$ (see Section 1.1 of this book). Therefore, at any distance r closer than $\lambda/2$:

1. The SE against magnetic field will be equal to its plane wave

value at $F_{NF\text{-}FF}$ multiplied by $2\pi r/\lambda$, i.e.:

$$SE_{(H)} = \frac{\lambda}{2g} \times \frac{2\pi r}{\lambda} = \frac{\pi r}{g}, \text{ for } r < \frac{\lambda}{2\pi} \quad (2.14)$$

and $r >> g$

2. The SE against electric field will be equal to its plane wave value at $F_{NF\text{-}FF}$ multiplied by $\lambda/2\pi r$, i.e.

$$SE_{(E)} = \frac{\lambda}{2g} \times \frac{\lambda}{2\pi r} = \frac{\lambda^2}{4\pi r g}, \text{ for } r < \frac{\lambda}{2\pi} \quad (2.15)$$

and $r >> g$

or,

$$SE_{(E)\,dB} = 77 - 40 \log F_{MHZ} - 20 \log (rg) \quad (2.16)$$

for r and g in meters

An interesting remark about Eq. (2.14) is that the near-field magnetic performance of a mesh becomes independent of frequency, a result confirmed by many practical measurements, as we will see further. Looking toward the left of Fig. 2.31, we see that, like for plane waves, the H-field performance is bounded by that of a metal sheet of the same thickness. More exactly, this lower bound can be predicted by taking the near-field SE of a fictitious metal plate having the same thickness as the wires forming the mesh and having a sheet resistance equal to that of the wire material (copper, steel, etc.) times the percentage of coverage of the wires. If the total dimension of the wire mesh panel is $D \times D$, and a total number of wires $N \times N$ with diameter d, the percentage of coverage is given by:

$$\text{percentage of coverage} = \frac{Nd(2D - Nd)}{D^2}, \quad (2.17)$$

for D and d in the same units

Or if, instead, the grid is designated by $K = N/D$, its number of wires per inch or wires per cm:

percent coverage = Kd(2 - Kd) for d expressed in same units as K (2.18)

For example, a typical tight mesh with 22 wire/inch, AWG 22 (0.6 mm or 0.024 in diameter) will have a coverage of:

$$22 \times 0.024\,(2 - 22 \times 0.024) = 0.77$$

Therefore, the conductivity of this mesh will be 0.77 that of a homogeneous copper sheet of the same thickness. The H-field SE of that fictitious metal would be, in dB:

	100 Hz	1 kHz	10 kHz	100 kHz
At 1 m distance	20	38	60	88
At 10 cm distance	6	22	43	70

These values would constitute the lower bound of the SE_H of our 22 wire/in copper mesh. The flat value of SE_H at 1 m will be, from Eq. (2.14):

$$20 \log \frac{\pi \times D}{(25.4/22)\, 10^{-3}} = 68 \text{ dB}$$

This model has been used to establish the series of curves in Fig. 2.31 and 2.32. They allow prediction of the SE_H of all practical grid sizes for several copper and steel meshes.

Some actual measurement results are also shown in Table 2.7. In the table, measured data come from 5 different sources, totaling 72 series of measurements. Near H-field test distances were 30 cm, 4.4 cm and 3.8 cm depending on the report. A phenomenon common to all measurements is that when the frequency reaches the near-far transition, a peaking of SE appears. This seems due to the fact that for $r = \lambda/2\pi$, all $1/r$, $1/r^2$, $1/r^3$ terms of H are equal but can be out of phase. Therefore, the wave impedance exhibits a momentary increase (E/H becoming large).

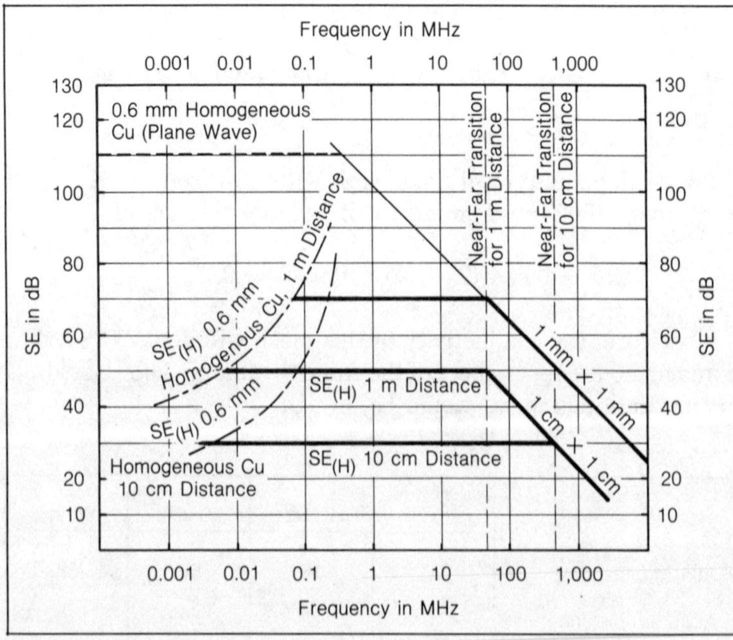

Figure 2.32—SE of AWG 22 (0.6 mm) Copper Wire Meshes

Table 2.7—Comparison of Measured Data vs. Calculated Values from Eq. (2.14)

Mesh Size	SE (Magnetic) Calculated	SE (Magnetic) Measured (Min-Max)
2/in (12.7 mm)	21 dB	15-19 dB
16/in, 7 mil wire (1.58 mm, 0.17 mm wire)	39 dB	33-41 dB
22/in 15 mil wire (1.15 mm, 0.35 mm wire)	45 dB	44-52 dB
60/in 7 mil wire (0.4 mm, 0.17 mm wire)	54 dB	40-62 dB

2.1.5.2 Panels with Holes

While not a member of the wire and mesh screen family, metal panels having punched holes are found in many applications. Convection air cooling of metal boxes containing electronic parts is one use in which the top cover plate frequently is perforated with small round apertures.

As early as 1940, many attempts had been made to model the leakage (or its reciprocal, viz., attenuation) of perforated metal plates. One might say that a perforated metal plate is just an extension of the metal meshes described before, except that in the latter case the hole area is simply larger. However, the mechanism of these leakages is fundamentally different: while a grid works primarily by reflection losses and provides no appreciable through losses (penetration), a perforated metal plate typically has holes with a depth-to-width ratio close to or larger than 1. Because of this, a "waveguide beyond cutoff" (WGBCO) effect is taking place as described in Chapter 7, and both reflection and penetration losses have to be accounted for.

On the other hand, with metal meshes, the radiation of adjacent sides of **each** hole are practically cancelling each other such that the leakage of the whole array corresponds to the leakage of one aperture only. Such mutual cancellation generally does not happen with punched holes since the metal width between each adjacent hole is equal to or larger than the hole diameter. Because of this, there is a cumulative effect of the multiple hole leakages. Depending on the solid angle under which the array is seen from both the source and the receiving antenna, N leakages can add up randomly (an approximation of their random phase addition would be $\sqrt{N}$ or, in dB, 10 log N) or coherently (like N, or in dB, 20 log N). This too is described more in detail in Chapter 7, "Aperture Leakages."

The works of Schelkunoff, Vasaka, Jarva, Kaden and Quine are often mentioned or reformulated in EMC literature for the calculation of perforated plates.[7, 8] Unfortunately, since some of the background is often omitted or truncated, the boundary and limiting conditions are overlooked or misunderstood, leading to errors which can be very large. Even more unfortunate, some typos and mistakes in the equations of the original works are carried over in many documents which add their own mistakes, and so on. For instance, a lot of the work by Quine deals with the **relative** improvement in SE of a naked aperture when this aperture is covered with a perforated plate. Nevertheless, his conclusions are often re-used to calculate the absolute shielding, i.e., the attenuation of the plate versus nothing, and the results can be grossly different.

The SE in decibels of a perforated metal plate is expressed as follows as long as the frequency of interest is less than the cutoff frequency of the apertures ($f < f_{cutoff}$):

$$SE = A_a + R_a + B_a + K_1 + K_2 + K_3 \tag{2.19}$$

where,

A_a = through attenuation (WGBCO) in dB, for f < f cutoff
= 27.3 D/g for rectangular holes
= 32 D/g for circular holes

with, D = depth of aperture
g = width of aperture if square or rectangular, (measured perpendicular to the E-field vector)

or, g = diameter for circular aperture

and, D and g expressed in the same units

R_a = aperture reflection loss in dB, due to mismatch between wave impedance and barrier impedance at the air-to-panel interface

$$= 20 \log \frac{(1 + K)^2}{4K} \approx 20 \log \frac{K}{4} \text{ if } K \geqslant 10 \text{ [from Eq. (1.82)]}$$

K = ratio of wave impedance to aperture characteristic impedance
≈ $\pi r/g$ for square or rectangular apertures and near H-field (r and g in same units)
= 3.68 r/g for circular apertures and near H-field
= $\dfrac{0.72 \times 10^4}{rF^2g}$ for square or rectangular apertures and near E-field (r and g in meters, f in megahertz
= 150/jgf for square or rectangular apertures and far field (g in meters, f in MHz)
= 173/jgf for circular apertures and far field (g in meters, f in MHz)

with,

r = distance from source to shield
$j = \sqrt{-1}$

B_a = correction for aperture reflection: similar to the B rereflection term with plain metal shields, negligible when absorption A_a is greater than 10 dB. Otherwise:

$$= 20 \log \left[1 - \frac{(K - 1)^2}{(K + 1)^2} \times 10^{-Aa/10} \right] \qquad (2.20)$$

K_1 = correction factor for number of openings per unit square, applicable when source (circuit or antenna) and victim (or receiving antenna) are far from the shield in comparison to center-to-center hole spacing

After Jarva, this term factors into the degradation caused by the ratio of the total open area to the total wall area illuminated by the radiator.[8] If the holes are evenly spread, this ratio is constant since any change in the area illuminated will cause a similar change in the total open area. This factor is always a negative number or, at best, equals 0 dB (only 1 hole).

$$= 10 \log (1 - an) \qquad (2.21)$$

where, a = area of each hole
n = number of holes per unit square (in same units as a)

K_2 = correction factor for metal penetration at low frequency, when the thickness of metal between holes is less than a skin depth. This accounts for the fact that each hole leakage is no longer independent of the metal intrinsic leakage, if the base metal itself shows a mediocre absorption loss

$$= - 20 \log \left(1 + \frac{35}{p^{2.3}}\right)$$

where,

$$p = \text{ratio of metal thickness to skin depth}$$

$$= t/\delta$$

$$= \frac{t/0.067}{\sqrt{f\sigma_r\mu_r}}$$

for t in millimeters and f in megahertz. σ_r and μ_r are conductivity relative to copper

Therefore,

$$K_2 = -20 \log \left[1 + \frac{0.067}{(t\sqrt{f\sigma_r\mu_r})^{2.3}} \right]$$

K_3 = correction factor for coupling between closely spaced shallow holes. K_3 increases (improves) when the holes are close and the opening depth is small compared to hole size. When attenuation A_a is small, Jarva states that there is a more or less pronounced neutralization of the radiation from the adjacent sides of two holes.

$$= 20 \log \frac{1}{\tanh\left(\frac{A_a}{8.7}\right)}$$

for $A_a < 10$ dB (negligible beyond 10 dB)

$$= 20 \log \frac{1}{\tanh(3.4\ D/g)}$$

(average for round or square holes)

Typical values of K_3 for A_a in dB are:

A_a	20	15	10	5	3	1
K_3	0.17	0.6	1.7	4.5	10	19

As it is, Eq. (2.19) is rather elaborate and applicable to many configurations of near or far field, including wire meshes. However, in most practical cases, simplifications occur which eliminate some of the correction terms of Eq. (2.19). With wire meshes for instance, only the reflection terms are significant, ending with the same results as the simple expression in Section 2.1.5.1. All the same, in many applications, the rereflection term B_a and correction terms K_1, K_2 and K_3 are nullified.

Illustrative Example 2.1

Determine the SE of a perforated plate when subjected to a near H-field at 0.1 m away from a 10 kHz source. There are 9 holes per cm^2 and each hole is a 0.1×0.1 cm square, leaving a gap of = 2 mm between holes. The metal is a 0.6 mm thick cold-rolled steel, with a $\sigma_r = 0.17$ and $\mu_r = 300$.

Thus,

$$A_a = 27.3 \frac{D}{g} = 27.3 \frac{0.6 \text{ mm}}{1 \text{ mm}} = 16.4 \text{ dB}$$

$$K = \frac{\pi r}{g} = \frac{\pi \times 10^{-1} \text{ m}}{10^{-3} \text{ m}} = 314 >> 10$$

Therefore,

$$R_a = 20 \log \frac{K}{4} = 37.9 \text{ dB}$$

$$B_a = 0 \text{ dB since } A_a > 10 \text{ dB}$$

$$K_1 = 10 \log (1 - an) = 10 \log (1 - 10^{-2} \text{ cm}^2 \times 10 \text{ holes/cm}^2)$$

$$= -0.46 \text{ dB}$$

$$K_2 = -20 \log \left[1 + \frac{0.067}{(g\sqrt{F\sigma_r\mu_r})^{2.3}} \right] =$$

$$-20 \log \left[1 + \frac{0.067}{(2\sqrt{0.01 \times 0.17 \times 300})^{2.3}} \right]$$

$$= -0.25 \text{ dB}$$

$$K_3 \cong 0 \text{ dB since } A_a > 10 \text{ dB}$$

$$SE = 16.4 + 37.9 - 0.4 - 0.25 = 54 \text{ dB}$$

Ultimately it should be checked that this number is not larger than the intrinsic SE of a plain metal sheet. From calculations in Chapter 1 or appendix curves, it is seen that at 10 cm from a magnetic source, the attenuation of 0.6 mm of steel is approximately 100 dB at 10 kHz. Therefore, it is concluded that the perforated plate gives 54 dB of attenuation compared to the 100 dB of an homogeneous plate.

2.1.5.3 Metalized Textiles

Metalized textiles represent an interesting development in recent years regarding what can be done with fabrics to convert them into electromagnetic shields for people, areas and equipment. Among the shielding applications of metalized textiles are partial shields for plastic equipment cases, shielding and grounding curtains, body contacts for electromedical sensors, static discharge wiper elements, latching devices, protective clothing for high-voltage linemen, flexible shielded shrouds and boots, and EMI-shielded accordion-type walkways.

Metalized textiles are perfect to "faradize" an ordinary unshielded facility. In less than a day, one can build a temporary shielded hut with modest performance (40 dB), wrap up cable harnesses, dress up unshielded windows in a metallic shelter, etc. These applications are described in practical terms in Chapters 13 and 14. Early conductive fabrics were simply a knitted mesh of thin copper wires, sometimes threaded with normal textile yarn. Other techniques have been developed more recently, such as:[9]

1. A knitted mesh of thin metal-coated fibers. The knitting results in an open-mesh, very compliant cloth.
2. Weaving of thin metal-coated fibers, resulting in a more or less dense cloth, from transparent to opaque
3. Metalization of conventional woven fabrics with nickel, silver or passivated copper

The type of metal and its thickness gives the basic fiber conductivity, while the density of the weave governs the overall surface resistivity. One family of conductive textile products is the silver metalized woven-nylon fabrics made by Technical Wire Products. The pure silver coating of the nylon cloth subtrates provides surface resistances of less than 10 Ω/sq. High strength, light weight and resistance to moisture and chemicals are qualities inherent in the nylon fabrics used as the substrates. The measured SE of the metalized textiles to plane waves follows a curve approximated by g = 0.7 mm (0.03"), as shown earlier in Fig. 2.30.

The maximum low-frequency, plane-wave shielding, however, ranges from about 65 dB for the heavy marquisette weave (Tecknit) to about 150 dB for the very fine rip-stop weave. The SE of other modern conductive textiles are shown in Figs. 2.33 and 2.34 from test results on TexGuard®, a fabric from U.S. manufacturer TW Trading, and on conductive fabrics made by DEVEX (Switzerland).[10] The latter company claims microwave power handling of 6 to 30 W/cm^2 at 12 GHz. Typical meshes have 20 to 300 μm openings, so their plane-wave or H-field attenuation can be approximated from Figs. 2.30 and 2.32, respectively. Figure 2.35 shows the TexGuard fabric applied as cable shielding, and Table 2.8 lists some of its principal characteristics.

Textile widths of 1.6 m (63") are available in rolls up to 300 m (1,000') long. Also available is 13 μm (0.5 mil) pressure-sensitive polyester fiber for easy installation and 23 μm (1 mil) clear nylon, to insulate or protect the textile surface. To maintain performance when exposed to the elements, a vinyl-coated fabric is available to weatherproof the fabric and bind the yarn together. The rated operating temperature range is from −51° to 121°C (−60 to 250°F).

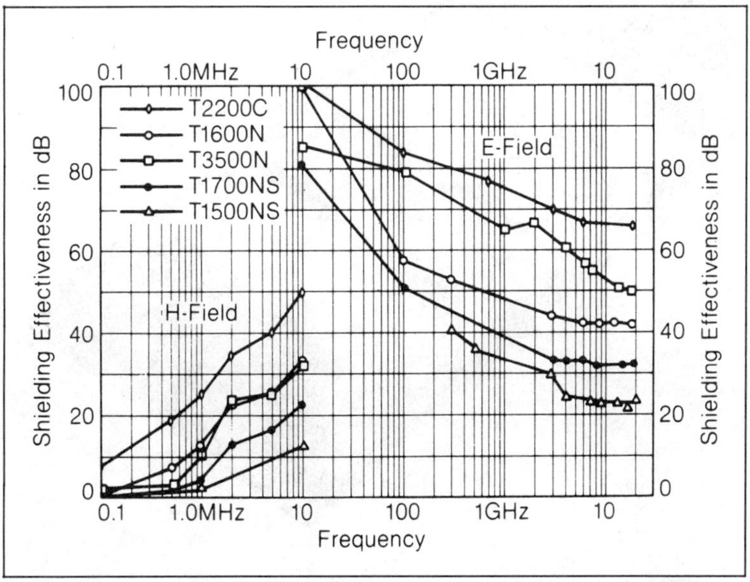

Figure 2.33—Shielding Effectiveness of Conductive Fabrics Listed in Table 2.8

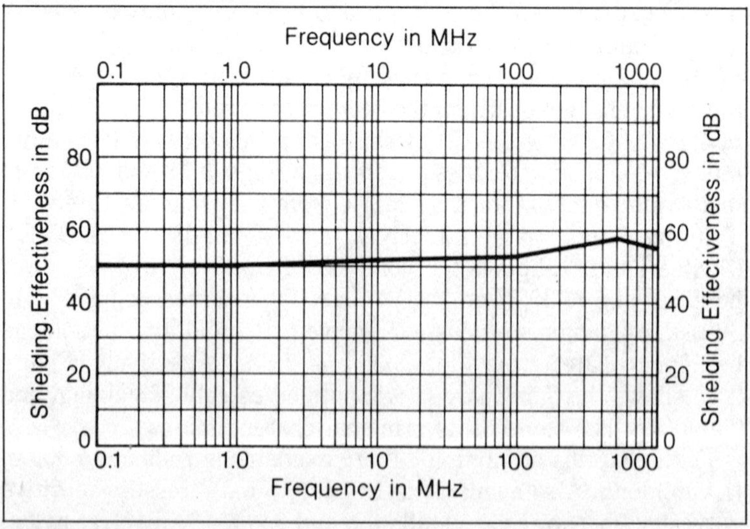

Figure 2.34—Plane-Wave Shielding Effectiveness of T1600N Material Measured by ASTM ES 7-83 Coaxial Test Method

Figure 2.35—Cable Shielding Using Conductive Fabric

Table 2.8—Characteristics of TexGuard® Conductive Fabrics

Product Name	Substrate Material	Metal Plating	Metal % by Weight	Fabric Weight g/sq. m	Surface Resistivy Ohms/Square	Description
T1600N	Polyester	Nickel	20	104.1	0.5	Silver gray color; no coatings
T2200C	Polyester	Copper	30	76.3	0.03	Copper color; no coatings
T1500NS	Polyester	Nickel	18	24.6	5.0	Silver gray color; very sheer fabric; transparent
T1700NS	Polyester	Nickel	20	44.8	1.0	Silver gray color; sheer fabric; almost transparent
T3500N	Polyester and Cotton	Nickel	35	245.3	0.5	Can be colored
T2100NA	Polyester	Nickel	8	59.4	2.5	Silver gray color; adhesive backed

*TexGuard is a Registered Trade Mark of TW Trading, International, Santa Cruz, CA 95060.

Another type of shielding fabric called Metshield (trademark of Allied Chemical Corp) was introduced around 1975.[11] This fabric is made of woven strands of Metglas®, an amorphous magnetic alloy (40 percent Fe, 40 percent Ni, 14 percent P and 6 percent Bi) which can be cut, bent and formed without loosing its high μr of 50,000 or more. The saturation level is about 7,800 gauss. This fabric, for a thickness of 58 μm (2.3 mil), is comparable to a 150 μm (6 mil) foil of commercial 80 percent nickel alloy. The low-frequency magnetic performance, for a source-to-shield distance of about 2.5 cm (Helmholtz-Coil method) is shown in Fig. 2.36. This fabric can be wrapped around cathode ray tubes (CRTs), sensitive sensors, medical instrumentation, etc. The only difficulty seems to be the lack of an easy bonding method to the supporting structure.

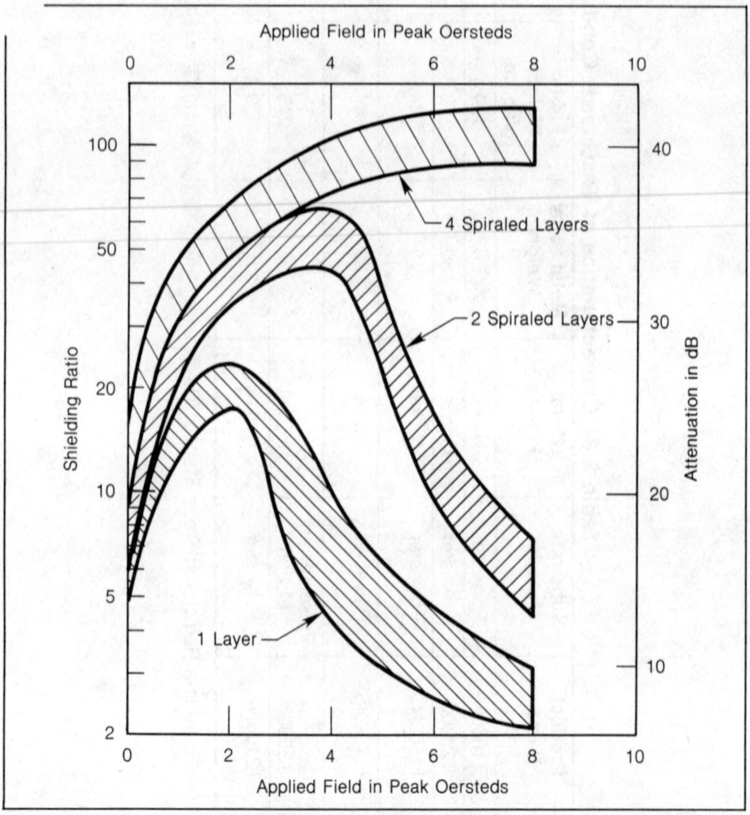

Figure 2.36—Magnetic Attenuation at 60 Hz of Single- and Multiple-Layer Cylinders (2'' I.D. × 7'') Made from METSHIELD® Magnetic Shielding Fabric

2.1.6 Effect of Protective Finishes on Shielding Performances

Many finishes are used for protecting metals against corrosion or simply for aesthetic reasons. They range from ordinary paint to enamel, varnishes, phosphate finishes (Parkerization), oxidation, plating, chromate, etc. Organic paint has no effect on SE since (except for a slightly different ϵ_r) it is transparent to electromagnetic (EM) fields and does not change the field-to-metal mismatch.

Some oxides are conductive, but most are not. For instance, the black anodized finish widely used on aluminum is resistive. Figure 2.37 shows the degradation in SE caused by some finishes. The reason why some chemical finishes affect the SE seems to lie in the transitional change they create in surface resistance by their progressive surface penetration. Instead of an abrupt air-to-metal interface, the EM wave sees a gradual change which reduces the impedance mismatch. Iridite and alodine are chromate treatments (non-oxide) that create good corrosion protection and high surface conductivity.

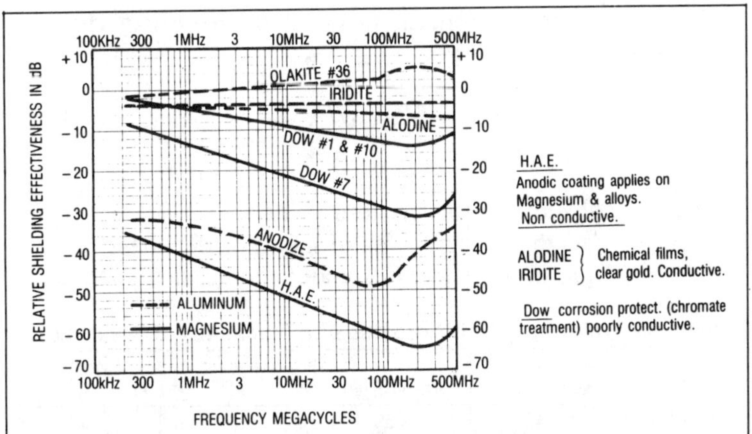

Figure 2.37—Effect of Various Finishes on Shielding Effectiveness of Aluminum and Magnesium (0 dB Reference is the SE of a Clean Surface)

2.2 Shielding Density for Weight-Sensitive Applications

For many applications, especially for aircraft and spacecraft, there exists a need for a **figure of merit** to relate the desired

SE to the required weight or volume of metal. Consequently, a concept of shielding density was devised by R. Schultz et. al. to relate the SE per unit weight per unit area of the shielding metal.[12]

The weight per unit area (W/A) of some of the most popular metals which may be considered for use in weight-sensitive applications is listed in Table 2.9. When transferred to an actual enclosure size, other considerations must also be factored in, as applicable, to determine the actual enclosure weight:

1. The weight of I/O connectors must be added and the weight of punched-out metal subtracted.
2. The weight of finger stock and other gasketing, extra interlocking lips and hinges must be added.
3. Copper plating and soldering must be added.
4. The reduction in weight due to bend radii over a square-cornered box must be subtracted.
5. The weight of excess seam metal due to overlaps, soldering, brazing, welding, etc., must be added. To simplify the present discussion, however, these weight additions and subtractions are not considered.

The desired figure-of-merit, designated **shielding density (SD)**, is defined as follows:

$$SD = SE_{dB}\ W/A \tag{2.26}$$

where,

SE_{dB} = shielding effectiveness in dB

W/A = weight per unit area of metal of thickness t required to yield SE_{dB} under the defined conditions

Table 2.9—Weight per Unit Area per Unit Thickness of Some Metals

Metal	oz/ft²/mil	gram/m²/μm	Relative σ_r	Relative μ_r
Copper	0.740	8.89	1	1
Monel	0.732	8.79	0.041	1
Brass	0.705	8.47	0.47	1
Steel	0.649	7.79	0.17	200
S3 Netic	0.642	7.71	0.02	300
Titanium	0.375	4.50	0.036	1
Aluminum	0.225	2.70	0.63	1
Magnesium	0.143	1.71	0.38	1

It is noted that identical SD values for different metals **do not** have identical electromagnetic SE_{dB} values. The SD values **do** indicate which metal will provide the greatest amount of shielding with respect to weight.

In considering SE, the worst case situation is selected, i.e., low-frequency magnetic fields with relatively little distance between emission source and areas to be protected. The rationale for this is that there will always be low-frequency magnetic field components from a source emission unless the source is carrier operated, such as a transmitter.

Furthermore, from a victim's standpoint, any electronic equipment (tuned or baseband) will have amplifiers with frequency responses extending down to either dc or low frequencies even though this may not be the most sensitive frequency region. In the low-frequency portion of the spectrum (approximately below 100 kHz), the shield metal thickness of nonmagnetic materials is usually a small fraction of a skin depth ($t/\delta << 1$), and absorption loss can be ignored. This was developed for Eqs. (1.100) and (1.102):

$$SE_{dB} \cong 20 \log (0.707\ Kt/\delta), \text{ for } t/\delta << 1 \text{ and } K >> 1 \quad (1.100)$$

$$= 20 \log (188.5\ k\sigma t) \quad (1.102)$$

$$= 20 \log (5.82 \times 10^{-3}\ r_m f_{Hz} \sigma_r t_{mil}) \quad (2.27)$$

$$= -44.7 + 20 \log (r_m f_{Hz} \sigma_r t_{mil}) \quad (2.28)$$

$$= -12.7 + 20 \log (r_m f_{Hz} \sigma_r t_{mm}) \quad (2.29)$$

where,

r_m = emission-source to metal-shield distance in meters
f_{Hz} = frequency in hertz
σ_r = conductivity of metal relative to copper
t_{mils} = metal thickness in mils
k = near field multiplier, from Eq. (1.4)

Equations (2.27) and (2.28) are for magnetic fields and $r_m \leqslant \lambda/2\pi$. Solving Eq. (2.28) for the metal thickness yields:

$$t_{mils} = \frac{\log^{-1}\left(\dfrac{SE_{dB} + 44.7}{20}\right)}{r_m f_{Hz} \sigma_r} \tag{2.30}$$

The weight associated with this metal required to give a shielding effectiveness of SE_{dB} is:

$$W/A = (W/A/mil)\ t_{mils} = \frac{(W/A/mil)\log^{-1}\left(\dfrac{SE_{dB} + 44.7}{20}\right)}{r_m f_{Hz} \sigma_r}$$

where, (W/A/mil) is obtained from Table 2.9

The conditions under which Eqs. (2.30) and (2.31) apply ($t/\delta << 1$ and $K >> 1$) are:

1. $r_m < \lambda/2\pi$ (in the near field)
2. $5{,}000 \geqslant t_{mil} \sqrt{f_{Hz}\ \sigma_r\ \mu_r}$ (absorption loss is negligible)
3. $SE_{dB} \geqslant 10$ dB

Equations (2.30) and (2.31) are plotted for aluminum in Fig. 2.38 versus frequency with SE_{dB} and r_m as parameters. The graph lines are coded in accordance with the listings in Table 2.10. This table is used to select the applicable line in the figure.

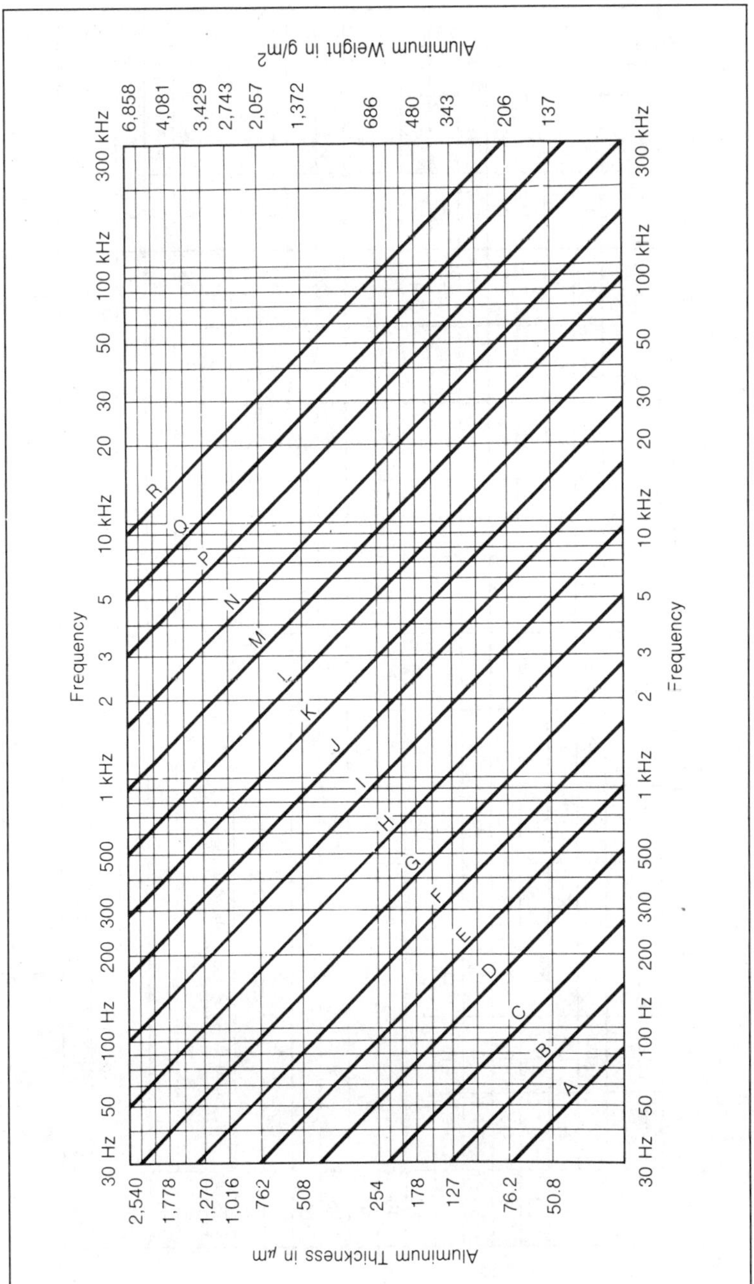

Figure 2.38—Aluminum Thickness and Weight vs. Frequency

Table 2.10—Applicable Line Selection for Use in Fig. 2.38

SE_{dB} \ r_m	1 cm	1.8 cm	3.2 cm	5.6 cm	10 cm	18 cm	32 cm	56 cm	1 m	1.8 m	3.2 m	5.6 m	10 m	18 m	32 m
10	M	L	K	J	I	H	G	F	E	D	C	B	A		
15	N	M	L	K	J	I	H	G	F	E	D	C	B	A	
20	P	N	M	L	K	J	I	H	G	F	E	D	C	B	A
25	Q	P	N	M	L	K	J	I	H	G	F	E	D	C	B
30	R	Q	P	N	M	L	K	J	I	H	G	F	E	D	C
35		R	Q	P	N	M	L	K	J	I	H	G	F	E	D
40			R	Q	P	N	M	L	K	J	I	H	G	F	E
45				R	Q	P	N	M	L	K	J	I	H	G	F
50					R	Q	P	N	M	L	K	J	I	H	G
55						R	Q	P	N	M	L	K	J	I	H
60							R	Q	P	N	M	L	K	J	K
65	Region of No. Validity:							R	Q	P	N	M	L	K	J
70									R	Q	P	N	M	L	K
75	viz, 5,000 is not $\geq t_{mil}\sqrt{f_{HZ}\sigma_r\mu_r}$									R	Q	P	N	M	L
80											R	Q	P	N	M
85												R	Q	P	N
90													R	Q	P
95														R	Q
100															R

Illustrative Example 2.2

Determine the thickness and weight of sheet aluminum stock required to give 30 dB of SE at 400 Hz when the magnetic emission source is: (1) 10 cm away, and (2) 3 m away from the location of the planned shield.

From Table 2.10, use line M for r_m = 10 cm and line G for r_m = 3 m in Fig. 2.38. At 400 Hz, Fig. 2.38 indicates:

1. At 10 cm distance: t >> 100 mils (2,540 μm) and W/A >> 22.5 oz/ft^2 (6,858 g/m^2)
2. At 3 m distance: t = 6.8 mils (175 μm) and W/A = 1.55 oz/ft^2 (475 g/m^2)

Thus, aluminum will not work for magnetic sources which are only 10 cm away since the shield would be too thick and heavy for weight-sensitive applications.

What remains is to compare the options in the above example for metals other than aluminum in order to determine if a lighter metal can be found with the same SE. The answer is obtained by using Eqs. (2.30) and (2.31) in which the ratio of t_{mil} is formed from the two (or more) candidate metals:

$$(t_{mil})\text{ ratio} = \frac{\sigma_{r1}}{\sigma_{r2}} \tag{2.32}$$

$$(W/A)_{ratio} = \frac{(W/A/mil)_2\ \sigma_{r1}}{(W/A/mil)_1\ \sigma_{r2}} \tag{2.33}$$

Equations (2.32) and (2.33) are computed for Table 2.11 relative to copper by using the information in Table 2.9.

Table 2.11—Relative Thicknesses and Weights of Some Metals for Yielding the Same Shielding Effectiveness

Metal	(t_{mil}) ratio	(W/A/mil) ratio	(W/A) ratio
Copper	1	1	1
Monel	24.4	0.989	24.1
Brass	2.13	0.953	2.03
Steel	50	0.877	43.9
S3 Netic	5.81	0.868	5.05
Titanium	27.8	0.507	14.1
Aluminum	1.59	0.304	0.483
Magnesium	2.63	0.193	0.509

Table 2.11 shows that steel is a poor low-frequency magnetic shield as long as it is operated under $t/\delta << 1$ conditions in which the absorption loss is negligible. Of course, if these conditions do not apply, then their SE will increase substantially (see Chapter 1). Monel gives poor shielding performance with regard to both relative thickness and weight requirements. The performance of titanium is not much better. While magnesium has a lighter specific weight than aluminum, a greater thickness is required due to its lesser conductivity. However, for the same SE, the net weight is greater than aluminum only by 5.4 percent. Thus, the preferred metal of the two should be chosen for a reason other than its shielding properties. The only metal not evaluated yet is a magnetic alloy like Netic with a μ_r of 300. In this case the skin depth at 400 Hz from Eq. (1.31) is:

$$\delta = \frac{0.066}{\sqrt{\mu_r \sigma_r F_{MHz}}} = 0.47 \text{ mm}$$

Therefore, the necessary thickness to provide the 30 dB at 400 Hz by absorption would be weight prohibitive as well.

Illustrative Example 2.3

Select the metal and compute the thickness and weight required to give 50 dB of SE at 10 kHz to a magnetic field source at a distance of 30 cm.

From Table 2.10, select line P. In Fig. 2.38, this requires at 10 kHz an aluminum thickness of 28 mils (0.7 mm) and a weight of 6.3 oz/ft^2 (1.9 kg/m^2).

Next, consider Mumetal®, which has a σ_r of 0.029 and a μ_r of 20,000. Since it violates the absorption loss constraint [see conditions in Eq. (2.31)] the above curves cannot be used. From Chapter 1 and the appendices, however, the thickness can be calculated as about 3 mils (0.08 mm). Based on a unit weight of 0.645 oz/ft^2/mil (0.75 kg/m^2/mm), the total weight of the Mumetall is 4 oz/ft^2 (1.2 kg/m^2) or about 63 percent of the weight of aluminum. This illustrates that only the very highly permeable metals will give a better magnetic shielding density than aluminum for the stated conditions.

Illustrative Example 2.4

Repeat Example 2.3 for a required attenuation of 30 dB with the other conditions remaining the same. Here, curve K is used in Fig. 2.38 to give an aluminum thickness of 2.8 mils (0.07 mm) and a weight of 0.63 oz/ft^2 (0.19 kg/m^2). From the equations in Chapter 1 or from the appendices, the required Mumetal thickness is 3.7 mils (0.09 mm) and the weight is 2.39 oz/ft^2 (0.72 kg/m^2) or about 280 percent of the weight of aluminum.

The reason Mumetal is heavier here is that Mumetal's performance is based less on absorption loss than in the preceding example where greater SE was required (50 dB versus 30 dB). Also, note the weight savings between the two examples: reducing the required SE_{dB} from 50 dB to 30 dB results in a weight reduction to about one-sixth of the original weight. This shows the importance of **not overdesigning for EMC, especially for critical weight situations such as aircraft and spacecraft.**

The preceding two examples clearly show the need for separate calculations for situations in which a single variable change results in conflicting metal choices. Generalizations cannot be made in which both nonmagnetic and magnetic metals may be used. Of the nonmagnetic family, aluminum and magnesium result in the least weight for the required shielding under all conditions.

2.3 Application Examples: Selection of Metal Type for a Given Performance

This chapter discusses how to use the shielding design graphs in this handbook with particular emphasis on the graphs presented in the appendices. A number of illustrative examples are given to familiarize the reader with the design graphs and to exercise the variables involved.

2.3.1 How to Use the Design Graphs

Chapters 1 and 2 contain several graphs of data pertinent to the topic of EM. For most shielding design situations, however, the

42 graphs presented in the appendices are of more direct use in computing SE versus:

1. Frequency
2. Conductivity of metal relative to copper
3. Permeability of metal relative to copper
4. Metal thickness
5. Distance between emission source and metal shield
6. E-field, H-field or plane-wave emission source

Each of the 42 graphs contains an enormous amount of data. Shielding effectiveness ranges from 0 to 300 dB, and frequency ranges from 10 Hz to 30 GHz. The variable parameters are metal thickness (ranging from 0.0001 mils to 1" (2.54 nm to 2.54 cm) and source-emission field type (E-field/plane wave or H-field/plane wave). Each graph corresponds to a metal class (seven different combinations of relative conductivity and permeability) and to a source-to-metal distance range (10 cm to 10 km). Thus, 7 × 6 (42) graphs in this handbook are required to define the SE of any metal.

It is necessary to classify each metal to determine which graph set to use. Thus, that metal closest to the conductivity and permeability of one of the seven sets is captioned by the set type. While this may result in an error in SE of several decibels in some instances, it otherwise precludes the requirement for hundreds of graphs corresponding to all specific metals. The metal class set (actually the appendix letter nomenclature) is defined in Table 2.12. Table 2.13 uses the metals listed in Table 2.1 and the definition in Table 2.12 to establish which appendix to use for any listed metal.

Table 2.12—Definition of Metal Class Based on Permeability and Conductivity Relative to Copper

Class or Metal Type (Appendix Letter)	Lower $\sqrt{\mu_r \sigma_r}$	Upper $\sqrt{\mu_r \sigma_r}$	Lower $\sqrt{\mu_r \sigma_r}$	Upper $\sqrt{\mu_r \sigma_r}$
A = Copper	1	1	1	1
Class A	0.45	> 1	1	2.24
B = Monel	0.2	0.2	5.0	5.0
Class B	0.05	0.45	2.24	20
C = Nickel	4.7	4.7	21.3	21.3
Class C	< 4.7	8	< 21.3	70
D = Iron	13.0	13.0	76.7	76.7
Class D	< 4.7	19.0	55	325
E = Hypernick	12.5	12.5	357	357
Class E	7.0	19	325	650
F = 78 Permalloy	29.4	29.4	272	272
Class F	19.0	40	140	650
G = High Permeability	56.6	56.6	1,414	1,414
Class G	24.0	> 56.6	650	> 1,414

As explained in Section 2.1, the permeability of a magnetic metal is frequency and flux-density dependent. Thus, it would seem to be impossible to define SE without knowing the operating conditions. Consequently, the applicable appendix in Table 2.13 shows a frequency split at 300 kHz (see Fig. 2.7) above which $\mu r \rightarrow 1$ for magnetic materials. Accordingly, above 300 kHz for the higher conductivity ($\sigma_r \geqslant 0.2$) magnetic metals, Appendix A (copper) is used; for $\sigma_r < 0.2$ magnetic metals, Appendix B (monel) is used.

After selecting the appendix letter from either Table 2.12 or 2.13, as applicable, select the correct design graph within the appendix. Each appendix subset is based on the distance between the EMI emission source and the metal barrier. Six such subsets are presented in each appendix as shown in Table 2.14 (the X corresponds to any appendix letter, that is, A, B, C, etc.).

Having selected the applicable graph within the appendix, apply the graph to the problem. The graph may be used to determine either: (1) the resulting SE given the metal thickness, operating frequency and E- or H-field problem, (2) the required metal thickness given the desired SE, operating frequency and E- or H-field problem, or (3) the upper and lower useful frequency given the desired SE, the metal thickness and E- or H-field problem.

2.3.2 Illustrative Examples

This section presents a number of examples which illustrate how the design graphs in the appendices can be used.

Illustrative Example 2.5

A sheet of 1.6 mm (1/16" or 62.5 mils) iron is considered for shielding a box from a strong magnetic field originating from a 60 Hz generator located 1.6 m (5') away. Determine the SE in dB.

Unless the type of iron is specified (see Table 2.13), Appendix D is selected as listed in Table 2.12. Since the distance is 1.6 m, Table 2.14 indicates that Fig. D.2, within Appendix D, is the closest to operational conditions. Select the design curve of t = 100 mils (to start) at 60 Hz for H-field conditions. Interpolating between t = 30 mils and t = 100 mils reveals that the expected SE is about 24 dB. Were the design objective to provide a SE of 50 dB at this distance, Fig. D.2 indicates that the thickness of the iron should be about 140 mils (3.6 mm).

Table 2.13—Metal Class for Choice of Appendix to Use

Metal	Appendix No. < 300 kHz	Appendix No. ⩾ 300 kHz	Metal	Appendix No. < 300 kHz	Appendix No. ⩾ 300 kHz
Alfenol, 16	E	B	Mumetal	G	B
Aluminum, soft	A		Nichrome	B	
Aluminum, tempered	A		Nickel	C	A
Beryllium	A		Permalloy, 45	D	A
Brass, (91% Cu, 34% Z_n)	A		Permalloy, 78	F	B
Brass, (66% Cu, 34% Z_n)	A		Permalloy, 4-79	G	B
Cadmium	A		Permendur	D	A
Chromium	A		Perminvar, 45-25	C	B
Conetic	G	B	Platinum	B	
Copper	A		Silver	A	
Gold	A		Sinimax	E	B
Hypernick	E	B	Steel, cold-rolled	C	B
Hiperco	D	B	Steel, hot-rolled	D	B
Iron, commercial	C	B	Steel, 50% Nickel	D	B
Iron, purified	F	B	Steel, stainless	D	B
Iron, Silicone 4%	D	B	Supermalloy	G	B
Lead	B		Tantalum	B	
Magnesium	A		Tin	B	
Manganese	B		Tungsten	A	
Mercury	B		Zinc	A	
Monel	B				
Monimax	D	B			

Table 2.14—Applicable Appendix Design Graph for Specified Distance

Figure Number	Nominal Distance	Applicable For
X.1	R_m = 10 cm	3 cm ≤ R_m ≤ 30 cm
X.2	R_m = 1 m	30 cm ≤ R_m ≤ 3 m
X.3	R_m = 10 m	3 m ≤ R_m ≤ 30 m
X.4	R_m = 100 m	30 m ≤ R_m ≤ 300 m
X.5	R_m = 1 km	300 m ≤ R_m ≤ 3 km
X.6	R_m = 10 km	R_m ≤ 3 km

Illustrative Example 2.6

A piece of sensitive electronic equipment is located near (100 m from) an AM broadcast station antenna transmitting at 1,250 kHz. Determine the SE of a 0.79 mm (0.03125" or 31.25 mils) sheet metal aluminum box enclosure to E-fields or plane waves, as applicable.

Table 2.13 indicates that Appendix A applies. For a distance of R_m = 100 m in Table 2.14, the closest graph is identified as A.4. Corresponding to a metal thickness of 30 mils and a frequency of 1.25 MHz, the SE for plane waves in the figure is about 196 dB. Note that 196 dB would never be obtained in practice because of the penetrations required through the box (see Chapter 7). Also note that the near/far-field interface ($R_m = \lambda/2\pi$) exists at 38.2 m. Thus, the electronic equipment box housing is located in the far field of the transmitter, and plane-wave conditions apply rather than E-field conditions.

Were the design objective to provide a SE of 100 dB at the 100 m distance, Fig. A.4 indicates that the thickness of aluminum should be about 18 μm (0.7 mils). Thus a 0.0254 mm (1 mil) foil covering the box would be adequate notwithstanding the fact that it would offer no mechanical strength. Note that graphs A.5 (1 km) and A.6 (10 km) also yield the same results since all are in the far field at 1,250 kHz.

Illustrative Example 2.7

A ground-level nuclear detonation produces a broadband EMP with most of its energy distributed in the 10 kHz band. From a distance of 5 km, the blast center looks like a magnetic source having billions of amperes. To protect electromagnetic equipment, specify the thickness of both aluminum and sheet steel to provide 300 dB isolation.

Tables 2.13 and 2.14 indicate that the graphs represented by Figs. A.6 (aluminum) and D.6 (steel) apply. When the frequency is 10 kHz, the thickness of aluminum should be about 18 mm (700 mils) and that of steel should be about 1.3 mm (50 mils). A separate calculation of magnetic flux density should be made to determine if the steel is saturated (i.e., the permeability is lowered).

Illustrative Example 2.8

A nearby (500 m away) L-Band weather radar operating at 1,300 MHz creates an electric field strength of 100 V/m in an internal room of a building where a computer is to be located. It is known that the computer, its peripherals and all interconnecting cables will not be susceptible to radiation below 3 V/m. Thus, determine a relatively inexpensive shield for the room.

The required SE is the ratio of the electric field strength [20 log (100/3)], which equals about 31 dB. Consider the SE of 25.4 μm (1 mil) household aluminum foil as one possible solution. Here the walls, ceiling, and flooring would be so lined with overlapping joints taped in place. For a frequency of 1,300 MHz, the SE is about 200 dB. Thus, the aluminum foil provides enormous SE if door seams, power entry ports and the like are treated such that they, too, provide $>$ 31 dB of attenuation to avoid parallel EMI entry modes.

Illustrative Example 2.9

A cable is carrying a number of 10 kb/s digital data lines at 5 V logic and 100 Ω impedance. The differential-mode current is 5 V/100 Ω = 50 mA. At a distance of 5 cm, the magnetic field intensity from a single wire is 0.16 A/m and the magnetic flux density is $B = \mu H = 2$ mG $= 2 \times 10^5$ pT $=$ 106 dBpT. If the differential-mode reduction due to reverse field gradients and twisted-pairs is about 70 dB, calculate the required thickness of a 10 cm pipe of permeable material to reduce the flux density to 1 milligamma (0 dBpT) on the outside. The required SE of the pipe is:

$$(106 \text{ dBpT} - 70 \text{ dB}) - 0 \text{ dBpT} = 36 \text{ dB}$$

This minimum attenuation must be provided from dc to about 10 kHz (first lobe of frequency spectrum for 50 μs pulses; $f \approx 1/\pi t$). Since less pipe attenuation exists at lower frequencies, a frequency of 10 Hz will be used, which is the lower limit on the appendix graphs.

For iron (Class D) and a distance of $R_m = 5$ cm, Tables 2.13 and 2.14 indicate that graph C.1 in the appendices should be used. The graph shows that a wall thickness of about 4 mm (150 mils) is required. Had Mumetal (graph G.1) been used, the required wall thickness would be reduced to about 0.25 mm (10 mils). Note that the SE of these metals is lowest at 10 Hz (about 36 dB); whereas, at 10 kHz it is greater than 300 dB for the iron and 88 dB for the Mumetal.

2.4 Bibliography

Achenback, B., "Particle Size Compaction," *EMC Technology*, October 1985.

Benson, P., "Effects of EPA Regulations on EMI Shielding Coatings," *ITEM*, 1985.

Birkin, J.A., "Advance Composite Aircraft Design," *1982 IEEE EMC Symposium Record* (New York: IEEE), Santa Clara.

Casey, K.F., "E.M. Penetration of a Composite Panel," *IEEE EMC Symposium Record* (New York: IEEE), Santa Clara.

Coniglio, J., "Nickel Coatings," *ITEM*, 1984.

CYCOM Co. catalog.

DEVEX, Inc., Chatel St. Denis, Switzerland, general catalog.

Donohue, R.K., and Eckersley, A., "Some Electromagnetic Compatibility Grounding and Shielding Considerations in Titanium Aircraft Structures" (Abstract), *IEEE Transactions on EMC*, Vol. EMC-10, no. 1, March 1968, p. 142.

Eckersley, A., "H-Field Shielding Effectiveness of Flame-Sprayed on Thin Solid Aluminum and Copper Sheets," *IEEE Transac-*

tions on EMC, Vol. 10, no. 1, March 1968, pp. 101-104.

"Electronics Installation and Maintenance Book-Radio Interference Reduction," Department of the Navy, Bureau of Ships, September 1963, pp. 10-3.

Epstein, M., and Schulz, R.B., "Magnetic-field Pickup for Low-Frequency Radio Interference Measuring Sets," *IRE Transactions, Electron Devices,* Vol. ED-8, January 1961, pp. 70-77.

Feber, R.R., and Young, F.J., "The Shielding of Electromagnetic Pulses by the Use of Magnetic Materials," *1969 IEEE EMC Symposium Record* (New York: IEEE), Vol. 69C3-EMC, June 17-19, 1969, Asbury Park, pp. 73-74.

Harrington, J.G., and Schulz, R.B., "Design of Minimum Weight and Maximum Effectiveness of Very-Low-Frequency Shielding," *IEEE Transactions on EMC*, Vol. EMC-10, no. 1.

Hornbostel, D.H., and Becker, S., et. al., "Study of Thin-film Interference Reduction Techniques," AIL Division Cutler-Hammer, Deer Park, New York, USAECOM Contract DA28-043-AMC-01970(E), 3rd Quarterly Report, July 1967. (Also, 1st and 2nd Quarterly Reports. Final Report, May 1966 to October 1967).

Jarva, W., "Shielding Efficiency Calculation Methods for Screening Waveguide Ventilation Panels and other Perforated Electromagnetic Shields," *Proceedings of the 7th Tri-Service Conference on Radio Interference Reduction and Electronic Compatibility*, Armour Research Foundation, November 1961, pp. 478-498.

Johnson, A., "New Conductive Fabrics for EMI Shielding," *EMC Technology*, October 1985.

Johnson, W.R., et. al., "Development of a Space Vehicle Electromagnetic Interference Compatibility Specification," NSA, Contract no. 9-73-5, TRW, 08900-60001-T000, June 28, 1968.

Lassiter, H.A., "Low Frequency Shielding Effectiveness of Conductive Glass," *IEEE Transactions on EMC*, Vol. EMC-6, July 1964, pp. 17-30.

Little, P., "Lightning Hazards with Composites," *1982 IEEE EMC Symposium Record* (New York: IEEE), Santa Clara, 1982.

Lodge, K.J., Brette, J., Loggie, N.J., "The Design of CFC Joints for Electromagnetic Compatibility," IERE EMC Symposium, 1982, Surrey, UK.

McCaskie, J., and Tsiamis, C., "A Gaseous Acid Etch System," 1981

Meeting of American Society of ElectroPlastics.

MIL-Handbook 419, p. 5.35.

Mooney, P.J., "Plastic EMI Shielding," *EMC Technology*, October 1985.

Patton, B.J., "Magnetic Shielding," Socony Mobil Oil Co., Dallas, Texas.

Patton, E.J., "Room-Size Enclosure for Geomagnetic Shielding," *1970 IEEE International EMC Symposium Record* (New York: IEEE), Vol. 70C28-EMC, Anaheim, July 14-16, 1970, pp. 89-96.

Pflub, D.R., "Composite Materials, EMC Perspectives," *EMC Technology*, July 1982.

Plantz, V.C., Schulz, R.B., and Goldman, R., "Shielding Density: An Electromagnetic Shielding Design Concept for Weight Sensitive Applications," *IEEE Transactions on EMC*, Vol. 10, no. 1, March 1968, pp. 126-129.

Quine, J.P., "Theoretical Formulas for Calculating Shielding Effectiveness of Perforated Sheets and Wire Mesh Screens," *Proceedings of the Third Conference on Radio Interference Reduction*, Armour Research Foundation, February 1957, pp. 315-329.

Schulz, R.B., "ELF and VLF Shielding Effectiveness of High-Permeability Materials," *IEEE Transactions on EMC*, Vol. 10, no. 1, March 1968, pp. 95-100.

Sellers, G., "Metglass Alloys," *1977 IEEE EMC Symposium Record* (New York: IEEE), p. 129.

Simon, R., "Aluminum Filled Polymers," Transmet Co., Tech. Notes.

Thomas, A.K., "Magnetic Shielded Enclosure Design in the DC and VLF Region," *IEEE Transactions on EMC*, Vol. 10, no. 1, March 1968, pp. 142-152.

Thompson, D., "Electroless Shielding," *EMC Technology*, October 1985.

Vaccari, J., "Design for Plated Plastics," *Design Engineering*, February 1981.

Walker, W.F., "Electrical Conductivity of Carbon Epoxy Composite," *1982 IEEE EMC Symposium Record* (New York: IEEE), Santa Clara.

Weck, R.A., "Advanced Shielding Techniques," U.S. Army Electronics Command, Fort Monmouth, New Jersey, Technical Report ECOM-2775, July 1966.

Weck, R.A., "Thin-Film Shielding for Microcircuit Applications and a Useful Laboratory Tool for Plane-Wave Shielding Evaluations," *IEEE Transactions on EMC*, Vol. 10, no. 1, March 1968, pp. 105-112.

Weck, R.A., and Lump, C.J., "Thin-Film Shielding Measurements," U.S. Army Electronics Command, Fort Monmouth, New Jersey, Technical Report ECOM-2798, January 1967.

Wills, A.P., "On the Magnetic Shielding of Tri-lamellar Spherical and Cylindrical Shells," Phys. Rev., Vol. 9, October 1899.

Chapter 3

Principal Test Methods to Characterize SE

Because SE is the accepted criterion to characterize the ability of a material to attenuate electromagnetic fields, it is important to have standardized test methods to measure it. Ideally, an SE test should give the following result: "The SE is the ratio of the power (or the field) which would be received at the measuring point if the shield was not there relative to the one actually received when the shield is in place."

As explained in the previous chapters, SE for a given material is highly dependent on the near- or far-field situation, whether the field is predominantly electrical or magnetic (if in the near field), the wave impedance, etc. It should be evident that an SE value measured under given conditions may be misleading if the application does not exactly match these conditions.

Several test methods have been devised to characterize the SE of a barrier or a sample of material while maintaining all possible versatility. They are:

1. MIL-STD-285
2. The Coaxial Holder Method (ASTM)
3. The Dual-Chamber Method (ASTM)
4. The Dual TEM Cell Method

These will be discussed individually.

3.1 MIL-STD-285

MIL-STD-285 (dated June 1956) is entitled "Attenuation Measurements for Enclosures, Electromagnetic Shielding, for Electronic Test Purposes, Method of." The setup is shown in Fig. 3.1. It is a substitution method, using two (electric or magnetic) antennas. Two readings are made, one with and one without the shield barrier in place. The SE is defined as the adjustment (increase) of the dB attenuation setting on the receiver (or the source) which produces the same reference reading when the shield wall is removed, **without changing the position of the antennas.** The rationale is that precision attenuators give better accuracy and repeatability than a receiver display. However, with the introduction of modern receivers with tracking generators, readout and operator errors have been eliminated by digital data processing.

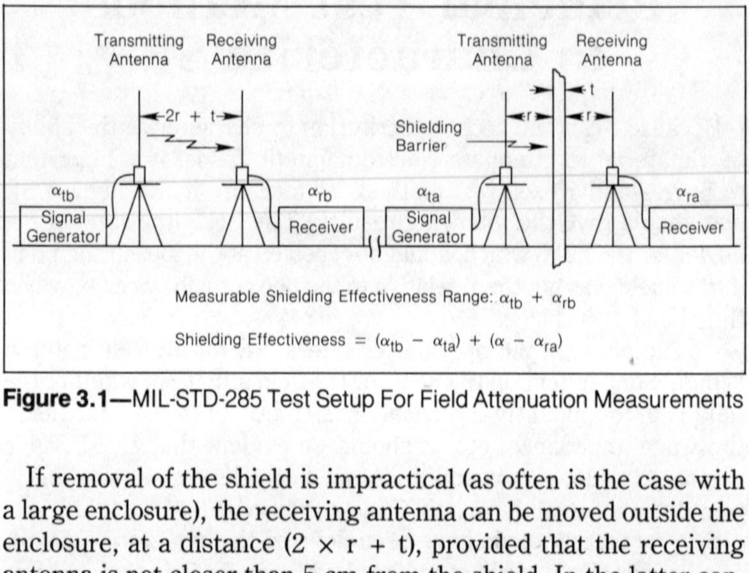

Figure 3.1—MIL-STD-285 Test Setup For Field Attenuation Measurements

If removal of the shield is impractical (as often is the case with a large enclosure), the receiving antenna can be moved outside the enclosure, at a distance (2 × r + t), provided that the receiving antenna is not closer than 5 cm from the shield. In the latter configuration, it must be assured that the shielding of the receiver case itself is larger by at least 20 dB than the SE being measured to ensure that a measurement error by signal entering through the receiver enclosure is not introduced. So the receiver generally is kept **inside** the enclosure, and the receiving antenna lead is brought in by a double-shielded cable or a tube via a shielded fitting. An optical fiber link can be used to avoid any parasitic coupling. There are several points that are mentioned here regarding MIL-STD-285 and its use:

1. E-field, or high-impedance waves, where $r < \lambda/2\pi$ and $Z_w \gg 377\ \Omega$, are measured at a distance of 30.5 cm (12") from the shielding-barrier interface. This results in much higher SE (optimistic values) than would exist at greater distances. As a minimum, measurements are made at least at 200 kHz, 1 MHz and 18 MHz and generally at other frequencies below 30 MHz.
2. H-field, or low-impedance waves, where $r < \lambda/2\pi$ and $Z_w \ll 377\ \Omega$, are measured at a distance of 30.5 cm (12") from the shielding-barrier interface. This results in much lower SE (pessimistic values) than would exist at greater distances. As a minimum, measurements are made at least at 200 kHz and generally at other frequencies below 1 MHz.
3. Plane waves, or far-field measurements, where $r > \lambda/2\pi$ and $Z_w = 377\ \Omega$ are made at a distance of 1.83 m (6 ft) or more (where $r \geqslant 2\lambda$) from the shielding-barrier interface. Within the available attenuation measurement range, distance is unimportant as long as $r > \lambda/2\pi$. Measurements are made at least at 400 MHz and generally at several other frequencies above 100 MHz.
4. Measurements are usually made in the open first, with no barrier between transmitting and receiving antennas. Then the measurements are repeated with the same orientation and separation of antennas but with the shielding barrier located in between. This gives erroneous readings for E-fields since the metal barrier detunes the antennas by capacitive loading. Antenna probes, whose dimensions are small compared to 30 cm (12"), will not have this problem. H-field antennas are Faraday-shielded.
5. The measurement range suggested in MIL-STD-285 can be substantially increased by using power amplifiers or power oscillators on the transmitting side and narrowband-tuned amplifiers or correlators with a fiber optic reference path on the receiving side. Manufacturers' E-field SE data often reflect inadequate measurement range.
6. For gaskets, wire screens, woven mesh and other metal forms whose thicknesses are very small compared to that of the shielded enclosure metal, a 45.7 cm (18") or 61 cm (24") square aperture is usually cut in the shield wall of an internal partition of the enclosure. Shielding effectiveness (insertion loss) measurements are made by taking the difference between field strength measurements when no test sample is used versus when the test sample is in place.

Detailed discussions regarding test methods and procedures of MIL-STD-285 are beyond the scope of this handbook. Some considerations regarding measurement accuracy are suggested in Vol. 6 of this EMI/EMC handbook series. However, the first and second points above regarding optimistic and pessimistic SE reported in manufacturers' literature deserve some emphasis here. For example, suppose there is a known source of interference in a real problem at a distance of 10 m against which one desires to shield. How much SE is needed when manufacturers' data sheets are reported in the language of MIL-STD-285 at a distance of 30.5 cm (12")?

To answer that question, one must develop correction data based on extrapolating the results of the MIL-STD-285 test data distance to the desired user distance. Such extrapolation data are a function of many variables and would require many graphs.

However, the problem may be simplified by making certain assumptions so that quantitative correction data can be presented here. For conditions under which the reflection loss is significant (i.e., $K \geqslant 10$), Eq. (1.64) becomes:

$$SE_{dB} \cong 20 \log (0.707) Kt/\delta) \text{ for } t/\delta << 1 \qquad (1.100)$$

$$SE_{dB} \cong 20 \log (e^{t/\delta}K/4) \text{ for } t/\delta >> 1 \qquad (1.101)$$

where,

$$K = Z_w/Z_m = k \times \text{constant [see Eq. (1.74)]}$$

$$k = \lambda/2\pi r \qquad \text{for E-fields}$$

$$= 2\pi r/\lambda \qquad \text{for H-fields}$$

$$= 1 \qquad \text{for plane waves}$$

When Eq. (1.100) or (1.101) is applied to MIL-STD-285 for any two distances, an error is developed which is a function of k alone. Thus, for any measurement distance r_m and any user-applied distance r_u, the correction in shielding efficiency, SE_{dB}, becomes:

$$\Delta SE_{dB} = 20 \log (r_m/r_u) \quad \text{for E-fields} \tag{3.1}$$

$$= 20 \log (r_u/r_m) \quad \text{for H-fields} \tag{3.2}$$

$$= 0 \quad \text{for plane waves}$$

in which it is understood that both r_m are in the near field.

When one of the distances is in the near field and the other is in the far field, Eqs. (3.1) and (3.2) become:

$$\Delta SE_{dB} = 20 \log (2\pi r_m/\lambda) \quad \text{for E-fields, and } r_m \text{ in near and } r_u \text{ in far field} \tag{3.3}$$

$$= 20 \log (\lambda/2\pi r_u) \quad \text{for E-fields, and } r_u \text{ in near and } r_m \text{ in far field} \tag{3.4}$$

$$= 20 \log (2\pi r_u/\lambda) \quad \text{for H-fields, and } r_u \text{ in near and } r_m \text{ in far field} \tag{3.5}$$

$$= 20 \log (\lambda/2\pi r_m) \quad \text{for H-fields, and } r_m \text{ in near and } r_u \text{ in far fields} \tag{3.6}$$

Equations (3.1) through (3.6) are plotted in Fig. 3.2 for a MIL-STD-285 distance of r_m = 30 cm (12") and any user distance r_u as shown.

Illustrative Example 3.1

A certain manufacturer's literature states that, per MIL-STD-285, a metalized silicone elastomer offers at least 90 dB of SE to E-fields at 10 MHz, 30 dB SE to H-fields at 10 MHz and 70 dB to plane waves above VHF. Determine the likely SE at a distance of 10 m.

From Fig. 3.2, the correction to E-fields at r_u = 10 m is −23 dB and to H-fields is +23 dB. Thus, for a 10 m distance, SE_{dB} (E-field) = 67 dB and SE_{dB} (H-field) = 53 dB. Note that the 10 m distance is in the far field ($r_u > \lambda/2\pi$) at 10 MHz, since the r_u = 10 m line in Fig. 3.2 ends at 4.8 MHz ($r_u = \lambda/2\pi$). Thus, there is no correction for SE_{dB} (plane waves) = 70 dB.

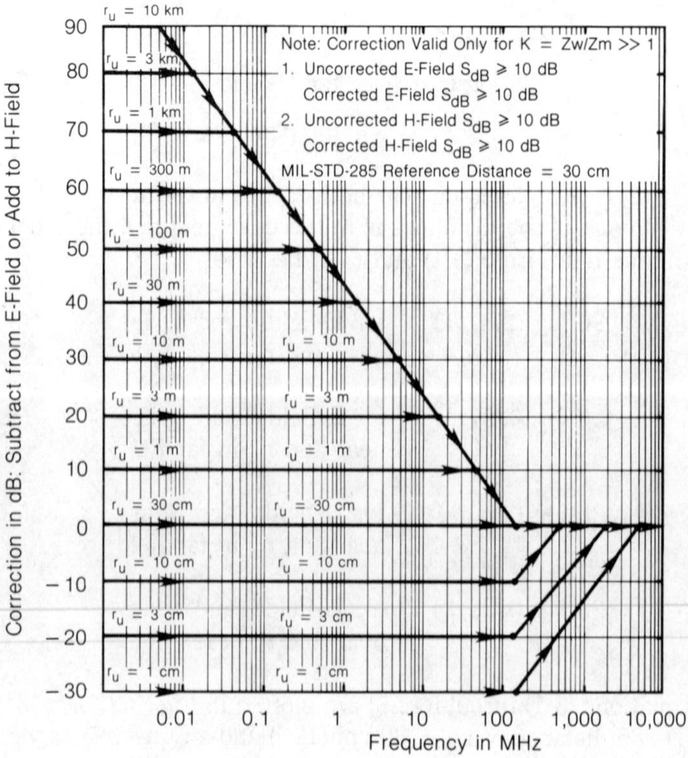

where,

α_{tb} = Reserve Signal Generator Attenuation in dB Before Shielding Measurement

α_{rb} = Reserve Receiver Attenuation in dB Before Shielding Measurement

α_{ta} = Signal Generator Attenuation in dB After Shielding Barrier in Place

α_{ra} = Receiver Attenuation in dB After Shielding Barrier in Place

r = 30 cm for Near E- or H-Field Evaluations (< 100 MHz)
r ≥ 1.80 m for Plane-Wave Evaluations (≥ 100 MHz)
For H-field evaluation, rod antennas are replaced by loops.

Figure 3.2—Correction in Shielding Effectiveness to Convert MIL-STD-285 Results to Another Distance

Illustrative Example 3.2

The SE to H-fields of 51 μm (2 mil) aluminum foil was measured at a 1 m distance at 42 kHz and found to be 50 dB. What protection would this foil offer at a 5 cm distance from a hostile magnetic field?

Since both conditions are in the near field, Eq. (3.2) applies:

$$\Delta SE_{dB} = 20 \log (5 \text{ cm}/100 \text{ cm}) = -26 \text{ dB}$$

Thus, SE_{dB} (H-field) $= 50 \text{ dB} - 26 \text{ dB} = 24 \text{ dB}$

More than 30 years after its issuance, MIL-STD-285 is still a widely used method for SE evaluation. Several national agencies in the United States and elsewhere in the world use this standard or similarly derived ones. The NSA Specification 65-6, for instance, uses a very similar set-up to check the SE of shielded rooms for communication security.

As its title states, MIL-STD-285 was primarily intended for the certification of shielded enclosures used for EMI testing. However, because of the dearth of other widely accepted standards, MIL-STD-285 has degenerated into a procedure which is used to test things like gaskets, conductive plastics, ventilation panels, etc. This introduces many variables that the standard cannot address, but people often blame their problems on MIL-STD-285 rather than their own misapplication of the standard (see Fig. 3.3)

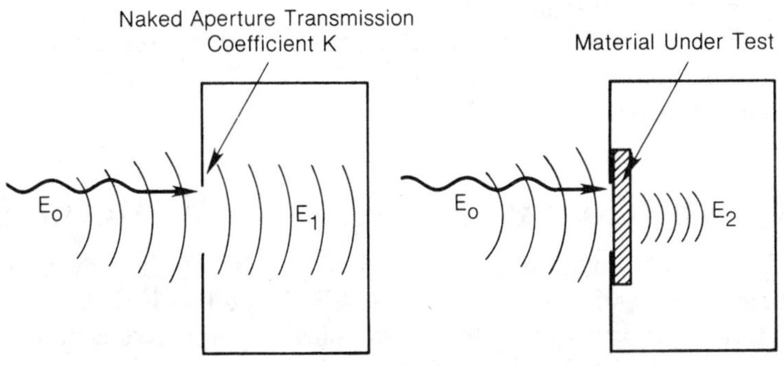

What is needed is E_0/E_2;
What is Actually Seen is E_1/E_2
Which Is Always $< E_0/E_2$ Since $K < 1$.

Figure 3.3—One Shortcoming of MIL-STD-285 Test Method

Typical problems encountered when trying to evaluate the SE of a material sample in a shielded enclosure include the following:

1. Since the test is usually performed by placing the test sample over an aperture in the shielded room wall, the size of the aperture or port influences the result; what is actually seen is the transmission coefficient of the aperture with and without the material. Therefore, two different test rooms will provide different results. Experience has shown that the larger the port, the more closely the data will agree with theoretical values.
2. Antenna size and distance influence the results. Although these phenomena are normal and their effects can be quantified, they complicate the issue since antennas and distances are not identical from one test facility to the next.[1]
3. Shielded room resonance causes peaks and nulls of E and H, giving apparent changes in SE.[2]
4. Below 30 MHz, the generic SE of materials in the far field is impractical to measure because it would require both antennas to be located at impractically large distances, with a quasi-infinite sheet of the material under test.
5. While it is generally agreed that the sample and its mounting port need to be as large as possible, the associated gasket or seam problems are growing. In many cases, what the experimenter actually sees is the joint attenuation, not the material attenuation.

To deal with these problems, other methods have been developed to evaluate shielding of materials.[3,4] These will be discussed in the following section.

3.2 The Coaxial Holder (ASTM) Method

The coaxial holder (Fig. 3.4) is recommended by the American Society for Testing and Materials (ASTM) Method ES7-83. This method's primary goal is to test samples of conductive coatings, composite materials and conductive loaded plastics. Since wave impedance is stabilized in a coaxial line, this method replicates a plane wave situation.

The method is basically an insertion loss measurement by substitution. One run is made with an empty holder, and another run is made with a standard size sample inserted in the midsection flange of the holder. The sample, in the form of a pierced

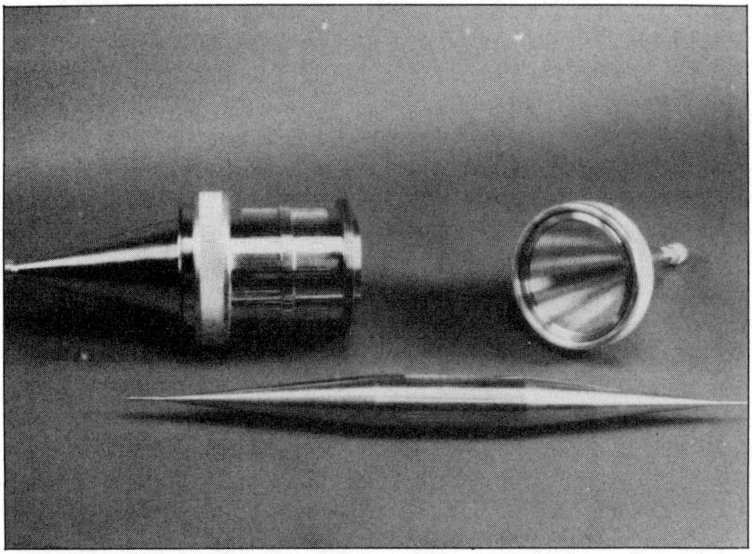

Figure 3.4—ASTM ES7-83 Coaxial Transmission Line Sample Holder (Far-Field Test)

wafer, introduces a low-impedance shunt. The resulting insertion is deemed to represent the plane wave (far field) SE of the material.

If the characteristic impedance of the holder is Z_0 and the impedance of the sample is Z_L, the insertion loss is:

$$IL_{dB} = 20 \log \left(1 + \frac{Z_0}{2\ Z_L}\right) \tag{3.7}$$

The principal limitation resides in the contact impedance and resistance between the sample and the holder. As seen in Fig. 3.5, this contact impedance Z_c is present in four places in the measurement circuit, and the insertion loss is actually:

$$IL = 20 \log \left[1 + \frac{Z_0}{2(Z_L + 2\ Z_c)}\right] \tag{3.8}$$

If Z_c approaches or exceeds the Z_L of the wafer, the measurement becomes meaningless. For instance, a wafer of aluminum having an actual SE of 100 dB at a few MHz, for an assumed contact resistance of 0 Ω, has been found to show an apparent SE of 54 dB if contact resistance is about 0.01 Ω and 14 dB if contact resistance is about 1 Ω. Since it is extremely difficult in practice to achieve a contact resistance less than 1 mΩ, this gives a practical dynamic range for the device of about 80 dB (for a 3 dB error).

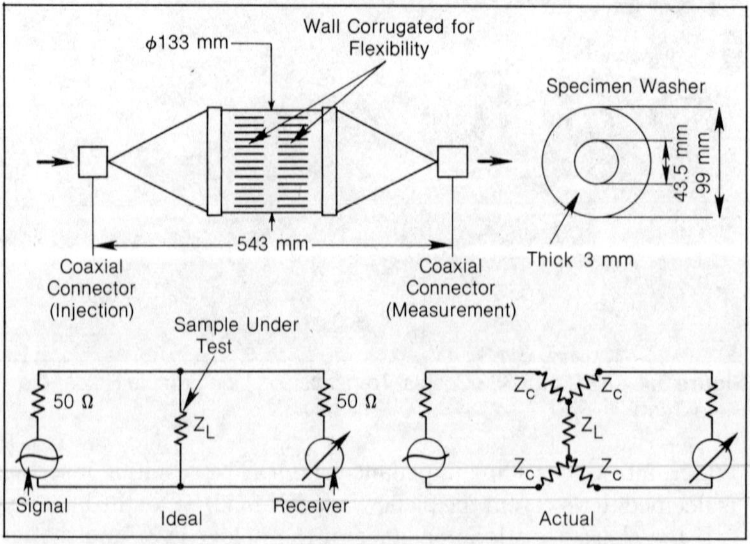

Figure 3.5—ASTM Coaxial Holder and Its Ideal/Actual Equivalent Circuit

The National Bureau of Standards (NBS) has developed an improved version (see Figs. 3.6 and 3.7) where the holder is equipped with enlarged flanges.[3] It is also recommended that the edges of the sample be coated with silver paint to decrease contact resistance. For composite materials, the resin-rich surface must be abraded at the edges before insertion. Provided the above limitations are known, the coaxial holder is a practical method for samples whose resistance is larger than a few milliohms.

The useful frequency range of the coaxial holder is 0.1 to 3,000 MHz (Fig. 3.8), with an accuracy of ± 3 dB. Calibration wafers with known SE can be used to check the condition of the holder. This repeatability is precious when testing for aging, humidity, salt spray, etc., since these environmental effects may cause only 3 to 10 dB changes of the material properties. These changes would probably go unnoticed in a shielded-room test like MIL-STD-285 because of the inherent repeatability problems of this method.

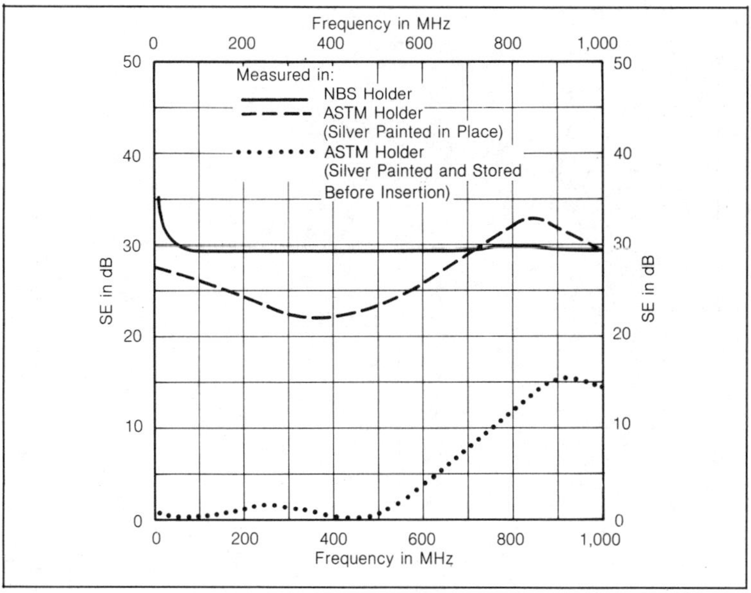

Figure 3.6—Measured Data of Gold-Film (Conducting) Sample Using Flanged and ASTM Holder

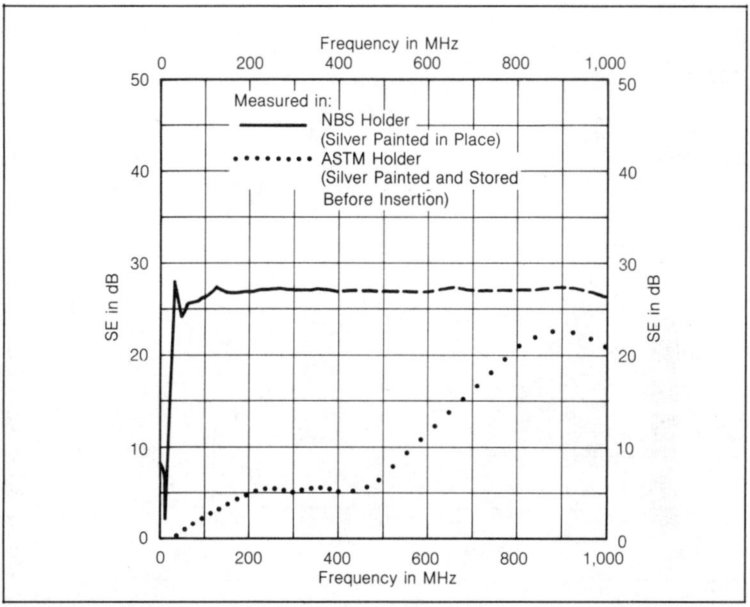

Figure 3.7—Measured Data of Laminated (Insulating) Sample Using Flanged and ASTM Holder

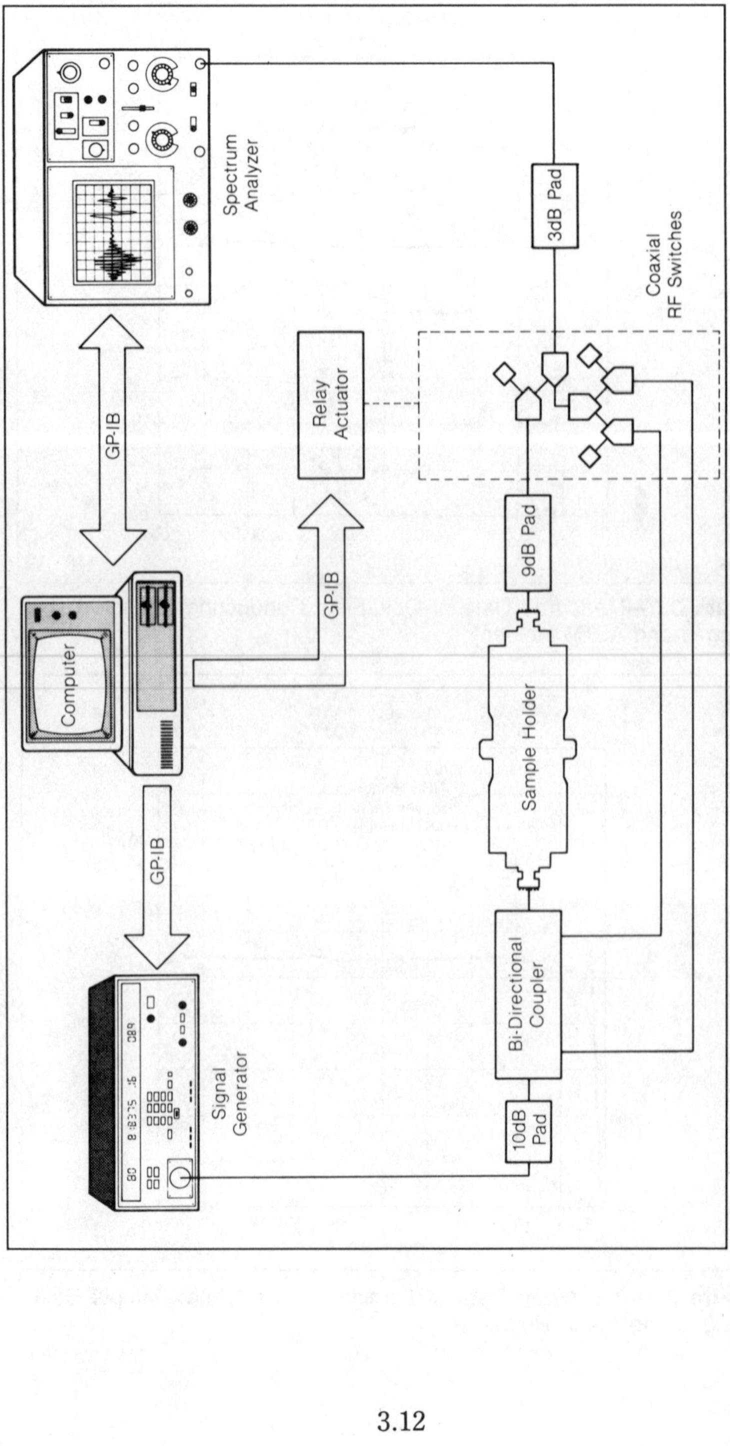

Figure 3.8—Block Diagram of Automated SE Measurement System that Uses Coaxial Holders

The coaxial holder is easy to use for inexpensive go/no-go quality control during the manufacture of metalized plastics (Fig. 3.9). It also allows for **time-domain** measurement of SE, where a single pulse is injected rather than a CW signal. Nevertheless, it does not provide near-field E- or H-field SE, which must be computed from the plane wave data or measured with either the conventional radiated method (MIL-STD-285) or the dual-chamber method as explained next.

Figure 3.9—The Dual-Chamber Box

3.3 The Dual-Chamber Method

This method, also described in the ASTM ES7-83 Standard, allows quick and easy near-field SE evaluation. The results depend on the size of the box, so to guarantee correlation and reproduceability, the box geometry is standardized to 18 cm (L) × 12 cm (W) × 16 cm (D) (Fig. 3.10). The specimen is a 7.5 × 15 cm plaque, up to 4 mm thick. Good contact pressure is applied through four rows of spring contacts (finger stock). Even more than the coaxial holder, this methods lends itself to benchtop measurements using a signal generator and a spectrum analyzer. The useful dynamic range is about 60 to 70 dB.

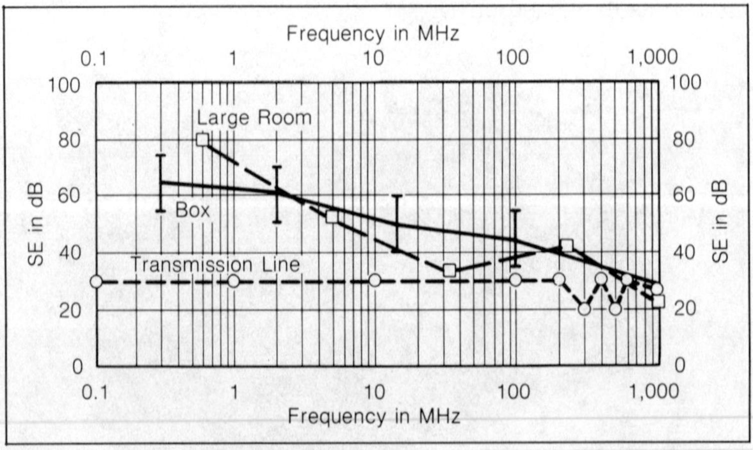

Figure 3.10—Transmission Line vs. Dual-Chamber Data Comparison, Conductive Composite.

3.4 The Dual Transverse Electromagnetic (TEM) Cell Method

The single TEM cell is a rectangular coaxial chamber developed by the NBS to evaluate EMC performances of equipment. The TEM cell can convert an RF voltage into a well defined plane-wave field and vice versa. Since it is a two-way transducer, it is assumed that two cells, when coupled by a common aperture, could be used to measure how much of the field in the first cell leaks into the second one, via the aperture. If a shielding material or accessory is placed over the aperture, the received field is reduced. The amount of this reduction is a direct measure of the sample's SE.

Figures 3.11 through 3.13 show the principle of the test set. The two cells share a common wall which is removable and serves as a calibrated aperture. The orifice is 12.7 cm × 12.7 cm, and four toggle clamps apply the mounting pressure. A dummy plate with calibrated spacers can also be used when evaluating gaskets.

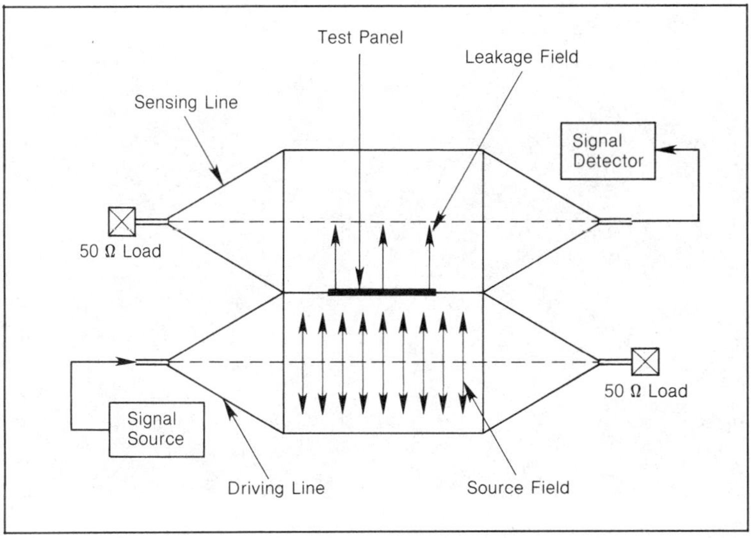

Figure 3.11—Common Aperture Coupling Inside Dual-TEM Cell Test Fixture

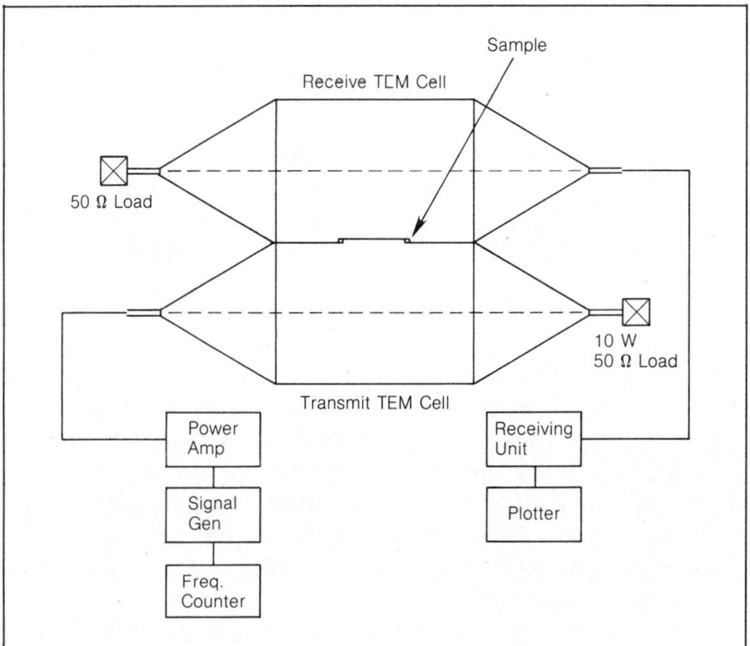

Figure 3.12—Block Diagram of Dual-TEM Cell Test System

Figure 3.13—Dual-TEM Cells

To avoid repeatability problems due to the imperfect symmetry of coupling between each end of the receiving TEM, a standard method has been adopted. In this method, the generation and detection equipments are connected at opposite ends. This also reduces the chances of parasitic stray coupling between the two instruments: if one expects a 100 dB dynamic range, one must be sure that the signal receiver itself is isolated by more than 100 dB from the power amplifier emanations. The tests can be performed either in the frequency domain using a CW signal or in time domain with a pulse generator or time-domain reflectometer (TDR).

An important difference of the dual TEM is that the wave **propagation is parallel to the sample and the E-field is perpendicular to it,** while in other methods the incidence is normal and the E-field vector is coplanar with the sample. Therefore the dual TEM cell may force a worst-case situation, giving more severe results than the other methods. The well defined field orientation in the cell enables testing of materials which are anisotropic, or sensitive to polarization.

Like the two previous methods, and unlike MIL-STD-285, the dual TEM cell eliminates the need for several antennas and shielded rooms. The lower frequency limit is limited by the poor

SE of the aluminum cell walls below 10 kHz, where magnetic leakages start to shunt the aperture leakage. Because the TEM cell is an efficient transducer, 10 W of power are enough to provide a dynamic range from 80 dB at 1 MHz to 100 dB at 500 MHz. This latter frequency is the upper limit above which a cell 30 cm in height will start to behave as a wave guide, no longer producing plane waves.

3.5 Evaluation of SE of Equipment Housings during Engineering Phases

Designers often need a quick evaluation of the SE of a prototype box without having to go to a fully equipped test facility. A typical case would be for equipment having to meet radiated emission limits of FCC Part 15, VDE 871 or CISPR 22. From previous engineering tests, the engineer knows that he needs a certain amount of attenuation, the burden of which is placed on the box shielding. A self-powered radiating source is needed, which can be either:

1. A broadband source in the form of a self-ringing relay, a car ignition coil and a spark plug. The relay is operated from a 6 V dry cell, so the whole set can be packed to occupy no more than 12 × 12 × 12 cm. When turned on, such a broadband source can generate a peak field in the 1 V/m at 1 m distance, with a spectrum expanding to more than 100 MHz.
2. A self-powered digital circuit based on a 10 MHz oscillator feeding a fast transistor-transistor logic (TTL) driver (Schottky or otherwise) connected to a 100 cm^2 loop terminated in few hundred ohms. At 1 m distance, such a device easily can produce 70 dBμV/m on its significant harmonics up to 200 MHz.

In either case, because these engineering measurements will be made in a noisy environment like a development lab, the engineer must determine the dynamic range required of his measuring equipment.

Assume for instance that the broadband incoherent ambient noise in the lab is 0 dBμV/m in a 1 kHz resolution bandwidth. Using a modern EMI receiver or spectrum analyzer, a sensitivity better than 0 dBμV can be expected. In the VHF range, a typical EMI measuring antenna can have an antenna factor (AF) ranging from 6 to 20 dBm^{-1}. Therefore, the minimum discernable field will be, at worst, equal to:

$$E\ dB\mu V/m\ (min) = 0\ dB\mu V + 20\ dBm^{-1}\ (AF) = 20\ dB\mu V/m$$

The ambient noise in this case is not a limiting factor for a 1 kHz bandwidth.

Let's assume the goal for SE is 40 dB in the 30 to 300 MHz span. Therefore, the radiating source which is needed to evaluate the housing must generate at the test distance and at the frequency of interest a minimum field equal to:

$$\begin{aligned} E\ source\ dB\mu V/m &= E\ min\ dB\mu V/m + SE\ min \\ &= 20\ dB\mu V/m + 40\ dB = 60\ dB\mu V/m \end{aligned}$$

Either one of the compact sources described above can produce such field strength at 1 m distance. The test procedure will then be:

1. With the antenna located at 1 m, measure the radiating source without any housing for both horizontal and vertical polarization, sweeping from 30 to 1,000 MHz.
2. Keeping the antenna at the same location, install the source in the prototype housing with all the vents, panels, etc., as realistically mounted as possible. Make another measurement scan. Check that during this second reading, the received levels are due to the source and not to the ambient.

For an **emission** situation, the difference between the two readings (see Fig. 3.14) is the SE of the box. It is not necessary to convert the reading into fields; a simple voltage comparison is sufficient. Besides its simplicity, the advantage of the method is that it gives the actual transfer function of the housing instead of expectations based on material performance only.

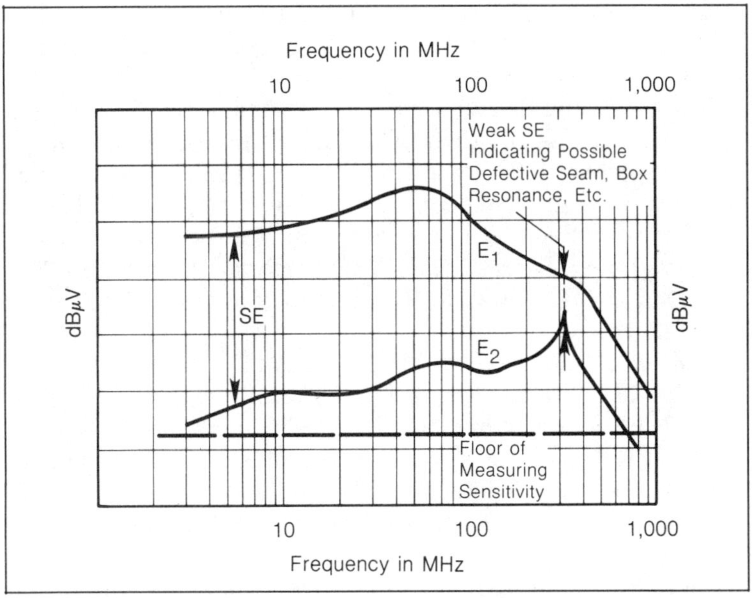

Figure 3.14—Fictitious Example of an Engineering Test of the Prototype of an Equipment Cabinet

3.6 Bibliography

Bridges, J.E., "Proposed Recommended Practices for the Measurement of Shielding Effectiveness of High-Performance Shielding Enclosures," *IEEE Transactions on EMC*, Vol. EMC-10, no. 1, p. 82, March 1968.

Cacace, R. and Hassett, R., "Investigation of Measurement Techniques for Transient Magnetic Fields," *Symposium Digest*, 7th National Symposium on Electromagnetic Compatibility, June 28-30, 1965, New York, New York.

"Conical Logarithmic Spiral Antennas," General Dynamics/ Astronautics, San Diego, California, U.S. Air Force AF33-(616)-7436, 1963.

Crawford, M.L. and Workman, J.L., "Using a TEM Cell for EMC Measurements," NBS, Boulder, Colorado, Technical Note #1013, July 1981.

Dinger, H.E. and Raudenbush, J., "A Technique for Measuring the Effectiveness of Various Shielding Materials," Naval Research Laboratory, NRL Report no. 4103, AD-9353, January 1953.

Dolle, W.C.; Van Steenberg, G.N.; and Jouffray, O.L., "Effects of Shielded Enclosure Resonances on Measurement Accuracy," *1970 IEEE International EMC Symposium Record* (New York: IEEE), Vol. 70C28-EMC, July 14-16, Anaheim, 1970, pp. 417-420.

Free, W.R., "Radiated EMI Measurements in Shielded Enclosure," *1967 IEEE EMC Symposium Record* (New York: IEEE), Vol. 27C80, Washington, D.C., July 18-20, 1967, pp. 43-53.

Free, W.R., et al., "Electromagnetic Interference Measurement Methods-Shielded Enclosure," Final Report on Contract no. DA 28-043 AMC-02381 (E), EES, Georgia Institute of Technology, 1967.

Kanellakos, D.P. and Schulz, R.B., "New Techniques for Evaluating the Performance of Shielded Enclosures," Fifth Conference on Radio Interference Reduction and Electronic Compatibility, Armour Research Foundation, IIT, Chicago, Ill., October 1959.

Klouda, J.C., "Practical Aspects in Evaluating Shielded Rooms," *Electro-Technology*, June 1961.

Lockwood, R.O., "New Technique for the Determination of the Integrity of Shielded Enclosures by the Measurement of the Perpendicular Magnetic Field," *1967 IEEE EMC Symposium Record* (New York: IEEE), Vol. 27C80, Washington, D.C., July 18-20, pp. 61-69.

Mendez, H.A., "A New Approach to Electromagnetic Field-Strength Measurements in Shielded Enclosures," *Convention Record of the 1968 WESCON*, Technical Papers Session 19, Los Angeles, California, August 20-23, 1968.

Mendez, H.A., "Meaningful EMC Measurements in Shielded Enclosures," *Record of the 1969 IEEE EMC Symposium Record* (New York: IEEE), Vol. 69C3-EMC, Asbury Park, June 17-19, 1969, p. 137.

MIL-STD-285, "Method of Attenuation Measurements for Enclosures Electromagnetic Shielding for Electronic Test Purposes."

O'Young, S.L.; Goldman, R. and Jorgensen, L., "Survey of Techniques for Measuring RF Shielding Enclosures," *IEEE Transactions on Electromagnetic Compatibility*, Vol. 10 no. 1, March 1968, pp. 72-81.

Scheps, R.D., "S.E. Measurements Using Dual TEM Cell Method," *EMC Technology*, July 1983.

Simon, R., "A New Standard Test Method," *EMC Technology*, January 1985.

Simon, R. and Stutz, D., "Testing Methods for Shielding Materials," *EMC Technology*, October 1983.

Stuckey, C.W.; Free, W.R.; and Robertson, D.W., "Preliminary Interpretation of Near-Field Effects on Measurement Accuracy in Shielded Enclosures," *1969 IEEE EMC Symposium Record* (New York: IEEE), Vol. 69C3-EMC.

Toler, J.C. and Evans, R., "Shielded Enclosure Specification in Perspective," *1970 Regional Electromagnetic Compatibility Symposium Record*, Vol. 70C64 - REGEMC, San Antonio, October 6-8, 1970, pp. III-A01-III-A-3.

U.S. Naval Civil Engineering Laboratory, "Proposed Specification for Electromagnetic Shielding of Enclosures and Buildings," Final Project Report, Contract N By-32220, Port Hueneme, California, 31 July 1963, pp. 97-113.

Wilson, P. and Ma, M., "Methods of Measuring Near-Field and Far-Field SE of Materials," EMC EXPO 86, Session T-28.

Chapter 4

Shielding Design Methodology and Procedures

This chapter sets the pace for the second part of this handbook by outlining the methodology and procedures of shielding design. The introductory section reviews shielding problems confronting design and retrofit engineers and identifies options for solving them. An overview of shielding design methodology and procedures is presented. The following chapters of the handbook cover the implementation of the methodology and procedures and include many illustrative examples.

4.1 Introduction

Design engineers responsible for the development of an EMI-shielded box or equipment enclosure usually proceed on an **osmosis-type** basis. This means that one or a combination of the following approaches is used:

1. Make the box design similar to earlier versions which are either known or believed to be EMI-free. Then test the box after design fabrication to confirm that it is, indeed, free of interference.
2. Starting anew, design and construct the box. Then EMI test it to determine whether it meets applicable legal emission specifications and recommended susceptibility immunity. Apply trial-and-error redesign techniques until the equipment finally complies.

3. Proceed according to number 1 or 2 above, but perform legally mandated emission tests only. Do not test for susceptibility unless a specific requirement (e.g. MIL-STD-461) applies.

For the most part, these traditional approaches allow test and evaluation to govern the outcome of the design. This seems like an antiquated approach in the twentieth century. The traditional approach often results in one or more of the following:

1. Substantial amounts of time and money are lost while the hit-or-miss process continues.
2. Ineffective components or techniques become part of the design.
3. EMC overdesign occurs with a resultant increase in cost and weight, reduction in reliability and maintainability, etc.
4. EMC underdesign occurs because the tests are limited in scope or type.

For example, the test setup may use a 1 m I/O cable length, whereas installation may require 10 m. Clearly, there must be a more scientific way to design an EMI-shielded box.

Even so, one must bear in mind the anxieties and frustrations of the traditional designer, especially when it comes to EMI control. He faces several dilemmas:

1. How much SE must the box/equipment design provide to meet one or more specifications or regulations?
2. How much SE is enough? Too much? Not enough?
3. Even if the answers to 1 and 2 are known, how can this be achieved in the design stage, before test confirmation or a retrofit cycle is attempted?
4. If the required SE is not known (as usually is the case), how can it be determined?
5. If an electromagnetic ambient threat must be met (required immunity to radiated susceptibility), how does this translate into SE requirements?

Clearly, what is needed is a **system approach** to shielded box design.

4.2 Shielding Strategy Principles for Susceptibility and Emission

Figure 4.1 is a flow diagram for determining SE objectives. The first question asks whether the required SE across the defined spectrum is already known. While SE requirements usually are not known, occasionally the designer may have the requirements dictated by an original equipment manufacturer (OEM) source, military procurement, etc. When the required SE is known, all that remains is to add in an appropriate margin of safety. See "Yes" choice in the first diamond of Fig. 4.1 (decision box).

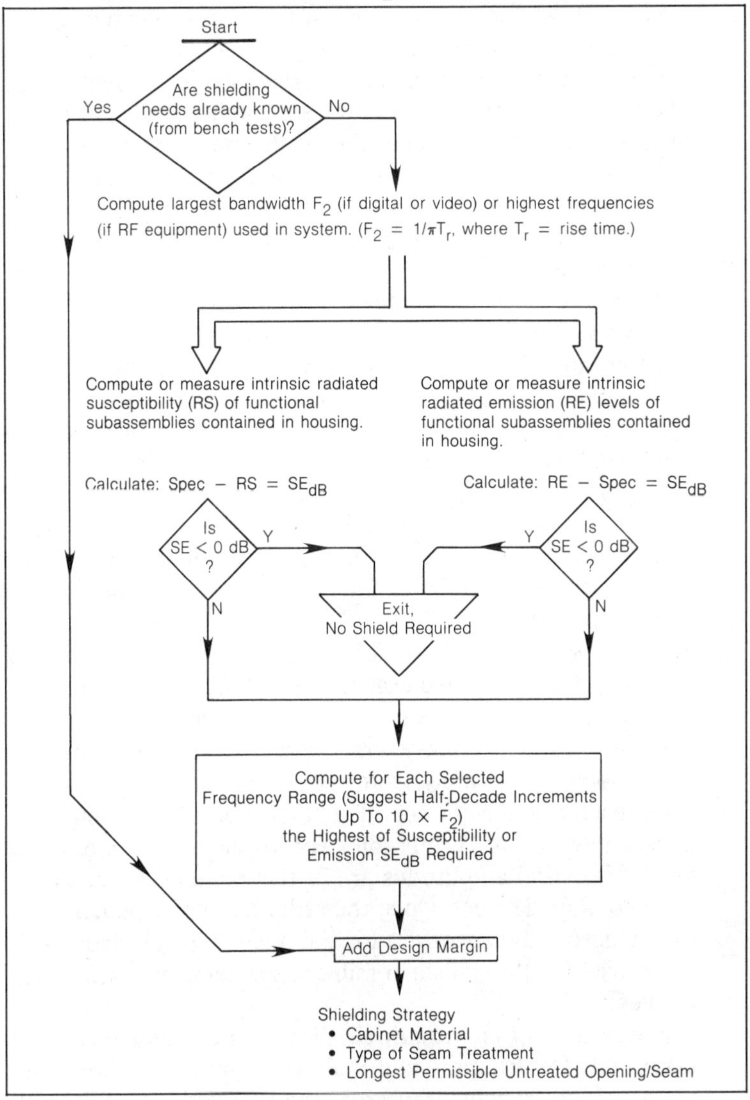

Figure 4.1—Routine for Setting SE Objectives

Since the SE needs usually are not known, the routine covers three cases:

1. **Case of susceptibility shielding** (left branch of diagram): Determine the electromagnetic ambient threats from 10 kHz to 10 GHz, based on the products intended applications and locations.

   Since most equipment and systems are located inside enclosures such as buildings or vehicles, it may be necessary to account for some attenuation of the outside ambient by the natural shielding properties of the surrounding walls, fuselage, bulkhead, etc. This reduces the electromagnetic ambient for indoor locations. In other cases, the equipment will not enjoy the shelter of a surrounding structure or the ambient threats will be **inside** the same structure. This results in an unattenuated field.

   Once the ambient has been determined, the next step is to compute the interference situation via the coupling of field-to-cable and PCBs based on a description of the victim's parameters. This includes its in-band and out-of-band analog sensitivity or logic noise immunity. Then the interference to noise (I/N) ratio is computed, giving the desired SE.

   Another possibility is that the radiated susceptibility of the bare (unshielded) equipment is known from previous EMI testing. In this case, the desired SE is simply the difference between the ambient level and the radiated susceptibility of the equipment.

2. **Case of emission shielding** (right branch of diagram in Fig. 4.1): Compute or measure the radiated emission levels for each subassembly to be contained in the future housing, excluding the I/O cables whose radiation has to be addressed and resolved separately from box shielding.

   For each frequency interval of at least one decade (one-half decade intervals are preferable), record the highest measured level. If several amplitudes are of the same order, compute their combined effect. Once the radiation envelope has been drawn across the spectrum for the unshielded electronics, it is compared to the civilian or military specifications, whichever applies.

3. **Case where both susceptibility and emissions are of concern:** Determine 1 and 2 above separately. Then compare in each frequency interval the susceptibility SE_1 with emissions SE_2 and select the toughest requirement of the

two. "Toughest" does not always mean the highest figure in dB. For instance, 20 dB of SE against a near-field 100 kHz magnetic source may be more difficult to achieve than 40 dB against a remote electric or plane-wave source in the same frequency range.

An abbreviated version of the prediction models for circuit emission and susceptibility will be given in following chapters as a prerequisite to calculating SE requirements. However, the reader who needs a more detailed coverage of the calculation process is invited to read the Volume 8, "EMI Control Methodology," of this handbook series. It contains the complete method, math models, graphs and tables to determine either the radiated emissions from a source or the radiated susceptibility of a victim.

4.3 Overview of the Shielding Design Procedure for the Susceptibility Case

The following chapters will first emphasize the susceptibility scenario because the field parameters it deals with are generally more easily defined. This will permit us to focus on the shield parameters alone. Chapter 9 will address the reciprocal problem (i.e., emissions case) where field parameters are more complex because of a wide range of frequencies and the problem of how the housing will function under near-field conditions. Figure 4.2 shows a flow diagram which illustrates how shielding design methodology and procedure works in a susceptibility situation. Figure 4.3 shows a suggested form to keep track of numerical results.

The first step is to determine if the SE requirements are known (Fig. 4.2, diamond 1). If they are not known ("N" decision), then the electromagnetic ambient threats (block 2) are determined with the help of a number of sources, including CBEMA publications, IEC-801, SAE AE4 Committee reports, SAMA documents and MIL-STD-461. Specific ambients may also be compiled (block 3) after discrete or representative radiators have been identified where the equipment in design will ultimately be located. Then the composite ambient may be computed from 10 kHz to 10 GHz (block 4) at several sample frequencies. These values are entered in Fig. 4.3 in row A.1.

Next, determine victim susceptibility (block 5) by first assuming that the box/case is transparent to the electromagnetic ambient environment. The victim is identified as either the most sensitive analog and/or the fastest (highest speed) logic family (block 6). The analog circuitry's signal-to-noise sensitivity or the logic family's noise immunity level (NIL) is determined along with the applicable victim bandwidth. Additional information should be gathered regarding discrete cable runs from box input connectors to printed circuit boards (PCBs) and single/multilayer board dimensions and types.

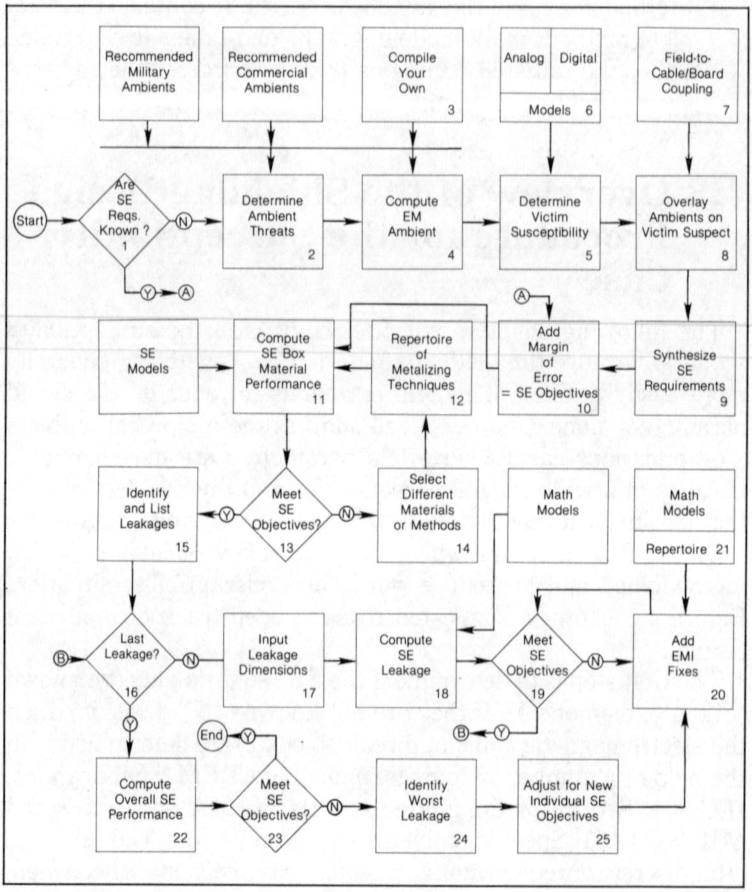

Figure 4.2—Flow Diagram for Shielding Design Methodology and Procedures (Susceptibility Case)

Topic Identification	EMI Fix	10 kHz	100 kHz	1 MHz	10 MHz	100 MHz	1 GHz	10 GHz
A. SE Objectives (1 + 2 − 3 + 4)								
1. EM Ambients (dBμV/m)								
2. Field-to-Cable/PCB Coupling (dB)								
3. Sensitivity or NIL (dBμV)								
4. Design Margin (dB)								
5. Box SE Modified Objectives								
6. Leakage SE Modified Objectives								
B. SE Box Performance								
C. Box Leakages*								
1. Viewing Window Apertures								
1.1 CRT Displays								
1.2 Alphanumeric Displays								
1.3 Panel Meter Displays								
1.4 Status Indicators								
2. Cooling Apertures								
2.1 Cooling Apertures								
2.2 Forced-Air Cooling								
3. Assembly and Maintenance Seams								
3.1 Box Housing Assembly								
3.2 Cover Panels								
4. Component Apertures								
4.1								
4.2								
D. Total SE* Performance								
E. Performance vs. Objectives (D − A)								

$$^{*}SE = -20 \log \left[\log^{-1}(-SE_{Box}/20) + \sum_{i=1}^{n} \log^{-1}(-\text{Leakages}/20) \right]$$

Figure 4.3—Worksheet for Recording SE Objectives and Performance

Using field-to-cable coupling models (block 7), the effect of the ambient on the victim wiring is computed to determine the differential-mode (DM) noise that will appear at the victim's input terminals (block 8). The difference between the coupled DM voltage and the in-band or out-of-band sensitivity constitutes the SE requirements at a number of sample frequencies across the spectrum (block 9).

Because there are errors in the prediction process (block 10), it is necessary to develop a margin of safety. A suggestion is to add 8 dB to the SE value below the frequency corresponding to half a wavelength for the discrete wire and PCB trace lengths and a 12 dB margin above this frequency. More generally, a minimum 10 dB default value may be used over the entire spectrum. The SE requirements plus the margins for error now equal the adjusted SE requirements or objectives (Box 10).

In the case of a wideband EMI situation (resulting from single or multiple pulses with low pulse repetition frequency), an additional margin may be necessary to account for the coherent addition of each frequency component shown in Fig. 4.3, a worksheet for recording SE objectives and performance. This is explained in more detail in Chapter 5, Section 5.2.3.2.

Next, compute the box material performance and the degradation due to the box leakages. Taking these in turn, the box material SE performance (block 11) is computed for the repertoire of shielding materials and metalizing techniques (block 12). These include sheet metal, conductive coatings, conductive composites and screen-lined boxes. The SE of the box is compared with the previously determined SE objectives (diamond 13) to see if performance meets objectives over the sampled frequencies. If it does not ("N" decision), then different shielding materials or methods are selected (block 14).

If the objectives are met ("Y" decision, diamond 13), the procedure then continues to address box leakages (block 15). One by one, the leakages are identified as listed in Fig. 4.3 (see items under C). The geometry of each aperture is identified to determine its leakage (block 17 of Fig. 4.2) and used to compute the SE (block 18). This is done before the installation of any EMI "fix." The results for each SE leakage are compared with objectives (diamond 19) to determine if the leakage is acceptable. If not ("N" decision), then an EMI fix to reduce the leakage (block 20) is selected from the repertoire of EMI fixes (block 21).

After addressing the last leakage (diamond 16, "N" decision), the overall SE is computed (block 22):

$$\text{Overall } SE_{dB} = -20 \log \left[\log^{-1}(-SE_{box}/20) + \sum_{i=1}^{i=n} \log^{-1}(-L_{idB}/20)\right] \tag{4.1}$$

where,

SE box = shielding effectiveness of box/case skin in dB

L_{idB} = ith aperture leakage in dB

This performance is compared with the objectives determined earlier (diamond 23) to see if the overall performance goal is met.

While the basic material and the individual leakages may meet the objectives, it does not necessarily follow that all leakages working in concert will still meet the overall objectives. When they do not (diamond 23, "N" decision), the worst-case leakage is readily identified (block 24). From this performance, an adjustment is made in the individual leakage SE objectives (not overall system objectives) to be met (block 25). Once this adjustment is made, it is rarely necessary to correct it again after the next iteration. Since the worst-case leakage is identifiable when Fig. 4.3 is filled out, the addition of an EMI fix (block 20) is readily achieved.

After the first EMI fix has been added (blocks 20 through 25), the process is repeated to determine if the EMI-suppressed leakage now permits the overall performance to meet the adjusted individual SE objectives (diamond 19). When this condition is met, there is no longer a need to EMI suppress further the box or its leakages further since the objectives have now been achieved.

4.4 References

1. MIL-STD-461C, "Electromagnetic Emission and Susceptibility Requirements for the Control of Electromagnetic Interference."
2. CBEMA/ESC5/75/31, Computer and Business Equipment Manufacturers' Association, 311 First St. NW, Washington, DC.
3. SAE AIR-1499, "Recommendation for Commercial EMC

Susceptibility Standards," Society of Automotive Engineers, Subcommittee AE4 on Electromagnetic Compatibility, Warrendale, PA.

4. Scientific Apparatus Makers Association (SAMA), 1101 16th Street NW, Washington, DC 20036.
5. White, D.R.J., *A Handbook on Electromagnetic Shielding Materials and Performance* (Gainesville, Virginia: Interference Control Technologies, Inc., formerly Don White Consultants, second edition, 1980), pp. 1.35-1.37.
6. CAD Program #5500, "EMC Design of Shielded Boxes, Cases, Cabinets and Enclosures to Control Electromagnetic Ambient Threats and Achieve Overall Shielding Effectiveness Requirements," Interference Control Technologies, Inc.
7. Milstead, F.C. and Cory, W.E., "The Effects of the Local Environment on Interference Measurements and Prediction," *1969 IEEE EMC Symposium Record* (New York: IEEE).
8. European Computer Manufacturers' Association (ECMA), "Limits and Measurement Methods for Radio Interference from Electronic Data Processing Equipment," standard ECMA-47.
9. White, D.R.J., "Eliminate the Guesswork from Shielding Enclosure Design—The Impact of Apertures," *EMC Technology*, April-July 1985, p. 29.

4.5 Bibliography

"Electro-Interference Control Requirements for Minuteman (WS 133A)," GM07-59-2617A, 20 October 1959.

Schulz, R.B. and Clapsaddle, R.L., "Electrocompatibility Aspects of Microelectronics," *IEEE Transactions on EMC*, Vol. EMC-6, January 1964, pp. 37-46.

Schulz, R.B.; Huang, G.D.; and Williams, W.L., "RF Shielding Design," *IEEE Transactions on EMC*, Vol. 10, no. 1, March 1968, pp. 168-175.

Sunde, E.G., "Switching Center Shielding Against Atmospheric Induction," Bell Telephone Labs (unpublished memorandum).

Chapter 5

Determining Shielding Effectiveness Requirements for Susceptibility

One major question about shielding that confronts both design engineers and equipment users is seldom addressed in EMC literature. This question is: "How much SE for any given product is enough, and how is this requirement determined?"

This chapter addresses this question and provides a methodology and procedure for getting the answers. Accordingly, this chapter discusses the determination of electromagnetic ambient threats, in-band and out-of-band susceptibility of victim devices, EMI coupling from field to cable into the victim's wiring and PCB, provisions for a design margin and the combined effects of all factors which determine SE objectives.

Figure 5.1 is a magnified excerpt of Fig. 4.2, where the ambient and ambient-to-victim coupling determination steps are stated. The objective of this first part is to derive a profile of the desired SE in decibels, including a predetermined safety margin.

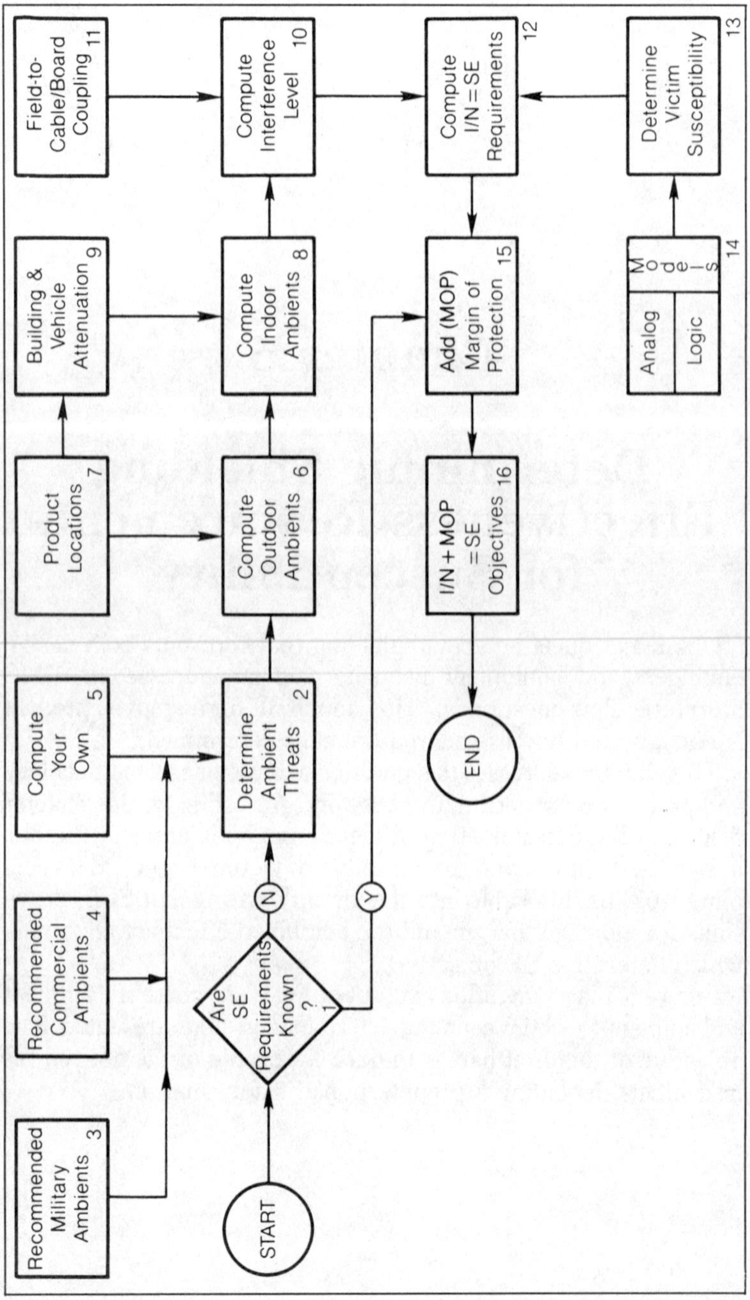

Figure 5.1—Methodology and Procedure Flow Diagram for Determining SE Objectives for the Susceptibility Case.

5.1 Electromagnetic Ambient Threats

The worst-case electromagnetic ambient environment is the greatest upper bound of spectrum radiation levels expected from different radiated emission threats and different typical product locations. For this methodology, the electromagnetic ambient spectrum covers the 10 kHz to 10 GHz spectrum for seven frequencies in one-decade steps. The ambients will include those from both licensed transmitters and lower power sources such as hand-held walkie talkies and CBs. Threats may also include lightning and EMP. The combined electromagnetic ambient spectrum profile will then be the upper boundary of the pertinent sources.

Some recommended electromagnetic ambients are stated as existing directly at the product location, where indoor and outdoor considerations are irrelevant since the building or vehicle attenuation has already been considered.

5.1.1 Predetermined Spectrum Profiles

From CBEMA,[1] SAMA[2] and SAE[3] AE-4 for commercial products and from MIL-STD-461[4] and MIL-Handbook-235[5] for military situations, the radiated electromagnetic ambient environment is usually specified from 10 kHz to either 1 GHz or 10 GHz. For example, for most commercial products, one recommended guideline is to use the following ambient environments:

E-Field = 1 V/m or 120 dB μV/m from 10 kHz to 25 MHz
E-Field = 10 V/m or 140 dB μV/m from 25 MHz to 1 GHz

On the other hand, for some military applications the recommended ambient is similar to commercial ambients below 1 GHz:

E-Field = 1 V/m or 120 dBμV/m from 10 kHz to 25 MHz
E-Field = 10 V/m or 140 dBμV/m from 25 MHz to 1,200 MHz
E-Field = 100 V/m or 160 dBμV/m from 1.2 GHz to 10 GHz

Thus, many standard electromagnetic ambients are already specified at the product location and therefore include the effect of any building or vehicle attenuation. Accordingly, they may be regarded as default electromagnetic ambients to be used wherever the defined product is located, unless a more specialized ambient environment has been determined (usually via measurement).[6]

For other product locations, such as automobiles and trucks, use higher recommended ambients at the outside of the vehicle. For example, the SAE-J1113[7] has recommended that field strengths of 200 V/m (166 dBμV/m) be used. The reason for these very high electromagnetic ambients is that vehicles will often wander off roads and may travel over field, hill and dale, ending up somewhere near a transmitting antenna array, broadcast tower, airport radar, etc.

For the heart pacemaker industry, where EMI ambients are very critical, field strengths of 400 V/m (172 dBμV/m) are recommended. These high levels represent, for example, a user holding an electric drill near the chest cavity containing his heart pacemaker.

5.1.2 Ambients from Licensed Transmitters

A licensed transmitter, e.g., communication-electronic (CE) transmitter, is an intentional radiator used for telecommunication from a carrier-operated transmitter and antenna to an intentional receiver containing its antenna. Thus, licensed transmitters include the whole field of low-frequency navigation, AM/FM and TV broadcast, land mobile and CB transmitters, point-to-point communications, radar and HF navigation radiators.

Radiation levels from the above sources may be determined with the help of Fig. 5.2. Note that the values are based on line-of-sight conditions (see the inset equation) and do not include the ground wave for low-frequency radiators. The X-axis represents the distance from the transmitter source to the victim location from 10 m to 100 km. The left Y-axis is the radiation level in dBV/m and dBμV/m, and the right Y-axis is in units of V/m. The parameter lines correspond to a number of specific radiation sources for either typical or maximum allowable effective radiated powers in the United States for broadcast service. For example, if a victim is located in an upper floor in a downtown office building 50 m away from an FM broadcast antenna farm on top of another building, then the electric field strength from Fig. 5.2 may equal 30 V/m.

To apply the data in Fig. 5.2, select the closest location for any given product or equipment to a transmitting source and then, by repeated use of Fig. 5.2, calculate a number of field strengths for the different frequencies corresponding to emitters in the defined region. This calculation is continued a number of times to cover all expected locations. For example, for microprocessor-based office equipment, the locations may include upper floors of commercial office buildings, street level floors, light or heavy industrial locations, shopping malls, airports and residential regions.

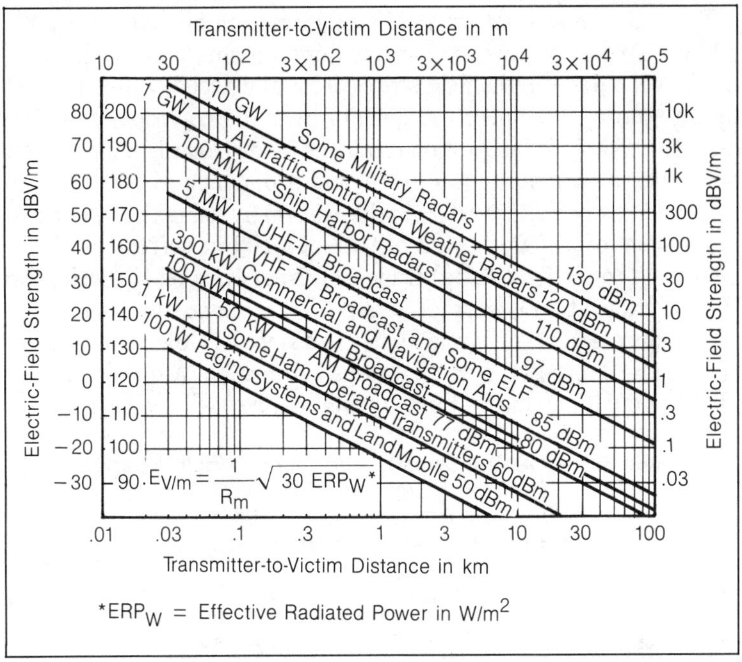

Figure 5.2—Free-Space Electric Field Strength vs. Transmitter Distance and Effective Radiated Power

Based on the spectrum profile for each location, the greatest upper bound is determined for all locations. This is the electromagnetic ambient environmental threat from licensed transmitters. Notice these values are based on the ambient located outside the building or vehicle and have not yet applied the attenuation for the building or vehicle as described later. Also, if a transmitter such as a hand-held walkie talkie is used inside a building, this will be added after the other fields have been corrected, taking structure attenuation into account.

5.1.3 Electromagnetic Ambient from Lightning Strokes

A lightning stroke typically has a rise time of 0.5 μs and a pulse duration of 30 μs. The designer may choose to use these as default parameters or to input his own rise time and pulse duration. The field strength from a lightning stroke also depends upon the amplitude of the stroke and the distance from it.

Table 5.1 shows field strength as a function of frequency for a 100 kA strike. Because the spectral increments used in this section cover a decade, the field strength for any decade is determined by integrating the spectrum density over the corresponding one decade band. This integration produces field strength values based on a stroke amplitude of 100 kA and a distance of 100 m.

E-field has been derived from the H term by using the free-space impedance of 377 Ω; i.e., 52 dBΩ. Since 100 m corresponds to a near-field-to-far-field ($\lambda/2\pi$) transition frequency of 477 kHz, the H-to-E translation is correct above this frequency but unreliable below since actual wave impedance is different. However, the field can still be expressed by a "fictitious" equivalent E term, and coupling calculations carried out in this manner will be correct.

Table 5.1—Spectral Density and Electric-Field Strength from a Lightning Stroke (I = 100 kA, Distance = 100 m, Rise Time = 0.5 μs; Pulse Duration = 30 μs)

Center Frequency	Frequency Range	Magnetic Spectral Density, in dBμA/m/MHz	Spectral Density in dBμV/m/MHz	Bandwidth in dB/MHz	Lightning in dBμV/m
10 kHz	1-32 kHz	192	244	−30	214
100 kHz	32-320 kHz	164	216	−11	205
1 MHz	0.32-1.7 MHz	147	199	9	208
3 MHz	1.7-5.8 MHz	123	175	19	194
10 MHz	5.8-17 MHz	103	155	29	184
30 MHz	17-58 MHz	83	135	39	174
100 MHz	58-170 MHz	63	115	49	164
300 MHz	170-580 MHz	43	95	59	154
1 GHz	0.58-1.7 GHz	23	75	69	144
3 GHz	1.7-5.8 GHz	3	55	79	134
10 GHz	5.8-20 GHz	−17	35	89	124

The designer may determine his own lightning electromagnetic ambient by selecting the values above in Table 5.1 and by adding the following adjustment:

$$\text{Adjustment} = 20 \log (I/d) \tag{5.1}$$

where,

I = lightning stroke current in kiloamperes
d = distance from stroke to victim in meters

5.1.4 Electromagnetic Ambient from an Electromagnetic Pulse

The procedure for determining the radiated ambient from an EMP is similar to that for a lightning stroke, except that distance does not enter the problem for an upper atmospheric explosion, and the rise time and pulse duration are much shorter. Table 5.2, the companion to Table 5.1 for lightning, gives the spectrum density values corresponding to a typical EMP with a rise time of 5 ns and a pulse duration of 50 ns.

Table 5.2—Spectral Density and Electric Field Strength from an Electromagnetic Pulse (A = 50 kV/m, Rise Time = 5 ns, Pulse Duration = 50 ns)

Center Frequency	Frequency Range	Spectral Density in dBμV/m/MHz	Bandwidth in dB/MHz	EMP in dBμV/m
10 kHz	1-32 kHz	194	−30	164
100 kHz	32-320 kHz	194	−11	183
1 MHz	0.32-1.7 MHz	188	9	197
3 MHz	1.7-5.8 MHz	184	19	205
10 MHz	5.8-17 MHz	173	29	202
30 MHz	17-58 MHz	160	39	199
100 MHz	58-170 MHz	149	49	198
300 MHz	170-580 MHz	129	59	188
1 GHz	0.58-1.7 GHz	109	69	178
3 GHz	1.7-5.8 GHz	89	79	168
10 GHz	5.8-20 GHz	69	89	158

Unlike the lightning stroke, for which there is an adjustment for distance, Table 5.2 for EMP requires no such distance adjustment for a high-altitude burst. For an amplitude and rise time other than given, the spectrum needs to be reconstructed.

5.1.5 Composite Electromagnetic Ambients

A composite electromagnetic ambient is made up of many licensed transmitters, a lightning stroke or an EMP. It is computed by taking the greatest upper bounds of all applicable electromagnetic ambient threats computed in earlier subsections of Section 5.1.

Illustrative Example 5.1: Computing the Electromagnetic Ambient

Consider an environmental threat based on licensed transmitters, a lightning stroke and an EMP threat. For the licensed transmitter, use the recommended SAE AE-4 profile of 1 V/m from 10 kHz to 25 MHz and 10 V/m from 25 MHz to 10 GHz. For the lightning stroke, consider the threat from a 50 kA stroke 200 m away. For the EMP threat, use the standard model given in Section 5.1.4

The separate transmitter, lightning and EMP threats are listed in Table 5.3 across from each frequency or range. The BB threats (Lightning & EMP) have been translated into equivalent NB fields by multiplying in each frequency sector the μV/m/MHz by the corresponding Bandwidth in MHz. The last column in Table 5.3 is the greatest upper bound of the three EM environmental threats.

Notice that, in this case, the levels from licensed transmitters are always lower than the composite threat. However, they do constitute the steady-state threat.

Table 5.3—Illustrative Example of Composite Electromagnetic Environmental Threats

Center Frequency	Frequency Range	Transmitters in dBμV/m	Lightning in dBμV/m	EMP in dBμV/m	Composite in dBμV/m
10 kHz	1-32 kHz	120	202	164	202
100 kHz	32-320 kHz	120	193	183	193
1 MHz	0.32-1.7 MHz	120	196	197	197
3 MHz	1.7-5.8 MHz	120	182	205	205
10 MHz	5.8-17 MHz	120	172	202	202
30 MHz	17-58 MHz	140	162	199	199
100 MHz	58-170 MHz	140	152	198	198
300 MHz	170-580 MHz	140	142	188	188
1 GHz	0.58-1.7 GHz	140	132	178	178
3 GHz	1.7-5.8 GHz	140	122	168	168
10 GHz	5.8-20 GHz	140	112	158	158

5.1.6 Building and Vehicle Attenuation

For electromagnetic ambients originating outside a building or vehicle, it is necessary to determine the attenuation offered by the natural shielding of the structure to calculate the ambient on the inside where products and equipments are to be located. Figure 5.3 shows a typical building attenuation profile corresponding to a steel-girdered building with a masonry and glass facade.[8] The X-axis ranges from 10 kHz to 10 GHz, and the Y-axis is the attenuation in dB offered by the structure. The parameter values correspond to the distance from the facade to the victim location.

For example, the expected building attenuation to an AM broadcast station at 1 MHz is about 50 dB at 2 m inside the facade. The corresponding attenuation to a 100 MHz, FM transmitter is 7 dB; whereas, attenuation to a 1,300 MHz, L-Band airport radar is 13 dB.

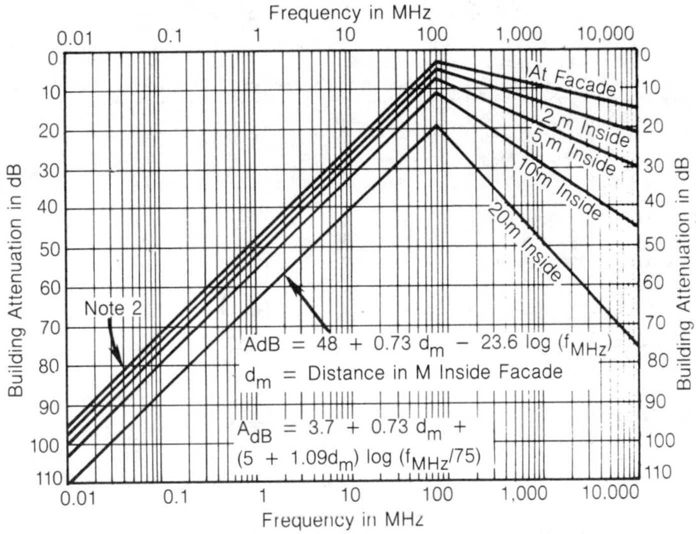

Notes: 1. Buildings surveyed are business-type, with metal girders.
2. Suggested extrapolation for E-Field Attenuation. H-Field attenuation decreases significantly below 1 MHz to approach 0 dB at around 10 kHz.

Figure 5.3—Building Attenuation to Outside Radiation vs. Frequency and Distance Inside Facade

5.1.7 Electromagnetic Ambient inside a Structure

To determine the electromagnetic ambient inside a building, subtract the building attenuation of Section 5.1.6 from the composite ambient determined in Section 5.1.5.

Illustrative Example 5.2: Computing the Electromagnetic Ambient inside a Structure

Using the composite electromagnetic ambient from Illustrative

Example 5.1 and Table 5.3, compute the ambient environment inside a building 2 m from the facade. The results are shown in Table 5.4. Notice that for lightning, the near H-field attenuation is considered at 10 and 100 kHz, making the lightning the biggest interior threat below 1 MHz.

Table 5.4—Illustrative Example of Composite Ambients on Outside and Inside of Building

Center Frequency	Frequency Range	Transmitters in dBμV/m	Lightng in dBμV/m	EMP in dBμV/m	Outside in dBμV/m	Attenuation in dB	Inside in dBμV/m
10 kHz	1-32 kHz	120	202	164	202	(75) 0*	202*
100 kHz	32-320 kHz	120	193	183	193	(55) 20*	173*
1 MHz	0.32-1.7 MHz	120	196	197	197	35	162
3 MHz	1.7-5.8 MHz	120	182	205	205	25	180
10 MHz	5.8-17 MHz	120	172	202	202	15	187
30 MHz	17-58 MHz	140	162	199	199	5	194
100 MHz	58-170 MHz	140	152	198	198	8	190
300 MHz	170-580 MHz	140	142	188	188	11	177
1 GHz	0.58-1.7 GHz	140	132	178	178	14	164
3 GHz	1.7-5.8 GHz	140	122	168	168	17	151
10 GHz	5.8-20 GHz	140	112	158	158	20	138

*For the lightning case, the building attenuation is the H-field SE

5.2 Coupling into Victim Wiring

The electromagnetic ambient, in terms of electric field strength versus frequency, has been determined across the spectrum from Section 5.2. The next step, shown in Fig. 5.1, is to couple this field to the victim wiring inside the box. It is assumed that the box is transparent; that is, it provides no electromagnetic shielding. The result from no shielding will be a coupled differential-mode voltage appearing at the analog/logic victim input terminals. This voltage then can be compared to the victim's in-band and out-of-band sensitivity, and the resulting I/N ratio will have been determined. This ratio is then the SE requirement which the box design must meet.

5.2.1 Coupling Relations

Figure 5.4 shows an example of internal box wiring originating at either an I/O connector or an internal device. The wiring progresses to a motherboard or backplane and then to a PCB containing the victim analog or logic device. The differential-mode coupling (DMC) defines the ratio of the DM-induced voltage to the arriving electric field strength:

$$DMC_{dB} = 20 \log (\pi S_m L_m / \lambda_m) \tag{5.2}$$

for all dimensions in meters and S_m, $L_m \leqslant \lambda_m/2$.

If, instead, all dimensions are expressed in cm:

$$DMC_{dB} = -40 + 20 \log [\pi(L_1S_1 + L_2S_2)/\lambda] \quad (5.3)$$

for L_1 and $L_2 < \lambda/2$

$$DMC_{dB} = -40 + 20 \log [\pi(\lambda S_1 + L_2S_2)/\lambda] \quad (5.4)$$

for $L_1 \geqslant \lambda/2$ and $L_Z < \lambda/2$

$$DMC_{dB} = -40 + 20 \log [\pi(L_1S_1 + \lambda S_2)/\lambda] \quad (5.5)$$

for $L_1 < \lambda/2$ and $L_2 \geqslant \lambda/2$

$$DMC_{dB} = -40 + 20 \log [\pi(S_1 + S_2)] \quad (5.6)$$

for L_1 and $L_2 \geqslant \lambda/2$

L_1 = length in cm of discrete wire from I/0 connector to PCB connector
L_2 = dimension in cm defined in Fig. 5.4
S_1 = dimension associated with L_1 in cm
S_2 = dimension associated with L_2 in cm
λ = wavelength in cm = $30{,}000/f_{MHz}$
f_{MHz} = frequency in MHz

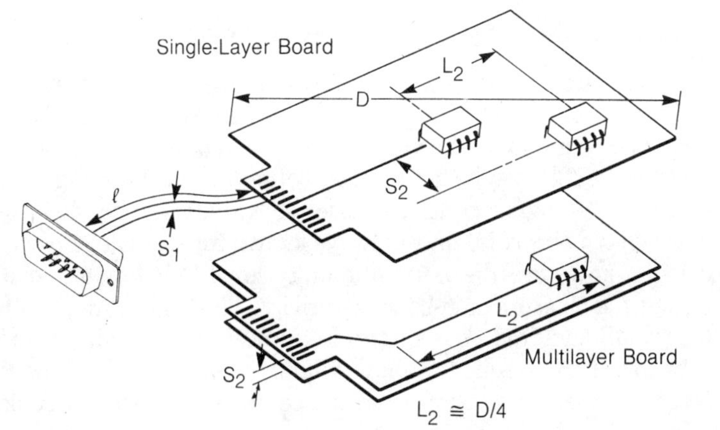

Figure 5.4—Internal Box Wiring Containing Cable from I/O Connector to Edge Connector and PCB Loop Pickup Area

Equations (5.3) through (5.6) are divided into two parts to account for the discrete wiring and PCB traces associated with either singlelayer or multilayer boards. The rationale and calculation details are covered in Volume 8, *EMI Control Methodology and Procedures*, of this handbook series.

Substituting Eq. (5.6) into Eq. (5.2) yields:

$$DMC_{dB} = -120 + 20 \log [f_{MHz} (L_1S_1 + L_2S_2)] \tag{5.7}$$

for, L_1 and $L_2 \leqslant \lambda/2$

Illustrative Example 5.3: Calculating DM Due to an Electric Field

Compute the DMC from an electric field at 10 MHz into an analog amplifier circuit with wire lengths (see Fig. 5.4) of $L_1 = 10$ cm, $L_2 = 20$ cm, $S_1 = 1.25$ mm and $S_2 = 5$ cm (singlelayer board).

From Eq. (5.7), the answer is:

$$DMC_{dB} = -120 + 20 \log [10 (10 \times 0.125 + 20 \times 5)]$$

$$= -120 + 60 \text{ dB} = -60 \text{ dB}$$

5.2.2 Sensitivity and Selectivity of Victims

The next step in the procedure shown in Fig. 5.1 is to compute the I/N ratio (block 12). Since the interference level (block 10) has already been computed, the victim in-band and out-of-band susceptibility (block 13) must be determined before I/N can be computed.

The S = N (signal equal to noise) sensitivity for bandwidth-limited white noise N is almost always known for an analog device. For logic, however, the noise-immunity level (NIL) or dc noise margin is the measure of in-band sensitivity. The bandwidth of both analog and logic devices is important in defining the limit of the passband and the frequency above which the sensitivity degrades at 20 dB/decade. These parameters are compiled and listed in Table 5.5 for logic.

Table 5.5—Table of Logic Families, NIL and Bandwidth

Logic Family	NIL in mV	NIL in dBμV	Bandwidth in MHz
1. PMOS/CMOS	1,000	120	1.6
2. CMOS	1,000	120	5
3. CMOS, High-Speed*	400	112	32
4. TTL	400	112	32
5. LP-TTL	400	112	20
6. LS-TTL	300	110	40
7. STTL	300	110	100
8. Schottky, Advanced	300	110	150
9. FAST	300	110	180
10. ECL-10k	100	100	160
11. ECL-100k	100	100	425

*for TTL-Compatible—otherwise, use 1,000 mV

However, as explained in Ref. 5.6, the 20 dB/decade increase in rejection is not constant and, especially with high-gain analog amplifiers, parasitic resonances and audio rectification can spoil the rejection at frequencies above 30 to 50 times the amplifier's bandwidth.

The noise sensitivity (N_S) of the victim is now defined:

$$N_S = N_{dB\mu V} + 20 \log (f_{MHz}/BW) \quad (5.8)$$

$$\text{for } 1 \leqslant f_{MHz}/BW \leqslant 10^3$$

$$= N_{dB\mu V} \text{ for } f_{MHz}/BW < 1 \quad (5.9)$$

$$= N_{dB\mu V} + 60 \text{ dB for } f_{MHz}/BW > 10^3 \quad (5.10)$$

where,

$N_{dB\mu V}$ = in-band analog S/N sensitivity or logic NIL (Table 5.5)

f_{MHz} = interfering frequency in megahertz

BW = victim bandwidth in megahertz

Illustrative Example 5.4: Calculating the Sensitivity of an Analog Device

Determine the sensitivity at 10 MHz of an analog device whose the in-band sensitivity is 120 μV (42 dBμV) and the bandwidth is 5 MHz. From Eq. (5.8):

$$N_S = 42 \text{ dB}\mu\text{V} + 20 \log (10/5) = 48 \text{ dB}\mu\text{V}$$

5.2.3 SE Requirements and Objectives

The electromagnetic ambient, the field coupling, the in-band and out-of-band analog sensitivity and logic NIL have been defined. Shielding effectiveness requirements now can be computed.

5.2.3.1 SE Requirements

The SE requirement is equal to the interference-to-noise ratio (without any margin):

$$SE = I/N \text{ or } SE_{dB} = I_{dB\mu V} - N_{sdB\mu V} \tag{5.11}$$

$$= E_{dB\mu V/m} + DMC_{dB} - N_{sdB\mu V} \tag{5.12}$$

where,

I = interference voltage at victim's input

$E_{dB\mu V/m}$ = field strength (see Table 5.4)

DMC_{dB} = DMC from Eqs. (5.2) through (5.6)

Illustrative Example 5.5: Calculating the SE Requirements

Using the illustrative examples 5.3 and 5.4, compute the SE requirements for a field strength of 1 V/m below 25 MHz and 10 V/m above 25 MHz. At 10 MHz, the SE is:

$$SE_{dB} = E_{dB\mu V/m} + DMC_{dB} - N_{sdB\mu V}$$

$$= 120 - 60 - 48 = 12 \text{ dB}$$

Since the SE requirement is needed across the 10 kHz to 10 GHz spectrum, Table 5.6 sums up the results of applying Eq. (5.12) several times (lines 5 and 6 will be discussed later).

Table 5.6—Summary of SE Requirement and SE Objective Computation

Variable/Parameter	10 kHz	100 kHz	1 MHz	10 MHz	100 MHz	1 GHz	10 GHz
1. Electric Field Strength dBμV/m	120	120	120	120	140	140	140
2. Field-to-Cable Coupling DMC_{dB}	−120	−100	−80	−60	−40	−28	−28
3. Sensitivity or NIL dBμV	42	42	42	48	68	88	102
4. SE Requirements (1+2−3)	−42	−22	−2	12	32	24	10
5. Margin of Error	10	10	10	10	10	10	10
6. SE Objectives (4+5)	0*	0*	8	22	42	39	19

5.2.3.2 Design Margin

Normally, the engineer designs to some signal-to-noise, S/N, objective. However, externally originating interference, I, combines with internal noise, N, so that in the real world S/N becomes:

$$\frac{S}{N + I} = \frac{S/N}{I + I/N} \tag{5.13}$$

One possible objective for I/N would be to set it equal to 1 so that I = N. However, the denominator in Eq. (5.13) would then equal 2 and S/N would already have degenerated by 3 dB. For I/N to become negligible, I/N must be less than 1 (perhaps 0.1) but not too small to avoid EMC overdesign (overkill). Also, remember that such a design margin must accommodate the fact that prediction models are not precise. There may be several dB of error, especially at higher frequencies. Thus, the useful design margin extremes are:

10 dB = light margin of error protection; may result in EMC underdesign
20 dB = recommended default margin of error protection
30 dB = heavy margin of error protection; may result in

In the case of a broadband EMI situation like a single pulse, a repetitive pulse with a low pulse repetition frequency (PRF) or several coexisting EMI frequencies (e.g, the composite threats of Tables 5.3 and 5.4), an additional margin is necessary because of the subdivided spectrum approach of the Prediction Worksheet (Figs. 4.3, 5.5, etc.). The reason for this is that, in a worst-case situation, the EMI voltages in each frequency band will finally add up coherently (or at least randomly) at the victim's input. One might run the risk of satisfying the I/N goal (−10, −20 or −30 dB) in each frequency band, but exceeding it on the whole. Therefore, and **for broadband EMI only**, an additional margin is recommended equal to 20 log X, where X is the number of frequency intervals used in the prediction. For instance, if seven frequency intervals are used as in the suggested worksheet, an additional margin of 20 log 7 = 17 dB should be included in each column for the SE requirements. So finally, a typical design margin will be:

1. Narrowband EMI (only one single-frequency EMI at a time) = 20 dB
2. Broadband EMI (EMI occupies the entire spectrum being analyzed) = 37 dB

5.2.3.3 Shielding Effectiveness Objectives

Section 5.2.3.1 showed how to determine SE requirements when not specified by a client or other procuring source. However, no margin of error was included. The last section indicated how to select an approximate margin of error.

Consequently, the SE objectives for the shielding box design are:

$$SE_{dB}\text{ (objectives)} = SE_{dB}\text{ (requirement)} + \text{design margin} \tag{5.14}$$

Using Illustrative Example 5.5 and its derivatives (Illustrative Examples 5.3 and 5.4), the SE objectives for a 10 dB design margin are shown in Table 5.6, lines 5 and 6. Figure 5.5 shows the worksheet for SE objectives and performance which was shown earlier in Chapter 4. Lines A.1 through A.4 have been filled in from Table 5.6. Line A is the resulting SE objective. This sheet will continue to appear in later chapters as more data become known.

Topic Identification	EMI Fix	10 kHz	100 kHz	1 MHz	10 MHz	100 MHz	1 GHz	10 GHz
A. SE Objectives (1 + 2 − 3 + 4)		0	0	8	22	42	34	20
1. EM Ambients (dBμV/m)		120	120	120	120	140	140	140
2. Field-to-Cable/PCB Coupling (dB)		-120	-100	-80	-60	-40	-28	-28
3. Sensitivity or NIL (dBμV)		42	42	42	48	68	88	102
4. Design Margin (dB)		10	10	10	10	10	10	10
5. Box SE Modified Objectives								
6. Leakage SE Modified Objectives								
B. SE Box Performance								
C. Box Leakages*								
1. Viewing Window Apertures								
1.1 CRT Displays								
1.2 Alphanumeric Displays								
1.3 Panel Meter Displays								
1.4 Status Indicators								
2. Cooling Apertures								
2.1 Cooling Apertures								
2.2 Forced-Air Cooling								
3. Assembly and Maintenance Seams								
3.1 Box Housing Assembly								
3.2 Cover Panels								
4. Component Apertures								
4.1								
4.2								
D. Total SE* Performance								
E. Performance vs. Objectives (D − A)								

$$*SE = -20 \log [\log^{-1}(-SE_{Box}/20) + \sum_{i=1}^{n} \log^{-1}(-\text{Leakages}/20)]$$

Figure 5.5—Worksheet for Recording SE Objectives and Performance

5.3 References

1. CBEMA/ESC5/75/31, Computer and Business Equipment Manufacturers' Association, 311 First St. NW, Washington, DC.
2. Scientific Apparatus Makers Association (SAMA), 1101 16th Street NW, Washington, DC 20036.
3. SAE AIR-1499, "Recommendation for Commercial EMC Susceptibility Standards," (Warrendale, PA: Society of Automotive Engineers, Subcommittee AE4 on Electromagnetic Compatibility).
4. MIL-STD-461B, "Electromagnetic Emission and Susceptibility (Also Requirements for the Control of Electromagnetic Interference," 1 April 1980, Notice 1 [Update], dated 5 January 1983).
5. MIL-Handbook-235-1A, *Electromagnetic (Radiated) Environment Considerations for Design and Procurement of Electrical and Electronic Equipment, Subsystems and Systems*, 5 February 1979.
6. White, D.R.J. and Mardiguian, M., *EMI Control Methodology and Procedures* (Gainesville, VA: Interference Control Technologies, Inc.).
7. White, D.R.J., "Building Attenuation Measurements," First Symposium and Technical Exhibition on EMC, May 20-22, 1975, Montreux, Switzerland.
8. White and Mardiguian, Chapter 7.
9. Ibid, Section 3.2.
10. Ibid, p. 4.4.
11. Ibid, p. 3.9.

Chapter 6

Shielding Effectiveness of Box Structure or Skin

Chapter 5 addressed how to determine SE requirements and objectives when they are not known in advance. By defining the electromagnetic ambient threat, the field-to-wire coupling and the victim sensitivity in-band and out-of-band, the methodology and procedures were presented and demonstrated. This chapter covers the first of two topics dealing with the actual shielding design of a box or housing enclosure: the metal or metalized plastic skin characteristics necessary to achieve the required SE. Compromises in basic shield performance which result from necessary apertures, plus methods of predicting and controlling resultant leakages, will be covered in detail in Chapter 7.

6.1 Box Skin Classification

Figure 4.1 in Chapter 4 presented the methodology to be used for designing a box or equipment shield. In that figure, blocks 11 and 12 identify the subject of this chapter; i.e., metalizing techniques and how to obtain the required SE from the box or case skin.

The box can be classified into one of six different types that exhibit some measure of SE:

1. Metal-framed plastic box
2. Metalized plastic box (paint, arc spray, deposition)
3. Conductive composite, particle-loaded plastic box
4. Conductive composite, filament-loaded plastic box
5. Wire mesh or screen-lined plastic box
6. Sheet metal box (or metal casting)

The first type, the metal-framed plastic box, is a six-sided, metal skeleton containing plastic panels. Some shielding is achieved by virtue of the bonded horizontal and vertical metal frame.

The second type is the metalized plastic box. Conductively coated plastic can be obtained by paint, arc spray, vacuum deposition or dipping (see chapter 2).

The third and fourth shielded box types are composites loaded with either conductive particle or filament. Particle-loaded types include conductive powder or flakes, whereas the filament-loaded type contains wire fibers smaller than the diameter of a human hair and ranging in length from a few millimeters to 1 cm.

The fifth type of shielded housing is a knitted-wire mesh or screen lining the inside of a non-metal box. It is seam bonded along all its mating edges. The SE of wire mesh and screen is described in Chapter 2.

The last of the six shielded housings is the more traditional sheet metal box, commonly used in military electronic equipment construction and some of the more rugged test instruments in the commercial and industrial world. Cast boxes may also be included in this category. The box is usually a five-sided configuration plus a top (or front) which can be removed for maintenance and access. Sheet metal construction is usually of steel or aluminum and is about 1 mm thick.

6.1.1 Metal-Framed Plastic Box

Figure 6.1 shows three metal-framed plastic (or other nonconductive material) boxes with skin which is transparent to electromagnetic fields. While the metal box is wide open, it does provide some low-frequency and mid-frequency shielding since the frame members have spacings of much less than one-half wavelength. This is assuming, of course, that the EMI threat is of plane wave nature, i.e., the source is in far field. Although this often is the case, there may be significant exceptions such as a nearby magnetic field. In

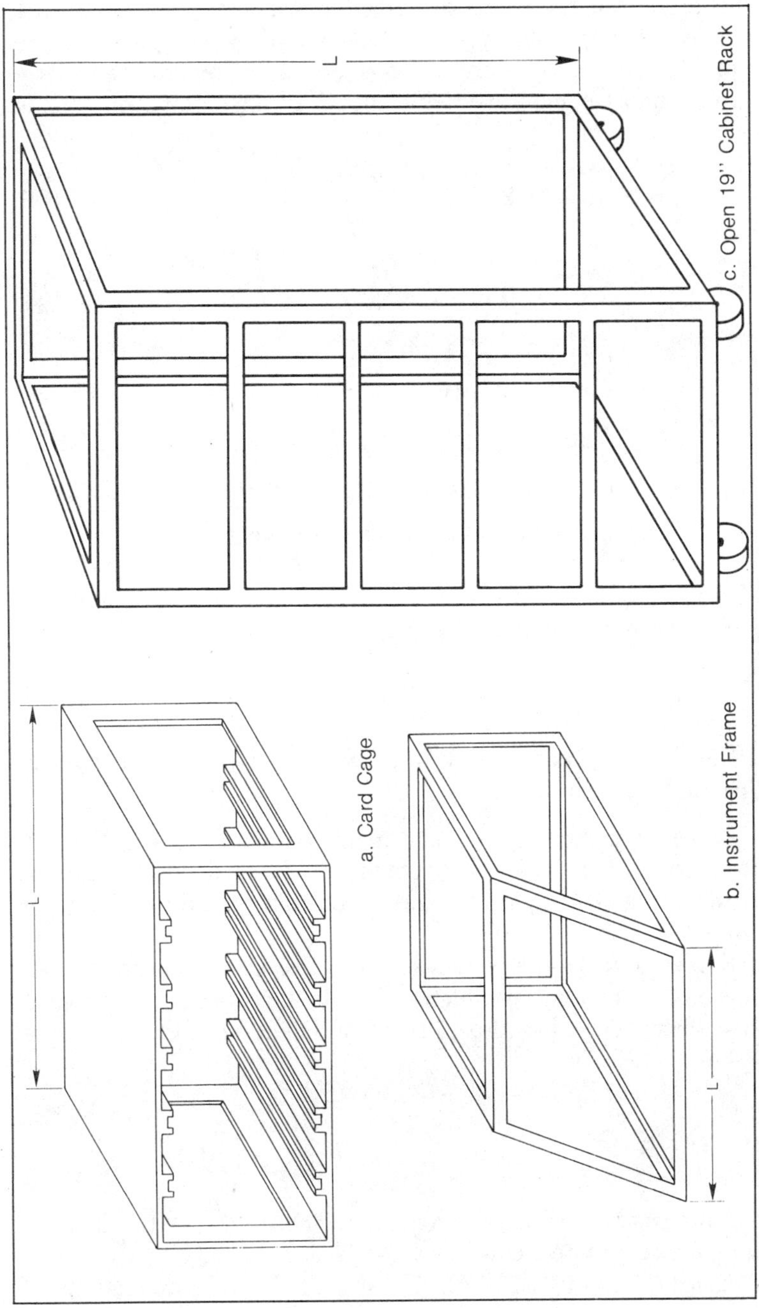

Figure 6.1—Three Examples of Open-Frame Fabrication with Metal Parts

the latter case, the model described in Chapter 2, Section 2.1.5.1, is perfectly applicable since the skeleton of the equipment frame, provided intersections are well bonded, can be regarded as a large grid. Against plane-wave fields, the SE of the metal frame box is:

$$SE_{box} = 20 \log \left(\frac{\lambda/2}{L} \right) \text{ for } L \leqslant \lambda/2 \qquad (6.1)$$

$$= 0 \text{ for } L \geqslant \lambda/2 \qquad (6.2)$$

where,

λ = wavelength in meters

= $300/f_{MHz}$

f_{MHz} = interference frequency in megahertz

L = largest dimension of frame in meters

When Eq. (6.1) is expressed in terms of frequency and the dimension L is in millimeters, Eq. (6.1) becomes:

$$SE_{box} = 104 - 20 \log (L_{mm} \, f_{MHz}) \text{ for } L < \lambda/2 \qquad (6.3)$$

Illustrative Example 6.1: Compute the SE of a metal-framed plastic box.

Consider an open-frame instrument box like the one in Fig. 6.1. It measures 40 cm wide × 30 cm high × 50 cm deep. It is covered with mating plastic panel members. Calculate the SE of the frame from 100 kHz to 1 GHz.

From Eq. (6.1), the worst-case shielding effectiveness will correspond to the longest dimension of the three sides; in this case, L = 50 cm. Thus, from Eq. (6.1) at 1 MHz, the SE of the box skin is:

$$SE_{box} = 104 - 20 \log (500 \times 1)$$

$$= 104 - 54 = 50 \text{ dB at } 1 \text{ MHz}$$

Notice that the SE goes to zero dB at 300 MHz, since the degradation increases with frequency at 20 dB/decade and 300 MHz is 2.5 decades (20 log 300/1 = 50 dB) above 1 MHz in frequency.

Substantial shielding can be achieved at low frequencies, but a typical open frame having a nonmetalized skin will offer little or no attenuation above 100 MHz. Parenthetically, the math model of Eq. (6.3) applies for building girder attenuation on the macro scale down to wire mesh on the micro scale. Both of these situations are discussed in Sections 5.1.6 and 2.1.5.1 of this handbook.

6.1.2 Metalized Plastic Box

This section addresses a popular form of box shielding: conductive surface coatings over plastic and other nonconductive housings. Several different processes are included in this category, such as conductive paints, arc spray, electroless and ion plating (including vacuum depositions). The pros and cons of each of these metalizing processes are covered in Chapter 2.

First, remember that in the conductive coating process, SE is achieved primarily by reflection loss and the absorption loss plays a smaller role, except at high frequencies. The measure of the performance of SE has been defined as:

$$SE_{box} = R_{dB} + A_{dB} \tag{6.4}$$

where,

R_{dB} = shielding effectiveness due to reflection loss

A_{dB} = shielding effectiveness due to absorption loss

1. The reflection loss has been defined as follows:

$$R_{dB} = 20 \log \left[\frac{(1 + K)^2}{4\,K} \right] \text{dB} \tag{6.5}$$

where,

$$K = Z_w/Z_B \tag{6.6}$$

Z_w = wave impedance

= 120π in the far field

Z_B = deposited metal surface impedance in ohms per square (this parameter being under the control of the person applying the process)

When Eq. (6.6) is substituted into Eq. (6.5) for K >> 1, the reflection loss becomes:

$$R_{dB} = 20 \log \left(\frac{K}{4} \right) = 20 \log \left(\frac{120\pi}{4\ Z_B} \right) \tag{6.7}$$

$$= 20 \log (94/Z_B) \cong 40 - 20 \log Z_B$$

2. The absorption loss has been defined as:

$$A_{dB} = 8.7\ t/\delta \tag{6.8}$$

$$= 0.13\ q\ t_{\mu m} \sqrt{f_{MHz}\ \mu_r\ \sigma_r} \tag{6.9}$$

where,

t = thickness of metal deposit

δ = skin depth in the same unit (mm, μm, etc.) as t

q = value varying from 0.1 to 1

μ_r = the permeability of deposited metal relative to copper

σ_r = conductivity of deposited metal relative to copper

The coefficient q is to derate the value of t (the actual coating thickness) relative to the performance of a pure metal with same thickness. Deposition must allow for captured air. Typical values of thickness of metal deposit compared to real homogeneous metal range from approximately 5:1 for arc spray (q = 0.2) to 1:1 for vacuum deposition (q = 1).

Equations (6.5) and (6.7) provide a minimum reflection loss corresponding to far-field conditions (distance, $R > \lambda/2\pi$) where the wave impedance is $120\pi = 377\ \Omega$. Additional reflection losses to electric fields exist where the interference source-to-box shield distance, r, is in the near field. This happens because the wave impedance, Z_w, is greater than 120π.

Thus, for near E-field conditions, Eq. (6.7) becomes:

$$R_{dB} = 20 \log \left[\left(\frac{120\pi}{4\ Z_B} \right) \left(\frac{\lambda/2\pi}{r_m} \right) \right] \quad \text{for } r_m \leqslant \lambda/2\pi \tag{6.7a}$$

$$= 20 \log [4{,}500/(Z_B r_m f_{MHz})] \quad \text{for } r_m \leqslant 48/f_{MHz}$$

For near H-field conditions, Eq. (6.7) becomes;

$$R_{dB} = 20 \log \left[\left(\frac{120\pi}{4\ Z_B} \right) \left(\frac{r_m}{\lambda/2\pi} \right) \right] \quad (6.7b)$$

$$\approx 20 \log 2r_m f_{MHz}/Z_B \quad \text{for } r_m < 48/f_{MHz}$$

Illustrative Example 6.2: Calculate the SE of an arc-sprayed conductive coating.

A manufacturer recommends that 0.5 mils (1 mil = 0.001" = 25.4 microns) of arc-sprayed zinc be used to achieve a surface impedance of 0.7 Ω/sq. Calculate the plane-wave SE of this conductive coating process from 10 kHz to 10 GHz.

Pure zinc (see Table 2.1) has a relative conductivity of 0.3. From Eqs. (6.7) and (6.9), the following values are computed, using q = 0.2:

R_{dB} = 20 log (94/0.7) = 43 dB (for all frequencies and plane-wave conditions)

$$A_{dB} = 0.13 \times 0.2 \times 0.5 \times 25.4 \sqrt{100 \times 1 \times 0.3}$$
$$= 1.8 \text{ dB at } 100 \text{ MHz}$$
$$= 18 \text{ dB at } 10 \text{ GHz}$$

where,

R_{dB} = reflection loss in decibels

A_{dB} = absorption loss in decibels

Thus, the absorption loss is insignificant up to about 100 MHz; the reflection loss is more important and results in an overall SE of 43 dB.

The example illustrates an earlier observation about conductive coatings and, to a lesser extent, metal-filled composites. This observation is that for electrically thin materials (thickness << skin depth), SE in free space is not a function of frequency and is only a function of the surface impedance (resistivity) since absorption loss is negligible for $t/\delta \ll 1$.

Illustrative Example 6.3: Compute the SE of an arc-sprayed conductive coating in near field:

Recompute Illustrative Example 6.2 for conditions corresponding to $r_m = 1$ m and a near electric field. The near-to-far field interface distance is $r_m = 48/f_{MHz}$ or $f_{MHz} = 48/r_m = 48$ MHz. Thus, above 48 MHz, Eq. (6.7) applies and $R_{dB} = 43$ dB. Below 48 MHz, at 1 MHz, for example, Eq. (6.7a) produces an SE of:

$$R_{dB} = 20 \log [4{,}500/(0.7 \times 1 \times 1)] = 77 \text{ dB}$$

If the electromagnetic ambient threat distance is unknown and varies over a wide range, use the more conservative Eq. (6.7) to compute SE.

Had the near field been magnetic, the reflection loss at 1 MHz would reduce to:

$$R_{dB} = 20 \log \frac{2 \times 1 \times 1}{0.7} = 9 \text{ dB}$$

6.1.3 Box with Conductive Particle or Filament Loading

An alternative to conductive surface coatings (see Section 2.1.4) is the use of metal-filled or conductive-filled plastics. Fillers in common use today include carbon black, graphite fiber (both metal-coated and uncoated), metal flakes, metal-coated glass and metal fiber.

Shielding properties of these materials are described in Section 2.1.4. Despite the occasional use of surface resistivity to rate their shielding properties, the SE of composites is more typically predicted by entering their volume resistivity in the curve of Fig. 2.26 or by calculating the resistance in ohms per unit square of the loaded plastic sheet using:

$$R = \rho L/S \tag{6.10}$$

where,

R = resistance in ohms per square

ρ = volume resistivity in ohm centimeters (Ωcm)

L = length of sample in centimeters

S = cross-sectional area of sample in square centimeters

Note that this resistance figure corresponds to current flowing in the plane of the specimen, not across the sample, because of the way in which the shielding if affected. If the sample is a square of side length L, we can replace S by the length times thickness product and obtain:

$$R(\Omega/\text{sq}) = \rho(\Omega\text{cm})/t_{cm} \tag{6.10a}$$

Then the SE can be found by using the surface impedance as previously explained in connection with the use of Eq. (6.7) for reflection loss. For applications above 1 MHz, use Eq. (6.8) to determine if absorption loss offers any additional SE contribution.

Particle-loaded and filament-loaded materials are two different manifestations of the same process. How, then, do filament-loaded composites differ from particle-loaded composites? The latter, typically made up of conductive powder or flakes, requires up to 40 percent additional weight to the plastic base carrier to attain an SE of about 35 to 40 dB. Contrast this with conductive filament fibers where about five percent by weight (1 to 1.13 percent by volume) is required to attain 35 to 40 dB of SE.

Conductive powder and flakes are more or less spherical in shape, whereas filaments are long, thin cylinders. For average aspect ratios of about 200 (aspect ratio = length/diameter), much more metal is in contact than with shorter ratios. This produces higher conductivity and better SE. This is why filament-loaded composites require only about 15 percent as much metal additives as particle-loaded plastics. Other advantages include:

1. Minimal alteration to base resin (plastic) properties
2. Shrinkage similar to unfilled resins
3. Low equipment and mold wear
4. Good colorability and surface finish
5. Good abrasion and corrosion resistance (long life)

See Section 2.1.4 for further discussion of these composites.

Notice that there is a serious safety concern with the handling and processing of thin filaments of carbon or steel. Being so small and light they easily can be airborne via chimneys and air conditioning systems. In this way they may reach human lungs and pose long-term hazards. Factories using this material must be equipped with special air filters, separators, etc.

Illustrative Example 6.4: Compute the SE of filament-loaded composites.

A plastic is loaded with stainless steel filaments which are 8 microns in diameter and have aspect ratios of 250. When filled with 5 percent by weight of the 2 mm fibers, the resistivity is 0.2 Ωcm. Determine the SE of the resulting 1 mm thick composite skin for a plane-wave field. From Eq. (6.10):

$$R = 0.2 \times L/Lt$$

$$= 0.2\ \Omega\text{cm}/0.1\ \text{cm} = 2\ \Omega/\text{sq}$$

From Eq. (6.7):

$$R_{dB} = 20 \log (94/2) = 33\ \text{dB}$$

Checking for absorption per Eq. (6.9) using $\rho = 0.2\ \Omega$cm compared to $\cong 2 \times 10^{-6}\ \Omega$cm for copper,

$$\sigma_r = 10^{-5}$$

and,

$$A_{dB} = 0.13 \times 10^3\ \mu m \sqrt{f_{MHz} \times 10^{-5}}$$

This gives 1.3 dB at 10 MHz and 4 dB at 100 MHz.

Notice that no q factor [of Eq. (6.9)] has been used since ρ already takes into account the low content of metal.

6.1.4 Wire Mesh or Screen-Lined Plastic Box

A wire mesh or screen-lined plastic shielded box is one in which the box skin is made of plastic, fiberglass, wood or other nonconductive material. A wire screen or knitted wire mesh is embedded in or lined inside all six sides. Except for the removable top, the eight seams joining the five sides are assumed to be seam bonded by soldering or another continuous-seam mating process. Like the open frames of Section 6.1.1, the SE of the wire mesh lined box follows the same principles of physics described in Section 2.1.5, except the dimension is that of the center-to-center spacing of the screen mesh as shown in Fig. 6.2

$$SE_{box} = 104 - 20 \log (g f_{MHz}) \tag{6.11}$$

for $g < \lambda/2$ and $r_m \geqslant 48/f_{MHz}$ (far-field condition)

where,

g = center-to-center wire spacing in millimeters

f_{MHz} = radiation frequency in megahertz

r_m = distance from radiating source in meters

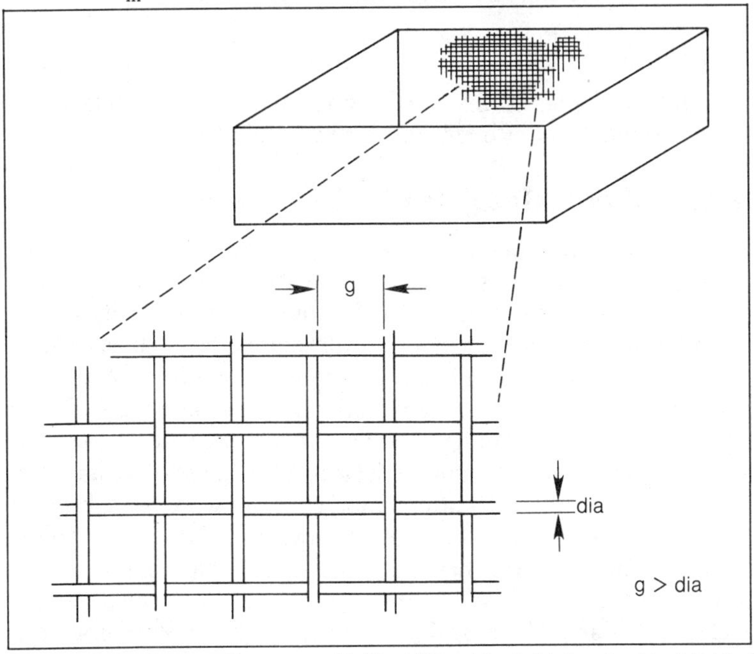

Figure 6.2—Square Screen or Knitted Wire Mesh Lining for Unshielded Box

Illustrative Example 6.5: Compute the SE of wire screen.

A spectrum analyzer is to be used outdoors in the presence of strong electromagnetic radiations from an airport L-band radar. An additional 30 dB of shielding beyond that of the spectrum analyzer housing is required at 1,300 MHz to ensure that a weak signal can be measured rather than the spurious responses of the spectrum analyzer resulting from shielding leakage. A mesh screen-lined box will surround the instrument. Determine if a 1 mm screen wire separation is adequate. From Eq. (6.11):

$$SE_{box} = 104 - 20 \log (1 \times 1{,}300) = 42 \text{ dB}$$

Thus, the screen will provide adequate (at least 30 dB) shielding provided each face's screen is bonded to the adjacent screens at intervals g' as follows:

$$30 = 104 - 20 \log (1{,}300 \text{ g'})$$

$$1{,}300 \text{ g'} = \log^{-1} [(104 - 30)/20] = 5{,}012 \text{ mm}$$

$$\text{or, g'} = 3.9 \text{ mm}$$

Such a short interval (3.9 mm) between bond points is tantamount to a continuous seam bond for physical realizability.

6.1.5 Sheet Metal Box

The traditional shielded equipment, instrument or other electronic product housing is fabricated from sheet metal. Most military equipments and many industrial and commercial product enclosures are still made from sheet metal, typically aluminum or steel, or more seldom, by casting. Such boxes are rugged and offer more SE to ambient threats from the base material than usually required, with the possible exception of low-frequency magnetic field threats. For example, household aluminum foil (with a thickness of 40 microns) offers more than 100 dB attenuation to plane waves and E-fields across the spectrum and to H-fields above 10 MHz. For magnetic fields below 1 kHz, however, household aluminum foil is transparent.

Chapters 1 and 2 have dealt extensively with the SE of practically all existing metals; therefore, these calculations will not be repeated here. Appendixes B through H also provide sufficient SE data for many metals, thicknesses and distances. Again, it is seen that virtually any metal with a thickness greater than 0.5 mm will exceed 120 dB above a few hundred kilohertz for any distance greater than 1 m. It is predictable that in practically all situations, the limiting factor in the SE performance of a finished box is not the metal, per se, but the various leakages, apertures and cracks, as explained in Chapter 7.

6.2 Entering the Box Skin Performance on the Worksheet

Figure 6.3 is the shielded box design objectives and performances worksheet first introduced in Chapter 4 and partially filled out in

Chapter 5 for line entries A.1 through A.4 (SE objective determination). This form will now be used to add the SE of a box skin shield on line B. The electromagnetic ambient threat on line A.1 corresponds to a 1 m distance in accordance with MIL-STD-461, RS03. Illustrative Example 6.3 produced the results entered on line B.

Figure 6.3—Worksheet for Recording SE Objectives and Performance

Top Identification	EMI Fix	10 kHz	100 kHz	1 MHz	10 MHz	100 MHz	1 GHz	10 GHz
A. SE Objectives (1 + 2 − 3 + 4)		0	0	8	22	42	34	20
1. EM Ambients (dBμV/m)		120	120	120	120	140	140	140
2. Field-to-Cable/PCB Coupling (dB)		-120	-100	-80	-60	-40	-28	-28
3. Sensitivity or NIL (dBμV)		42	42	42	48	68	88	102
4. Design Margin (dB)		10	10	10	10	10	10	10
5. Box SE Modified Objectives								
6. Leakage SE Modified Objectives								
B. SE Box Performance	.2 Ω/SQ.	117	97	77	57	43	48	61
C. Box Leakages*								
1. Viewing Window Apertures								
1.1 CRT Displays								
1.2 Alphanumeric Displays								
1.3 Panel Meter Displays								
1.4 Status Indicators								
2. Cooling Apertures								
2.1 Cooling Apertures								
2.2 Forced-Air Cooling								
3. Assembly and Maintenance Seams								
3.1 Box Housing Assembly								
3.2 Cover Panels								
4. Component Apertures								
4.1								
4.2								
D. Total SE* Performance								
E. Performance vs. Objectives (D − A)								

$$*SE = -20 \log \left[\log^{-1}\left(-SE_{Box}/20\right) + \sum_{i=1}^{n} \log^{-1}\left(-\text{Leakages}/20\right)\right]$$

6.3 References

1. Duncan, D., "A Low Cost Method of RFI/EMI Shielding," *EMC Technology*, October 1984, p. 75.
2. Furman, Stan, "Conductive Coating Considerations for Electronic Engineers," *EMC Technology*, October 1984, p. 49.
3. Mauriello, A.J., "Selection and Evaluation of Conductive Plastics," EMC Technology, October 1984, p. 59.
4. Nablo, J.C., Tolokan, R.P., "Stainless Steel, the Invisible Conductive Filler for Plastics," *EMC Technology*, October 1984, p. 43.
5. Schweizer, R.A., "Glass-Fiber Length Degradation in Thermoplastics Processing," *Plastics Compounding*, November, 1981.
6. Seymour, R.B., *Conductive Polymers*, (Plenam Press, 1981), pp. 49-57.
7. Simon, R.M., "EMI Shielding with Aluminum Flake Filled Polymer Composites," *Proceedings of the 1984 International Symposium on EMC*, Tokyo, Vol. 1, p. 732.
8. Simon, R.W., "EMI Can Be Made of Conductive Plastics," *Industrial Research and Development*, June 1982, pp. 104-109.
9. Simon, R.W. and Stutz, D., "Test Methods for Shielding Materials," *EMC Technology*, October 1983, p. 39.
10. Simpson, R.M., "Flexible Foil Laminates: A Cost Effective Alternative for EMI Shielding," *EMC Technology*, October 1984, p. 79.
11. White, D.J.R., "Electromagnetic Shielding Materials and Performances," (Gainesville, VA: Interference Control Technologies, Inc., 1980).
12. MIL-STD-461B, RS03, "Electromagnetic Emission and Susceptibility" (1 April 1980), Notice 1 (update), dated 5 January 1983 RS03, "Radiated Susceptibility Field Strengths."
13. EMC Technology, special issue on Conductive Coatings for Shielding Boxes, October-December 1985.

Chapter 7

Evaluation and Control of Aperture Leakages

Figure 7.1 shows a metal or metalized plastic box with many aperture leakages. The leakages compromise the shielding integrity of the basic box metal/metalized skin. In general, the SE of the aperture depends on the aperture size and the frequency. Addi-

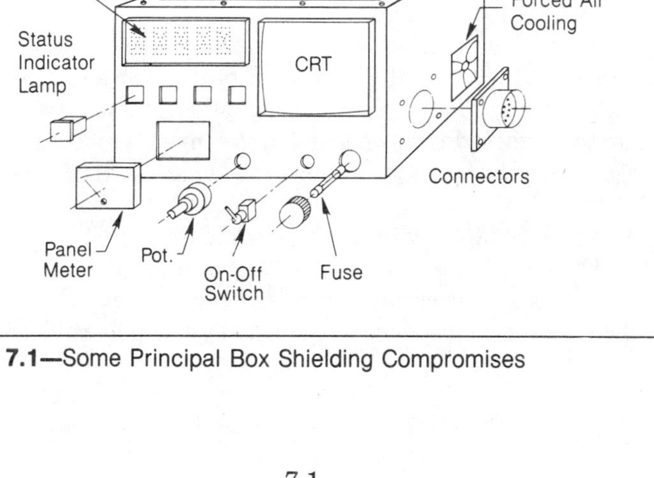

Figure 7.1—Some Principal Box Shielding Compromises

tionally, the SE with leakages can never be better (higher dB) than the SE of the basic skin material. At the low end, the aperture leakage is limited by the aperture height-to-length ratio and the shadow effect (i.e., SE does not go to exactly zero). In addition, when the depth of a hole exceeds the aperture dimensions, the WGBCO effect occurs. This chapter covers the analytical study and quantitative prediction of aperture leakages and their reduction. Qualitative aspects and guidelines on available hardware are covered in Chapter 10.

7.1 Classification of Aperture Leakages and Their Control

Figure 7.2, first introduced in Chapter 4, depicts the worksheet for recording input data and computed results on SE requirements and performance. Lines C and C.1 through C.4 address the subject of aperture leakages and classify them as follows:

1. Viewing window apertures
2. Cooling apertures
3. Mating panel assembly and maintenance seams
4. Component apertures

7.1.1 Viewing Window Apertures

Viewing windows are those apertures through which one views CRT, alphanumeric, panel meter and similar displays (see Fig. 7.1). Here the display is inside the box, beyond the box skin surface. Some displays may be surface mounted, such as status indicators and push-button indicators.

Usually, viewing windows have the largest area and, hence, the greatest leakage, i.e., the lowest SE of the four aperture leakage classifications. Thus, with some exceptions, viewing windows require the highest degree of aperture treatment for interference control. Among the leading options for viewing aperture leakage control are:

1. Finely knitted wire mesh or screens
2. Optically transparent RF reflective covers
3. EMI-suppressed indicators and buttons

The first two EMI controls usually include a clear plastic

Figure 7.2—Worksheet for Recording SE Objectives and Performance

Topic Identification	EMI Fix	10 kHz	100 kHz	1 MHz	10 MHz	100 MHz	1 GHz	10 GHz
A. SE Objectives (1 + 2 − 3 + 4)								
1. EM Ambients (dBμV/m)								
2. Field-to-Cable/PCB Coupling (dB)								
3. Sensitivity or NIL (dBμV)								
4. Design Margin (dB)								
5. Box SE Modified Objectives								
6. Leakage SE Modified Objectives								
B. SE Box Performance								
C. Box Leakages*								
1. Viewing Window Apertures								
1.1 CRT Displays								
1.2 Alphanumeric Displays								
1.3 Panel Meter Displays								
1.4 Status Indicators								
2. Cooling Apertures								
2.1 Cooling Apertures								
2.2 Forced-Air Cooling								
3. Assembly and Maintenance Seams								
3.1 Box Housing Assembly								
3.2 Cover Panels								
4. Component Apertures								
4.1								
4.2								
D. Total SE* Performance								
E. Performance vs. Objectives (D − A)								

$$*SE = -20 \log [\log^{-1}(-SE_{Box}/20) + \sum_{i=1}^{n} \log^{-1}(-\text{Leakages}/20)]$$

substrate which sandwiches the wire mesh (see Fig. 7.3) or on which the vacuum-deposited film (see Fig. 7.4) is applied. Continuous peripheral electrical bonding between the treated window and the metal or metalized box is important.

Figure 7.3—Representative Shield Screen Windows for Viewing (courtesy of Tecknit)

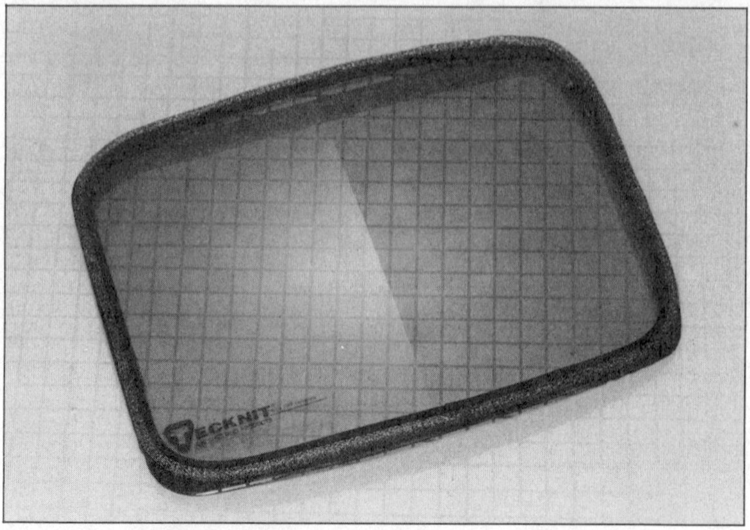

Figure 7.4—Optically Transparent, RF Reflective Windows

Another option, though expensive, is to build a "dog house" behind the display. The display itself no longer contributes to SE degradation if the auxiliary box is properly mounted to make it RF-tight and if wiring to the display is properly treated (filtered).

7.1.2 Cooling Apertures

Equipment cooling is one of many requirements of box and equipment design. Accordingly, the second aperture classification involves convection and forced-air cooling. Convection cooling usually consists of an array of slots or circular holes arranged such that warm air can pass through the top or upper portion of the box. Forced-air cooling provides for at least one large fan aperture for exhausting or drawing air. There may or may not be a corresponding hole designed specifically for air movement because air can pass through other apertures.

Several techniques are commonly used for leakage control of cooling apertures:

1. Break large slots into more slots of smaller dimensions.
2. Add depth to small apertures to effect WGBCO behavior.
3. Place louvers over the fan hole.
4. Place a screen over the fan hole.

7.1.3 Mating Panel Member Seams and Gaskets

To permit component installation, maintenance, etc., equipment boxes must have either one or more of the six sides (or portions thereof) removable or hinged, or a telescoping "drawer" which fits inside the rectangular receptacle (see Fig. 7.5). Either approach results in mating panel members and associated seams, fasteners and slit apertures where leakage can occur.

Among the leading options for leakage control of seams are:

1. More or different fasteners
2. Panel member stiffeners
3. Depth (thickness) added to panel members where they meet to perform in accordance with WGBCO effect
4. Electrical gaskets
5. Combinations of the above

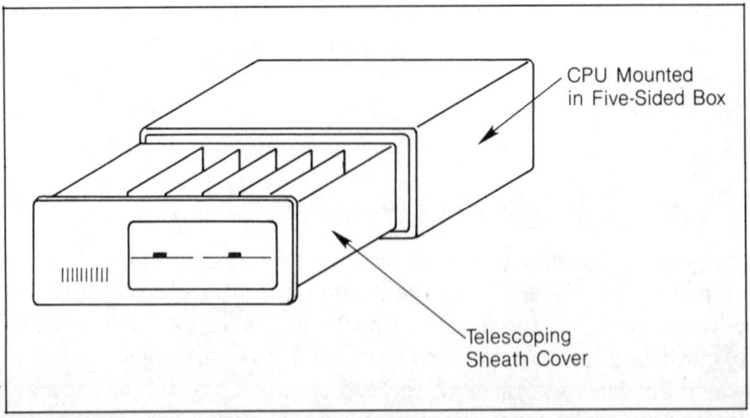

Figure 7.5—Typical Five-Sided Box Using a Telescoping Sheath Cover

7.1.4 Component Apertures

Figure 7.1 shows a number of component apertures not covered in the previous discussion, i.e., fuses, switches, potentiometers, connectors, etc. These apertures (without components in place) use math models similar to those for larger apertures. Among the leading options for control of component aperture leakages are EMI-shielded components filling the aperture area, and standard (non-EMI-suppressed) components located within a shielded compartment having miniature filtered access lines through the bulkhead.

7.2 Calculation and Control of Aperture Leakages

Expanding on the previous section, the following discussion provides math models, performance charts and graphs for predicting and controlling EMI aperture leakages.

Numerical prediction of the leakage of an aperture in a metal barrier is a subject which has been addressed in various ways. The complexity of the topic is such that the EMC engineer or equipment designer is caught between a bevy of complex and often contradictory mathematical expressions on the one hand and a set of qualitative rules of thumb on the other. This section describes a simplified model which cannot aspire to rigorous accuracy but provides an acceptable approximation of the SE of an aperture.

First, we need to revisit the principle of reciprocity between an aperture in a thin wall and its complementary dipole (Fig. 7.6). Babinet's principle states that a slot in a plane behaves like a dipole of the same length and width as the slot, with simply a 90° rotation in the orientation of the E-fields and H-fields. In other words, a horizontal slot behaves as vertical dipole and vice versa. We know that a dipole becomes an efficient radiator when its length reaches half-wave resonance. Therefore, we can deduce that a $\lambda/2$ slot will also be an efficient antenna; i.e., an inefficient shield barrier.

But a numerical expression is more reliable than an intuitive feeling. Since we have established that reflection shielding at the wave-to-barrier interface is calculated from the ratio $K = Z_{wave}/Z_{barrier}$, we can calculate the impedance of a slot to a normally incident field, and thus the reflection coefficient will be known. The theory of slot antennas (slots as radiators) can be used here. The impedance of a slot is found from:

$$Z_{slot} \times Z_{dip} = \frac{Z_w^{\ 2}}{4} \tag{7.1}$$

where,

Z_w = the impedance of the electromagnetic field in the first media, i.e., generally air. In the air, for far-field conditions, $Z_w = 120\ \pi\Omega$.

Z_{slot}, Z_{dip} = the impedances of the slot and its complementary dipole, i.e., a dipole whose length is equal to the slot length, and whose width (L >> W) is equal to the slot height.

Because the impedance of a thin dipole at resonance equals 75 ohms, the impedance of its complementary slot will be:

$$Z_{slot} = \frac{(120\pi)^2}{4 \times 75} = 473\ \Omega$$

Using the formula of Eq. (1.64) for reflection loss:

$$R = \frac{(1 + K)^2}{4K} = \frac{[1 + (377/473)]^2}{4 \times (377/473)} \approx 1 \equiv 0 \text{ dB}$$

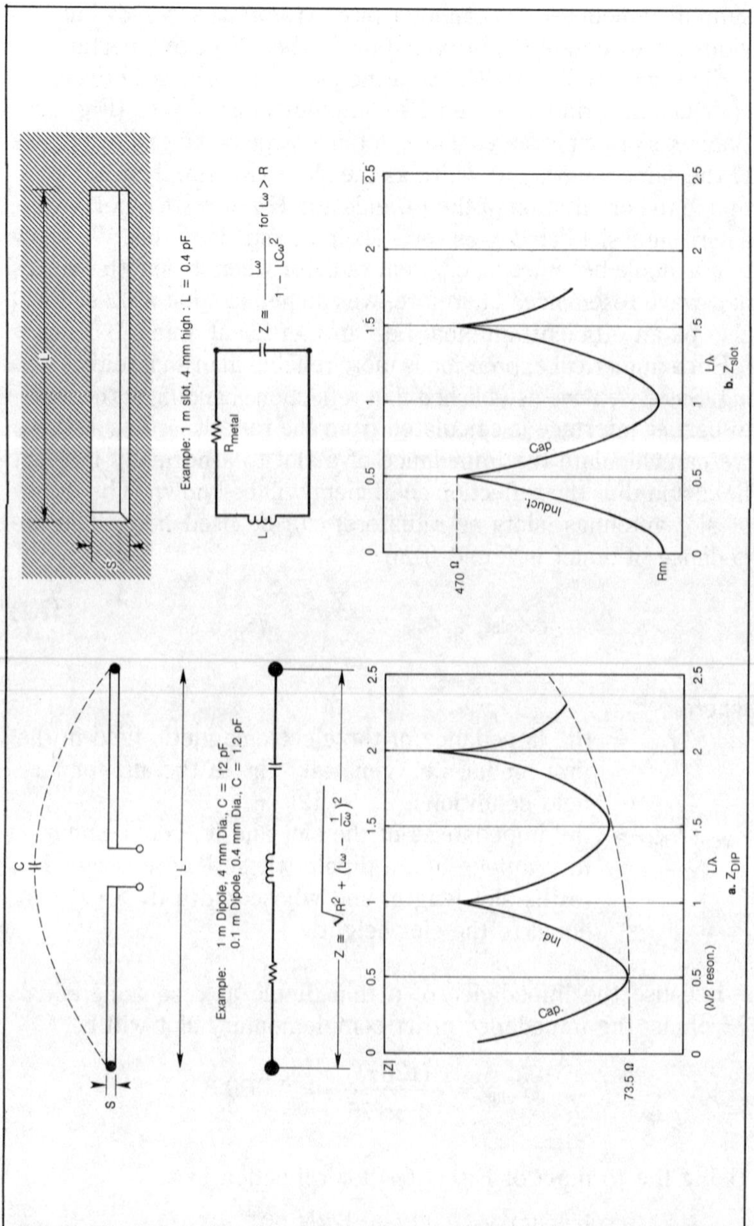

Figure 7.6—Reciprocity between a Dipole (a) and Its Complementary Slot (b). Approaching resonance, Z of the dipole decreases, making an efficient field transducer, and Z_{slot} increases, making an inefficient shunt attenuator.

This confirms that, at resonance, the reflection loss of a slot is virtually 0 dB.

Next, it is more interesting to compute how this reflection (the slot impedance) behaves below resonance. Bearing in mind that (1) if we know how to calculate a dipole impedance at any frequency, we can derive the impedance of its complementary slot by Eq. (7.1), and (2) the same equation shows that below resonance, since Z_{dip} increases, the complementary Z_{slot} will decrease. Figure 7.6 shows how both impedances evolve for a thin dipole and slot.

The last parameter we need to address is the effect of slot height. Is the length the only influencing parameter in calculating Z_{slot}, or does the height also count? If so, to what extent? Here again, considering the equivalent dipole, we know that length is the predominant factor and that the thickness plays a secondary role. The difference between a "fat" and a "thin" dipole resides in its selectivity or quality factor "Q" near resonance (more or less, how "sharply" tunable it is). Formulas state that the impedance at any frequency of a dipole of length L and diameter S is approximately:

$$Z_{dip} = -j\ 120\ l_n\ (2L/S)\ \cot(\pi L/\lambda) \tag{7.2}$$

In the capacitive region (when $L < 0.5\lambda$), a dipole 10 times thicker has only double the capacitance (half the impedance) as derived from the natural logarithm in the equation. By reciprocity, a slot 10 times thinner will have half the inductance and hence half the impedance, Figure 7.6 and Table 7.1 show some calculated values

Table 7.1—Some Typical Impedance Values for a Dipole and its Complementary Slot

	L = 0.16 λ	L = 0.32 λ	L = 0.5 λ ≡ Resonance
Dipole diam = $10^{-3} \times$ L	X = −1,000 Ω R = 10 Ω	X = −500 Ω R = 30 Ω	|Z| = R + jX = 73.5 Ω
Slot (complementary)	X = 36 Ω	X = 72 Ω	|Z| = 473 Ω

of Z_{dip} and its complementary slot for a typical ratio of L/λ.

Another approach, giving basically the same results, appears in Ref. 1 where the slot impedance is derived from the characteristic impedance of a line made of two parallel plates side by side.

Based on the calculations summarized above, the reflection loss of a thin slot in an **infinite** metal baffle for far-field conditions is:

$$R = \frac{2l_n\ 2L/S}{\pi \tan \pi L/\lambda}$$

and,

$$SE_{apertdB} = \{97 - 20 \log L\, f_{MHz}\} + \{20 \log (1 + l_n \frac{L}{S})\} \quad (7.3)$$

$$\text{for } L < \lambda/2$$

where, L = slot length in millimeters
S = slot height in millimeters
l_n = natural log

In the above equation, the first element enclosed in braces is the square or round aperture reflection term, and the second is the "fatness factor" for S < L. The latter equals 0 dB when L = S (square aperture) and allows for a progressive improvement of SE as the slot becomes thinner and thinner. For instance:

L/S	Fatness factor
10	+10 dB
100	+15 dB
1,000	+18 dB

7.2.1 Leakage from a Single Aperture

Figure 7.7 shows a rectangular aperture in a metal panel or metalized plastic shield. An incident electric field is vertically polarized to hit the horizontal edge of the aperture for maximum coupling. As shown, at low frequencies (L << λ/2) the induced surface currents are interrupted by the slot and flow around it.

To complete a general equation for the aperture leakage, we must account for the fact that the box wall is not an infinite plane, nor is the volume behind the slot uniformly illuminated. Finally, if the thickness of the metal is an appreciable fraction of the length L, an additional "through" attenuation (WGBCO effect, similar to the absorption term AdB in the SE of metal) exists. This is explained in Section 7.2.5.

Thus, the complete, generally applicable equation for slot SE is:

$$SE_{dB} = 97 - 20 \log (Lf_{MHz}) + 20 \log (1 + l_n L/S) + SE_{shad} + 30\ d/L \qquad (7.4)$$

where,

d = depth of the slot = metal thickness unless laps, louvers or tubes are installed

L, S = slot length and height in millimeters

SE_{shad} = shadow effect in dB, explained below

For a circular hole in which L = S = the diameter of hole, add 2 dB to the 97 dB to obtain 99 dB. This number comes from 20 log (π/4) = 2 dB because the area of a hole versus that of a bounded square is 0.79 or 2 dB less. Also note that this case is for matched polarization between E-field and the slot. For crossed polarization (E-field and longer slot dimension align), about 20 to 30 dB improvement may be expected, but this is not realistic because the arriving E-field may take on any polarization.

In Eq. (7.4), "97 − 20 log L_{fMHz} + 20 log (1 + l_n L/S)" is the total reflection term, while "30 d/L" represents the absorption term.

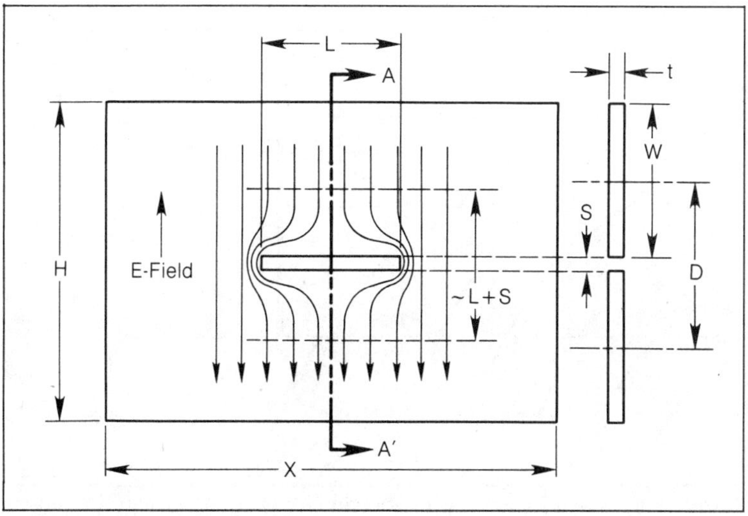

Figure 7.7—Front Cover Panel with Slot Aperture Dimensions S × L

Note that nothing prevents SE from becoming larger than 97 dB since the term "– 20 log Lf_{MHz}" can be a positive number when $Lf_{MHz}< 1$. However, SE_{ap} is bounded on the high side by $SE_{boxskin}$ since the slot attenuation in dB cannot exceed that of the surrounding metal. Translated in terms of the simple circuit, this means that in Fig. 7.6, the slot impedance can never become smaller than R_{metal}. When L reaches $\lambda/2$, tan $\pi L/\lambda$ becomes ∞ and the reflection term equals zero.

So, Eq. (7.4) is bounded on the low-value side by:

$$SE_{low} = SE_{shad} + SE_{WGBCO}$$

The shadow effect term appearing in Eqs. (7.4) and (7.5) may be explained with the assistance of a few sketches. Figure 7.8 shows a six-sided metal box with a leaky aperture. Areas inside the box will be shadowed by the oncoming radiation as a result of the shape of the radiation pattern. Thus, there must be some reduction (shadow effect) in aperture leakage in these regions.

Figure 7.9 shows the antenna pattern inside the box, where the box is assumed to have a very low Q-factor (the box is not sharply resonant) due to absorption and scattering of its internal components. For slot dimension conditions in which $L < \lambda/2$, the pattern is shown in Fig. 7.9a. When $L >> \lambda/2$, the number of lobes

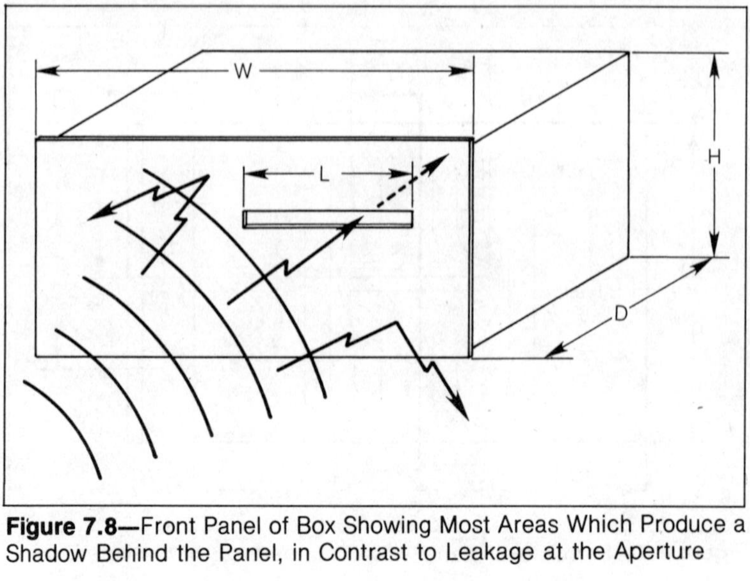

Figure 7.8—Front Panel of Box Showing Most Areas Which Produce a Shadow Behind the Panel, in Contrast to Leakage at the Aperture

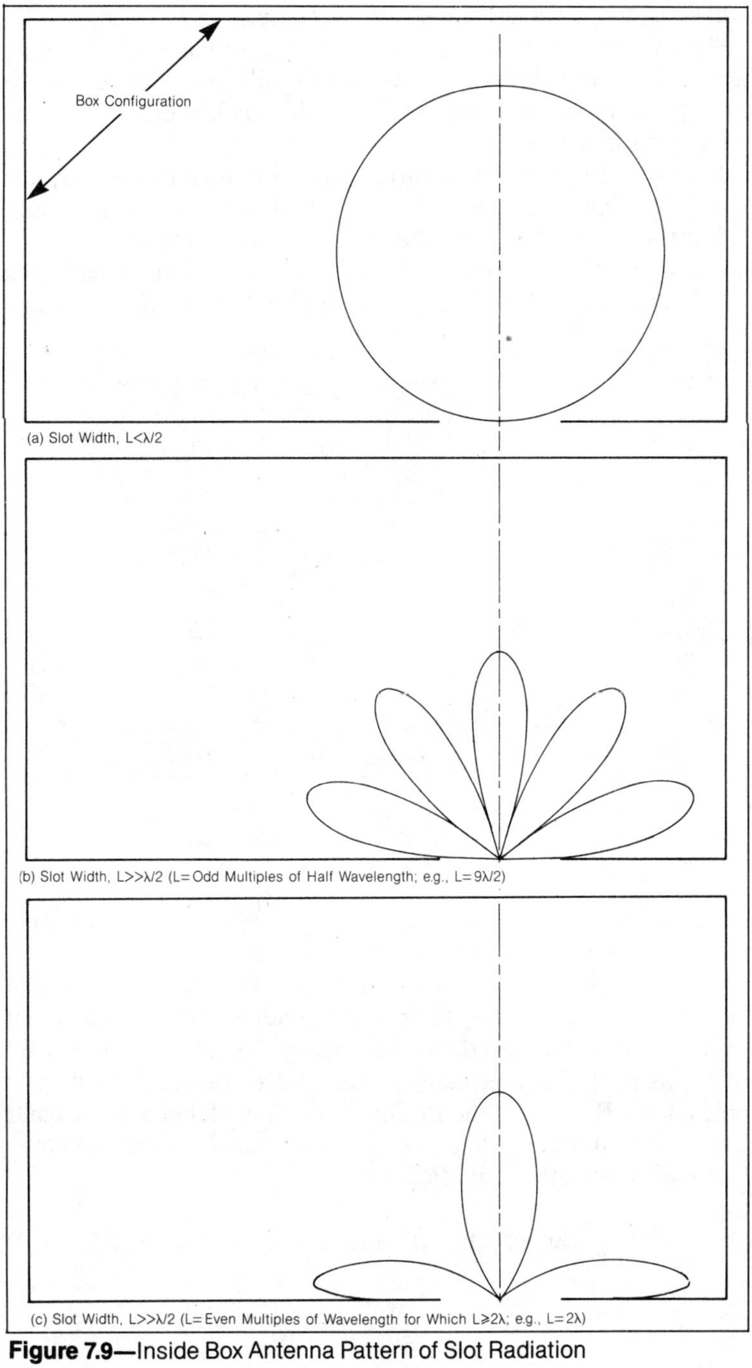

Figure 7.9—Inside Box Antenna Pattern of Slot Radiation

will equal the slot-width dimension in multiples of wavelength (Fig. 7.9b). Above about $3(\lambda/2)$, the lobe structure subtends an angle greater than the slot beamwidth and the shadow effect begins to lose its effectiveness. Above $L = 2\lambda$, the shadow effect is lost as suggested in Fig. 7.9c.

Figure 7.10 shows the three slot gain pattern regions for $L < 1.5\lambda$. Table 7.2 shows the results when the total inside box volume is integrated over the three regions. Note that shallow boxes (low values of D or D/W ratios in Fig. 7.1) and small slots (low values of L or L/W and S or S/H ratios) produce shadow

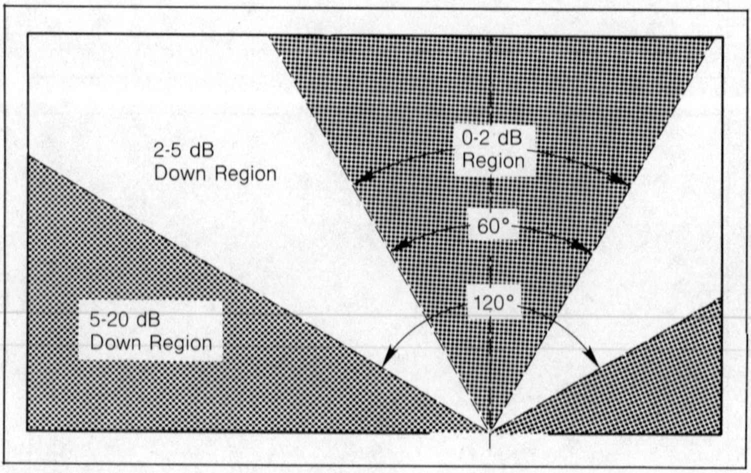

Figure 7.10—Inside Box Illumination from Slot-Coupled Radiation for $L \leqslant 3\lambda/2$

shielding increases on the order of 10 to 15 dB, while the converse produces only a few dB of shadow shielding.

Most box housing configurations have a D/W ratio in the range of 0.5 to 1.5. Therefore, their corresponding shadow protection contribution will be less than 5 dB. Where D/W ratios approximate unity, as in the case of many boxes, the shadow contribution is only a few dB. Thus, if the shadow effect is not taken into account, 3 dB can be used as a default in Eqs. (7.4) and (7.5) and the equation can be greatly simplified:

$$SE = 100 - 20 \log Lf + 20 \log \left(1 + l_n \frac{L}{S}\right) + 30\ d/L \qquad (7.4a)$$

Illustrative Example 7.1: Compute SE corresponding to slot leakage

The front panel of a metalized plastic box (SE_{box} = 75 dB) measures 15 cm high by 40 cm wide. It contains a single-slot aperture measuring 1 mm high by 10 cm long. The box is 30 cm deep. Calculate the SE corresponding to the slot leakage.

Equation (7.4) produces a slot leakage at 10 MHz of:

$$SE_{apt} = 97 - 20 \log (100 \times 10) + 20 \log (1 + l_n \frac{100}{1}) + SE_{shad}$$

$$= 97 - 60 + 15 + 4 = 56 \text{ dB}$$

Since it is metallic plastic, there is virtually no WGBCO.

Table 7.2—Additional Shielding Effectiveness in dB due to Cover Panel Shadow Effect (See Fig. 7.8 for Dimensions Identification)

L/W	S/H	D/W														S/H	L/H
L/W	S/H	0.05	0.10	0.15	0.20	0.25	0.30	0.40	0.50	0.60	0.75	1.0	1.3	1.5	2.0	S/H	L/W
0.01	0.01	13	10	8	7	6	6	5	5	4	4	3	2	2	1	0.01	0.01
0.01	0.2	13	9	8	7	6	6	5	5	4	4	2	2	2	1	0.32	0.01
0.01	0.4	5	6	6	5	5	5	5	4	4	4	2	2	1	1	0.4	0.01
0.01	0.6	3	2	2	3	3	3	3	3	3	3	2	1	1	1	0.6	0.01
0.01	0.8	1	1	1	0	0	0	1	1	1	2	1	1	1	1	0.8	0.01
0.01	1.0	0	0	0	0	0	0	0	0	0	0	0	0	0	0	1.0	0.01
0.2	0.01	15	10	8	7	6	6	5	5	4	4	3	2	2	1	0.01	0.2
0.2	0.2	15	9	8	7	6	6	5	5	4	4	2	2	2	1	0.2	0.2
0.2	0.4	6	7	6	6	5	5	5	4	4	4	2	2	2	1	0.4	0.2
0.2	0.6	3	2	2	3	3	3	3	3	3	3	2	1	1	1	0.6	0.2
0.2	0.3	1	1	1	1	0	0	1	1	2	2	1	1	1	1	0.8	0.2
0.2	1.0	0	0	0	0	0	0	0	0	0	0	0	0	0	0	1.0	0.2
0.4	0.01	15	10	8	7	6	6	5	5	4	4	3	2	2	1	0.01	0.4
0.4	0.2	10	8	7	6	6	6	5	5	4	4	2	2	2	1	0.2	0.4
0.4	0.4	6	6	6	6	5	5	5	4	4	4	2	2	1	1	0.4	0.4
0.4	0.6	3	3	4	4	4	4	4	4	3	3	2	2	1	1	0.6	0.4
0.4	0.8	1	1	1	1	1	1	2	2	2	2	1	1	1	1	0.3	0.4
0.4	1.0	0	0	0	0	0	0	0	0	0	0	0	0	0	0	1.0	0.4
0.6	0.01	15	10	8	7	6	6	5	5	4	4	3	2	2	1	0.01	0.6
0.6	0.2	6	6	6	6	5	5	5	4	4	4	2	2	2	1	0.2	0.6
0.6	0.4	3	4	4	4	4	4	4	4	4	3	2	2	1	1	0.4	0.6
0.6	0.6	3	2	2	3	3	3	3	3	3	3	2	1	1	1	0.6	0.6
0.6	0.6	2	2	1	1	2	2	3	3	3	3	2	1	1	1	0.3	0.6
0.6	1.0	0	0	0	0	0	0	0	0	0	1	0	0	0	0	1.0	0.6
0.8	0.01	15	10	8	7	6	6	5	5	4	4	3	2	2	1	0.01	0.8
0.8	0.2	3	5	5	5	5	5	4	4	4	3	2	2	1	1	0.2	0.8
0.8	0.4	3	2	2	3	3	3	3	3	3	3	2	1	1	1	0.4	0.8
0.8	0.6	2	1	1	1	2	2	2	2	2	2	2	1	1	1	0.6	0.8
0.8	0.3	1	1	1	1	0	1	1	2	2	2	1	1	1	1	0.8	0.8
0.8	1.0	0	1	0	0	0	0	0	1	1	1	1	1	0	0	1.0	0.8
1.0	0.01	15	10	8	7	6	6	5	5	4	4	3	2	2	1	0.01	1.0
1.0	0.2	3	3	3	4	4	4	4	4	3	3	2	2	1	1	0.2	1.0
1.0	0.4	2	1	1	1	2	2	2	3	2	2	2	1	1	1	0.4	1.0
1.0	0.6	1	1	1	1	0	0	1	1	2	2	1	1	1	1	0.6	1.0
1.0	0.8	0	1	0	0	0	0	0	0	1	1	1	0	0	0	0.8	1.0
1.0	1.0	0	0	0	0	0	0	0	0	0	0	0	0	0	0	1.0	1.0
L/W	S/W	0.05	0.10	0.15	0.20	0.25	0.30	0.40	0.50	0.60	0.75	1.0	1.3	1.5	2.0	S/W	L/
L/W	S/W	D/W														S/H	L/

Code And Notes: W = Box Width in mm; D = Box Depth in mm; L = Slot Width in mm. Where an Array is Used, L = End-to-End Dimension in mm. Table Valid for <5λ. Values Approach Zero at L = 2λ.

The upper limit of 75 dB is achieved when:

$$75 = 97 + 19 - 20 \log (100 \times f_{MHz})$$

or $$100 f_{MHz} = \log^{-1} [(116 - 75)/20]$$

$$f_{MHz} = 0.01 \times 112 = 1.1 \text{ MHz}$$

From Eq. (7.5) the lower limit is:

$$SE_{low} = SE_{shad} = 4 \text{ dB}$$

The above calculations are plotted in Fig. 7.11. The 4 dB is obtained from Table 7.2, where L/W = 10 cm/40 cm = 0.25, S/H = 1/150 = 0 and D/W = 30 cm/40 cm = 0.75. The 4 dB ap-

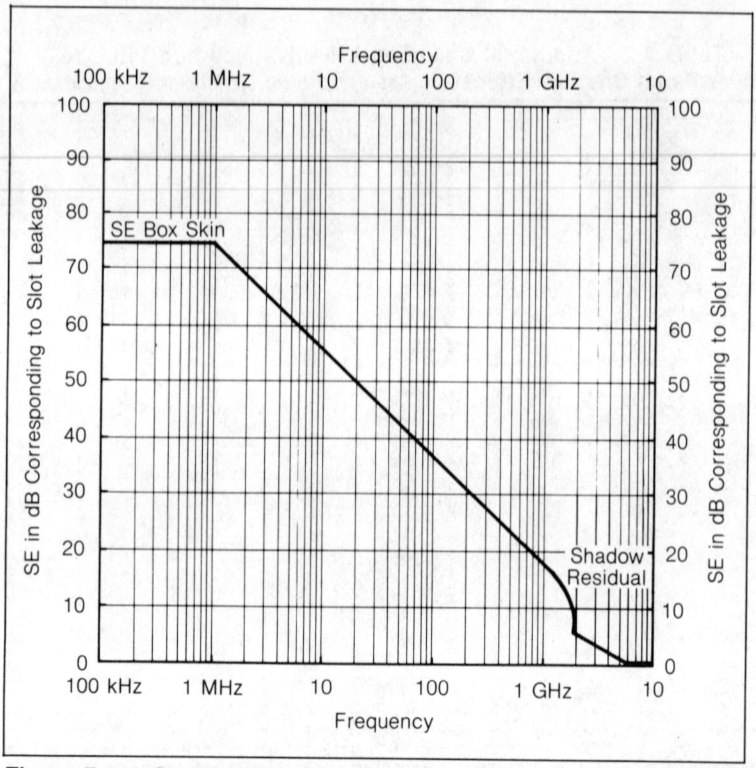

Figure 7.11—Shielding Effectiveness of 1 mm × 10 cm Slot Leakage in 40 cm Wide × 15 cm High × 30 cm Deep Box (Far Field, Plane Waves)

plies at L = λ/2 or 1.5 GHz. Above, SE_{shad} progressively decreases, reaching 0 dB above 2λ, or 6 GHz.

Figure 7.12 is a graph of the same box having several different slot configurations for a skin material offering a basic SE of 100 dB. Table 7.3 provides some of the important data points for plotting the graph.

Notice in Fig. 7.12 how a five-sided box (one side entirely open) still offers some attenuation against plane-wave fields. This attenuation corresponds to the limited intercept area bounded by box sides, corresponding to a hooded or "shrunken" antenna. For instance, looking at our 15 cm × 40 cm box, the open area corresponds to 0.06 m². The area of an ideal intercept antenna is

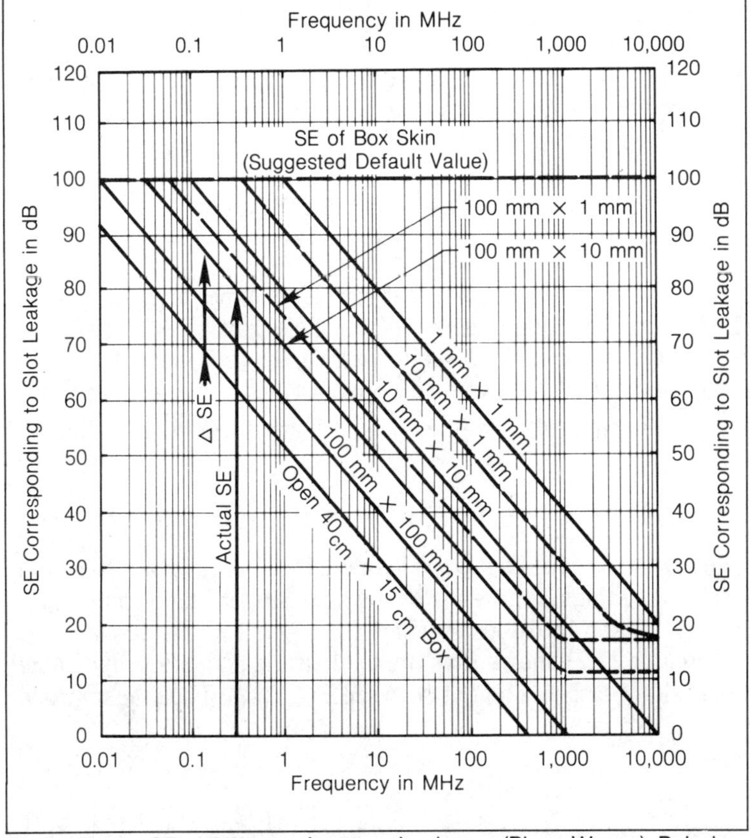

Figure 7.12—SE of Different Aperture Leakages (Plane Waves), Polarization Perpendicular to Slot Length SE ≅ 100 − 20 log L_{mm} + 20 log (1 + l_n L/S) − 20 log F_{MHz}

Table 7.3— Principal Data Points for Plotting Fig. 7.12 for 40 cm Wide × 15 cm High × 30 cm Deep Box

Slot Configuration	(No Shadow) SE at 1 MHz	Shadow	Total
A. 1 mm × 1 mm long	97 dB	4 dB	100* dB
B. 1 mm × 1 cm long	87 dB	4 dB	91 dB
C. 1 mm × 10 cm long	72 dB	4 dB	76 dB
D. 1 cm × 1 cm long	77 dB	4 dB	81 dB
E. 1 cm × 10 cm long	67 dB	4 dB	71 dB
F. Open Box	49 dB	0 dB	49 dB
G. No Slots	100* dB	N/A	100 dB

*For values of SE_{dB} > 100 dB, replace by SE_{box} skin. If unknown, a default value of 100 dB can be entered except for near H-field conditions.

1.6 $\lambda^2/4\pi$, i.e., about 10^4 m^2 at 1 MHz. The radiated power intercepted by this open side of the box versus the power intercepted by an ideal antenna is, therefore:

$$10 \log \frac{10^6}{0.06} = 52 \text{ dB}$$

It is important to recognize that very often experimenters mistake (1) the actual SE of an aperture (i.e., the vertical value of any slot SE in Fig. 7.12), for (2) the relative SE of the same aperture, (i.e., the improvement seen when an open box side or a large aperture is blinded by a panel having a smaller slot (i.e., Δ SE in Fig. 7.12).

7.2.2 Leakages from Subdivided Apertures

One way to reduce aperture leakage is to break an aperture into smaller openings. Figure 7.13 shows the slot of Illustrative Example 7.1 broken into four smaller slots.

Equation (7.7) shows that the aperture leakage is proportional to the slot length and height (also see Fig. 7.12). Thus the improvement in SE is:

$$\Delta \text{SE} = 20 \log \left[\frac{L_b}{L_a} \right] - 20 \log \left[\frac{1 + l_n (L_b/S_b)}{1 + l_n (L_a/S_a)} \right] \quad (7.6)$$

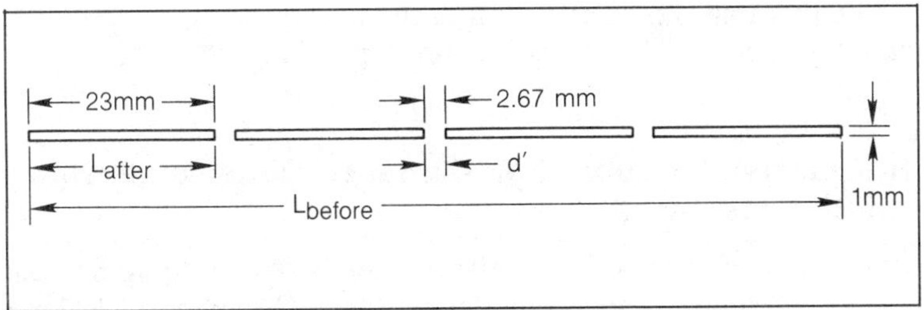

Figure 7.13—1 mm × 10 cm Slot Divided Into Four Slots

where,

SE = SE improvement ($SE_{before} - SE_{after}$)
S_b = slot height before
S_a = slot height after

This corresponds to the fact that breaking the length into shorter segments creates a less leaky slot. But at the same time, this improvement may be slightly offset by a change in the fatness factor.

If, as is frequently the case, $S_b = S_a$ 30, Eq. (7.6) becomes:

$$\Delta SE \approx 20 \log \left(\frac{L_b}{L_a} \right) - 20 \log \left[\frac{L_b}{L_a} - \frac{1 + l_n L_{b/s}}{1 + l_n L_{a/s}} \right]$$

$$= 20 \log (100_{mm}/23_{mm}) - 20 \log \left(\frac{1 + l_n 100/1}{1 + l_n 23/1} \right)$$

$$= 10 \text{ dB} \qquad (7.7)$$

Equation (7.7) will produce the value of the leakage, which is that of curve B in Fig. 7.12 + 10 dB with a ceiling of 100 dB if SE_{skin} = 100 dB.

What happens when a bigger aperture is divided into an array of smaller apertures? Again Eq. (7.6) applies, but this time both the L and S dimensions are involved. If the partitioning is even, i.e. if L and S are divided by the same number, $L_b/L_a = S_b/S_a$, the second term of Eq. (7.6) equals zero and:

$$\Delta SE \cong 20 \log \sqrt{N} = 10 \log N \qquad (7.8)$$

where, N = total number of smaller apertures

For this improvement to apply, the apertures must be similar and contiguous. This aspect is developed in Section 7.3.

Illustrative Example 7.2: Compute the leakage from a cooling aperture.

A shallow (10 cm deep × 40 cm wide × 20 cm high) box has a forced-air cooling hole diameter of 10 cm. Calculate the leakage at 1 GHz. If SE is less than 40 dB, divide the hole into smaller squares to ensure a 40 dB SE performance.

At 1 MHz, Eq. (7.4) (note the 99 dB figure for holes) produces:

$$\begin{aligned} SE_{apt} &= 99 - 20 \log (100 \times 1^2) + SE_{shad} \\ &= 99 - 40 + SE_{shad} \\ &= 59 \text{ dB} + SE_{shad} \end{aligned}$$

From Table 7.1, D/W = 10/40 = 0.25, L/W = 10/40 = 0.25 and S/H = 10/20 = 0.5. Thus:

$$SE_{shad} = 4 \text{ dB and } SE_{apt} = 59 + 4 = 63 \text{ dB at 1 MHz}$$

The lowest value for SE_{apt} is obtained from Eq. (7.8):

$$SE_{low} = 0 + 4 \text{ dB} = 4 \text{ dB}$$

since, for a round hole, the fatness factor = 0 dB. This value is valid up to a frequency corresponding to L = 1.5λ. Since L = 0.1 m, λ = 0.067 m and f_{MHz} = 300/0.067 = 4.5 GHz. The above SE of 63 dB at 1 MHz corresponds to 3 dB at 1 GHz, 3 decades higher in frequency. To ensure a 40 dB performance at 1 GHz, 40 dB – 3 dB or 37 dB improvement is necessary. Thus, applying this to Eq. (7.6) yields:

$$\Delta SE = 20 \log \left(\frac{L_b}{L_a} \right), \text{ for } L = S \tag{7.9}$$

Thus,

$$37 \text{ dB} = 20 \log \left(\frac{100}{L_a} \right)$$

or,

$$\begin{aligned} 100/L_a &= \log^{-1} (37/20) = 70.8 \\ L_a &= 1.4 \text{ mm} \end{aligned}$$

A circular array of 1.4 mm square apertures will provide the necessary performance. Keep the separation distance between the 1.4 mm squares small to ensure that little reduction of the forced-air movement occurs. A square wire grid having less than 10 percent optical coverage will work, but some forced-air movement reduction will result. Optical coverage is a measure of the "opacity" of the mesh, which depends on how "tight" the mesh is.

7.2.3 Performance of a Screen Mesh

Figure 7.14 shows a square screen mesh, or knitted-wire mesh of wire diameter (dia.) and center-to-center mesh spacing of g. Since g is usually much greater than dia., the optical coverage area is small (typically less than 10 percent).

The behavior of this type of barrier has been described in Sections 2.5.1.1 and 6.1.5, where Eq. (6.11) established that:

$$SE_{dB} = 104 - 20 \log (g_{mm} \times f_{MHz})$$

$$\text{for } g \leqslant \lambda/2 \text{ and far-field conditions} \quad (6.11)$$

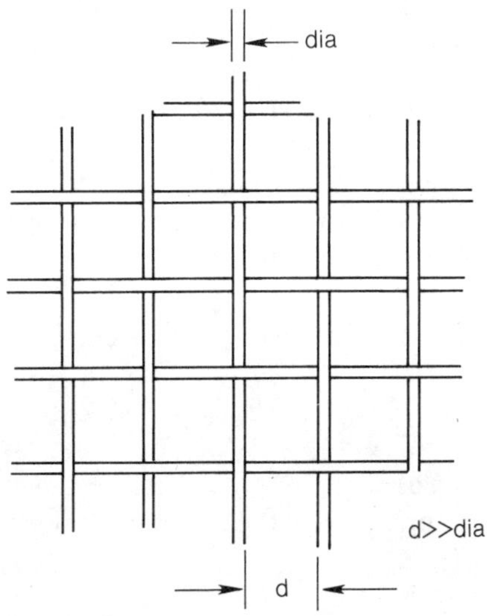

Figure 7.14—Square Screen or Knitted-Wire Mesh Covering a Leaky Aperture

For a near H-field source at distance r, Eq. (2.14) established that:

$$SE_{(H)dB} = 20 \log \frac{\pi r}{g}, \qquad \text{for r and g in the same units}$$

Remember that this value, although constant, is bounded by the low-frequency magnetic SE of the metal itself, which generally is poor.

A similarity exists between Eqs. (6.11) and (7.4). If we make g = L = S, and accounting for a typical shadow effect, Eqs. (7.4) and (6.11) become nearly identical.

Illustrative Example 7.3: Compute aperture leakage from a box with a wire screen.

There is a 25 cm wide × 19 cm high, 4:3 aspect ratio, rectangular display area on a CRT. Compute the aperture leakage over the 10 kHz to 10 GHz spectrum for the metalized plastic box (front panel = 40 cm wide × 30 cm high) (1) having no EMI screen protection, and (2) with a d = 1 mm screen protection. The box is 40 cm deep and has a metalized plastic SE skin of 60 dB.

From Eq. (7.4), the leakage at 1 MHz from the rectangular CRT hole, assuming far field condition exists, is:

$$SE_{apt} = 97 - 20 \log (250 \times 1) + 20 \log (1 + l_n \frac{250}{190}) + SE_{shad}$$

(see Table 7.2)

$$= 97 - 48 + 2 + 0 = 51 \text{ dB}$$

Lower-bound SE at higher frequencies is obtained from Eq. (7.5):

$$SE_{low} = SE_{shad} = 2 \text{ dB}$$

Figure 7.15 is a plot of the SE of the CRT rectangular hole. When a 1 mm screen is applied over the rectangular aperture at 1 MHz, Eq. (7.4) yields:

$$SE_{apt} = 97 - 20 \log (1_{mm} \times 1_{MHz}) + SE_{shad}$$

$$= 97 + 0 + 2 = 99 \text{ dB}$$

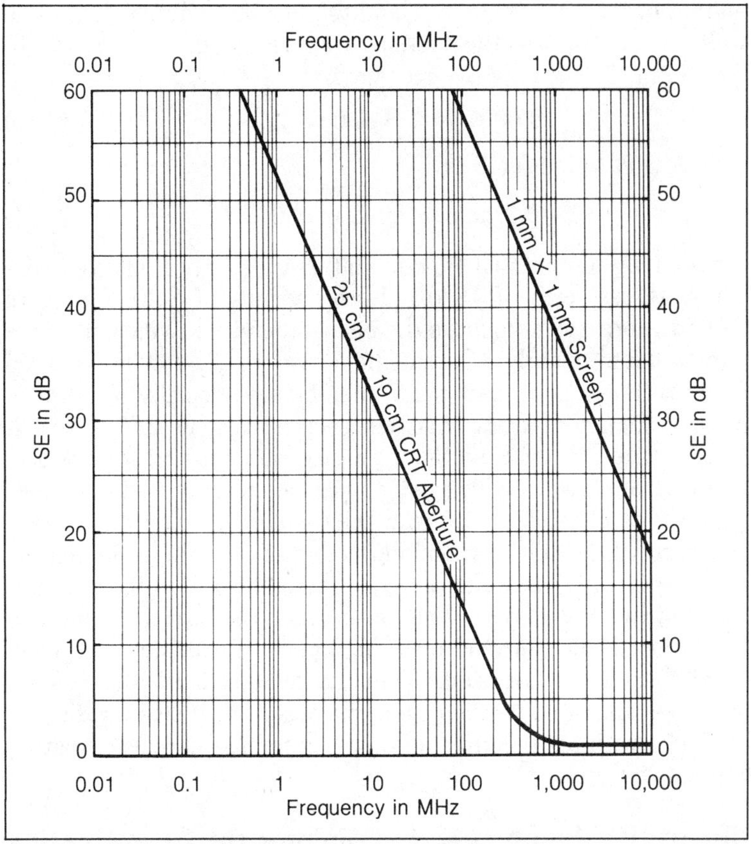

Figure 7.15—SE Performance of CRT Aperture Before and After Adding a 1 mm × 1 mm Knitted-Wire Mesh Screen

Since the SE of the skin is only 60 dB, however, it overrides the aperture SE at 1 MHz up to 80 MHz. At 1 GHz the leakage is 39 dB (99 dB − 60 dB).

7.2.4 Performance of Mating Panel Members

Figure 7.16 shows examples of mating panel members used for access to the side (front, back, left/right side, top or bottom or portions thereof) of a box. In each case, the easily definable (larger) dimension is L. The smaller dimension S, however, may have wide variations, depending upon the thickness, rigidity and type of mating of the panel members and the fastening dimension, L, as well as how tightly the fasteners are secured. For Fig. 7.16, therefore, S may vary over a range:

$$0.0001\ L \leqslant S \leqslant 0.05\ L \tag{7.10}$$

The default value for S is:

$$S = 0.3 \text{ percent of L or } 0.003\ L \tag{7.11}$$

Incidentally, notice that this default value corresponds to a constant fatness term of 19 dB. Lower values of S (0.0001 L to 0.001 L) correspond to more rigid panel members, which are more tightly and frequently secured. Higher values of S (0.005 L to 0.05 L) correspond to thinner panel members, especially if made of plastic with large spacings between securing fasteners. The default value suggested for S (without an electrical gasket) may depart readily from the real value, which can be measured with a feeler gauge. For instance, a 1.5 mm thick steel cover secured every 100 mm will exhibit a typical deflection of 0.6 mm in the middle of the seam, i.e., an **average** S = 1/2 × 0.6/100 = 0.3 percent.

Along with the above definitions for L and S, Eqs. (7.4) and (7.5) can be used to define the resulting aperture leakage when electrical gaskets are not used. When gaskets are used, the leakage becomes that rated by the gasket manufacturer when used in accordance with his recommendations, including minimum mating and closing pressures.

Illustrative Example 7.4: Compute the SE of mating panel members.

A five-sided box is opened at the top as shown in Fig. 7.16, where it is lapped over by 5 mm to accept screw fasteners for the cover which is 1 mm thick. Screws are located around the periphery every 15 cm on all four sides. The SE objective is 60 dB at 1 MHz and 40 dB at 1 GHz. Determine if this objective is achieved; if not, determine what to do.

From Eqs (7.4) and (7.11), at 1 MHz:

$$SE_{apt} = 97 - 20 \log (150) + 20 \log \left(1 + l_n \frac{150}{0.003 + 150}\right)$$

$$+ 0 + SE_{WGBCO}$$

$$= 97 - 43 + 16 + SE_{WGBCO} = 70 \text{ dB}$$

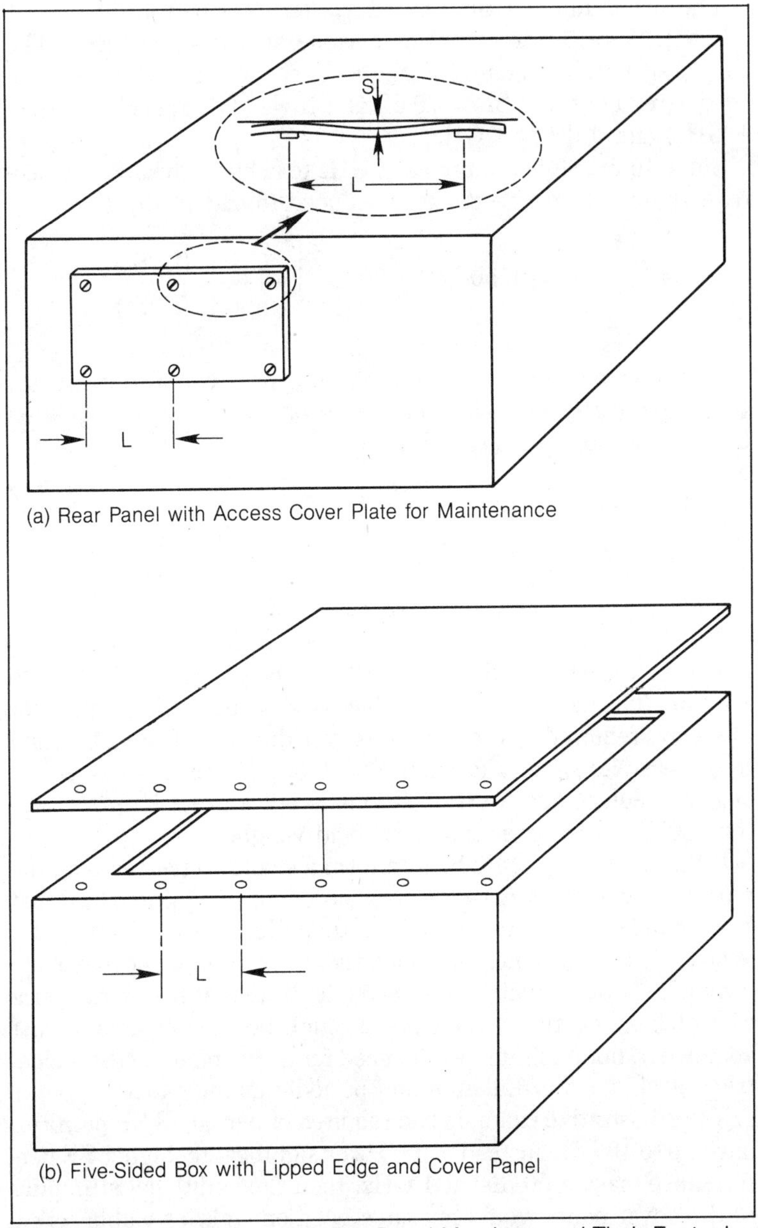

Figure 7.16—Examples of Mating Panel Members and Their Fastening Dimensions

The lower limit of SE, the SE_{shad} term, is 0 from Table 7.1 when L/W = 1 for complete peripheral seam apertures. The SE_{WGBCO} term, discussed in the next section, is assumed to be zero here. Thus, the SE = 70 dB at 1 MHz (3 decades lower than 1 GHz) and 0 dB at 1 GHz.

Since 40 dB SE is required at 1 GHz to achieve the above objective, the aperture size must be reduced to satisfy Eq. (7.6):

$$40 = 20 \log (150/L_a) - 20 \log \left[\frac{1 + l_n (L_b/S_b)}{1 + l_n (L_a/S_a)} \right]$$

As the screws get closer, the gap height S will also decrease, such that $L_b/S_b = L_a/S_a$ and the second term in the equation becomes 20 log 1 or zero. So:

$$40 = 20 \log \left(\frac{150}{L_a} \right)$$

$$\frac{150}{L_a} = 10^2 \text{ or } L_a = 1.5$$

Therefore, we would have to reduce the seam length L_{after} to 1.5 mm. This solution is not viable because of the large number of screws required. Increasing the cover thickness from 1 to 2 mm to increase its rigidity (assume $S = 0.001\ L_a$) results in a screw spacing requirement of 6 mm. This is not a great improvement for doubling the cover thickness (and weight).

If the cover edge were bent to form a shallow inverted pan, the stiffness would increase by an order of magnitude so that $S \approx 0.0001\ L_a$. For this situation, then, due to the creation of a WGBCO effect (see next section), the screw spacing becomes approximately 5 cm, which may be viable. Note that the extra operation of bending the cover edge through 90° increases the cost somewhat, but it eliminates the need for an even more costly electrical gasket, its installation and periodic maintenance.

If the illustrative example had required either 40 dB SE performance up to 10 GHz instead of 1 GHz or significantly higher SE performance (say, > 50 dB) at 1 GHz, then the additional stiffening of the cover panel would not have been enough for viable screw fastener spacing. One can then determine a general guideline for an upper electromagnetic ambient threat frequency of 1 GHz:

SE Requirement	Comments
⩽ 30 dB (up to 1 GHz)	Stiffen cover and avoid electrical gaskets.
30 - 50 dB (up to 1 GHz)	Twilight Region (stiffen and lap over lip flange discussed in next section).
50 - 60 dB (up to 1 GHz)	Same as 40 - 50 dB region but requires excessive number of screw fasteners and extremely rigid cover.
⩾ 60 dB (up to 1 GHz)	Use EMI gasket.

See Chapter 10 for a complete discussion of EMI gaskets.

7.2.5 Effects of Aperture Gap Depth

In Eq. (7.4), the WGBCO effect appeared as the SE_{WGBCO}. However, in situations examined so far, the aperture size was large compared to metal thickness, so waveguide ("tunnel") attenuation did not exist. The shielding was almost entirely due to the reflection mismatch at the air-to-skin boundary between the wave impedance Z_w and the slot impedance Z_s (see Fig. 7.17). There was no additional through attenuation.

In contrast, when the depth, d, of the opening approaches or exceeds its largest dimension, WGBCO effect takes place. The term **beyond cutoff** means that a tube can only perform as a waveguide if the incident wavelength is less than a critical value, λ_c, corresponding to half the aperture width.

For longer wavelengths, i.e., lower frequencies, the guide practically blocks the transmission of the electromagnetic energy. The transition between the waveguide propagation mode and the waveguide attenuation is an abrupt function such that below one-third of its cutoff frequency, the attenuation of the tube becomes practically steady, as seen in Fig. 7.17. The attenuation of a deep hole is given by:

$$A_{dB} = 0.0018\, df_x \sqrt{\left(\frac{f_c}{f_x}\right)^2 - 1} \tag{7.12}$$

$$= 0.0018\, df_x \sqrt{\left(\frac{1.5 \times 10^4}{g\, f_x}\right)^2 - 1}$$

where,

d = hole depth in cm
g = largest transverse dimension of hole in centimeters
f_x = EMI frequency in megahertz
f_c = hole $\lambda/2$ cut-off frequency, in MHz = 15,000/g

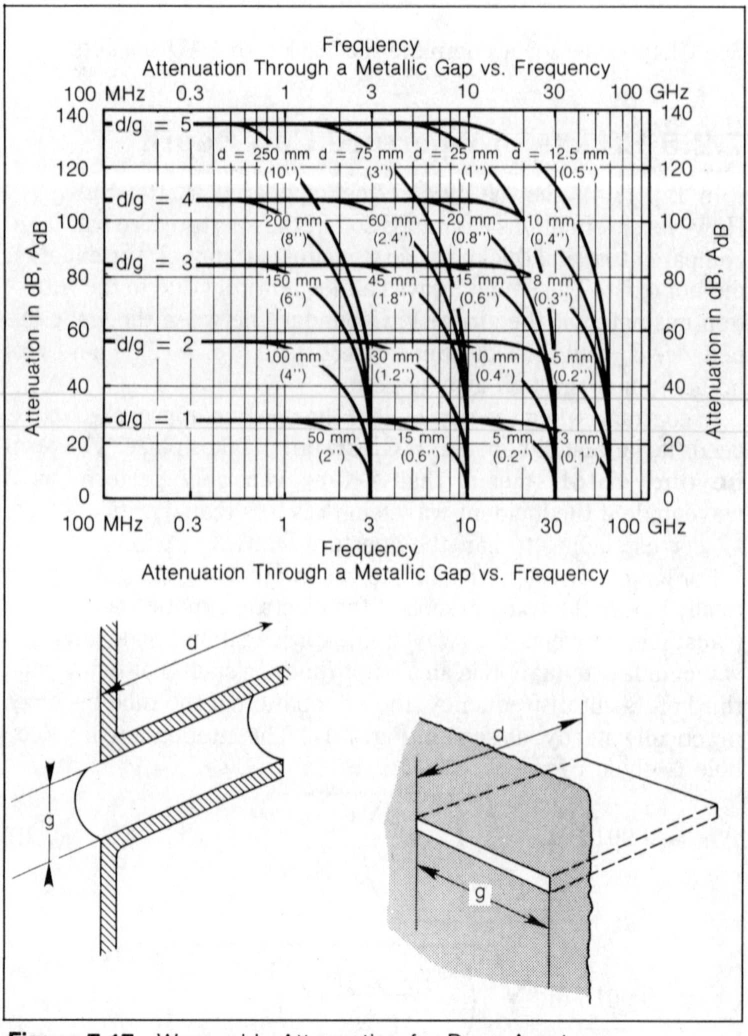

Figure 7.17—Waveguide Attenuation for Deep Apertures

or in English units:

$$A_{dB} = 0.0046\ df_x\sqrt{(f_c/f_x^2) - 1} \quad \text{for d in inches} \qquad (7.12a)$$

This equation is plotted in Fig. 7.16 for d in inches for several gap depths and widths. When $f_c/f_x > 3$ (corresponding to a 5 percent error), Eq. (7.12) reduces to:

$$A_{dB} \approx 27.3\ d/g \text{ for square or rectangular holes} \qquad (7.13)$$

$$A_{dB} = 32\ d/g \text{ for round holes} \qquad (7.14)$$

This characteristic is used advantageously in many applications such as artificially lengthening the depth of a seam by widening and bending the edges of a flange, using honeycomb structures for air vents, brazing a tube for cable entries, etc.

Figure 7.18 shows two examples of apertures whose depth, d, may be comparable to the larger dimension of the aperture: (1) a lapped box/case lip and lid secured by screws or other fasteners, and (2) convection cooling slots or holes in a ribbed, metalized plastic box. When either of these types of apertures is used, we include a WGBCO factor in addition to the SE provided solely by reflection. To simplify further calculations, an average value has been retained which is half-way between the WGBCO of a round hole and that of a square hole:

$$S_{(WGBCO)dB} \approx 30 \text{ depth/width} \qquad (7.15)$$

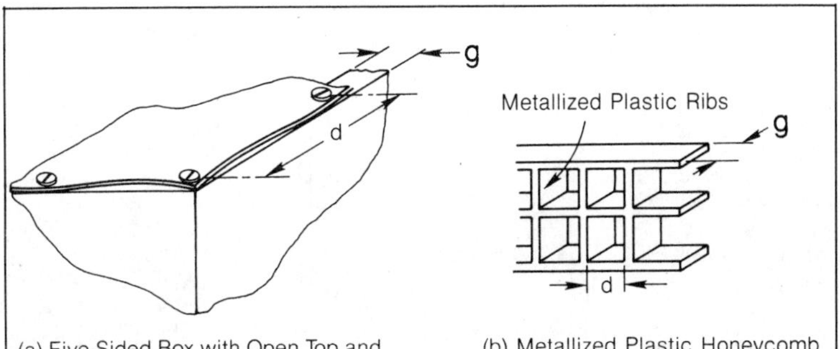

(a) Five-Sided Box with Open Top and Lipped Flange to Accept Cover Plate

(b) Metallized Plastic Honeycomb Having Square Apertures

Figure 7.18—Application of Waveguide-Beyond-Cutoff in Aperture Leakages

This approximation is acceptable for $d > g$ and for $d \leqslant \lambda_{EMI}/6$. It is the approximation made in Eq. (7.4).

Illustrative Example 7.5: Compute the SE of a feed-through hole.

Figure 7.19 shows a 12.7 mm hole in a thin housing, originally intended for mounting a fiber optic plastic connector. Assuming that metal skin always provides more than 100 dB of SE and that the box SE shadow is 3 dB, ensure that at least 40 dB of SE still exist at 1 GHz. Using Eq. (7.11):

$$SE_{dB} = 97 - 20 \log 12.7\ 20 \log f_{MHz} + 3$$

$$= 78 - 20 \log f_{MHz} = 18 \text{ dB @ 1 GHz}$$

Replacing the plastic connector with a metallic fiber optic fitting or connector will create a WGBCO effect, provided the metal of the tube makes good contact with the box skin (tightly screwed,

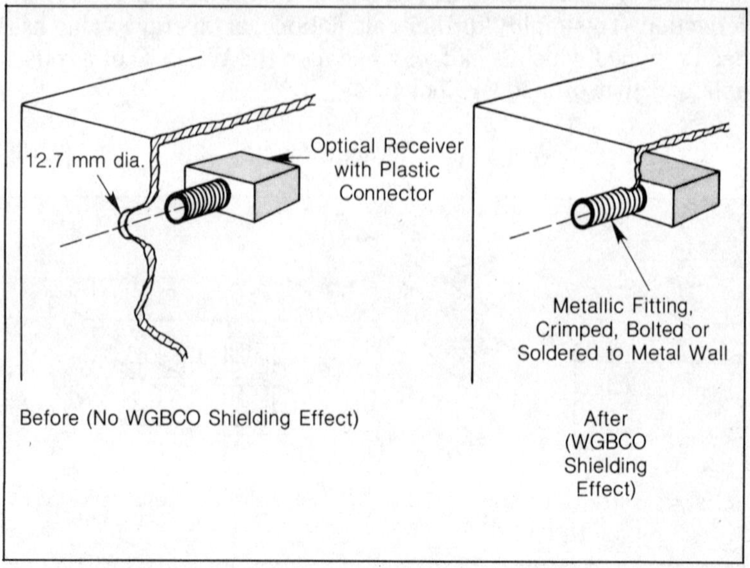

Figure 7.19—Taking Advantage of Waveguide Attenuation for a Feed-Through Hole

crimped or soldered). Assuming the connector length is 25 mm and the inner diameter is 8 mm, the new SE at 1 GHz is:

$$SE_{dB} = 100 - 20 \log 8 - 20 \log f_{MHz} + 30 \times 25/8$$

$$= 115 \text{ dB}$$

The new SE is equal to or better than the metal skin, so the lesser number will dictate the SE of the whole box.

We must also check that 1 GHz is still below the cutoff frequency of the 8 mm hole. This cutoff frequency is:

$$F_{co} = \frac{15{,}000}{g_{cm}} = 18{,}750 \text{ MHz or } 18.7 \text{ GHz}$$

7.3 Cumulative Leakages of Multiple Apertures

A question often arises as to whether, in an array of N apertures, the leakages of all apertures are cumulative. Literature dealing with slot antenna arrays and the like provide elaborate math models for slotted plates. However, they generally address the intentional radiation of slots rather than SE. The complexity of the equations and the number of boundary assumptions make them inappropriate for the quick approximations that the EMC engineer needs.

Common sense suggests that an array of identical slots radiate more than a single slot since each aperture is a small antenna. However, this seems to contradict the following observation that: if N identical holes radiate N times more than one, a wire mesh would provide no improvement since breaking a large aperture into N apertures (each one being just N times smaller) would produce no benefit. Since experience shows that screen meshes work, a deeper examination of this phenomenon is in order.

7.3.1 Contiguous Apertures in a Thin Wall

As was briefly discussed in Section 2.1.5.1, when identical holes are separated by a narrow gap and the metal wall is thin (no through attenuation), the resultant fields at the inner edges of adjacent holes cancel each other since they are created by opposing currents (Fig.

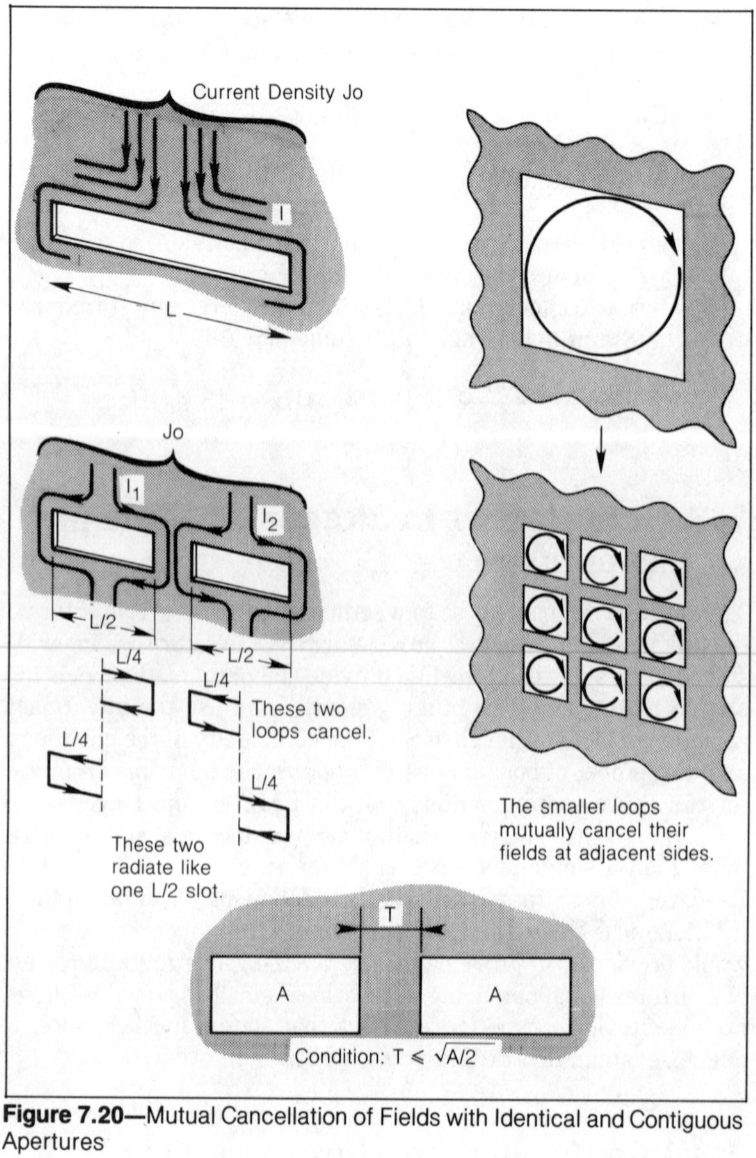

Figure 7.20—Mutual Cancellation of Fields with Identical and Contiguous Apertures

7.20). Because there is no attenuation through the shallow aperture, any cancellation that takes place on the outboard side of the slot also occurs on the inboard side.

As a result, the SE performance of the array is equal to that of a single element alone. The width of the metal rib which separates the two slots, beyond which this neutralization ceases to exist, is approximately:

$$T = \sqrt{A/2}$$

$$= 0.7 \text{ g for square apertures}$$

where, A is the area of one slot

Therefore, any array of same holes in a thin plate will exhibit:

1. The SE equal to that of one hole if $T \leqslant \sqrt{A/2}$, i.e. mutual cancellation exists
2. The SE equal to that of one hole minus 20 log N if $T \geqslant \sqrt{A/2}$, i.e., mutual cancellation does not exist because the holes are not contiguous

7.3.2 Several Deep Apertures

If either the metal is thick or the gap depth artificially increased (overlapping edges, feed-through tube, etc.), we know from Section 7.2.5 that both reflection **and** through-loss (WGBCO) exist. If there is a waveguide effect (especially with an array, the number of coherent sources will degrade the overall SE since there will be a vectorial addition of the field from each individual element. This exists as long as the separation between adjacent elements does not exceed λ/2, where periodic vectorial subtraction (destructive interference due to opposite phase combinations) will appear. Therefore, the SE of an array of N deep holes will be:

$$SE_{TOT} = (97 - 20 \log g - 20 \log F) + (30\, d/g - 20 \log N) \tag{7.16}$$

In the equation, the first element in parentheses represents the reflection mechanism and the second is absorption. Absorption must be ⩾ 0 dB; it cannot be negative.

To a certain extent, the absorption is not affected by near-field or far-field conditions. The limitations are explained next.

7.3.3 Contribution of the Number of Holes When the Victim (or the Source) Is in the Near-Field of the Array

In the above example, the effect of multiple holes was found to be −20 log N. This factor is accurate if all the holes behave as coherent, synchronous sources and if both the emission source and the victim antenna are in the far field with respect to the perforated barrier. Usually, **at least** one of the two reference points (the source or the victim) is within the housing, so it will be in the Fresnel (near-field) region of the array of holes. In this case, the wave front of the impinging field is curved and three factors contribute in reducing the cumulative effect of the N holes:

1. The phase shift
2. The distance gradient
3. The vectorial addition (polarization angle)

These will be discussed in turn.

7.3.3.1 The Phase Shift

When the observation point is at an infinite distance (Fig. 7.21a), the E-field impinging on the array of holes is:

$$E_{TOT} \text{ (for the array)} = \Sigma E \text{ (1 slot)}$$

However, in the near field (Fig. 7.21b), the augmented path length, Δ, causing a phase lag is:

$$\Delta = \sqrt{r^2 + (h/2)^2} - r$$

where,

r = distance between the victim and the middle of the array
h = distance from the middle of the array to the outermost edge of the array

Therefore, for the N holes of an array, the total received field at the victim, if it were for phase lag only, would be:

$$E_{TOT} = \int_{i=1}^{i=N} E_o \cos\left(\frac{\Delta i}{\lambda} \times 2\pi\right) \tag{7.17}$$

where,

E_o = field due to one slot only, oriented in the plane of the array holes

Δ = additional path length = $\sqrt{r^2 + h_1/2} - r$

h_1 = incremental distance from middle of array to the ith slot

Integrating Eq. (7.17) over a vertical array of N equidistant slots gives the following results as a function of the largest value of Δ_N (i.e. kλ) which is the increase in path length to the outermost edge of the array:

for $\Delta_N = \lambda/8$ $\quad E_{TOT} = E_O + 20 \log N - 0.3$ dB

for $\Delta_N = \lambda/4$ $\quad E_{TOT} = E_O + 20 \log N - 1$ dB

(corresponding to $r = \dfrac{2h^2}{\lambda} N$)

for $\Delta_N = \lambda/2$ $\quad E_{TOT} = E_O + 20 \log N - 4$ dB

If we fix 1 dB as the maximum objective for the pessimistic error, we see that using 20 log N to compensate for a number of holes remains an acceptable assumption within the following conditions:

Frequency	100 MHz	300 MHz	1 GHz	3 GHz	10 GHz
λ in m	3	1	0.3	0.1	0.03
h_N = 5 cm, r_{min} =	0.15 cm	0.5 cm	1.5 cm	5 cm	15.0 cm
h_N = 15 cm, r_{min} =	1.50 cm	5.0 cm	15.0 cm	50 cm	1.5 m
h_N = 50 cm, r_{min} =	15.00 cm	50.0 cm	1.5 m	5 m	15.0 m

r_{min} = minimum source-array separation

In other words, if an array of holes has a dimension of 15 cm, provided the victim is at least at 5 cm of the array, the 20 log N adding factor provides an estimate (with less than 1 dB error) of the combined effect of the N holes, up to 300 MHz. Therefore, it can be concluded that in most practical cases, the phase lag is not a significant reduction factor.

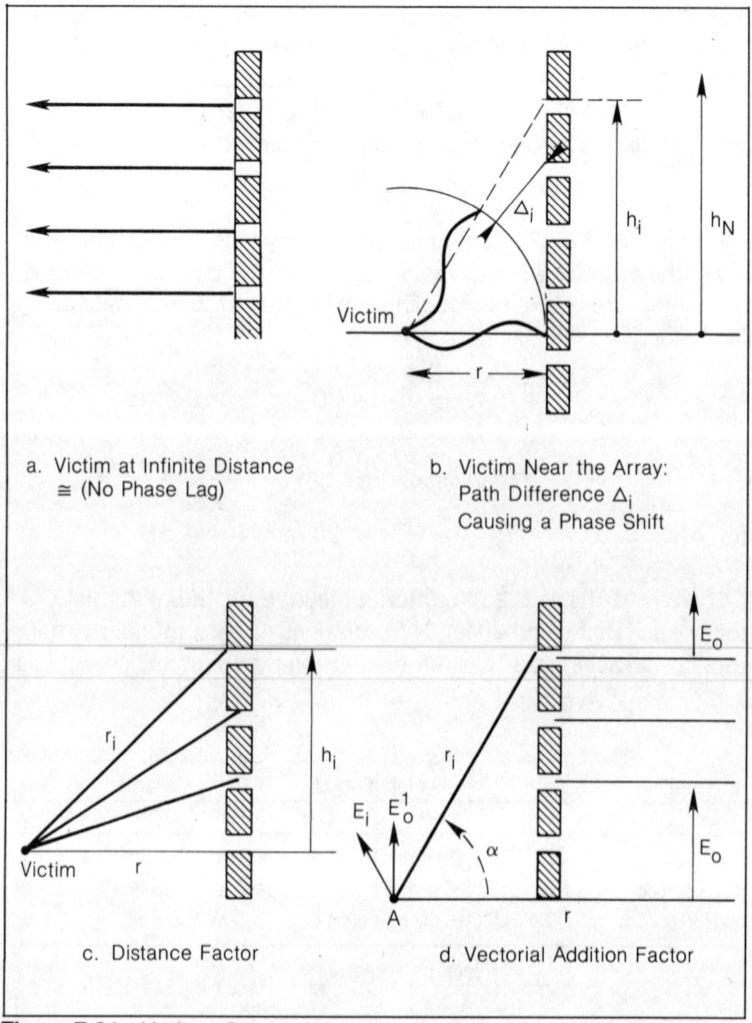

Figure 7.21—Various Contributors to the Additive Effect of N Holes in Line

7.3.3.2 The Distance Gradient

This field strength error (Fig. 7.21c) is due to the fact that the radiation received from the i_{th} slot differs from the radiation received from the nearest slot (i = 1) by a factor equal to:

$$20 \log (r/r_i)^P$$

with,

p = 1 when the victim is in the far field (the amplitude of the far field is proportional to r^{-1})

p = 3 when the victim is in the near field of all the array (the amplitude of the field is proportional to r^{-3} according to Maxwell's equations)

For instance, if $r = h_N$, radiation from the most remote slot will reach the victim with a magnitude reduced by 3 dB (compared to the nearest slot) for far-field conditions, and decreased by 9 dB for near-field conditions. Of course this factor for all slots in between will be less than the reduction in radiation from the most remote slot.

7.3.3.3 The Vectorial Addition (Polarization Angle)

This last factor is easy to understand when looking at Fig. 7.21d. The contribution of the field vector E_i to the vertically polarized field E'_O received by the victim is simply equal to $E'_O \cos \alpha$, where E'_O = field at point A due to E_O axially illuminating the nearest slot. The vectorial factor applies to E– or H–fields, but not to both field components simultaneously because:

$$\text{cox}\ \alpha = r/r_i$$

$$E_i = E'_O \frac{r}{\sqrt{r^2 + (h_i/2)^2}} \qquad (7.18)$$

$$\text{and } E_{TOT} = \sum_{i=0}^{i=N} E_i$$

7.3.3.4 Some Practical Results

When combining the contribution of Figs. 7.21c and 7.21d (given that phase shift is negligible) and entering some typical values for r, N and h, the curve of Fig. 7.22 results. The vertical axis corresponds to the degradation of SE in dB due to N holes relative to the degradation due to one individual hole. The curves illustrate the fact that for a victim-to-array distance r larger than the full array dimension D, the degradation factor is close to 20 log N. Conversely, for closer distances, like r = 0.1 D, the degradation factor is significantly less, approaching 10 log N (i.e., proportional to $\sqrt{N}$) for a large number of holes. This probably explains why some data from actual SE measurements of perforated barriers show an N dependency while others indicate a $\sqrt{N}$ dependency.

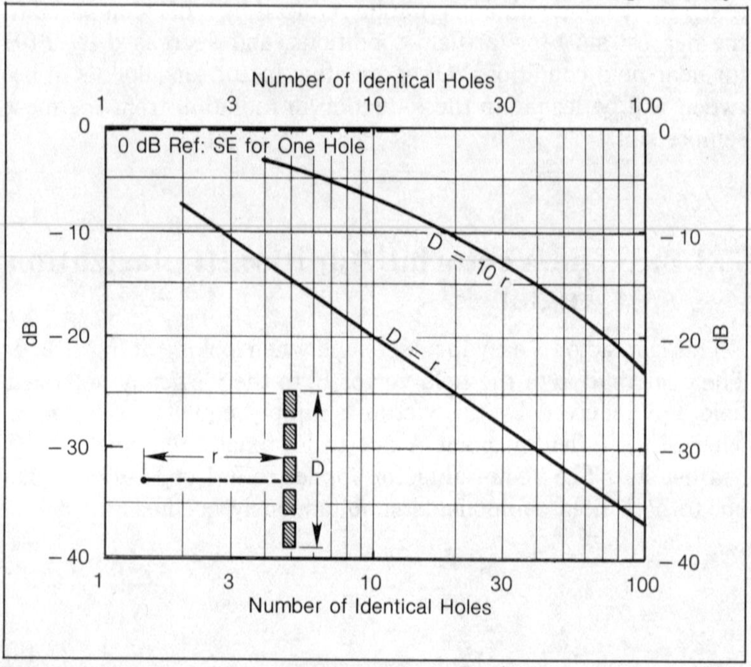

Figure 7.22—Total Degradation Caused by N Idential Holes vs. the SE of a Single Hole

7.3.4 Summary of Recommendations for Simplified Calculation of Leakage from N Apertures

Although the problem is not trivial and an infinite number of particular cases may exist, the following can be recommended at least to obtain a satisfactory prediction or, at the worst, result in a conservative design (see Fig. 7.23):

1. Thin metal barrier (no WGBCO effect), identical and contiguous holes:

$$SE_{(Nholes)} \equiv SE_{(1hole)}$$

2. Thin metal barrier (no WGBCO), identical noncontiguous holes:

$$\begin{aligned} SE_{(Nholes)} &= SE_{(1hole)} - 20 \log N, \text{ if } r \geqslant D \\ &= (SE_{(1)} - 20 \log N) < SE_N < SE_{(1)}, \text{ if } r < D \end{aligned}$$

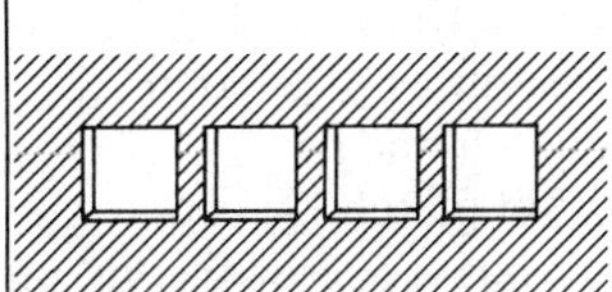

1. Multiple, Contiguous Apertures, Thin Wall: $SE_{TOT} = SE_{1\ hole}$

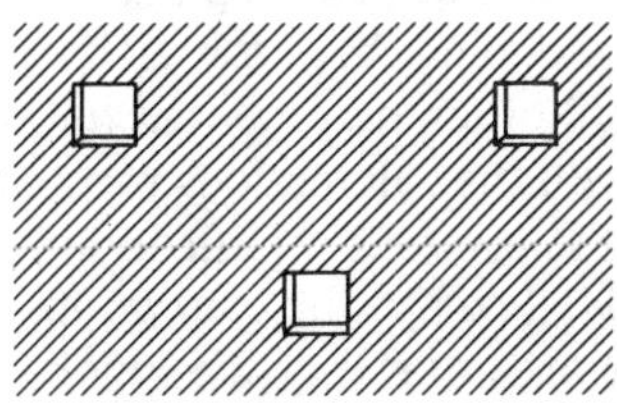

2. Similar Apertures, Noncontiguous: $SE_{TOT} \cong SE_{1hole} - 20 \log N$ (Worst Case). Cannot be < 0 dB

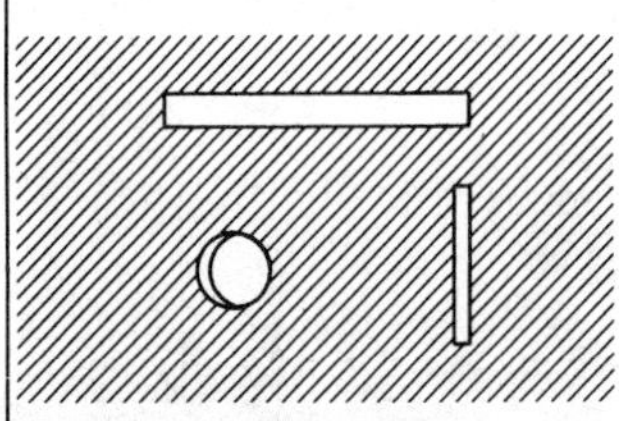

3. Dissimilar Apertures: $SE_{TOT} \leqslant$ SE of the Largest Aperture

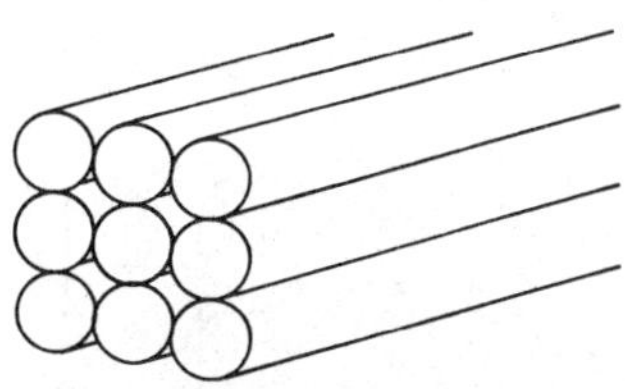

4. Similar Deep Apertures, Contiguous or Not: $SE_{TOT} \cong SE_{1\ hole} - 20 \log N$ $= \text{Refl. Loss}_{(1\ hole)} + \left[\text{Abs. Loss}_{(1\ hole)} - 20 \log N\right]$ (Always $\geqslant 0$ dB)

Figure 7.23—Summary of Rules for Multiple Apertures

3. Thick barrier or deep aperture ($d \geqslant g$):

$$SE_{(Nholes)} = SE_{(1hole)} - 20 \log N \text{ if } r \geqslant D$$
$$= (SE_{(1)} - 20 \log N) < SE_N < SE_1, \text{ if } r \leqslant D$$

4. Any barrier, dissimilar holes:

Take the SE of the largest aperture, or combine if both SEs are nearly equal.

An additional simplification is suggested in Fig. 7.24, where the hole array area is compared to the actual victim intercept area. Figure 7.24a corresponds to the assumption made so far of a point-like victim. Figure 7.24b corresponds to a victim intercept area equal to the array, and Fig. 7.24c corresponds to a victim intercept area larger than the array. For situations b and c, the problem could be treated by a method of finite areas where a summation would be made of the coupling of any single aperture A_i of the array to all elements V_j of the victim's area, successively. This summation would then be repeated for each aperture A_i from $i = 1$ to $i = N$. Given the calculations presented at the beginning of Section 7.3.3, the following simplification is recommended for an array of similar holes (except for contiguous holes in a thin plate): Consider only the area of the array which is within a radius $3 \times r$ of the victim (or source) location and take the reduced number N_r of holes in this perimeter to calculate the $-20 \log N$ correction.

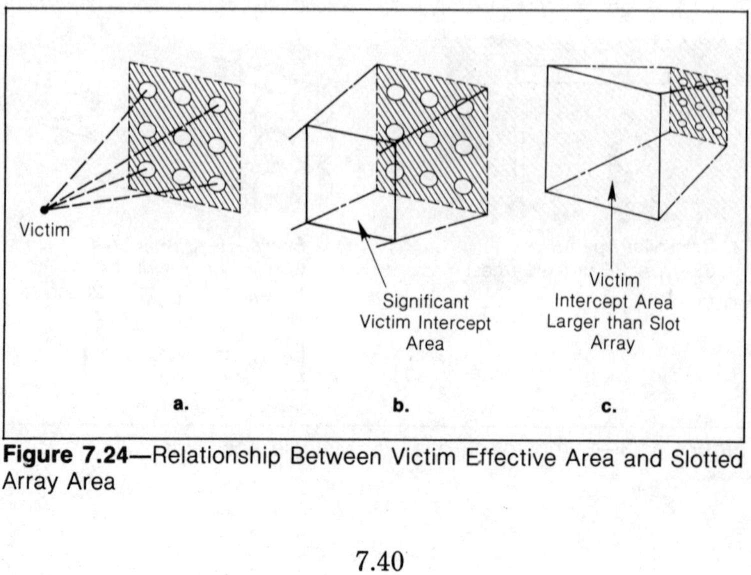

Figure 7.24—Relationship Between Victim Effective Area and Slotted Array Area

Illustrative Example 7.6: Compute the SE for convection cooling slots.

Fig. 7.25 shows the convection cooling slots in the top of a metalized plastic box offering a skin SE_{dB} of 50 dB. Determine the SE up to 10 GHz and ensure that an SE of at least 40 dB will be provided. Note that worst-case polarization will exist for horizontal E-fields. From Eq. (7.4), the SE at 1 GHz is:

$$SE_{apt} = 97 - 20 \log (80 \times 10^3) + 20 \log (1 + l_n L/S) + SE_{shad}$$

$$= 97 - 98 + 12 + 3 = 14 \text{ dB}$$

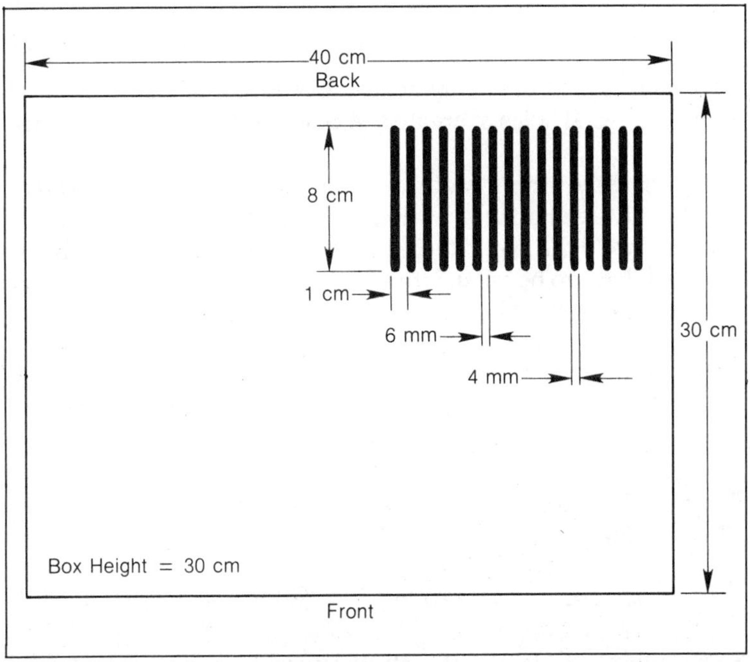

Figure 7.25—Top of Metalized Plastic Box with Convection Cooling Slots

The lower limit, up to $l = 1.5\lambda$, is 3 dB. Above this, SE = 0 dB. Since the slots are close and the ribs are narrow, the condition for contiguous slots is met and the N holes do not add up, since: T = 6 mm, while $\sqrt{A/2} = 12.6$ mm, so $T < \sqrt{A/2}$. Thus, the cooling slots fail at both 1 GHz and 10 GHz.

Assuming *a priori* that we would break the array into small square apertures, the L/S fatness would become 0 dB and the required objective can be expressed as:

$$40 \text{ dB (objective)} = 97 - 20 \log (\text{Lf})$$

So at the highest frequency of interest, 10 GHz:

$$20 \log (\text{L} \times 10^4) = 57 \text{ dB}$$

$$20 \log \text{L} = 57 - 80 = -23 \text{ dB}$$

$$\text{L} = 10^{-23/20} = 0.07 \text{ mm}$$

This value is, of course, impractical. Let us now increase the depth of the metalized holes to create a WGBCO effect, as shown in Fig. 4.18b.

Assuming the former width, S, of the slot will be kept, if we break the slot into even 4 mm × 4 mm squares, the WGBCO alone will account for 30 d/L, i.e., 30 d mm/4 mm. However, the number of holes will have to be taken into account since the criterion for contiguous holes in a thin wall is no longer satisfied. So, finally:

$$\text{SE} = 97 + 3 - 20 \log 4 - 20 \log \text{F} + \underbrace{30 \times \text{d}/4 - 20 \log \text{N}}_{\geqslant 0 \text{ dB}}$$

Breaking the array into regular holes will result in about 10 × 16 holes so:

$$\text{SE} = 88 - 20 \log \text{F} + (30 \text{ d}/4 - 44)$$

Solving for 10 GHz and the 40 dB objective produces:

$$40 = 88 - 80 + (30 \text{ d}/4 - 44)$$

$$32 = 30 \text{ d}/4 - 44$$

or $$\text{d} = \frac{76 \times 4}{30} = 10 \text{ mm}$$

This attenuation will hold up at frequencies for which d = λ/6 (limit of validity for the WGBCO simplified formula). This wavelength is:

$$\lambda = 6d = 6 \times 4 \text{ mm} = 24 \text{ mm}$$

which corresponds to a frequency of:

$$f_{MHz} = 300 \times 10^3/24 = 12.5 \text{ GHz}$$

Thus, a 10 × 16 array of 4 mm × 4 mm convection cooling holes having a depth of 10 mm will produce the required 40 dB at 10 GHz.

7.4 References

1. *Reference Data for Radio Engineers*, 6th edition (Indianapolis: Howard Sams & Co., Inc.).

7.5 Bibliography

Boothman, G., "Designing Cabinets for Shielding," *1982 IEEE EMC Symposium Record* (New York: IEEE), Santa Clara, 1982.

Chou and Adams, "Coupling through Slots," *IEEE Transactions on EMC*, May 1977.

Cowdell, R., "Simplified Computation of Perforated Shields," *1968 IEEE EMC Symposium Record* (New York: IEEE).

Hoeft, L., "H Field Shielding of Generic Equipment Cases," *1982 IEEE EMC Symposium Record* (New York: IEEE), Santa Clara, 1982.

Hoeft, L., "How Big a Hole in an Enclosure," *1986 IEEE EMC Symposium Record* (New York: IEEE).

Honig, "EM Coupling through Slots," *IEEE Transactions on EMC*, November 1977.

Jarva, W., Mirsky, M., *"Development of Shielding Effective Test Methods,"* Filtron Report FR-1351-10, November 1967.

Kadden, H., "Wirbelstrome and Schirmung in der Nachrichtentechnik," published by Springer, Berlin, 1959.

Madle, J.P., "Introduction to Field Penetration," *1978 IEEE EMC Symposium Record* (New York: IEEE).

MIL Handbook 419, Chapter 5.

Quine, J.P., "Theoretical Formulas for Calculating S.E. of Perforated Sheets and Wire Mesh," Third Conference on Radio Interference-Armour Research Foundation, February 1957, p. 315-329.

Rostek, P., "Techniques of Shielding," *1974 IEEE EMC Symposium Record* (New York: IEEE).

"S.E. of Apertures," *IEEE Transactions on EMC*, February 1986.

Chapter 8

Examples of Shielding Design and Retrofit

Previous chapters established shielding design methodology and procedures. Those chapters detailed the three main building blocks of shielding design:

1. Determining SE requirements and objectives
2. Achieving SE of box skin
3. Identifying and controlling aperture leakages

This chapter puts the previously covered concepts together in the form of examples of shielding design and retrofit. The results will be compiled on the worksheet used in the previous chapters.

8.1 Susceptibility Example: Initial Performance

Figure 8.1 shows a typical box requiring sufficient shielding to permit EMI-free operations when exposed to EMI radiations at levels recommended by CBEMA, SAE AE-4 or other standards.[1-2] For instance:

10 kHz – 25 MHz: 1 V/m (120 dB μV/m)
25 MHz – 1 GHz: 10 V/m (140 dB μV/m)

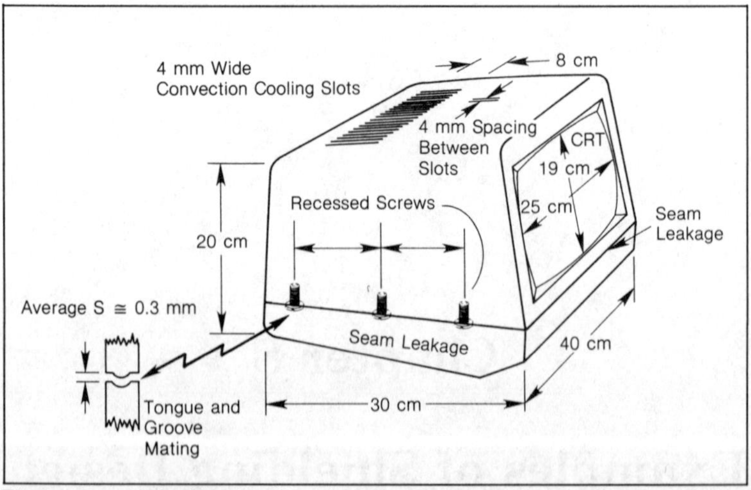

Figure 8.1—Video Display Terminal Showing Three Leaky Aperture Types

Three aspects of shielding will be addressed in filling out the worksheet in Fig. 8.2:

1. Determining the required SE
2. Indicating the basic box or case material selection and design
3. Controlling three EMI aperture leakages in the box:
   a. Leakage from a CRT window
   b. Leakage from convection cooling slots
   c. Leakage from mating of box cover with base assembly

8.1.1 Determining SE Requirements

Chapter 5 summarized the technique for determining the required SE (also see blocks 2 to 16, Fig. 5.1). The specified 1 and 10 V/m fields are coupled to the equipment's most sensitive analog and logic circuits via the PCB traces and discrete input wires going to and from the input/output (I/O) port. To determine the required SE, no box shielding is used at first. The most sensitive analog circuit in Illustrative Example 5.4 is a disk-read amplifier of 5 MHz

Figure 8.2—Worksheet for Recording SE Performance Objectives

Topic Identification	EMI Fix	10 kHz	100 kHz	1 MHz	10 MHz	100 MHz	1 GHz	10 GHz
A. SE Objectives (1 + 2 − 3 + 4)		0	0	8	22	42	34	20
1. EM Ambients (dBμV/m)		120	120	120	120	140	140	140
2. Field-to-Cable/PCB Coupling (dB)		−120	−100	−80	−60	−40	−28	−28
3. Sensitivity or NIL (dBμV)		42	42	42	48	68	88	102
4. Design Margin (dB)		10	10	10	10	10	10	10
5. Box SE Modified Objectives								
6. Leakage SE Modified Objectives								
B. SE Box Performance		117	97	77	57	43	46	61
C. Box Leakages*		92	72	52	32	12	0	0
1. Viewing Window Apertures								
1.1 CRT Displays		93	73	53	33	13	3	1
1.2 Alphanumeric Displays								
1.3 Panel Meter Displays								
1.4 Status Indicators								
2. Cooling Aperatures								
2.1 Cooling Apertures		115	95	75	55	35	15	11
2.2 Forced-Air Cooling								
3. Assembly and Maintenance Seams								
3.1 Box Housing Assembly		120	100	80	60	40	20	0
3.2 Cover Panels								
4. Component Aperatures								
4.1								
4.2								
D. Total SE* Performance		92	72	52	32	12	0	0
E. Performance vs. Objectives (D − A)		92	72	44	10	−30	−34	−20

$$^{*}SE = -20 \log \left[\log^{-1}(-SE_{BOX}/20) + \sum_{i=1}^{n} \log^{-1}(-\text{leakages}/20) \right]$$

bandwidth having an in-band sensitivity of 0.12 mV (42 dBμV), and the highest speed logic family (TTL) having a noise immunity level of 400 mV (112 dBμV).

From the calculations in Table 5.6, Fig. 8.2 shows that the greatest SE objectives exist at 100 MHz (see line A). This 42 dB SE also includes a 10 dB design margin at this frequency (line A.4).

8.1.2 Determining Box Skin SE

Chapter 6 presented some details on selecting a suitable metalizing process, metalized composites, or sheet metal for equipment cases. Whatever the approach, the SE performance (line B, Fig. 8.2) must equal or exceed that determined for the SE requirements in line A.

As described in Chapter 6, the basic SE of sheet metal boxes (with no leakages) is well in excess of 100 dB—more than enough to meet the worst-case requirements (e.g., EMP threat). For all other nonhomogenous cases, however, the SE achieved is a function of the metal type, the metalizing process, its deposition thickness, and homogeneity (size, length and percentage loading of metal particles or fibers for composites).

From Illustrative Examples 6.2 and 6.3, where a deposited metal surface impedance of 0.7 Ω/sq on a plastic surface was calculated, Fig. 8.2 (line B) shows that the worst-case SE box case performance (43 dB) is at 100 MHz, where it barely meets the overall SE requirements (42 dB on line A).

8.1.3 Determining SE Box Leakage

Lines C.1 through C.3.2 in Fig. 8.2 identify several different types of aperture leakages. In general, leakages listed earlier tend to be the more important aperture leakages. Their importance results from the observation made in Chapter 7 that the larger the single aperture area, the greater the leakage; i.e., the SE of the leakage has a lower dB value. Also, any aperture attenuation in dB can never be higher in value than the SE_{dB} of the basic box skin material.

As will be shown, there is no guarantee that even if each aperture leakage in dB exceeds the overall SE objective at any frequency, the combination of leakages will meet the overall objective. For example, from Eq. (8.1), assume that each aperture leakage is identical. They will combine (worst-case) to produce an overall aperture leakage:

$$SE = -20 \log \left[\sum_{i=1}^{n} \log^{-1} (-L_{idB}/20) \right] \tag{8.1}$$

$$= L_{idB} - 20 \log N \text{ for } L_i = L_n \tag{8.2}$$

where,

SE = Shielding effectiveness of all apertures
L_{idB} = any single aperture attenuation in dB
N = number of other noncontiguous apertures

In N above, apertures in an array are excluded. Noncontiguous means that the edge separation distance is at least one-half the larger aperture dimension (see Fig. 8.3)

Thus, for up to 10 equal but distinct aperture leakages, for example, the combined attenuation leakage will be below that of any single aperture by an amount shown in Table 8.1.

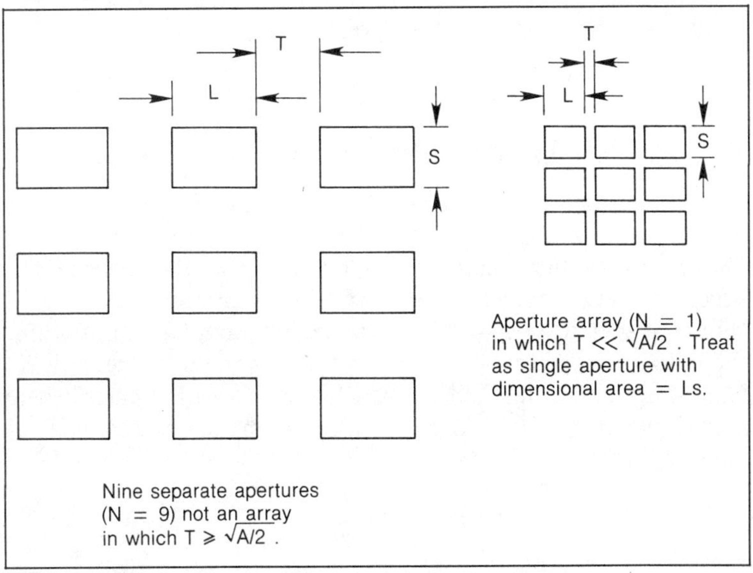

Figure 8.3—Criteria for Distinguishing Between Discrete Apertures in Defining N Form in Eq. (8.1)

Table 8.1—Combined Leakage of Several Equal, Noncontiguous Apertures

Number of Apertures	Degradation Due to Combination Leakage
2	6 dB
3	10 dB
4	12 dB
5	14 dB
7	17 dB
10	20 dB

When designing an overall shielded box for the first computation run or iteration process (see blocks 20 through 25 in Fig. 4.2), it is neither necessary nor desirable that Eq. (8.1) be used in advance to calculate the required individual SE for each leakage. Usually only one or two leakages tend to dominate; to prevent EMC overdesign, it is best to defer any use of Eq. (8.1) until the end of the first run.

Another obvious simplification exists when there are only a few leakages which are distinct and unequal. If, in a given column of the Fig. 8.2 worksheet, a given leakage predominates over the others (i.e., less SE in dB) by at least 20 dB, only this leakage needs to be considered in the first calculation of combined SE. If the margin is only 10 dB, the others can be neglected, but this will result in an approximate 2 dB error.

8.1.3.1 SE Box Leakage for CRT Displays

Based on the computations from Illustrative Example 7.4, the leakage from a large CRT aperture gets progressively greater (lower in dB) with an increase in frequency, since the aperture size becomes a larger percentage of a wavelength. Figures 8.2 and 7.15 show the aperture leakage performance without any attempt (for the initial run or iteration) to EMI-suppress the rectangular hole. Here it is seen on line C.1 that the CRT aperture offers nearly zero attenuation at 1 GHz (it becomes 3 dB at and above 250 MHz as obtained from Fig. 7.15).

8.1.3.2 SE Leakage for Convection Cooling Slots

Illustrative Example 7.6 shows aperture leakage from convection cooling holes and slots. When applied to the illustrative example in Fig. 8.1, which involves twelve 8 cm × 4 mm slots, the leakage on line C.2.1 of Fig. 8.2 is shown. It is also observed that the leakage gets worse (lower SE) as the frequency increases and that it is less than the overall SE requirements at frequencies above 100 MHz (200 MHz by interpolation). Again, remember that no attempt is made during the first run to reduce any slot leakage, since this will be done on second or subsequent iterations.

8.1.3.3 SE Leakage for Cover Plates, Mating Members and Seams

As shown in Fig. 8.1, the third aperture leakage comes from a horizontal seam formed by the mating metalized members of a box and its cover or pedestal base. Where screws or fasteners are used, the principal leakage dimension is the distance between the screws or fasteners. Separation (assuming there are no electrical gaskets) between securing screws and fasteners can vary widely depending on the rigidity of both secured members, their finish and other variables. This gap, however, is less sensitive than the fastener separation.

Based on the dimensions in Fig. 8.1 and the calculations in Illustrative Example 7.5, the leakage due to mating edges with no EMI fix is shown on line C.3.1 of Fig. 8.2.

8.1.4 Combined SE Performance

Once all three aperture leakages in Fig. 8.2 have been computed, their values in lines C.1.1, C.2.1 and C.3.1 may be combined in accordance with Eq. (8.3), which is also shown at the bottom of Fig. 8.2, to produce the overall leakage entry shown in Fig. 8.2, line C.

$$SE = -20 \log \left[\log^{-1}(-SE_{BOX}/20) + \sum_{i=1}^{n} \log^{-1}(-\text{leakages}/20)\right] \quad (8.3)$$

Similarly, both the SE box performance (line B) and the total leakage performance (line C) are combined in accordance with Eq. (8.3) to produce the total SE performance appearing in line D. The surplus of SE performance, or its deficiency in meeting SE objectives (line A), is entered on line E (line E = line D – line A).

No dB entry on line E should be negative, and the least positive entry should not be greater than 10 dB to prevent box overdesign. Figure 8.2 shows that first-run box design performance starts to fail (line E becomes negative) at 100 MHz, i.e., at some frequency between 10 and 100 MHz (16 MHz by interpolation) and continues to fail up to 10 GHz. The worst leakage frequency (most negative number on line E) exists between 100 MHz (–30 dB) and 1 GHz (–34 dB).

Also note that the worst aperture leakage (CRT display on line C.1.1) will require at least 31 dB of EMI suppression (SE objectives – performance = lines A – C.1.1 at 1 GHz). In fact, all three leakages (lines C.1.1, C.2.1 and C.3.1) fail, or nearly fail, the SE objectives (line A) between 100 MHz and 10 GHz.

8.2 Strategy for Adjusted SE Performance for Box and Aperture Leakage Requirements

To ensure that individual box leakages will not collectively fail after the first EMI fix and during the second iterative run, use a modified version of Eq. (8.3). The new collective box leakages should not be greater than:

$$\sum_{i=1}^{3} \log^{-1}(-L_{idB}/20) = \log^{-1}(-SE/20) - \log^{-1}(-SE_{box}/20) \tag{8.4}$$

Thus, at the worst-case frequency of 100 MHz, Eq. (8.4) becomes (see Fig. 8.2, lines A and E):

$$\sum_{i=1}^{3} \log^{-1}(-L_{idB}/20) = \log^{-1}(-42/20) - \log^{-1}(-43/20)$$

$$= 0.00794 - 0.00708$$

$$= 0.00086$$

Thus, SE for combined leakage = $-20 \log (0.00086) = 61$ dB. However, since the SE requirements of each leakage or leakage combination cannot be greater than the integrity of the box-skin (43 dB), it is necessary to increase the box skin SE. This will also reduce the 61 dB leakage requirements which are generally more costly to control. To determine the box skin SE performance, consider the following: There are N leakages and one box for a total of N + 1 items contributing to overall SE. Determine the increase in SE of the box from the **overall** objectives (42 dB at 100 MHz per line A, Fig. 8.2). Then the new SE requirement for box SE performance becomes only:

$$SE\ BOX_{newreq} = SE_{old} + 20 \log (N + 1)$$

$$= 42 \text{ dB} + 12 \text{ dB} = 54 \text{ dB} \tag{8.5}$$

for N = 3 aperture leakages

Thus, in the next iteration, the box SE performance will be designed to provide at least 54 dB for the worst-case frequency of 100 MHz. Repeating this process [Eqs. (8.4) and (8.5)] for the other frequencies produces a box SE requirement of 34 dB and 46 dB at 10 and 1,000 MHz, respectively. Equation (8.4) is again applied to get the new combined aperture leakage SE required now that the box-skin requirements have been increased:

$$\sum_{i=1}^{3} \log^{-1}(-L_{idB}/20) = \log^{-1}(-42/20) - \log^{-1}(-54/20)$$

$$= 0.00794 - 0.00200 = 0.00616$$

The new combined SE due to aperture leakages, $L_{newc} = 45$ dB. By increasing the box SE by 12 dB (54 dB − 42 dB), the combined leakage requirements have been reduced by 16 dB (61 dB − 45 dB). The leakage requirement for each individual leakage [see Eq. (8.2)] is:

$$L_{idB} = L_{newc} + 20 \log N$$

$$= 45 \text{ dB} + 9.5 \text{ dB} = 54 \text{ dB} \quad (8.6)$$

In summary, the first run or iteration produced the SE objectives, box performance, individual leakage performance, combined leakage performance, combined box-leakage performance and performance-objective results. From the above critique, a secondary box requirement and leakage requirement are synthesized for subsequent iterations. This is summarized in Fig. 8.4. Remember, the original overall SE objectives never change unless the user has a new requirement.

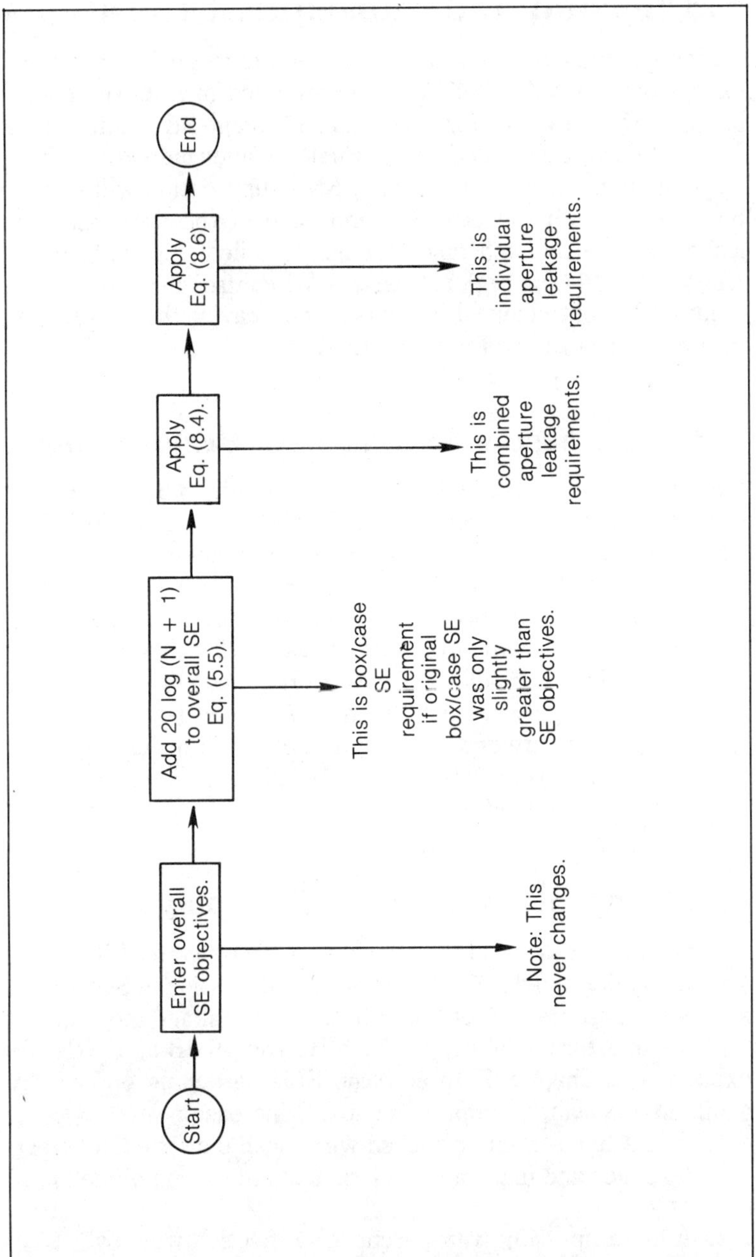

Figure 8.4—Methodology for Determining Box/Case SE Requirements and Aperture Leakage Requirements to Achieve Overall SE Objectives

8.3 Second and Subsequent Iterations

Figure 8.5 is a repeat of lines A and A.1 through A.4 in Fig. 8.2 except that lines A.5 and A.6 are now filled in from the results of applying Section 8.2 for new box and leakage SE requirements. Two actions are now necessary. First, tougher individual SE requirements must be met. Second, EMI suppression will have to be done on both the box skin and each of the three aperture leakages because each failed, or nearly failed (Fig. 8.2), in the 100 MHz to 10 GHz region. Special EMI-control measures will be required for controlling CRT leakage since, having the largest aperture area, it is the worst of the three.

8.3.1 Increasing SE Box Skin Performance

As explained in Section 8.2 and shown in Fig. 8.5, the new box skin SE performance requirement is 54 dB at 100 MHz and 1 GHz. The 12 dB additional improvement performance may be readily achieved, as explained in Illustrative Example 6.2, by simply reducing the surface resistance of the metalized plastic by a factor of four. Thus, 0.7 Ω divided by 4 produces a 175 mΩ/sq impedance (use 0.2Ω). This essentially achieves the required box skin SE. The results from re-applying far-field Eq. (6.7) and near-field Eq. (6.7a) for Z_b = 0.2 Ω/sq are entered on line B of Fig. 8.6. As previously mentioned, this assumes that the EMI radiating source is at 1 m, and that the near field is an E-field.

8.3.2 Improving SE of CRT Displays

Figure 8.2 showed that the attenuation through the CRT display was totally degraded at 1 GHz. From the discussion in Section 8.2 and as seen on line A.6 of Fig. 8.5, the new leakage requirement for each aperture is 54 dB at 100 MHz and 46 dB at 1 GHz. As explained in Chapter 7, three basic EMI hardening options are available for leakage improvement (dB increase):

1. Add a fine screen or knitted-wire mesh over the CRT aperture face and make a peripheral bond to the metal/metalized case.
2. Add an optically transparent RF reflective window.
3. Bond the backshell of the CRT to the metalized plastic to form an "electrical funnel." This decreases the effective aperture

Figure 8.5—Same as Fig. 8.2 after modified box skin and leakage requirements (lines A.5 and A.6) are computed.

Topic Identification	EMI Fix	10 kHz	100 kHz	1 MHz	10 MHz	100 MHz	1 GHz	10 GHz
A. SE Objectives (1 + 2 − 3 + 4)		0	0	8	22	42	34	20
1. EM Ambients (dBμV/m)		120	120	120	120	140	140	140
2. Field-to-Cable/PCB Coupling (dB)		−120	−100	−80	−60	−40	−28	−28
3. Sensitivity or NIL (dBμV)		42	42	42	48	68	88	102
4. Design Margin (dB)		10	10	10	10	10	10	10
5. Box SE Modified Objectives		0	0	20	34	54	46	32
6. Leakage SE Modified Objectives		0	0	20	34	54	46	32
B. SE Box Performance								
C. Box Leakages*								
1. Viewing Window Apertures								
1.1 CRT Displays								
1.2 Alphanumeric Displays								
1.3 Panel Meter Displays								
1.4 Status Indicators								
2. Cooling Aperatures								
2.1 Cooling Apertures								
2.2 Forced-Air Cooling								
3. Assembly and Maintenance Seams								
3.1 Box Housing Assembly								
3.2 Cover Panels								
4. Component Aperatures								
4.1								
4.2								
D. Total SE* Performance								
E. Performance vs. Objectives (D—A)								

$$*SE = -20 \log \left[\log^{-1}(-SE_{BOX}/20) + \sum_{i=1}^{n} \log^{-1}(-\text{leakages}/20) \right]$$

Figure 8.6—Final shielded box performance after box skin and aperture leakage have been EMI suppressed.

Topic Identification	EMI Fix	10 kHz	100 kHz	1 MHz	10 MHz	100 MHz	1 GHz	10 GHz
A. SE Objectives (1 + 2 − 3 + 4)		0	0	8	22	42	32	20
1. EM Ambients (dBμV/m)		120	120	120	120	140	140	140
2. Field-to-Cable/PCB Coupling (dB)		−120	−100	−80	−60	−40	−28	−28
3. Sensitivity or NIL (dBμV)		42	42	42	48	68	88	102
4. Design Margin (dB)		10	10	10	10	10	10	10
5. Box SE Modified Objectives		0	0	20	34	54	46	32
6. Leakage SE Modified Objectives		0	0	20	34	54	46	32
B. SE Box Performance	.2 Ω/sq	127	107	87	67	53	58	71
C. Box Leakages*								
1. Viewing Window Apertures								
1.1 CRT Displays	.3mm MESH	152	132	112	92	72	52	32
1.2 Alphanumeric Displays								
1.3 Panel Meter Displays								
1.4 Status Indicators								
2. Cooling Aperatures								
2.1 Cooling Apertures	4mm x 4mm	155	135	115	95	75	55	37
2.2 Forced-Air Cooling								
3. Assembly and Maintenance Seams								
3.1 Box Housing Assembly	EMI GASKET	121	101	81	61	50	45	40
3.2 Cover Panels								
4. Component Aperatures								
4.1								
4.2								
D. Total SE* Performance		117	97	77	57	45	38	26
E. Performance vs. Objectives (D—A)		117	97	69	35	3	4	6

$$^{*}SE = -20 \log \left[\log^{-1}(-SE_{BOX}/20) + \sum_{i=1}^{n} \log^{-1}(-\text{leakages}/20) \right]$$

size to that of the diameter of the CRT neck.

As shown in Illustrative Example 7.4, the second approach, a fine wire mesh with 0.3 mm center-to-center wire spacing, is selected. This mesh gives a new SE performance to the CRT shown on line C.1.1 of Fig. 8.6.

8.3.3 Improving SE Box Leakage for Convection Cooling Slots

Figure 8.2, line C.2.1 shows that 15 dB leakage exists at 1 GHz. To achieve the new 46 dB SE leakage requirement, an additional 31 dB of EMI protection is required.

As explained in Illustrative Example 7.6, the 4 mm × 8 cm convection cooling slots shown in Fig. 8.1 are broken up into multiple smaller 4 mm × 4 mm slots. It is also necessary to increase the depth of the metalized plastic in this area to 5 mm to achieve additional attenuation based on the "waveguide beyond cutoff" (WGBCO) effect. The results of this are shown on line C.2.1 in Fig. 8.6.

8.3.4 Improving SE Box Leakage for Covers and Mating Seams

Line C.3.1 of Fig. 8.2 shows that the worst-case seam performance is at 10 GHz where SE = 0 dB. Based on the requirement, 32 dB EMI suppression is required. As shown in Illustrative Example 7.5, this suppression is achieved by adding an electrical gasket around the periphery of the mating base cover and top sections. The new performance obtained by adding this gasket (60 dB at 10 kHz, 50 dB at 1 MHz and 100 MHz and 40 dB at 10 GHz) is illustrated on line C.3.2 of Fig. 8.6.

8.3.5 New Overall Performance for Second Run

Figure 8.6 summarizes the changes for the second run. On line D, the new total SE performance has been computed using Eq. (8.3) to combine lines B and C. The required SE has now been achieved, and no further design is necessary. If the second iteration had produced any negative entries on line E, it would have been necessary to go into a third iteration.

Conversely, if more than 20 dB had been indicated on line E for the worst case, an overdesign situation would have resulted. Neither of these two situations exists in Fig. 8.6. It is always important to review or analyze the EMC solution for the shielding problem. EMI control techniques to achieve mission objectives are not unique. It is desirable to repeat one or two additional complete runs of the procedures. This repetition provides multiple options for selecting the best of EMI fixes. The final decision will be based on other than EMI considerations such as cost, reliability, maintainability, aesthetics, etc.

8.4 References

1. MIL-STD-461C, "Electromagnetic Emission and Susceptibility Requirements for the Control of Electromagnetic Interference."
2. CBEMA/ESC5/75/31, Computer and Business Equipment Manufacturers' Association, 311 First St. NW, Washington, DC.
3. SAE AIR-1499, "Recommendation for Commercial EMC Susceptibility Standards," (Warrendale, PA: Society of Automotive Engineers, Subcommittee AE4 on Electromagnetic Compatibility).
4. Standard ECMA-47, "Limits and Measurement Methods for Radio Interference from Electronic Data Processing Equipment," (European Computer Manufacturers' Association).
5. MIL-Handbook-235-1A, *Electromagnetic (Radiated) Environment Considerations for Design and Procurement of Electrical and Electronic Equipment, Subsystems and Systems*, 5 February 1979.

8.5 Bibliography

Milstead, F.C. and Cory, W.E., "The Effects of the Local Environment on Interference Measurements and Prediction," *IEEE EMC Symposium Record*, 1969.

Various documents, Scientific Apparatus Makers Association (1101 16th Street NW, Washington, DC 20036).

Chapter 9

Reciprocity: Shielding Design for Emission Control

The quantitative approach for shielding a radiating source housed inside a box is similar in principle to the susceptibility case treated in earlier chapters. However, several noteworthy differences exist:

1. The maximum allowable emission is now a civilian or military limit for radiated EMI emissions, generally defined at 1, 3 or 10 m from the radiating equipment.[1-3]
2. The radiation of the circuit inside the shielded box (the source) must be calculated for all the significant spectrum, covering all intended and (probably) unintended frequencies.
3. The shielded housing will be, for a more or less significant part of this spectrum, in the near field of the source, which complicates the issue, i.e., it is now necessary to know the wave impedance to calculate the SE of box skin, apertures, etc.

This chapter will follow the same methodology as the design of shields for susceptibility. The design sequence will be the right-hand branch of Fig. 4.1, which is the mirror image of the left-hand, susceptibility branch. However, since the general routine is very similar to the one described in Chapters 5 through 8 for susceptibility, the different steps of emission shielding are concentrated in this single Chapter 9.

9.1 Determining Shielding Effectiveness Requirements for Emissions

This section reviews how to determine the attenuation needed to comply with a given emission limit, given a set of HF sources within the equipment being designed. The methodology is summarized in the flowchart of Fig. 9.1.

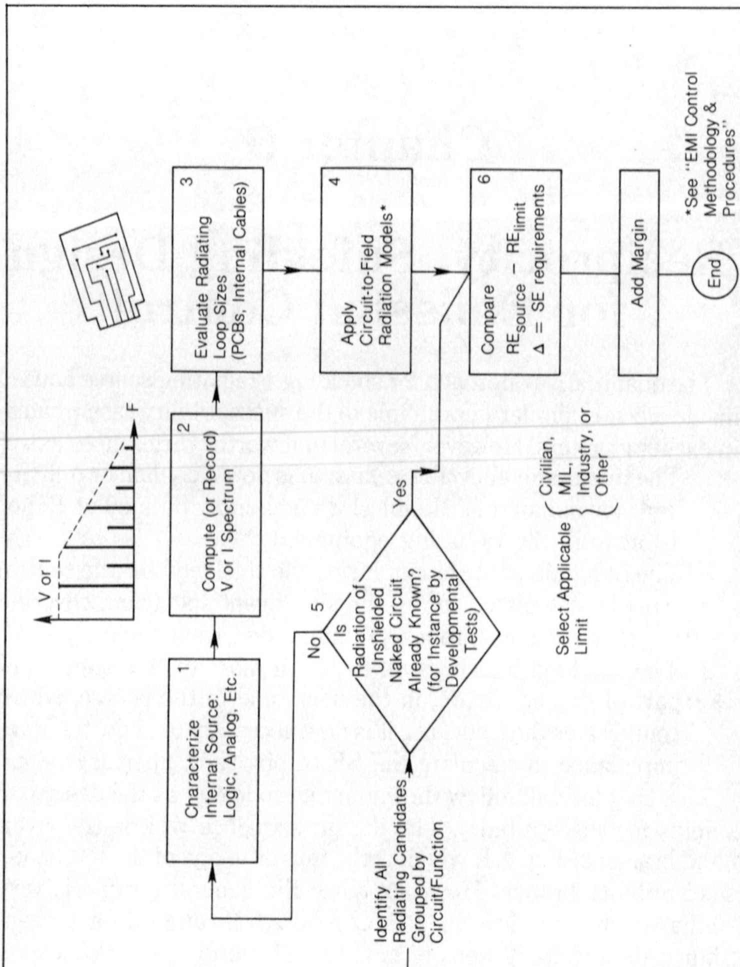

Figure 9.1—Flow Diagram for Shielding Design in the Case of Radiated Emission Control

First, the designer identifies the predominant radiating sources within the equipment (box 1). This may require a tedious listing of all the functions and all their corresponding circuits one by one. However, a simplification can be made by selecting the circuits which have the highest I × A × F product, where I is the current through the circuit, A is the geometric area and F is the frequency of the component of interest. This product expresses, using the same units for all circuits (e.g. mA, cm^2 and MHz) the relative propensity to radiate beyond the quasi-static region.

For those circuits selected in the previous screening, the voltage spectrum (or current spectrum for predominantly magnetic sources) is plotted, either by the calculation of a Fourier transform of the time domain waveform or by actual bench measurements (box 2). The circuits' loop areas are also identified (box 3). Using radiation models described briefly in the next paragraphs, and more deeply covered in Volume 8 of this handbook series, the radiated E-field (or eventually H-field) at a given distance r is calculated (box 4). If the radiation levels of the bare circuits (e.g., a breadboard prototype) are already known ("Yes" option in box 5) from an early EMI measurement, the previous steps are, of course, unnecessary.

Finally, in the absence of any shield, the radiated EMI spectrum is compared to the applicable radiated EMI limit (FCC Parts 15 and 18, CISPR 22, VDE 871, MIL-STD-461, TEMPEST, etc.). Unless the radiation levels are already below the limit (a lucky occurrence), the difference RE_{source} - RE_{limit} is the required shielding (box 6). Ultimately, as for susceptibility, a safety margin of 10 to 30 dB is added and the SE requirements across the spectrum of concerns are then determined.

9.2 Performing Spectral Analysis of the Radiated EMI Sources

Many types of radiated EMI sources exist but, at the risk of oversimplifying, they can be classified into two categories: **analog** and **digital,** or more accurately, **sinewave** and **pulsed.** Sinewave sources include radio transmitter circuits of all kinds (AM, FM, TV, radar), receivers, local oscillators, industrial, scientific and medical HF equipment using powerful tuned oscillators, microwave ovens, etc. The spectral profile of such sources is fairly simple since it is basically the source's continuous-wave (CW) frequency F_0 widened by the modulation bandwidth. However, since absolute spectral purity does not exist, it is prudent to account for spurious emissions up to the tenth through thirtieth harmonics. The amplitude of these spurious emissions is typically 30 to 60 dB below the amplitude of the fundamental, but the higher frequencies leak more readily through the box apertures and cracks.

Pulsed sources include digital circuits, digital clocks, switched-mode power supplies, ac-to-dc or dc-to-dc converters, etc. As digital data processing replaces analog processing, these sources become numerically prevelant.

The purpose of digital equipment is to process time-varying waveforms, generally represented by zero volts for a logical "0" (or "false") and a few volts for logical "1" (or "true"). On the other hand, EMI specifications (FCC, CISPR/VDE, MIL-STD-461, etc.) are defined in the frequency domain, as are performance data of filters and shielding materials. So, one way or another, time-domain signals must be translated into the frequency domain or vice versa.

Fourier theory states that a periodic signal can be expressed in terms of an infinite series of sine and cosine components, with frequencies which are multiple integers of the pulse period. However, commercial specifications extend to 1 GHz. If one were to take a complex time-domain waveform and perform a rigorous Fourier analysis, it could take hours or days to compute the amplitudes and phases of every harmonic from the fundamental to number 1,000 or number 100,000. Fortunately, some simplifications exist which, providing some assumptions are made, allow for a quick calculation process with acceptable results.[4, 5] Therefore, the problem can be broken into two steps:

1. Identify the waveform characteristics and period of the time-varying signal.
2. Perform a time-to-frequency-domain conversion so it can be compared to the specification limit once the proper coupling coefficients have been applied.

Step 1 is relatively easy: connect an oscilloscope to the source and then take a picture of the display. Be sure that the bandwidth of the oscilloscope is wide enough to reproduce the rise times of the logic family without distortion (use bandwidth $> 0.3/\tau_r$ as a rule of thumb). Then approximate the time-domain waveform by one of the following waveforms:

A. Trapezoidal waveforms have the following general shape:

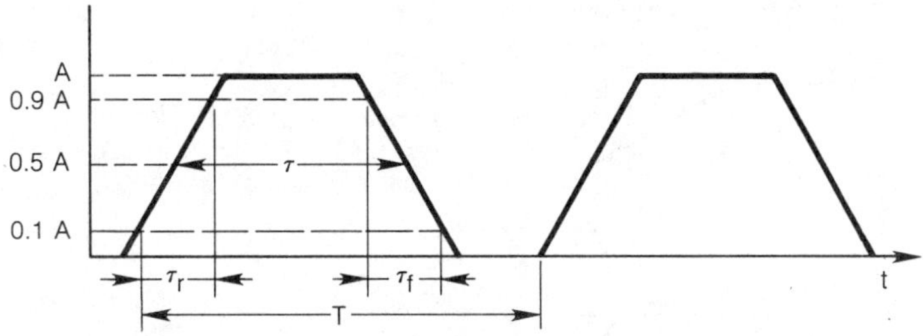

B. Triangular pulses have the following general shape:

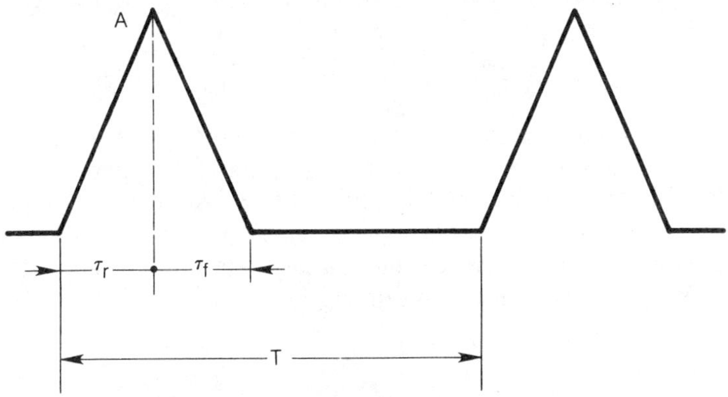

Once the waveform has been approximated by the shape shown in A or B above, the following approximations are required to perform step 2:

1. If $\tau_r \neq \tau_f$, only the smallest of the two will be considered.
2. Only the envelope of the maximum amplitudes will be considered; i.e., the individual frequency components are not discerned.

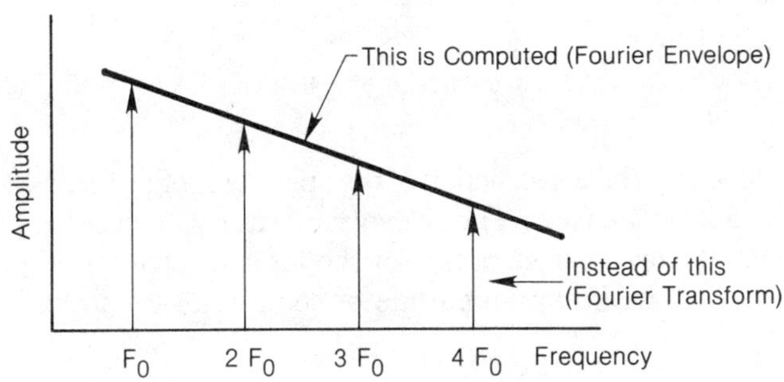

The approximate frequency domain representation for trapezoidal waveform A is shown below:

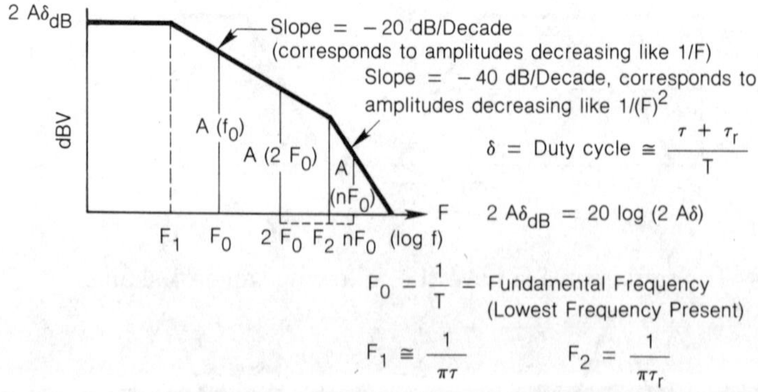

For a periodic waveform, the frequency spectrum is composed of discrete frequency components consisting of the fundamental (f_0) and integral multiples (harmonics) of f_0. The units of 2A are volts, millivolts, amps, etc. If $\tau_r \neq \tau_f$, select the smaller of the two to calculate F_2. The approximate Fourier spectrum representation for triangular waveform B, assuming $\tau_r \cong \tau_1$, is:

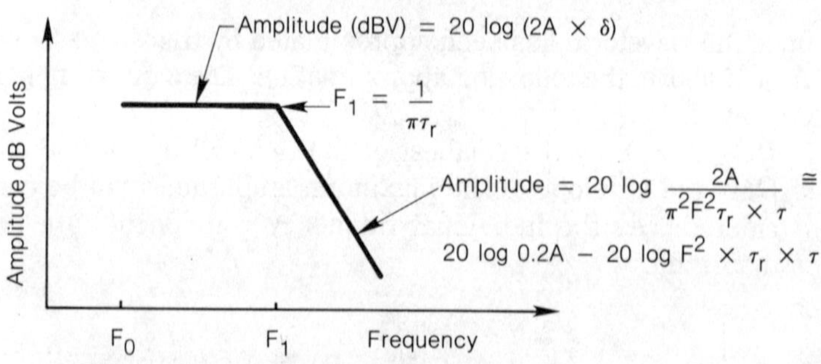

Note: F_0 = 1/T can be greater than F.

To speed up the construction of the spectral profile, the nomogram of Fig. 9.2 can be used. The nomogram was constructed to provide a fast and easy approximation of the total spectrum occupied, as well as the maximum amplitude of some typical harmonics.

Illustrative Example 9.1

To understand how to use Fig. 9.2, thereby producing Fig. 9.3, consider a clock signal having the following characteristics:

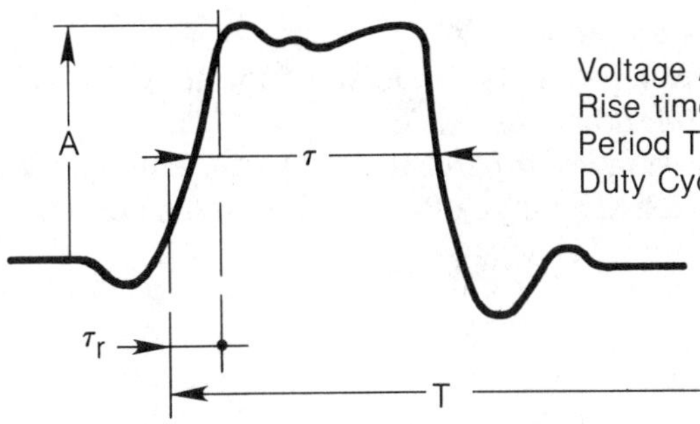

Voltage Amplitude = 4 V
Rise time = 10 ns
Period T = 100 ns
Duty Cycle = 50%

First, identify the three key frequencies of the spectrum. The fundamental, F_0, is obviously 1/100 ns = 10 MHz.

The first corner frequency, F_1, is computed as follows:

$$F_1 = \frac{1}{\pi\tau} = \frac{1}{\pi \times (100/2) \times 10^{-9}} \approx 6 \text{ MHz}$$

Note that at this frequency, the spectrum has no components yet because the fundamental is greater than F_1, but F_1 is necessary to construct the envelope.

The second corner frequency, F_2, is computed by:

$$F_2 = \frac{1}{\pi\tau_r} = \frac{1}{\pi \times 10 \times 10^{-9}} \approx 30 \text{ MHz}$$

Note how critical this frequency is, as it is related to the rise time. The shorter the rise time, the greater the spectral occupancy.

What remains is to calculate the amplitude of the reference line:

$$20 \log 2 \text{ A} \times \tau/\text{T} = 20 \log 2 \times 4 \text{ V} \times \frac{50}{100} = 12 \text{ dBV}$$

The frequency F_1 (6 MHz) is plotted to the top 0 dB reference, then a negative 20 dB/decade slope is drawn, using the parallel grids, to the frequency F_2 (30 MHz). From F_2, a negative 40 dB/decade slope is drawn. The envelope has now been constructured; it is the upper bound of the amplitudes of the frequency spectrum. The form below the nomogram allows one to compute the amplitude of any significant harmonic in dBV. To convert from dBV to dBμV, one should simply add 120 dB, since 1 V = 10^6 μV = 120 dBμV.

A few clarifications are necessary:

Question: The envelope shows the locus of the maxima. What about the other harmonics?

Answer: With a simple 50 percent duty cycle, only odd harmonics exist, and they follow this envelope exactly as represented below:

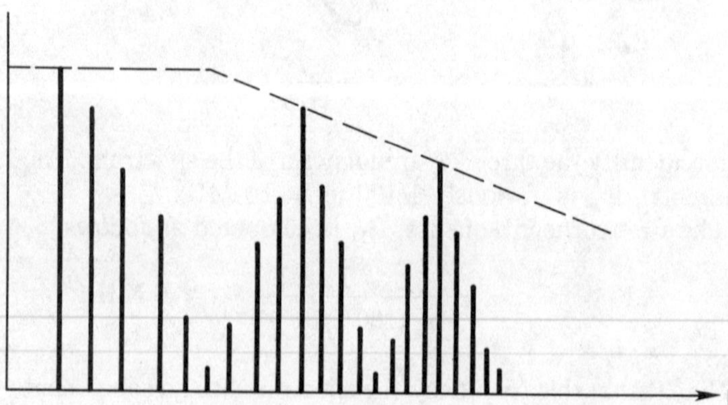

For smaller duty cycles, i.e., when the period becomes significantly long compared to the pulse duration, only a few harmonics in the series reach the upper bound. However, experience shows that these harmonics can be sufficient to create an EMI problem, precisely because they have the highest amplitude. **Murphy's Law** (which perhaps has found its most impressive field of application in EMC) would state that if any susceptible circuit has some especially critical resonance, this is exactly where the EMI sources will be the strongest.

Question: Should not the envelope show phase or polarity?

Answer: Yes. At every multiple of $1/\tau$, there is a 180° phase reversal due to the fact that harmonics follow a sine or cosine function of frequency (see below). However, EMI receivers used to measure EMI emissions (for FCC or CISPR) and spectrum analyzers are insensitive to the phase of the emission and would display the absolute value.

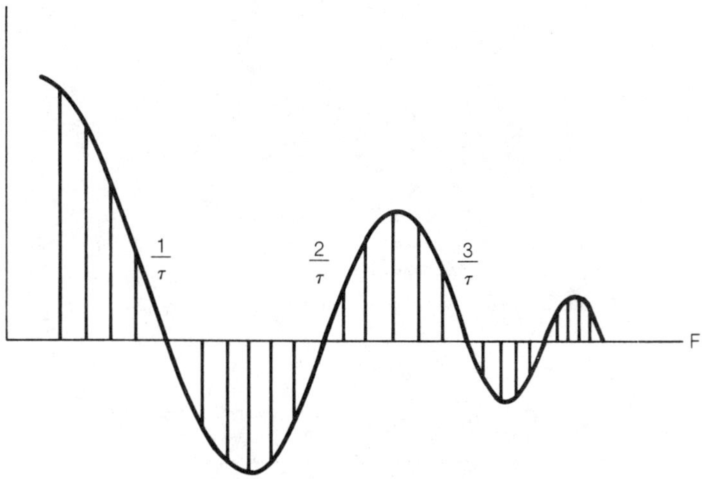

Question: When harmonics are closely packed together in the frequency spectrum (i.e., the period is relatively long and there are many harmonics within a frequency interval) should they be added in some manner?

Answer: For digital signal processing systems, the majority of repetitive signals have repetition rates higher than 10 kHz. An example is switched-mode power supplies using switching rates of 20 kHz or more. In addition, digital clock frequencies are always greater than a few hundred kilohertz. Since the receivers used to measure EMI emissions have a bandwidth of 10 kHz for conducted EMI and 100 kHz for radiated EMI, the noise will generally appear **narrowband,** i.e., **the receiver will only see one harmonic at a time within its resolution bandwidth.**

However, it is true that in some situations (e.g., for MIL-STDs), the broadband addition of spectral components should be considered. In this case, the spectral amplitude is expressed in microvolts/kHz or microvolts/MHz. This topic is beyond the scope of our book, but the reader will find more details on this subject in Refs. 5 and 6.

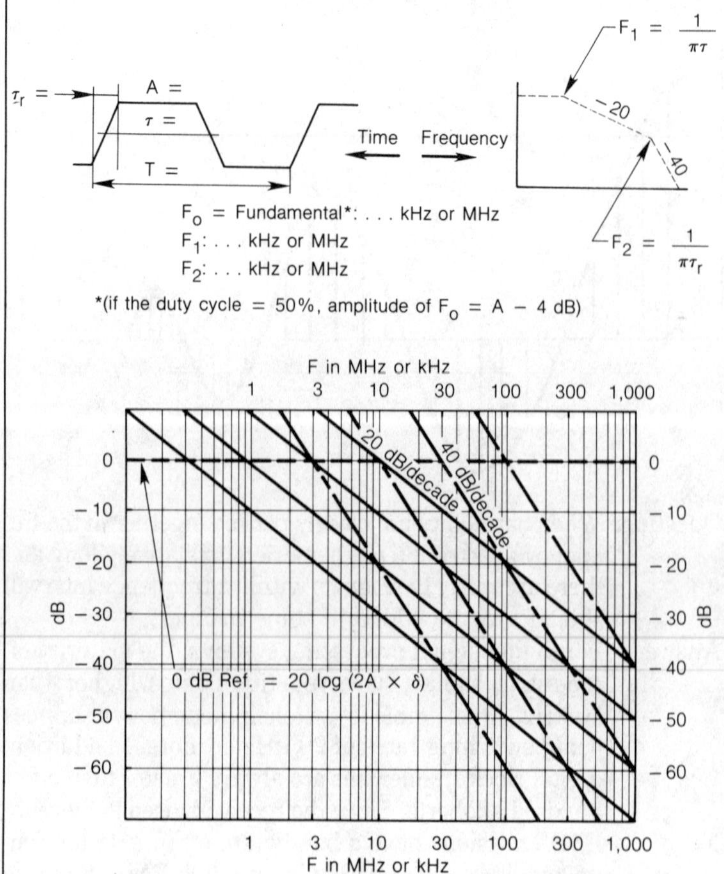

Harmonic #	Frequency	Ref. Amplitude dB Volt (or dB Amp)	Slope decrease	Amplitude of Harmonic

Figure 9.2—Nomogram to Predict Pulse Train Spectrum

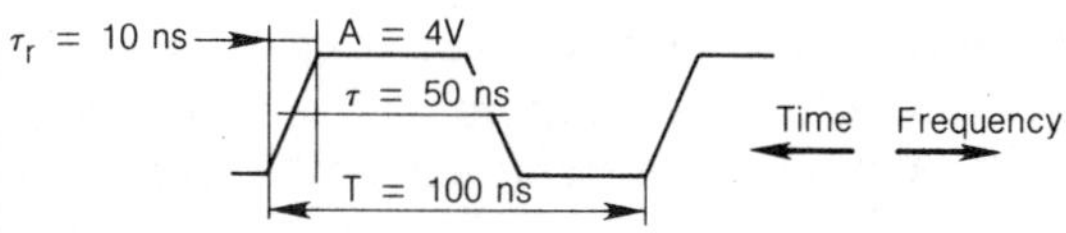

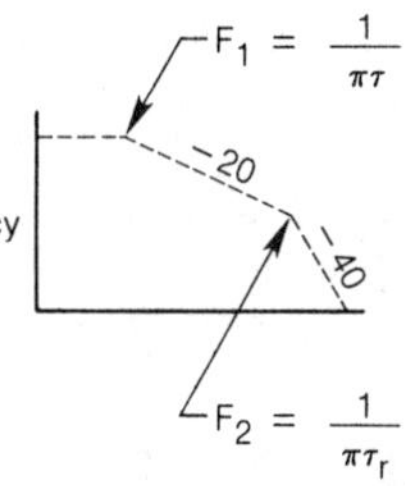

F_o = Fundamental*: . . . 10 MHz
F_1: 6 MHz
F_2: 30 MHz
*(if the duty cycle = 50%, amplitude of F_o = A − 4 dB)

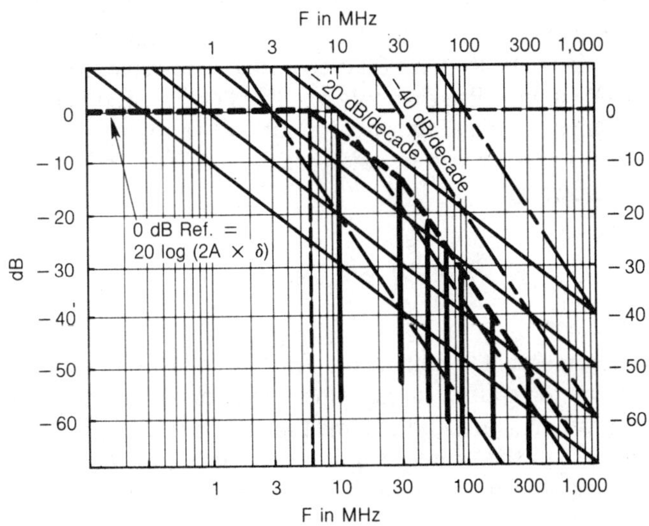

Harmonic #	Frequency MHz	Ref. Amplitude dB Volt (or dB Amp)	Slope decrease dB	Amplitude of Harmonic dBV
Fundamental	10	12 dBV	−4	8
#3	30	12 dBV	−14	−2
#5	50	12 dBV	−22	−10
#7	70	12 dBV	−28	−16
#9	90	12 dBV	−32	−20
#15	150	12 dBV	−42	−30
#30	300	12 dBV	−54	−42

Figure 9.3—Nomogram to Predict Pulse Train Spectrum (with Illustrative Example 9.1)

9.3 Prediction of EMI Radiation from a Given Circuit

In the days when radio frequency interference was primarily a matter of radio communication transmitters and receivers, very sophisticated analytical approaches were developed to predict the interference level given all the parameters of the transmitting source. More recently, a multitude of unintentional RF sources have appeared due to the proliferation of high-performance computing devices which use faster and faster logic technologies with higher clock and bit rates. All computers and digital equipment, therefore, have become threats to radio communication. A simple calculation serves to illustrate the problem: assume a minicomputer consisting of a large planar board with 60 chips, each chip consuming about 250 mW. Assuming also that only one-third of the circuits resident on these chips are toggling synchronously at an internal clock frequency of 50 MHz, it can be said that the total switched power at a given instant is:

$$1/3 \times 60 \times 0.250 = 5 \text{ W}$$

Now assume that a minuscule fraction of this power is not dissipated by Joule effect in the chips, wiring or various resistances, but is radiated instead. And assume that at the 50 MHz fundamental, only 10^{-6} of the switched power is radiated (5 μW). A simple formula gives the field strength of a given radiator:

$$E \text{ V/m} = 1/r \sqrt{30 \, P_r} \tag{9.1}$$

where,

r = distance from the source in meters

P_r = radiated power (including antenna gain)

At 3 m distance, the 5 μW radiated from the PC board will result in an electric field:

$$E = 1/3 \sqrt{30 \times 5 \times 10^{-6}} = 4 \text{ mV/m or } 72 \text{ dB}\mu\text{V/m}$$

This is more than the minimum field strength received by TV and FM listeners in areas distant from the broadcast station. Therefore, in case of frequency coincidence (co-channel EMI), the computer clock and its harmonics may seriously affect radio reception in the vicinity.

The FCC and other regulatory bodies worldwide have set RF emissions limits such as FCC Docket 15J which stipulates that no more than 100 μV/m (40 dBμV/m), measured at 3 m, should radiate above 30 MHz from personal computers.

Compared to a CW transmitter where the characteristics of the radiation source (transmitter power, antenna gain and pattern, spurious harmonics, etc.) are well identified, a digital electronic assembly is much more difficult to model. So the traditional approach is to design the PCB layout, motherboard and interconnects using the best available knowledge. In the area of EMI, this may include a blend of rules-of-thumb, company recipes and the philosophy of, "We do it this way because we always did it this way and it worked." The manufacturer then runs a test to see if it passes FCC regulations. Needless to say (and the example above shows why), in many cases it does not pass. This means redesign, engineering changes and retrofits.

The following method allows quick prediction of radiation from printed circuits and associated wiring. The prediction is based on solutions of Maxwell's equations for a small loop or doublet. The far field electric term from a short doublet is given by:

$$E\ \text{V/m} = \frac{I \times l \times 60\pi}{r\,\lambda} \qquad (9.2)$$

where,

I = current in the short wire in amperes
l = length of short wire in meters
r = distance in meters
λ = wavelength in meters

Since the radiating circuit is made of two more or less parallel wires carrying opposite currents, the net field would be null if they were infinitely close. But in fact, their separation, S, creates a phase difference and therefore a non-zero net field. Taking into account this phase shift gives the radiated far-field expression for a pair of parallel wires:

$$E\ V/m = \left(\frac{I \times l \times 60\pi}{r\,\lambda}\right)\left(\frac{\sin 2\,\pi s}{\lambda}\right) \tag{9.3}$$

$$\approx \frac{I \times l \times 120\pi^2}{r}\left(\frac{s}{\lambda^2}\right)$$

The simplification is based on the fact that the sine of a small number is approximately the number itself. Given that $m = 300/F_{MHz}$, replacing $l \times s$ by the area of the circuit and using more convenient units results in:

$$E\mu V/m = 1.3 \times A \times I/r \times (F)^2 \tag{9.4}$$

where:
- A = area of the radiating circuit in cm^2
- F = frequency in MHz
- I = circuit current in amps

Interestingly, the result is the same whether we use a loop or two parallel wires.

However, these equations are based on current, which is generally not what the designer uses the most. It is often more convenient to use the drive voltage. When I is replaced by V/Z in the above equations, the following expression results which directly gives the radiated field as a function of the driving voltage:

$$E\mu V/m = \frac{1.3\ A\ V(F)^2}{r\ Z} \tag{9.5}$$

where,
- V = driving source voltage
- Z = circuit impedance

The curves in Fig. 9.4 gives the radiated field strength at 1 and 3 m from a circuit normalized as a 1 cm^2 loop area, driven by 1 V, and terminated by loads from 10 to 377 Ω. Above the far-field transition, if loads are larger than 377 Ω, the field becomes essentially that of a voltage-driven dipole, i.e., it depends on the driving voltage only. What remains is to compute (or measure) the Fourier series of the digital signal.

The results in Fig. 9.4 are fairly accurate as long as the radiating element is electrically short ($l << \lambda$). When l approaches a fraction of a wavelength, the limitations are:

1. At frequencies where l exceeds $\lambda/4$ (i.e., the wire and its return path exceed $\lambda/2$) the geometry of the loop entered in the model is clamped to $\lambda/4$ of the frequency used in the calculation.

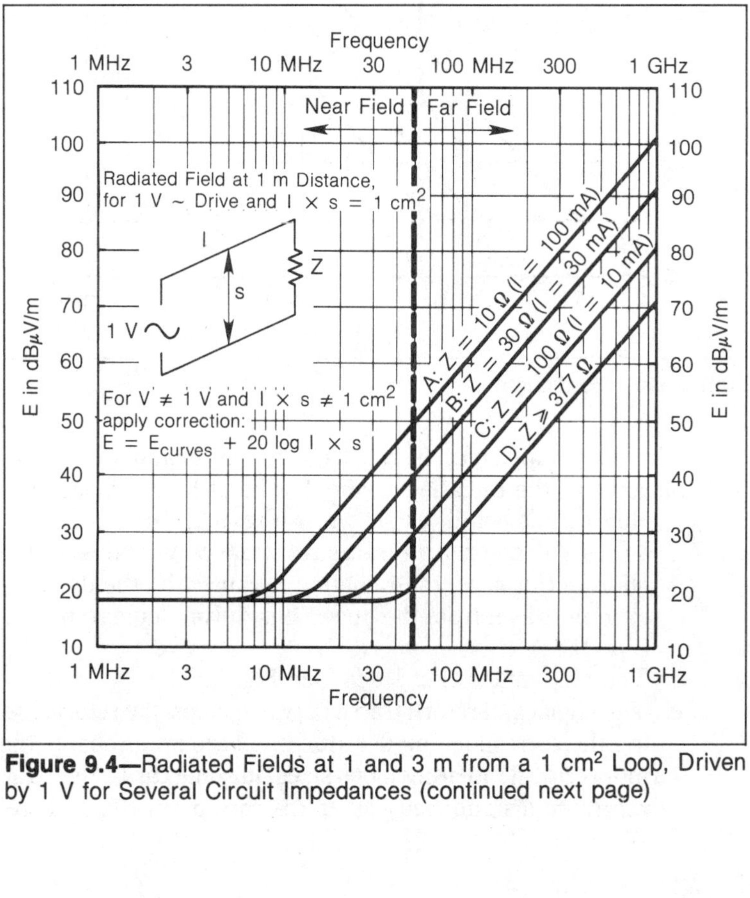

Figure 9.4—Radiated Fields at 1 and 3 m from a 1 cm^2 Loop, Driven by 1 V for Several Circuit Impedances (continued next page)

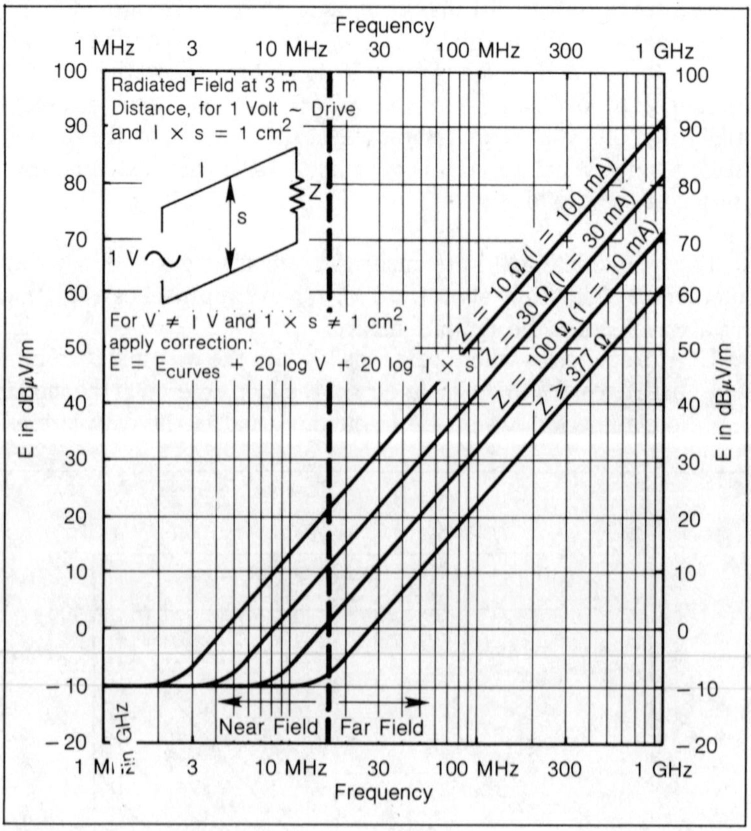

Figure 9.4—(continued)

2. The load impedance is replaced by the characteristic impedance of the radiating element. The radiating loops on a printed circuit board, as shown in Fig. 9.5, can consist of:
   a. The +Vdc and return traces from the power buses to the chip. In this case, the radiation is caused by the dynamic current drawn from the power bus during logic transition (Fig. 9.5a).
   b. The chip and socket leads (see Fig. 9.5b)
   c. The signal and return traces between chips, the returns being the corresponding 0 V traces. These are probably the most critical radiating loops since, in some cards, they can represent a significant part of the card perimeter and are

often the most active lines. In Fig. 9.5a, for instance, the run EFGH reveals a very large loop.

d. The signal and return wires in a flat cable, etc.

If the PCB is multilayer or double-sided with a ground plane, an immediate reduction of the radiating loop occurs since the spacing S now is merely the height of the traces above the ground plane, which is the thickness of the board dilectric (see Fig. 9.5c).

The following is a practical example of how to predict the radiation from a circuit board.

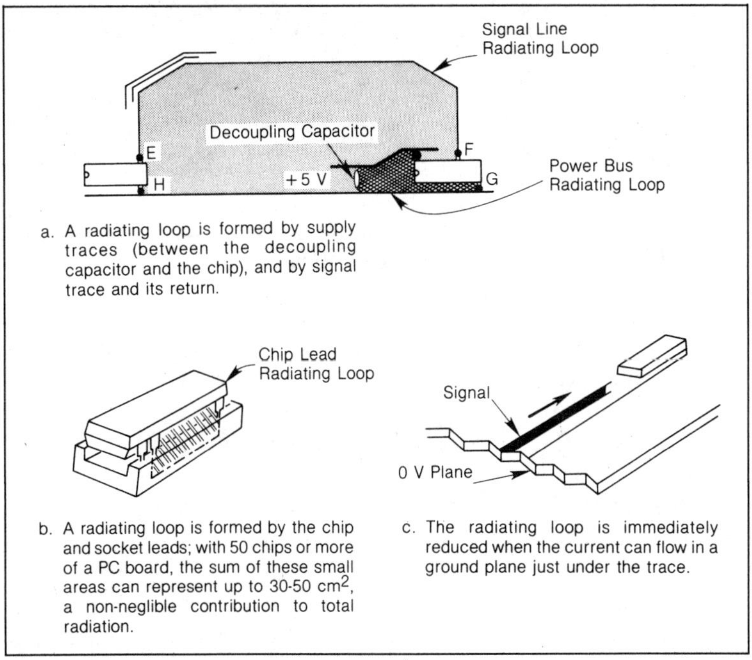

Figure 9.5—Radiating Elements in PCBs

Illustrative Example 9.2: Radiation vs. FCC Class B Limits (at 3 m)

Using the same digital pulses as the example 9.1, the radiated emission is estimated from the large card (mother board) shown in Fig. 9.6. The critical (radiating) portion of the circuit consists of:

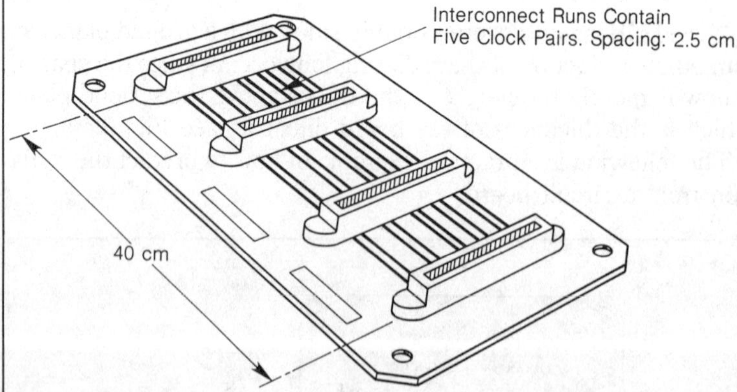

Figure 9.6—Motherboard for Radiation Example

1. Five clock traces and their returns, carrying 10 Mhz clock
2. Voltage swing: 4 V
3. Rise time: 10 ns
4. Load: 500 Ω (corresponds roughly to a TTL input with 10 pF of capacitance)
5. Trace run: l = 40 cm
6. Pair separations: S = 2.5 cm

The amplitude of each harmonic has already been calculated in Illustrative Example 9.1 and entered in the form at the bottom of Fig. 9.7. Since the curves are normalized for 1 V, 1 loop and 1 cm^2, the following dB correction factors must be applied:

1. The actual voltages, in dBV
2. The area of one loop = 20 log $l \times$ s cm^2
3. The number of synchronous loops = 20 log N

One should remember that when the largest dimension l reaches λ/4, the value entered for l in the area correction must be clamped to the λ/4 for the frequency under consideration. In this case, 40 cm represents one-quarter of the wavelength at F = 1/4 × 300 MHz/0.4 m = 188 MHz. Therefore, beyond 188 MHz (harmonic

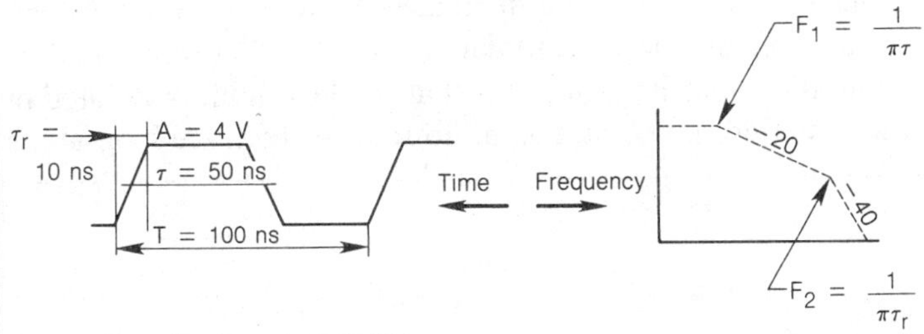

F_o = Fundamental:* 10 MHz

F_1: 6 MHz

F_2: 30 MHz (*If the duty cycle = 50%, amplitude of F_o = A − 4 dB)

Harm. #	Freq. MHz	Ref. Ampl. dB Volts	Slope decrease (dB)	5 Amplitude of Harmonic	6 Correction for Loop Area	7 Correction for N Synchronous Loops	8 E_o, dBμV/m from	Ttl Field dBμV/m	Spec Limit	Off-spec?
Fund.	10	12	−4	8	40	14	−10	52	NA	
3	30	12	−14	−2	40	14	2	54	40	14
5	50	12	−22	−10	40	14	10	54	40	14
7	70	12	−28	−16	40	14	16	54	40	14
15	150	12	−42	−30	40	14	28	52	43	+9
30	300	12	−54	−42	36*	14	41	49	46	+3
45	450	12	−63	−50	32	14	48	44	46	−2

Total Field = 5 + 6 + 7 + 8

*Above 188 MHz, since the trace length exceeds λ/4, the effective I × s area has been recalculated using λ/4 for I, at each discrete frequency.

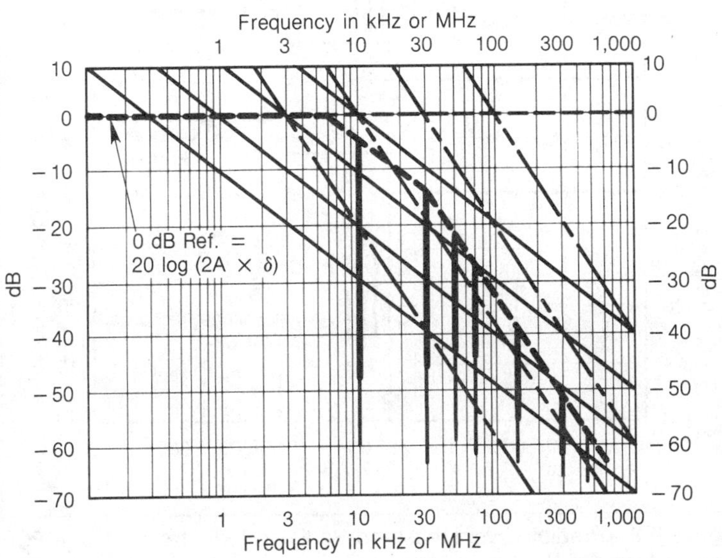

Figure 9.7—Radiation Prediction Form (Use for Illustrative Example 9.2)

no. 18) the area correction factors show a gradual reduction. Thus, the efficiency of the antenna decreases as does the voltage spectrum. As a result, the total radiation profile collapses.

The prediction clearly indicates that the FCC limit is violated by 14 dB, so action has to be taken. To reduce the trace loop areas,

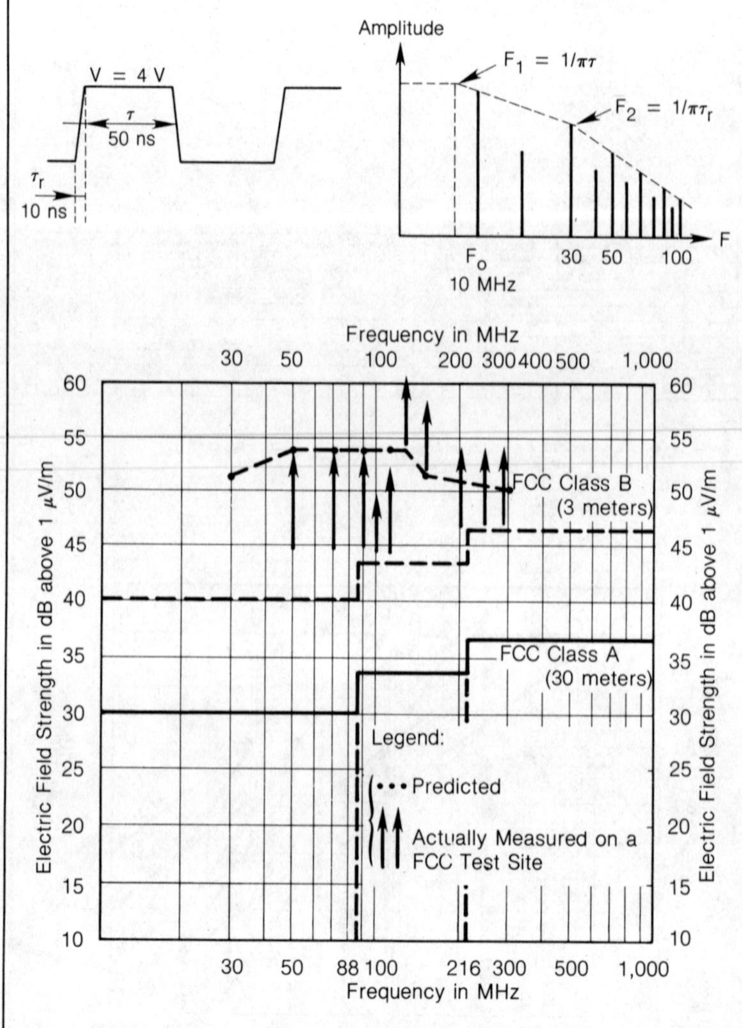

Figure 9.8—Predicted vs. Measured Radiated Field from a Large PCB with 10 MHz Clock

install a ground plane or shield the equipment housing.

By comparison, Fig. 9.8 shows the actual measurements of this PCB (the tested card used 7404 TTL inverters) on a 3 m FCC test site. Another precise validation of this model is found in Ref. 7. For the sake of precision, it must be stated that:

1. The Fourier series gives the peak value of each harmonic while the EMI receiver measures in RMS values. This results in a −3 dB difference.
2. The FCC procedure calls for scanning the antenna height to search for the maximum reading. Up to 6 dB increase in measured radiation can occur when the ground reflected wave arrives at the antenna in phase with the direct radiated wave. This analysis does not account for common-mode currents on external cables.
3. In actual equipment, the radiation from common-mode currents leaking by the external cables (I/Os, etc.) may contribute significantly to the total radiation.[6]

This section has only described methods to calculate radiated emission from rather simple circuits. A complete method for calculating the radiated E-fields and H-fields from a wide variety of circuits is presented in the Volume 8 of this handbook series.

9.4 Principal Radiated Emission Standards, Civilian and Military

In the civilian arena, maximum levels have been set by the International Special Committee on Radio Frequency Interference (CISPR, abbreviated from French) to protect radio communications from electrical interference by various sources. Since the CISPR limits were generated by national committees of the CISPR, they are now in use or being progressively adopted as national regulations in Europe, the U.S.A. and most other industrial countries. Small differences may exist in some countries; however, if a product is designed to meet CISPR interference limits, it is likely that it will comply with limits of most countries.

Although CISPR has addressed several categories of noise sources (electric appliances, fluorescent lights, vehicle ignition systems, etc.), CISPR Publication 22 is the most widely applicable for EMI suppression in digital data processing equipment, and Publication 11

Table 9.1—A Comparison of Principal Radiated Emission Limits, in dBμV/m

Frequency	FCC 15.J (1)		CISPR Publ-11		VDE (4)		MIL-STD-461 (5)	Frequency
	Class B 3 m	Class A 30 m	ISM 30 m	ISM 100 m	0871-B 3 m	0871-B 10 m	RE02 1 m	
10 kHz	NA	NA	NA	NA	91	NA	NA	10 kHz
14 kHz	NA	NA	NA	NA	88	NA	NA/35	14 kHz
20 kHz	NA	NA	NA	NA	85	NA	34	20 kHz
30 kHz	NA	NA	NA	NA	82	NA	33	30 kHz
50 kHz	NA	NA	NA	NA	78	NA	32	50 kHz
70 kHz	NA	NA	NA	NA	75	NA	32	70 kHz
100 kHz	NA	NA	NA	NA	72	NA	31	100 kHz
150 kHz	NA	NA	NA	NA/34	68	NA	30	150 kHz
200 kHz	NA	NA	NA	34	65	NA	30	200 kHz
300 kHz	NA	NA	NA	34/48	62	NA	29	300 kHz
500 kHz	NA	NA	NA	48/34	78	NA	28	500 kHz
700 kHz	NA	NA	NA	34	75	NA	27	700 kHz
1 MHz	NA	NA	NA	34	51	NA	26	1 MHz
1.6 MHz	NA	NA	NA	34/48	50		26	1.6 MHz
2 MHz	NA	NA	NA	48	49	NA	25	2 MHz
3 MHz	NA	NA	NA	48	48	NA	24	3 MHz
4 MHz	NA	NA	NA	48/34	47		24	4 MHz
5 MHz	NA	NA	NA	34	46	NA	23	5 MHz
7 MHz	NA	NA	NA	34	45	NA	23	7 MHz
10 MHz	NA	NA	NA	34	44	NA	22	10 MHz
20 MHz	NA	NA	NA	34	42	NA	20	20 MHz
25 MHz	NA	NA	NA	34		NA	20	25 MHz
30 MHz	NA/40	NA/30	NA/54	34/NA	41/NA	NA/34	21	30 MHz
41 MHz	N40	30	54/30	NA	NA	34	23	41 MHz
50 MHz	40	30	30	NA	NA	34	25	50 MHz
68 MHz	40	30	30/54	NA	NA	34	27	68 MHz
70 MHz	40	30	54	NA	NA	34	27	70 MHz
87 MHz	40	30	54/30	NA	NA	34	28	87 MHz
88 MHz	40/43	30/34	30	NA	NA	34	28	88 MHz
100 MHz	43	34	30	NA	NA	34	29	100 MHz
108 MHz	43	34	30/54	NA	NA	34	30	108 MHz
162 MHz	43	34	54/30	NA	NA	34	32	162 MHz
200 MHz	43	34	30	NA	NA	34	34	200 MHz
216 MHz	43/46	34/37	30	NA	NA	34	34	216 MHz
230 MHz	46	37	30/54	NA	NA	34	35	230 MHz
300 MHz	46	37	54	NA	NA	34	37	300 MHz
470 MHz	46	37	54/40	NA	NA	34/40	40	470 MHz
500 MHz	46	37	40	NA	NA	40	40	500 MHz
700 MHz	46	37	40	NA	NA	40	42	700 MHz
960 MHz	46	37	40/54	NA	NA	40	44	960 MHz
1 GHz	46/NA	37/NA	54/NA	NA	NA	40	45	1 GHz
2 GHz	NA	NA	NA	NA	NA	40	49	2 GHz
3 GHz	NA	NA	NA	NA	NA	40	52	3 GHz
5 GHz	NA	NA	NA	NA	NA	40	55	5 GHz
7 GHz	NA	NA	NA	NA	NA	40	58	7 GHZ
10 GHz	NA	NA	NA	NA	NA	40	60	10 GHz

Note:
1. CISPR Publication 22, international limits for computer emissions are similar.
2. NA = Not Applicable
3. Where two numbers appear, left number corresponds to frequency increment below indicated frequency and right number corresponds to frequency increment above indicated frequency.
4. As per 1982 update of German regulations. Below 30 MHz the VDE 871 measures the H-field radiation and translates it into equivalent E-field.
5. MIL-STD-461 limit is shown for the most severe category; other classes have relaxations.

addresses shielding of industrial, scientific and medical HF generators (ISMs). In turn, these CISPR recommendations have been translated by the governments of member countries into national laws or standards, such as FCC (U.S.A.), VDE (Germany), BS (United Kingdom), etc.

The left part of Table 9.1 summarizes the levels of the principal civilian limits, including the distance at which the limit is stipulated. When the radiation has been calculated (Section 9.3) or measured at a distance other than the specification test distance, an inverse-distance correction factor of 20 log (D_{test}/D_{specif}) can be applied.

For military equipment, the most famous and universally adopted specifications are contained in MIL-STD-461 (currently in version C), issued by the U.S. Department of Defense (DoD). Military standards are usually more stringent in the sense that they cover frequencies from 30 Hz to 10 GHz. The measurement of radiated EMI is made at 1 m. The MIL-STD-461 levels have also inspired several agencies to adopt specifications for controlling emissions in critical, though nonmilitary, equipments such as medical electronics, aircraft instrumentation, public telecommunications systems, etc. The levels of MIL-STD-461 are shown on the right-hand column of Table 9.1.

9.5 Box Skin SE in the Emission Case

The categories of box construction and the corresponding calculation routine are the same as described in Section 6.1. However, since the radiating source is inside the box, near-field conditions will likely exist for all or part of the spectrum. Therefore, given the average distance r from the source to the closest wall, SE is calculated for four types of boxes:

1. For a metal-framed box (as was shown Fig. 6.1), the box SE is calculated as follows:
   a. If $r < \lambda_{EMI}/2\pi$, i.e. $r_{meters} < 48/F_{MHz}$, the box SE $= \pi r/L$ for magnetic field sources and $r > L$. For electric field sources and $R > L$, box SE $= \lambda^2/4\pi rL$.
   b. If $r > \lambda_{EMI}/2\pi$, i.e. $r_{meters} > 48/F_{MHz}$, box SE $= 104 - 20 \log (L_{mm} \times F_{MHz})$ for $L \leqslant \lambda/2$ [from Eq. (6.3)].

Since in many cases (e.g., welded tubes or struts construction) the condition $r > L$ is not met, this type of frame provides virtually no barrier against LF magnetic emissions from inside.

2. For metalized plastics or metal-loaded plastics:
   As discussed in Section 6.1.2, this type of housing offers little or no absorption and works primarily by the reflection losses R_{dB}, i.e., field and barrier mismatch. Therefore, given the barrier impedance Z_b in ohms per square, the reflection loss is calculated as follows:
   a. If $r_m < \lambda_{EMI}/2\pi$, i.e. $r_m < 48/F_{MHz}$, against a magnetic-field source:

$$R_{dB(H)} = 20 \log \frac{2r_m F_{MHz}}{Z_b} \qquad \text{[from Eq. (6.7b)]}$$

   Against an electric source:

$$R_{dB(E)} = 20 \log \frac{4{,}500}{(Z_b\, r_m\, f_{MHz})} \qquad \text{[from Eq. (6.7a)]}$$

$$= 73 \text{ dB} - 20 \log (Z_b r_m f_{MHz})$$

   b. If $r_m \geqslant \lambda_{EMI}/2\pi$, i.e. $r_m > 48/f_{MHz}$:

$$R_{dB} = 20 \log 94/Z_b$$

[from plane wave reflection, Eq. (6.7)]

3. For wire meshes or screen-lined plastics, the calculations are identical to case 1 above, except that the condition $r > L$ is replaced by $r > g$, where g is the mesh size, and always will be satisfied. Therefore some attenuation to H-field can be expected within the limits described in Section 2.1.5.
4. For a sheet metal box, the SE_{dB} calculated in Chapter 2 and shown in the appendix applies, provided the near-field condition is accounted for.

9.6 Aperture Leakages in the Case of Emission Shielding

As previously mentioned, the SE of an aperture in the near field departs significantly from its plane wave value given by Eq. (7.4a). (incorporating an average shadow effect of 3 dB):

$$SE_{dB} \approx 100 - 20 \log (L_{mm} f_{MHz}) + 20 \log (1 + l_n \frac{L}{S}) + 30 \, d/g$$

In the above equation, 30 d/g is the WGBCO absorption term, and the rest of the equation represents the reflection term.

Since in near field the wave impedance $Z_w \neq 377 \ \Omega$, the reflection term will be affected. The mismatch between the wave impedance and the slot impedance will be higher for the case of a near electric field (more mismatch = greater SE) and lower for the case of a near magnetic field (lesser mismatch = lower SE). Chapter 1, Section 1.2 describes the near-field wave impedances for ideal E-field or H-field cases. The more pragmatic "in-between" cases established that:

$$Z_w = \frac{18{,}000}{r_m f_{MHz}} \text{ for } Z_{circuit} \geqslant \frac{18{,}000}{r_m f_{MHz}}$$

(high-Z case)

$$Z_w = Z_{circuit} \text{ for } \frac{18{,}000}{r_m f_{MHz}} > Z_{circuit} > 7.9 \, r_m f_{MHz}$$

(general near-field case)

$$Z_w = 7.9 \, r_m f_{MHz} \text{ for } Z_{circuit} < 7.9 \, r_m f_{MHz}$$

(ideal low-Z case)

This concept appeared in Chapter 1, Fig. 1.4, and Eqs. (1.18), (1.19) and (1.20), showing that the wave impedance cannot be higher than the "ideal" Hertzian electric dipole (perfect E-field) nor lower than the "ideal" magnetic loop (perfect H-field) and in most cases is bound by the radiating circuit impedance Z_c.

Setting aside the case of pure E-field, which is generally academic for virtually all EMI situations, it is established that when an emission source within a shielded enclosure is in a near-field case i.e. $r_m < 48/f_{MHz}$ two conditions may apply:

1. If $Z_c \geqslant 7.9\ r_m\ f_{MHz}$,

$$SE_{(near)\ dB} = 100 - 20 \log \frac{377}{Z_c} - 20 \log (L_{mm}\ f_{MHz}) + 20 \log \left(1 + l_n \frac{L}{S}\right) + 30 \frac{d}{g} \tag{9.6}$$

$$\approx 48 + 20 \log Z_c - 10 \log (L_{mm}\ f_{MHz}) + 10 \log \left(1 + l_n \frac{L}{S}\right) + 30 \frac{d}{g} \tag{9.6A}$$

2. If $Z_c < 7.9\ r_m\ f_{MHz}$,

$$SE_{(near)\ dB} = 20 \log \frac{\pi r}{L} + 20 \log \left(1 + l_n \frac{L}{S}\right) + 30 \frac{d}{g} \tag{9.7}$$

In case 1, the SE value found will be valid for any wave impedance low or high, since a circuit impedance Z_c smaller than 377 Ω, forcing a predominant H-field mode, will result in lower SE than plane wave while a $Z_c > 377$ Ω will result in a higher SE. Notice that in case 2 for the "ideal" H-field loop, the SE becomes independent of both the frequency and Z_c, as long as condition 2 is met.

Illustrative Example 9.3

A 5 V, 20 A power supply operates at a switching frequency of 100 kHz. In the secondary loop (formed by the transformer output, the rectifying diodes and the capacitor bank), the peak half-wave current is 50 A with a repetition frequency of 100 kHz. The impedance of the loop is very low (merely the impedance of the few turns of the transformer secondary, and the capacitor equivalent series resistance), estimated at 0.1 Ω. What attenuation is offered by a 50 mm diameter hole, located 250 mm from the center of the source?

First, we must check if we are in the near field and which of Eqs. (9.6) and (9.7) applies. Because 0.25 m is $< 48/0.1$, the near field is indicated. We can also compute that $7.9\ r_m \times f_{MHz} = 7.9 \times 0.25 \times 0.1 = 0.197$. Since $Z_c = 0.1\ \Omega$, which is < 0.197 as per case 2, Eq. (9.7) applies. Thus,

$$\text{SE} = 20 \log \frac{\pi \times 250}{50} + 0 \text{ dB (round hole, L = S)} +$$

$$0 \text{ dB (no WGBCO)}$$

$$= 24 \text{ dB}$$

This value is valid until either the impedance condition or the near-field condition (i.e., $f \leqslant 48/0.25$ or 192 MHz) no longer applies.

Illustrative Example 9.4

A 13 MHz RF signal is generated by the output section of a transmitter amplifier having 75 Ω of circuit impedance. What is the attenuation offered by the same aperture as in the previous example, i.e., a 50 mm hole, 250 mm away from the source?

Because 0.25 m is $< 48/13$, we are again in near field. We compute that $7.9\ r_m \times f = 7.9 \times 0.25 \times 13 = 26$. Since $Z_c = 75\ \Omega$, which > 26, case 1 and Eq. 9.6 apply. Thus,

$$\text{SE} = 48 + 20 \log 75 - 20 \log 50 \times 13 + 0 \text{ dB} + 0 \text{ dB}$$

$$= 29 \text{ dB}$$

This equation is applicable until the Z_c condition is not met (occurring at 37 MHz). Here SE will rejoin the value of 24 dB found in the previous example and keep this value until the near-field condition no longer applies, i.e. for $f \geqslant 48/0.25 = 192$ MHz, above which the general equation for aperture SE applies.

9.7 Degradation of Reflection Shielding Due to Cavity Resonance

When the largest dimension of a metal or metalized plastic box reaches $\lambda/2$, the internal source excites the first cavity resonance of the box.[8] Ironically, the more conductive the metal, the more pronounced the resonance can be since a high-Q resonant circuit is formed, with a small resistive part (compared to the reactive component). Quality factors of 3, 5, or more have been measured in empty metal boxes with smooth internal walls and a source located in the middle. This is particularly troublesome when the only attenuation expected is by reflection. The case of a thin metal coating on a plastic box is shown in Fig. 9.9a where the first incident wave is reflected and, after a two-way trip, comes back in phase on the same wall, practically doubling the first amplitude and so on. Because of this, the actual SE at these resonances can be less than expected.

Figure 9.9b shows an external field coupling through an aperture in the box and exciting the resonance of the box. The result is peak field regions inside the box, increased coupling and lower SE than would be expected. Several factors can counteract this effect:

1. Boxes are not empty; rather, they are populated with a mix of conductive and nonconductive items (cards, cables, components, etc.) which degrade the cavity Q.
2. One should design the equipment to have many inner compartments, ribs, etc. which will resonate only at higher frequencies and increase multipath attenuation.
3. In the microwave region, the resonance can be decreased by absorptive, lossy coatings based on ferrites, corrugated, graphite-coated layers, etc.

9.8 Importance of Aperture and Seam Polarization with Near-Field Sources

The orientation and polarization of seams and slots is more critical in the case of a nearby H-field source. Since the attenua-

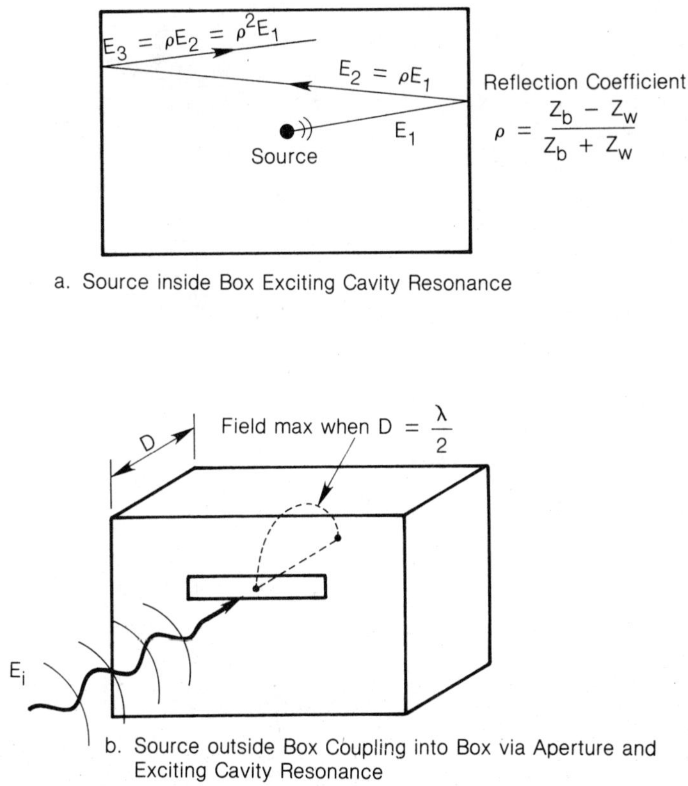

Figure 9.9—Degradation of SE due to Box Internal Reflections

tion of slots is generally mediocre, the designer should take advantage of the directivity of the magnetic field lines, especially when ferrous or other magnetic shields are employed.

Figure 9.10 shows some precautions concerning the orientation of aperture and seams against an H-field. Typically, cross-polarization provides at least 20 dB of improvement over the worst-case, parallel orientation. Since the magnetic field lines can exhibit highly distorted paths, defeating any optimization of slot orientation, it is preferable to replace long slots with arrays of smaller round or square holes wherever possible.

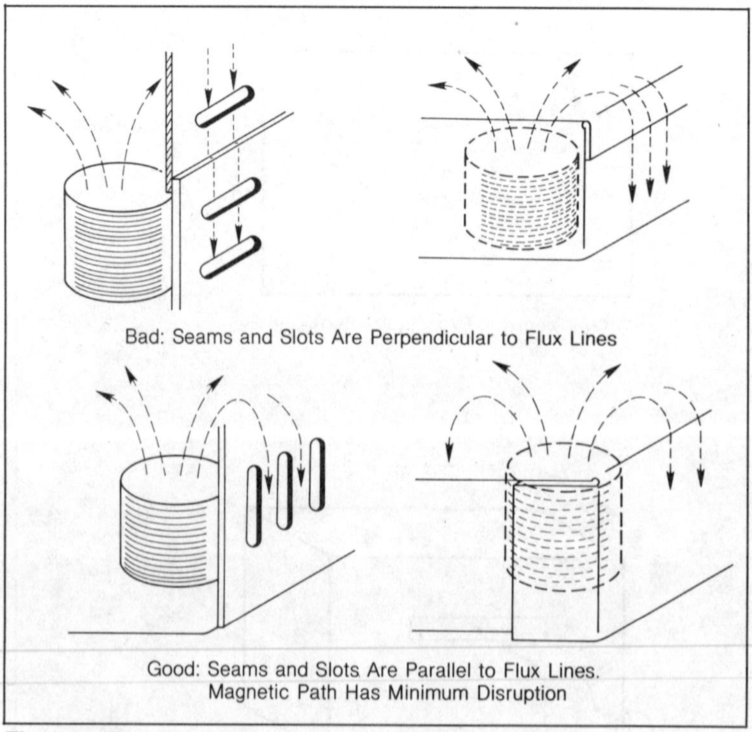

Figure 9.10—Precautions to Provide Box Shielding against Nearby H-Field Sources

A peculiar polarization case occurs when the field propagates parallel to the shield surface with the E-field perpendicular to the slot. The usual SE equations do not apply since they assume normal incidence. Absorption is low and reflection on the metal surface is potentially large (reflection factor improves with large incidence angles), but the presence of the aperture creates a direct capacitive coupling path between the E-field source and the victim (or antenna) on the other side of the shield.

9.9 Examples of Shielding Design and Hardening for Emissions

Radiated EMI from clock pulses has been a growing concern ever since electronic data processing (EDP) equipment has invaded the military and industrial/commercial applications. Clock emissions always emerge from the broad radiated spectrum produced by EDP equipment because, in contrast to the random nature of a typical NRZ message, clock emissions appear as a stable, coherent signal with a corresponding narrowband spectrum. Civilian and military organizations (CISPR, VDE, FCC) are controlling the problem by putting adequate limits on those radiated emissions, such that a compliant equipment will not cause co-channel interference to a reasonably close radio receiver.

To fully implement our SE methodology in the emission case, we will use the form of Fig. 9.11, which is derived from the form

Figure 9.11—Shielded Box Design Requirements and Performance Worksheet (Emission Case)

Topic Identification	EMI Fix	10 kHz	100 kHz	1 MHz	10 MHz	30 MHz	100 MHz	300 MHz	1 GHz	3 GHz	10 GHz
A. SE Objectives (1 + 2 − 3 + 4)**											
1. Circuit Voltage (dBV)											
2. Voltage-to-Field Coeff. (dBμV/m/V)											
3. Specification Limit (dBμV/m)											
4. Design Margin (dB)											
5. Box SE Modified Objectives											
6. Leakage SE Modified Objectives											
B. SE Box Performance											
C. Box Leakages*											
1. Viewing Window Apertures											
1.1 CRT Displays											
1.2 Alphanumeric Displays											
1.3 Panel Meter Displays											
1.4 Status Indicators											
2. Cooling Aperatures											
2.1 Cooling Apertures											
2.2 Forced-Air Cooling											
3. Assembly and Maintenance Seams											
3.1 Box Housing Assembly											
3.2 Cover Panels											
4. Component Aperatures											
4.1											
4.2											
D. Total SE* Performance											
E. Performance vs. Objectives (D − A)											

$$^{*}SE = -20 \log \left[\log^{-1}(-SE_{BOX}/20) + \sum_{i=1}^{n} \log^{-1}(-\text{leakages}/20) \right]$$

**SE cannot be negative

in Fig. 4.3 as used in Chapters 5 through 8. The "A" items in the upper portion have been modified to correspond to predicted emission levels.

Line A.1 is the circuit voltage at each frequency (unless the source is a CW, whereas only one frequency would be entered). These data come from the Fourier analysis of Section 9.2 and appear in column 5 of Fig. 9.3.

Line A.2 is the combination of coefficients (loop area × number of loops × field from a 1 cm^2 loop) found by adding the values in columns 6, 7 and 8 of Fig. 9.7. That is, the figures entered on line A.2 equal:

$$20 \log N + 20 \log A + E_0 \text{ dB}\mu\text{V/m found in Fig. 9.4a or 9.4b}$$

where,

N = number of loops

A = area in cm^2

E = electric field strength in dBμV/m

The B, C and D sections are identical to the susceptibility worksheet. A finer resolution is desirable in the VHF-UHF range, and the decade intervals are divided into half-decades since more problems in meeting radiated EMI standards are found in these frequency bands.

Illustrative Example 9.5: Shielding Design for the 10 MHz Clock of Example 9.1 vs. FCC Class B Limits

We will continue with the 10 MHz clock of Example 9.1 and the PCB of Example 9.2. We assume the same unmodified box as shown in Fig. 8.1, with a zinc-arc coating that provides surface resistivity of 0.7 Ω/sq.

The average PCB trace-to-shield distance (mean distance between the closest and the most remote) is 15 cm. This corresponds to a near-far transition frequency of:

$$f_{NF-FF} = 48/0.15 = 320 \text{ MHz}$$

Thus, all emissions below 320 MHz will be in a near-field case.

To determine the SE of the box skin, use the routine of Section 8.1.2 except for the wave impedance below 320 MHz. SE is due to reflection losses:

$$SE_{\text{skin (near)}} = \text{R losses} = 20 \log \frac{Z_w}{4Z_b}$$

In our case, Z_w is bound by the circuit impedance Z_c of 500 Ω, so:

$$SE_{\text{skin}} = 20 \log \frac{500}{4 \times 0.7} = 45 \text{ dB}$$

Figure 9.12—Shielded Box Design Requirements and Performance Worksheet (Emission Case)

Topic Identification	EMI Fix	10 kHz	100 kHz	1 MHz	10 MHz	30 MHz	100 MHz	300 MHz	1 GHz	3 GHz	10 GHz
A. SE Objectives (1 + 2 − 3 + 4)**					NA	24	21	13	3	NA	NA
1. Circuit Voltage (dBV)					8	−2	−22	−42	−62	−82	−102
2. Voltage-to-Field Coeff. (dBμV/m/V)					44	56	76	91	101		
3. Specification Limit (dBμV/m)					NA	40	43	46	46	NA	NA
4. Design Margin (dB)						10	10	10	10		
5. Box SE Modified Objectives											
6. Leakage SE Modified Objectives											
B. SE Box Performance						45	45	48	50	55	63
C. Box Leakages*											
1. Viewing Window Apertures											
1.1 CRT Displays					37	27	17	7	3	2	2
1.2 Alphanumeric Displays											
1.3 Panel Meter Displays											
1.4 Status Indicators											
2. Cooling Aperatures											
2.1 Cooling Apertures					56	46	36	26	12	3	3
2.2 Forced-Air Cooling											
3. Assembly and Maintenance Seams											
3.1 Box Housing Assembly					58	48	38	28	18	8	3
3.2 Cover Panels											
4. Component Aperatures											
4.1											
4.2											
D. Total SE* Performance					55	25	15	5	0	0	0
E. Performance vs. Objectives (D − A)					NA	1	−6	−8	−3	NA	NA

$$*SE = -20 \log \left[\log^{-1}(-SE_{BOX}/20) + \sum_{i=1}^{n} \log^{-1}(-\text{leakages}/20) \right]$$

**SE cannot be negative

*SE Box Skin Improves at 100 MHz due to Emergence of Absorption Loss.

Next, determine aperture leakages for the CRT, cooling slots and assembly seams. Above 320 MHz, the figures are identical to those of the initial run in Chapter 8, Fig. 8.2. Below 320 MHz, Eq. (9.7) applies; the SE of the three apertures are calculated accordingly and entered in lines C.1, C.2 and C.3 of Fig. 9.12.

The first iteration shows that the housing fails to meet the objective by −6 dB and −8 dB between 100 and 300 MHz. The cause of this failure is leakage through the CRT. Other leaks do not exceed our dB budget. Since in the second iteration of the susceptibility study, the CRT leakage was reduced (SE improved) by about 60 dB, this fix will be more than adequate to satisfy the emission shielding needs.

Illustrative Example 9.6: Determining the Necessary SE Hardening for Higher Clock Frequency and Tougher Emission Limits (MIL-STD-461)

In the two decades that RFI regulations have existed, the typical clock rates used by CPUs have jumped from a few megahertz to few tens of megahertz, i.e., one order of magnitude. Without debating over the linear or logarithmic increase of the clock speeds over the years of technological advancement, it is already foreseeable (and the technology is available) that by the early 1990s, clock rates in the 100 to 500 MHz range will be common.

Since the radiated EMI levels will increase with frequency (as explained in Section 9.3) and the specifications will certainly not be relaxed, a greater amount of EMI suppression will be needed. This example is to forecast the additional "dB budget" needed when going from 10 to 100 MHz clock rates.

When changing from a 10 MHz to a 100 MHz clock, we assume the following changes will occur in the PCB design:

1. Because of the high speeds involved, all PCB traces carrying high-speed signals will be terminated in matching resistances approximately equal to their characteristic impedance Z_0, for which we will assume to be 100 Ω.
2. Multilayer boards will be used (for the same speed concern), resulting in lower radiation.
3. The 10 ns transition time of the logic pulses will be reduced to 1 ns.

4. Due to the highly critical aspect of military equipment and the cost and difficulty in performing several development trials, a 20 dB margin is desired. This is also to account for the deterioration of some SE figures under vibration and harsh climatic conditions (salt spray, thermal shock, etc.).

The limit to be met is the most stringent category of MIL-STD-461 RE02. The result of a first run, with a bare housing simply coated with the 0.7 Ω/sq zinc-arc discussed in Chapter 8, appears in Fig. 9.13.

Notice that the radiation efficiency (line A.2) has increased compared to the former 10 MHz clock against the civilian (FCC)

Figure 9.13—100 MHz Emission Example: Military Equipment, First Run

Topic Identification	EMI Fix	10 kHz	100 kHz	1 MHz	10 MHz	30 MHz	100 MHz	300 MHz	1 GHz	3 GHz	10 GHz
A. SE Objectives (1 + 2 − 3 + 4)**							52	51	33	16	0
1. Circuit Voltage (dBV)							8	-2	-22	-42	-62
2. Voltage-to-Field Coeff. (dBμV/m/V)							53	70	80	90	100
3. Specification Limit (dBμV/m)							29	37	45	52	60
4. Design Margin (dB)								20	20	20	20
5. Box SE Modified Objectives											
6. Leakage SE Modified Objectives											
B. SE Box Performance							45	48	50*	55*	63
C. Box Leakages*											
1. Viewing Window Apertures											
1.1 CRT Displays							17	7	3	2	2
1.2 Alphanumeric Displays											
1.3 Panel Meter Displays											
1.4 Status Indicators											
2. Cooling Aperatures											
2.1 Cooling Apertures							36	26	12	3	3
2.2 Forced-Air Cooling											
3. Assembly and Maintenance Seams											
3.1 Box Housing Assembly							38	28	18	8	3
3.2 Cover Panels											
4. Component Aperatures											
4.1											
4.2											
D. Total SE* Performance							17	7	3	2	2
E. Performance vs. Objectives (D − A)							-35	-44	-30	-14	2

$$*SE = -20 \log \left[\log^{-1}(-SE_{BOX}/20) + \sum_{i=1}^{n} \log^{-1}(-\text{leakages}/20) \right]$$

**SE cannot be negative

*SE Box Skin Improves at 100 MHz due to Emergence of Absorption Loss.

specification. This is because the measurement is made at 1 m and because the clock frequency and second corner frequency have increased (the rise time being 10 times shorter than before). This increase in radiated levels is compensated to some degree by the use of a multilayer board.

The first run shows an increase in the SE requirements of about 30 dB versus those of Illustrative Example 9.5. Using the same rationale as in the susceptibility example, Chapter 8 and Fig. 8.6 indicate that:

1. For the CRT opening, the 0.3 mm mesh provides adequate

Figure 9.14—100 MHz Clock Emission Example: Military Equipment, Second Run

Topic Identification	EMI Fix	10 kHz	100 kHz	1 MHz	10 MHz	30 MHz	100 MHz	300 MHz	1 GHz	3 GHz	10 GHz
A. SE Objectives (1 + 2 − 3 + 4)**							52	51	33	16	0
1. Circuit Voltage (dBV)											
2. Voltage-to-Field Coeff. (dBμV/m/V)											
3. Specification Limit (dBμV/m)											
4. Design Margin (dB)											
5. Box SE Modified Objectives											
6. Leakage SE Modified Objectives											
B. SE Box Performance	BETTER CONDUCTIVE PAINT .07 Ω/□						65	68	70	75	83
C. Box Leakages*											
1. Viewing Window Apertures											
1.1 CRT Displays	.3 MESH						72	62	52	42	32
1.2 Alphanumeric Displays											
1.3 Panel Meter Displays											
1.4 Status Indicators											
2. Cooling Aperatures											
2.1 Cooling Apertures	4×4 mm SLOTS						75	65	55	45	37
2.2 Forced-Air Cooling											
3. Assembly and Maintenance Seams											
3.1 Box Housing Assembly	EMI GASKET						60	60	45	30	30
3.2 Cover Panels											
4. Component Aperatures											
4.1											
4.2											
D. Total SE* Performance							54	51	39	27	23
E. Performance vs. Objectives (D − A)							2	0	6	11	23

$$*SE = -20 \log \left[\log^{-1}(-SE_{BOX}/20) + \sum_{i=1}^{n} \log^{-1}(-\text{leakages}/20) \right]$$

**SE cannot be negative

*SE Box Skin Improves at 100 MHz due to Emergence of Absorption Loss.

attenuation, with a provision for the combinational leakage (from several apertures).

2. For cooling holes, the division of long 80 mm × 4 mm slots into smaller 4 × 4 slots plus the WGBCO effect provide adequate attenuation, with a provision for the combinational leakage.
3. For assembly seams, the existing SE provided by the 150 mm screws spacing falls short by about 20 dB. This would require a 10 times reduction of the seam length, i.e. using a screw spacing of 15 mm, which is quite unpractical. Therefore a gasket must be installed. The one selected in Section 8.3.4 is inadequate, and an EMI gasket providing at least 60 dB at 100 MHz and 300 MHz, 45 dB at 1 GHz and 30 dB at 3 GHz should be selected. To provide even gasket pressure, an increase in the number of screws will be required (see Chapter 10).
4. The zinc arc spray does not provide sufficient SE. To meet the objectives with the reserve for combinational leakage, about 20 dB of improvement in box skin SE is needed at 100 MHz. This means that a process with a surface resistance of 0.07 Ω/sq or less has to be found. This value is in the range of the best industrially available coatings like silver or copper paints or electroless plating (see Chapter 2). Otherwise, a box made of homogenous metal should be selected.

Figure 9.14 shows the worksheet with the second run and the applied fixes. This example, performed on a simple box, shows that a one-decade increase in clock rate, even when accounting for some PCB improvements, results in a significant burden in box shielding (about 30 dB).

9.10 References

1. FCC Rules and Regulations, Vol. 2, Part 15.
2. CISPR Publications 11, 14 and 22.
3. MIL-STD-461C, "Electromagnetic Emission and Susceptibility Requirements for the Control of Electromagnetic Interference."
4. Di Marzio, A., "Graphical Solutions in Harmonics Analysis," *IEEE Transactions on Aerospace*, September 1968.
5. Audone, B., "Graphical Harmonics Analysis," *IEEE Transactions on EMC*, May 1979.
6. White, D.R.J. and Mardiguian, M., *EMI Control Methodology*

and Procedures (Gainesville, VA: Interference Control Technologies, Inc., 1985).

7. Bush, D., (technical paper presented at the International EMC Symposium, Zurich, 1985).
8. Wu, C.Y., "Field Distribution Inside a Box with Apertures," *1977 IEEE EMC Symposium Record* (New York: IEEE), p. 296.

Chapter 10

Shielding Integrity Hardware Performance and Installation Guidelines

The previous chapters explained that, since any practical enclosure has apertures and seams, the theoretical metal skin SE of 100 dB or more is never obtained due to leakages through apertures. This chapter discusses the available hardware by which shielding integrity can be restored, typical performances and practical applications for shielded boxes, chassis and equipments, cabinets, rooms and vehicles.

Any practical shielded enclosure has necessary penetrations and apertures of one kind or another. Some examples of such penetrations and apertures include:

1. Cover plates and access cover members
2. Meter windows
3. Windows for viewing digital or other displays (CRTs, LCD, plasma, etc.)
4. Potentiometer shafts
5. Cooling apertures
6. Power line and signal lead connectors
7. Indicator lamps
8. Pushbuttons
9. On-off switches
10. Fuses

Thus, as calculated in Chapters 7 and 8, it is not uncommon to find the plane-wave attenuation of a basic shield material to be greater than 100 dB, for example, while the actual enclosure will exhibit 50 dB, particularly in the VHF/UHF portion of the spectrum. Here leakage of one or more of the above types compromises the integrity of the basic shielding material.

The previous calculations allowed to predict the leakage reduction obtained via meshes, waveguide effect, and so on. This chapter addresses the practical aspects of commercially available shielding integrity parts, their performances, mounting precautions and other mechanical, packaging and environmental guidelines.

10.1 Bonding Seams and Joints

The means of preserving the SE integrity of seams can depend on the type of service required of the seam. In terms of service, there are basically three types of joints (see Figs. 10.1 and 10.2):

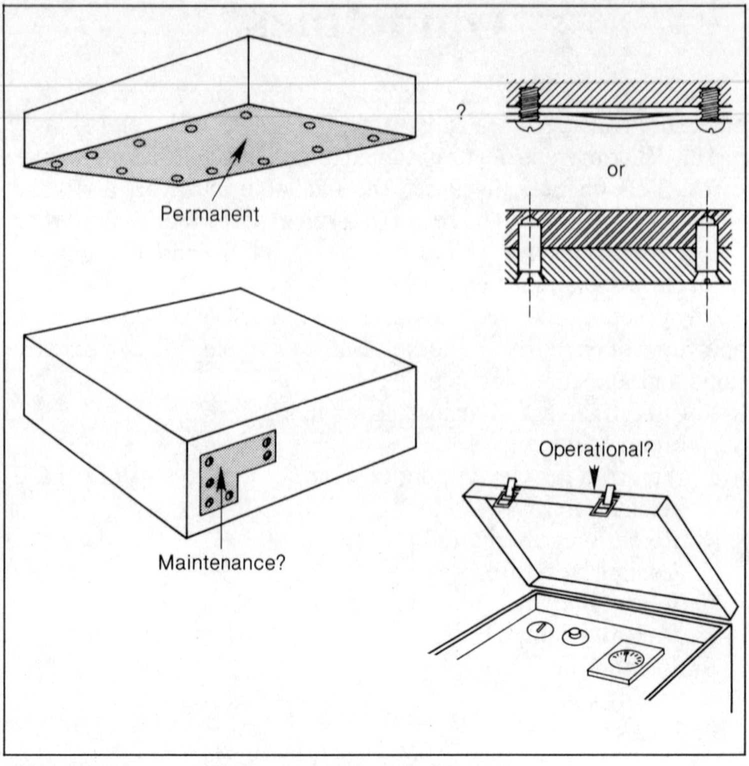

Figure 10.1—Identification of Types of Closures

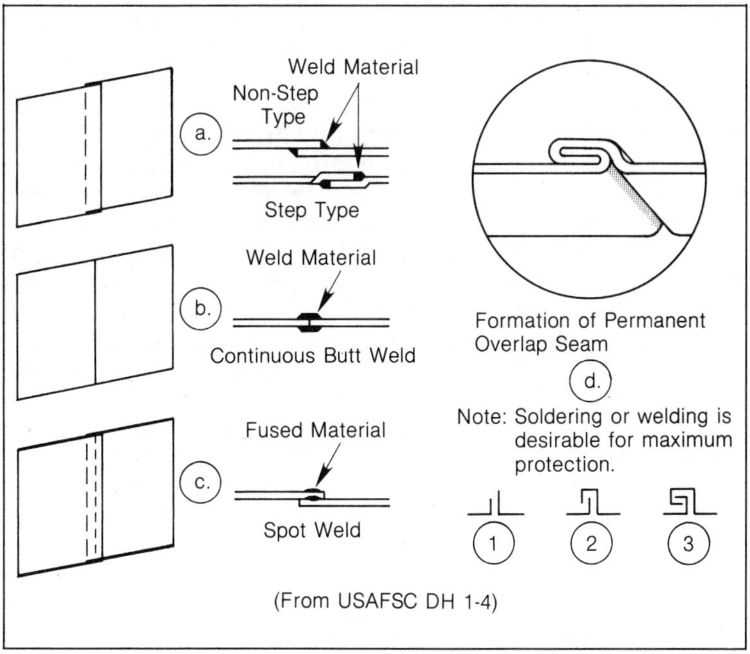

Figure 10.2—Permanent and Semipermanent Shield Seam Configurations

1. **Permanent joints** are factory-made and never disassembled except for a major overhaul. These can be welded, brazed, riveted, etc. These joints are discussed in this section.
2. **Semipermanent joints** typically include maintenance panels, covering parts which are only serviced or replaced by a technician qualified to work on that equipment (and not by the maintenance crew of the host system which, in this case, should be regarded as users). These joints are typically bolted, screwed or crimped. For this type of joint, special tooling and alignment precautions can be specified, and the replacement of a gasket, cleaning of gasket seat or application of conductive caulking are feasible measures.
3. **Operational openings** are expected to be more or less frequently handled by the user. Therefore, no special care or skill can be assumed as far as gasket handling or replacement are concerned. This is typically the domain of EMI gaskets discussed in Section 10.2.

The loss of RF shielding integrity across the **interface** of mating surfaces is a main reason why shielding effectiveness is compromised. Here, the conductivity of the interface may be much higher, and the permeability may be much lower because of the type of interface bond used. Thus, resulting material interfaces may be classified into two types: physically inhomogeneous and physically homogenous.

A physical inhomogeneous interface bond results when shielding members are directly connected by screws, rivets, spot welds, and the like. The interface connection is not continuous and there results a bowing or waviness effect between connected members. This in turn develops slits or gaps which leads to complete radiation or penetration at frequencies where the length of the slit approaches $\lambda/2$. The exact calculation of the SE for this type of thin slot was carried out in Chapters 7 and 8.

There are several techniques available for reducing electromagnetic emission or susceptibility leakage of mating members. If members are joined by screws or rivets, Eq. (7.4) shows that A_{dB} may be significantly increased by using more screws or rivets per linear dimension of the interface due to the reduction in the gap length g. Figure 10.3 shows joint shielding effectiveness as a function of screw spacing for the indicated parameters.

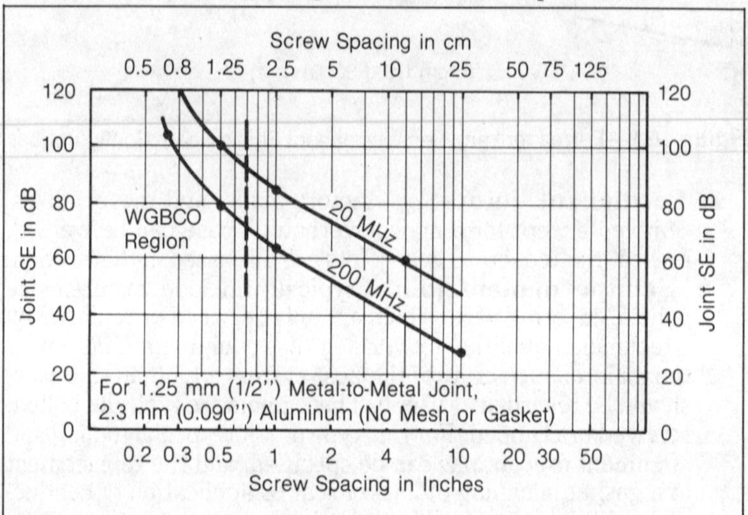

Figure 10.3—Minimum Shielding Effectiveness for Screw-Secured Joints (Plane Wave)

Other techniques available for reducing the leakage in a physical homogenous mating member bond, involve attempting to eliminate or reduce the inhomogeneity. Figure 10.2 illustrates some of these approaches. Where members do not have to be disengaged or separated, a continuous seam weld around the periphery of the mating surfaces is preferred. This type of weld is not critical provided it is continuous and has no weld **pinholes**. One exception involves using a weld filler with different electrical characteristics from those of the basic shield member material. Hence, either the conductivity or permeability of the weld filler may be much lower resulting in degradation of shielding effectiveness. The seam weld technique is of questionable value when used with the more exotic magnetic materials ($\mu r > 1{,}000$; see Table 2.1) which must be annealed before assembly. Here, welding will destroy the specific properties that the annealing produced.

An alternative technique shown in Fig. 10.2 is the overlap seam. All nonconductive material (e.g., paint, rust, coatings, etc.) must be removed from the mating surfaces before they are crimped. Crimping must be performed under sufficient pressure to ensure positive contact between all mating surfaces.

For shield members that may have to be separated from time to time, such as cover and access plates for equipment alignment or maintenance, none of the above techniques is acceptable. A temporary but good bond is required, and this is the role of RF gasketing materials such as finger stock, resilient mesh and conductive elastomers. The topic of gaskets is covered in Section 10.2.

10.2 EMC Gaskets

This section discusses another very important class of techniques used to restore shielding integrity at seams and joints where other than permanent fastening methods are used. Gaskets are employed for either temporary or semipermanent sealing applications between joints or structures, such as:

Temporary RF Sealing Applications

1. Securing access doors to enclosures, cabinets or equipments
2. Mounting cover plates or removal panels for equipment maintenance, alignment or other purposes

Semipermanent RF Sealing Applications

1. Mounting either screen or conducted glass windows to housings containing electrical or electronic test equipment
2. Mounting honeycomb and other ventilation covers to enclosures, cabinets and equipment

3. Securing parallel members of an equipment housing to a frame structure using machine screws

Except for the finger stock and spiral types, all gaskets (whether they seal EMI or high-pressure fluid, make a container waterproof or simply keep forced ventilating air from escaping at a door-to-cabinet joint) conform to the unavoidable irregularities of the mating surfaces of a joint. Some examples are:

1. The joint at the lid of a fluid container
2. Housing for an emergency radio or beacon to be dropped into the sea
3. The joint between the cover and enclosure of a radar pulse modulator

In each example the joint has two relatively rigid mating surfaces, and neither surface is perfectly flat. When the surfaces are mated without a gasket, closing forces may not be sufficient to cause the two surfaces to mutually seal. Resultant gaps will allow leaks to exist. A gasket resilient enough to mate to both surfaces under reasonable force, however, will eliminate these leaks. In a garden-hose coupling, for example, trying to prevent a leak by force alone without a gasket will be fruitless.

With a gasket (washer) placed in the hose fitting against a faucet, even hand torque results in a water-tight joint. To try to get the same water-tightness by accurate machining of both surfaces would be prohibitively expensive. Thus, in most cases, the least expensive way to obtain a tight joint (water-tight, oil-tight or EMI-tight) is to make the mating surfaces to normal tolerances in terms of flatness, rigidity and roughness, and then to add a gasket to compensate for the resulting poor fit between the two surfaces.

The performance of a gasket used to maintain or restore the SE of a given enclosure depends as much on the design of the box, especially around the mating member, as on the gasket itself. Once the SE objectives for closing edges (lines 3.1 and 3.2 of the SE worksheet) have been determined, several parameters have to be considered for proper gasket selection:

1. Environmental considerations
2. The shape and material of mating members
3. Gasket material and compression
4. Type and number of fasteners

Since two or more of the above parameters typically are interactive, more than one iteration may be necessary before reaching the optimum configuration. The next sections will guide the reader through the most logical sequence of selection steps:

1. Listing the environmental constraints
2. Determining the proper style and material of mating members
3. Selecting an appropriate gasket
4. Designing and dressing mating faces for gasket mounting

10.2.1 Environmental Considerations

In addition to the desired SE figures, the following constraints may also be applicable to the joint:

NEMP or Other Unusual Current Flow (Lightning) Survivability

Not all gaskets need to withstand electromagnetic pulse currents, but if the equipment falls into this category, the gasket material must survive the high current density. Some gaskets are damaged with a single-pulse current as low as 200 A per cm of gasket for about 100 ns duration. Other gaskets can withstand more than 1 kA per centimeter of gasket.

Inside and Outside Pressure (Air, Gas, Vapor, Etc.)

The inside and outside of the box may not be at the same pressure, being exposed to a casual or permanent vacuum, or high pressure. In this case, the gasket will have to be a combination of an electrical and a pressure seal. Otherwise, two gaskets (one electrical, one for pressure) will be needed.

Salt Spray, Corrosive Atmosphere

Any joint between dissimilar metals (or even similar ones if there is some regular current flow) will corrode in the presence of an electrolyte such as salt water. Also, some environments are basically corrosive, containing chemicals, fumes, etc. (e.g., an engine compartment, the proximity of a rocket nozzle, the environment in chemical or other industrial operations). The two metal-to-gasket interfaces in an EMI seal must resist this type of environment if the application calls for it. For instance, some ordinary wire mesh gaskets lose 30 to 50 dB of their initial SE after extended exposure to salt spray.[1–3] As for pressure, this threat may eventually require a combined gasket. Or as a minimum, the galvanic combination of the gasket material with edges material must be checked, and a proper selection of both flange/cover finish and gasket material is required.

Unusual Temperatures and Thermal Shock

In certain applications, the gasket will have to stand extremely high (280°C) or low (−60°C) temperatures, or both alternately. Neglecting these aspects may cause a loss of SE during certain periods of time, compromising the mission of the host system. Although not related to EMI, the tragedy of the U.S. space shuttle "Challenger" in 1986 was traced to a disregard of gasket performance under low-temperature conditions.

Vibrations

In spite of historical evidence that two adverse conditions may occur jointly to create a failure that neither would have created alone, there is still an obscure belief that "Murphy will never strike twice." Consequently, EMI testing is almost invariably done under static mechanical conditions.

In real life, some equipment may be subject to severe vibrations, causing temporary flexing of covers, flanges, chassis, etc. This flexing tends to reduce contact pressure during the vibration. Some gaskets may lose 20 to 40 dB of SE under only 3 g of vibration. Figure 10.4 shows how the contact pressure decrease degrades the shielding effectiveness of one type of EMI gasket. Therefore, if the application implies vibration, the flange-fastener-gasket combination must be designed for the worst-case gap.[2]

Number of Reclosures

Openings requiring frequent access will typically have poorer mechanical tolerances, and also a lesser number of screws or will have quick-release fasteners or latches. Also, the gasket selection may need to account for openings and closures which occur in harsh

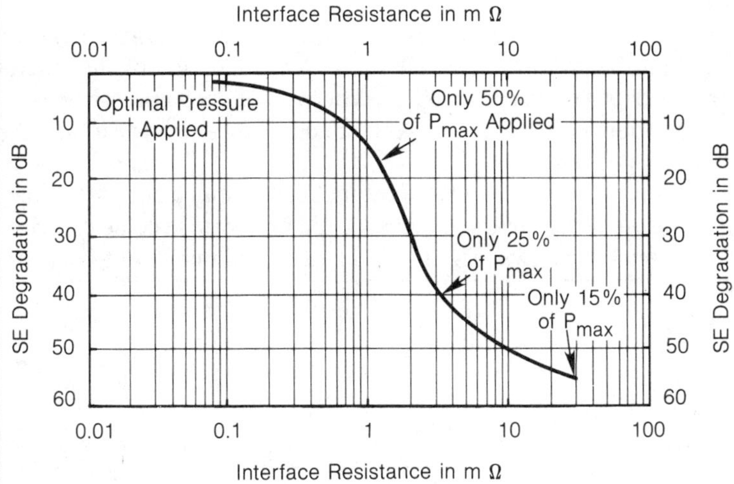

Figure 10.4—Typ. Degradation in SE of Elastomer Gaskets Caused by Insufficient Pressure, Resulting in an Increase in Resistance of the Gasket-Metal Joint; Measured Δ dB about 100 dB of SE when Correctly Compared (Source: Chomerics)

environments (salt fog, moisture) when some corrosive agent may stay trapped under the covers. A large number of reclosures in the product shelf life may also prohibit the use of some gaskets which exhibit too large a compression set, where the metal retains a variable "memory" of its previous cross section.

Repeated Flexures

Some applications require that the part equipped with the gasket be folded, bent over, etc., or the gasket path may be tortuous. This generally is specified as the gasket "flex range" and given in terms of degrees such as 0 to 90° or 0 to 180°.

Sunlight, Ultraviolet (UV) Radiation, Etc.

The action of sun or other light may accelerate the aging of some conductive elastomer gaskets by causing hardening and cracking.

10.2.2 Design of the Mating Members

The mating sections (edges, flanges, door frames, etc.) should be as rigid and stiff as possible. If enough thickness cannot be provided, then ribs, folded edges or other reinforcing features must be added.

Installing the gasket in a groove in the mating member is usually preferable, allowing a well controlled deflection of the gasket (groove depth ranges from 50 to 90 percent of uncompressed gasket height, depending on recommended closure pressure for the selected gasket).

Whatever the box material (cast metal, sheet metal or particularly plastic), the design of the flanges should minimize leakage contributions from the screws (Fig. 10.5). The size and number of fasteners

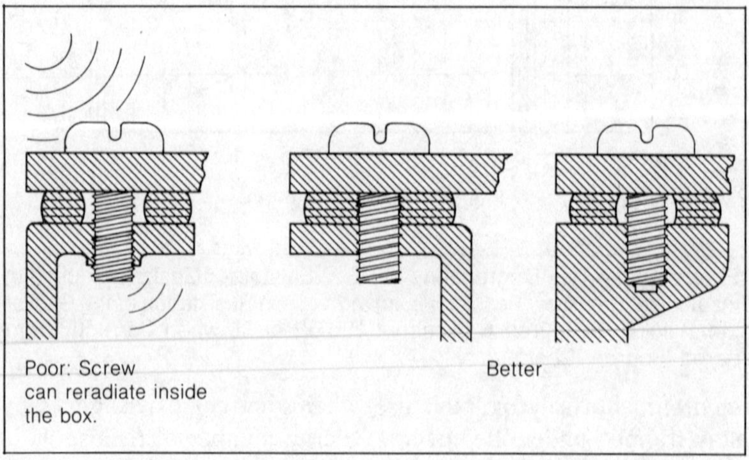

Figure 10.5—Incorrect and Correct Design of Metal Flanges to Prevent Screw-Leakage Inside the Box

cannot be finalized until the gasket itself is selected, and this choice itself can depend on the compression available from the flange. To solve this typical "chicken-and-egg" situation, a first iteration is generally made with a typical number of screws of fasteners (Fig. 10.6), then the validity of the design is rechecked once a gasket choice is made.

With plastic boxes, tongue-and-groove or other self-adjusting profiles can be used for holding the gasket (Fig. 10.7). If a box is painted for aesthetics, the manufacturer must take care (1) not to extend the decorative paint too far inside the gasket groove or seat and (2) to extend the conductive coating (paint, plating, etc.) well inside the gasket groove or seat. Figure 10.8 illustrates improper and correct applications.

For low SE requirements (below 20 to 30 dB) a gasket may not

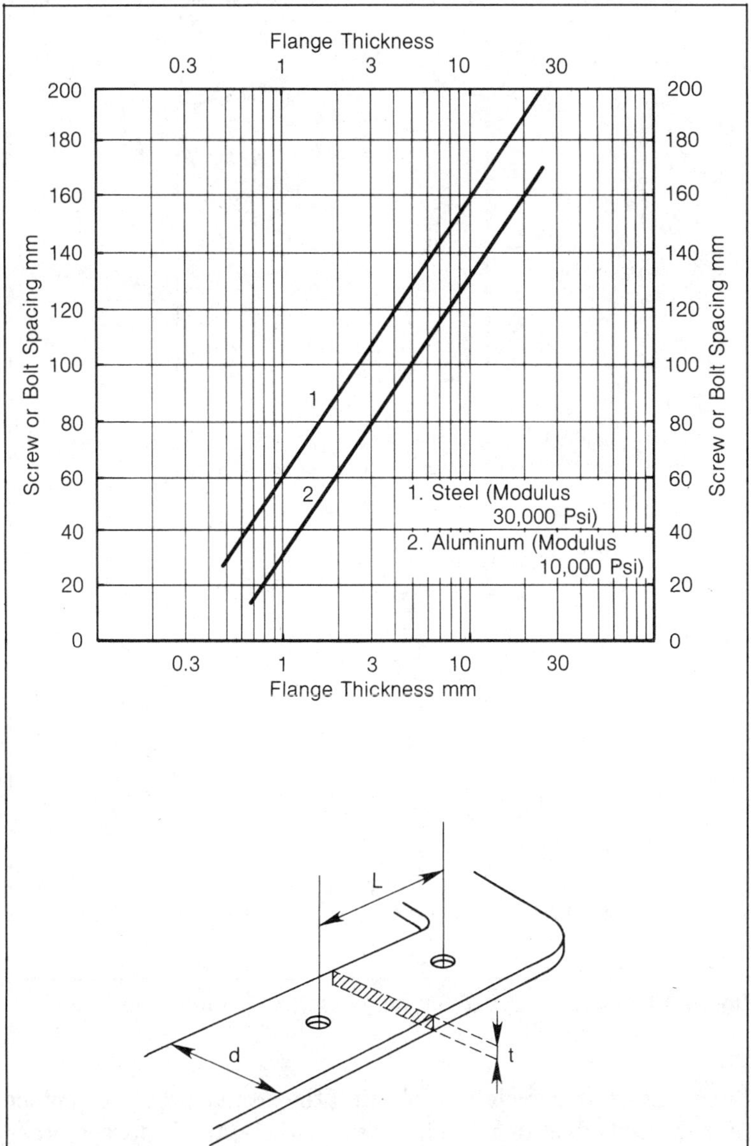

Figure 10.6—Recommended Maximum Spacing to Minimize Buckling of Cover Flanges—Valid for Width-to-Thickness Ratios d/t of ≅ 10 to 30

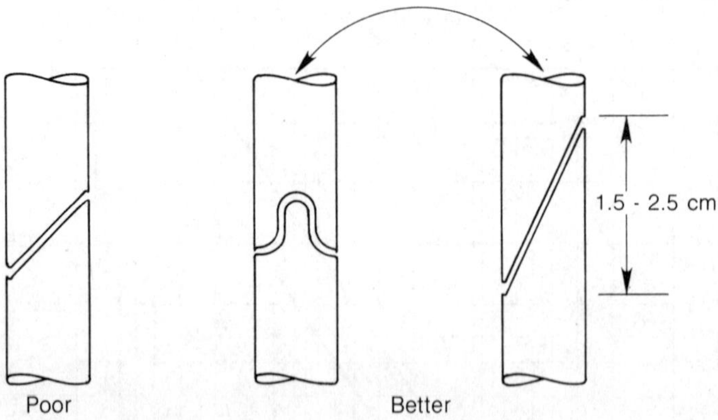

a. A tongue-and-groove design of plastic edges will provide a longer path against EMI reradiation or a longer creepage against arcing.

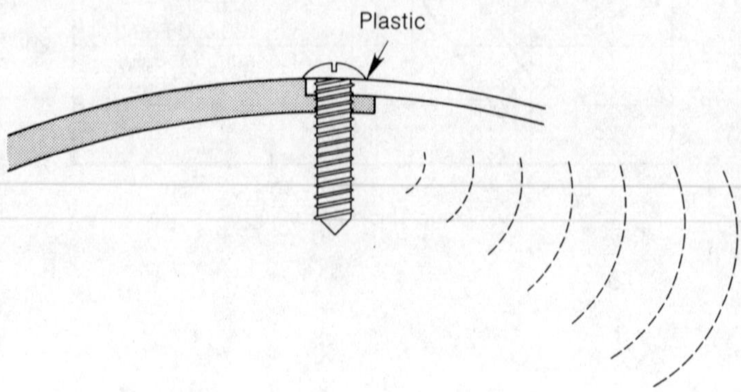

b. Long protruding ungrounded or poorly grounded screws will act as reradiators inside the housing. They should be embedded in cover thickness; otherwise, fillets should contact the metallized portion.

Figure 10.7—Plastic Cover Design for Better EMI Immunity

be necessary with metalized plastics; the normal pressure contact of metalized edges or SE enhancement via regularly spaced brass springs may be sufficient.

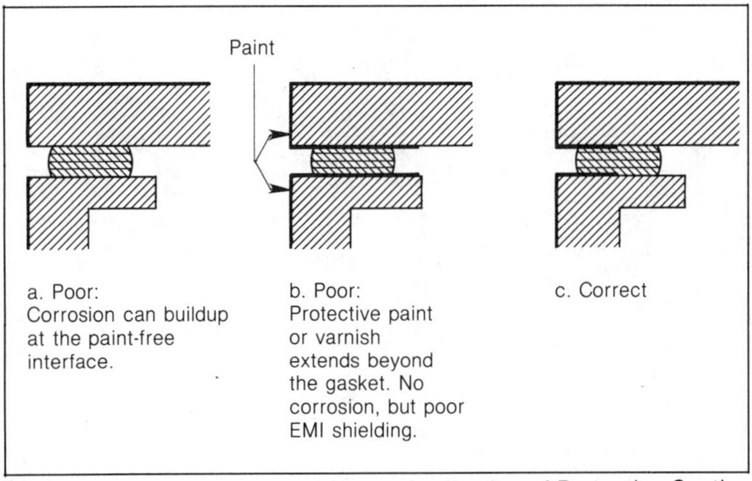

Figure 10.8—Inadequate and Proper Application of Protective Coating for Avoiding Penetration of Salt Spray in Marine Type Environments

10.2.3 Gasket Selection: Material, Dimensions and Performances

There exists a plethora of EMI gasket types, shapes, binders and materials. In fact, the profusion of gaskets is so great that it is likely to be confusing to all but those who regularly specify or use them. This is recognized by the many suppliers who have responded by producing credible application notes and design and order guides.

10.2.3.1 Types of Gaskets and Materials

For the convenience of this discussion, EMI gaskets are divided into four types: (1) knitted-wire mesh, (2) oriented immersed wires, (3) conductive plastics and elastomers, and (4) spring-finger stock. The last type is different from the first three and operates on a significantly different principle. A brief summary of the four types is presented below followed by a comparison of all four. Several examples are shown in Fig. 10.9.

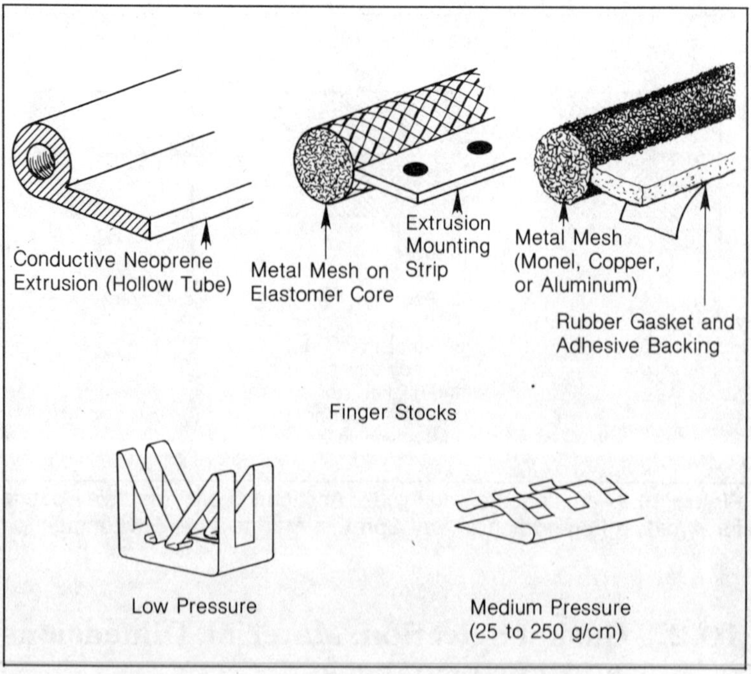

Figure 10.9—Examples of Cover Gasketing Materials

10.2.3.1.1 Knitted-Wire Mesh Gaskets

Figure 10.10 shows some examples of knitted-wire mesh gaskets.They are made from resilient, conductive, knitted-wire and somewhat resemble the outer jacket of a coaxial cable. Nearly any metal that can be produced in a fine-wire form can be fabricated into these EMI gaskets. Typical materials used are monel, aluminum, silver-plated, copper-clad steel. They should not be used in a corrosive or salt spray environment. These gaskets may employ either an air core or, for maximum resiliency, they may use a spongy neoprene or silicone core. Cross sections may be round, rectangular, or round with mounting fins. They are generally applied to shielding joints having a periphery greater than 100 mm and cross sections between 1.5 mm and 20 mm. The all-metal (no core) type has deflection range up to 30 percent. The spongy core type has a large deflection range up to 70 percent and can provide limited sealing against dust or air.

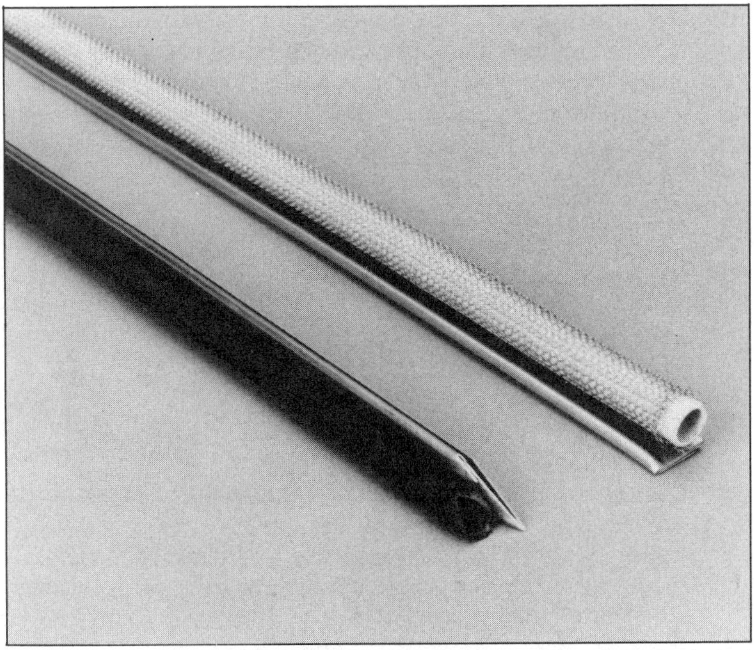

Figure 10.10—Typical Knitted Wire-Mesh Gaskets (courtesy of Technical Wire Products)

10.2.3.1.2 Oriented-Immersed Wire Gaskets (Porcupine or Expanded Mesh)

Figure 10.11 shows some examples of oriented immersed-wire gaskets. They are made with a myriad of fine, parallel, transverse-conductive wires whose parallel impedance across the gasket interface is very low. Each wire may be insulated from its neighbors or interweaved with them in the expanded version. A typical density is about 150 wires per cm^2 (1,000 wires per in^2). Standard materials used are monel or aluminum embedded in either a solid silicone (hard gasket) or a sponge silicone (soft gasket) elastomer. As such, **this gasket provides a simultaneous EMI and pressure seal**. The ends of the embedded wires protrude a few mils on either side of the elastomer to assist in piercing any residual grease/oil film and oxide on the surface of the mating members. These minuscule protruding "hairs" are usually tin plated. This feature is especially good where aging and subsequent maintenance may result in a panel member no longer being clean and grease-free. The gaskets should not be used where top covers are closing just above bare electrical parts, PCBs, etc. because, with aging and vibration, few filaments may work loose and fall into the electronics.

Figure 10.11—Typical Oriented Immersed-Wire Gaskets (courtesy of Chomerics)

Available cross sections range from 3 mm × 3 mm to 15 mm × 15 mm and come in any length. Since they are very stiff, their compression range is typically 5 to 15 percent. The material provides efficient EMI sealing for small, even joints but cannot absorb large joint irregularities. However, it is the most tolerant gasket type to pressure irregularities. Figure 10.12 shows the influence of pressure on SE of porcupine gaskets.

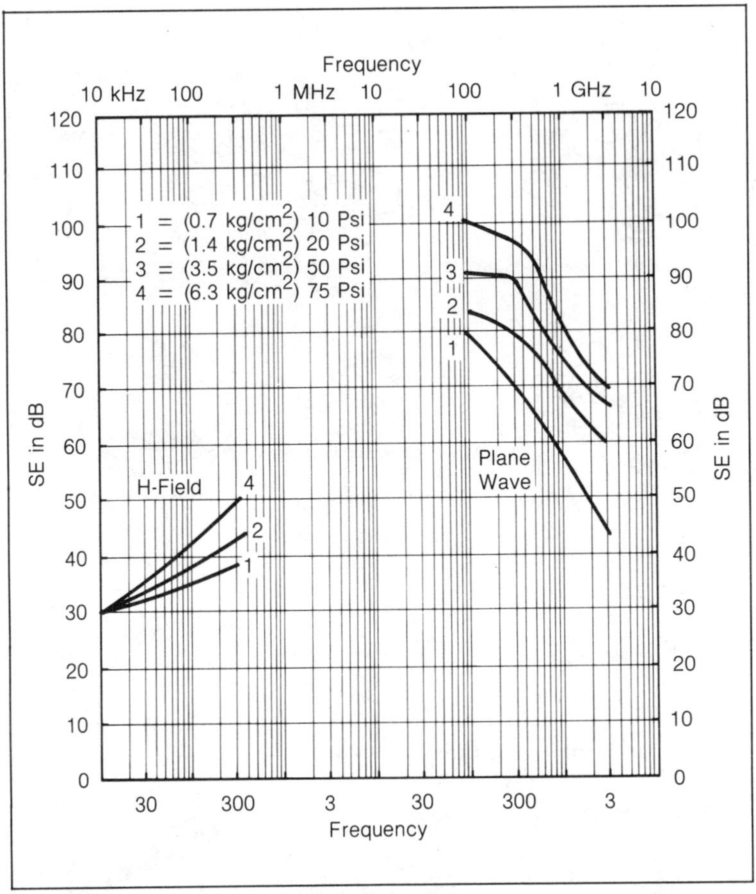

Figure 10.12—Influence of Pressure on SE of Porcupine-Type Gaskets

10.2.3.1.3 Conductive Elastomer Gaskets

Figures 10.13 and 10.14 show some examples of conductive plastic and elastomer gaskets. The most efficient (and expensive) are made with a myriad of tiny silver balls immersed in a silicone rubber or vinyl elastomer binder and carrier. As such, this gasket

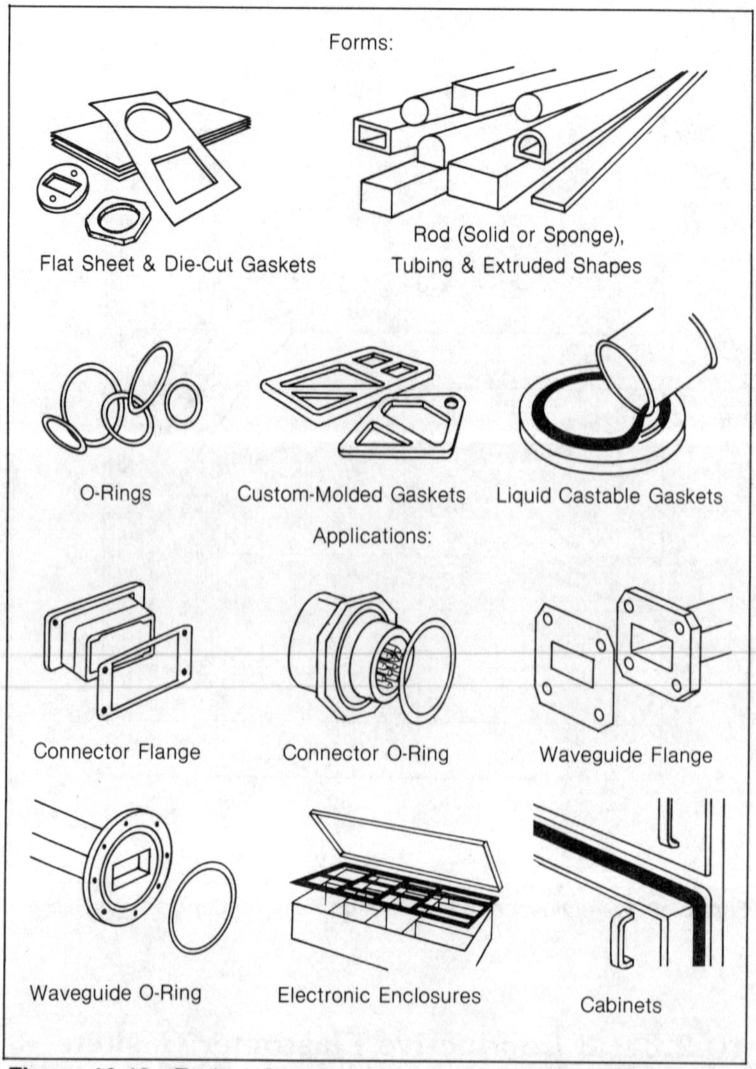

Figure 10.13—Typical Shapes of Conductive Plastic and Elastomer Gaskets

provides a simultaneous EMI and hermetic seal. Offering volume resistivities from 0.1 to 0.01 ohm centimeters (Ωcm) and useful over a wide range of temperatures, these gaskets are provided in sheets, die cuts, molded parts and extruded shapes. Some versions are

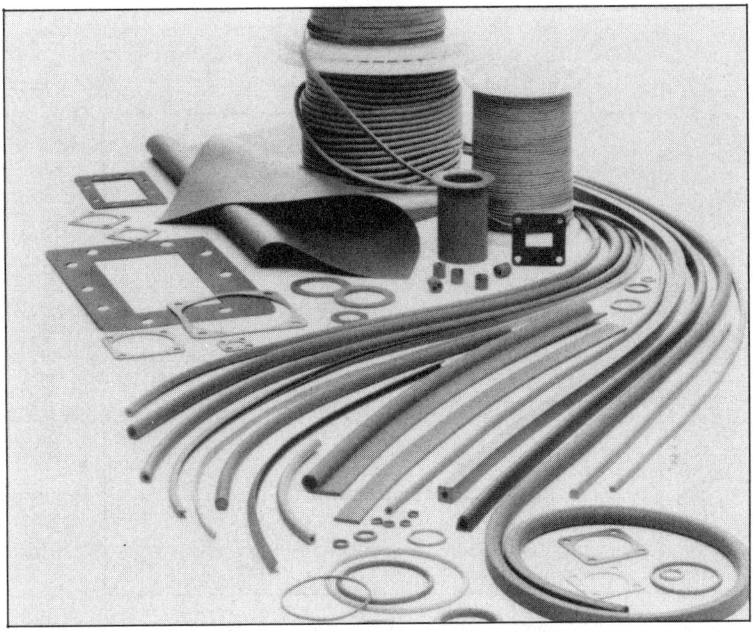

Figure 10.14—Typical Conductive Elastomer Gaskets

operable down to cryogenic temperatures. They offer low closing pressures, low compression set and maintenance, and long life.

For this type of elastomers, one of the largest manufacturers (Chomerics) uses the following criteria:

Compression set (reusability criteria) ≈ 30 percent maximum
Elongation = 80 percent minimum, 300 percent maximum
Volume resistivity = 0.01 Ωcm after heat cycling and tensile test

A more recent gasket material consists of small mercury particles trapped in the elastomer core. Performances are close to silver loading, and compression set is excellent.

The lower-cost versions use carbon filled elastomers. Due to their higher resistivity, they can offer only to 20 to 30 dB of plane-wave SE at best.

Elastomer gaskets are available with solid or hollow cores. The hollow type is very pliable and can be deflected by more than 50 percent to absorb large joints unevenness. Figure 10.15 illustrates gasket deflection vs. pressure for two common elastomer gaskets.

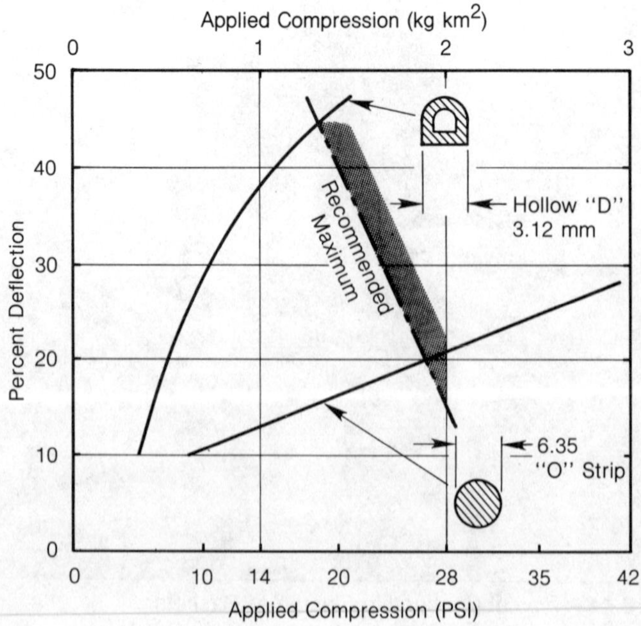

Figure 10.15—Gasket Deflection vs. Pressure for Two Elastomer EMI Gaskets

10.2.3.1.4 Molded-in-Place EMI Gaskets

A few manufacturers offer custom vulcanized EMI gasketing which consists of a conductive elastomere gasket plus an extra hermetic gasket which is factory-molded onto a cover (Fig. 10.16). This is especially attractive for covers with complex shapes or many internal compartments. Compression stops can be included easily, and the need for conductive epoxy is eliminated. The hot vulcanization allows several ribs (typically 2 or 3) of gasket to be put in parallel to reduce the closure force for a given performance. Typical resistance of such gaskets is approximately 1 mΩ per cm length, providing a plane-wave SE of better than 90 dB for a double row with a typical 5 cm joint between screws.

Figure 10.16—Molded-in-Place EMI Gasket from Chomerics

10.2.3.1.5 Spring-Finger Gaskets and Spiral Gaskets

Figure 10.17 shows some examples of beryllium copper spring finger gaskets stamped into different configurations. Gaskets similar to these were introduced over 40 years ago and were the first EMI gaskets on the market. Since there existed little elastomer technology in the 1940s, it is natural that joint unevenness (see Section 10.2.3.3) could be accommodated by a series of individual **fingers**, each capable of flexing a different amount. Thus, shielded enclosures, cover plates and other heavy-duty applications used, and still use, this type of gasket.

Forty years after their first appearance, this type of gasket still has unmatched SE performances, often beyond the measuring instrumentation range (>120 dB, as shown in Fig. 10.18).[4,5] When properly maintained, the self-cleaning action provided by the con-

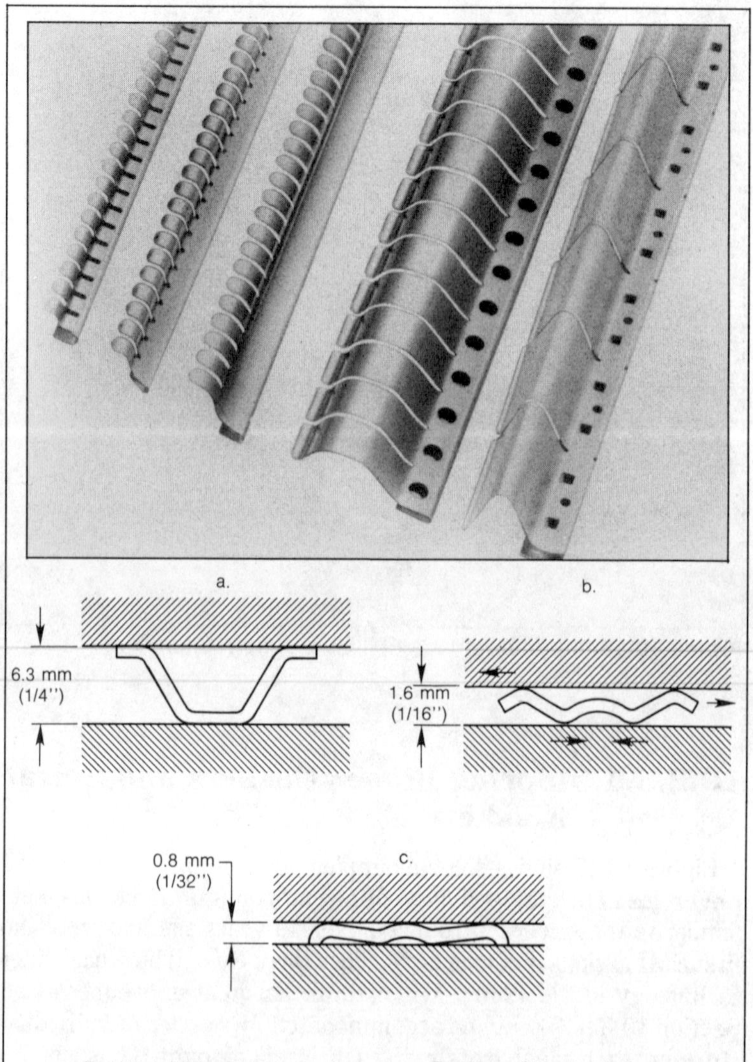

a. End View, Minimum Compression (Three Lines of Contact)

b. End View, Medium Compression (Four Lines of Contact Due to Arching at Center). Notice the Wiping Action of Edges.

c. End View, Close to Maximum Compression (Seven Lines of Contact—Three at Top, Four at Bottom). The gasket could be compressed further down to 0.4 mm (1/32") and yet recover.

Figure 10.17—Typical Spring-Finger Strip Gaskets

stant wiping of the fingers maintains an extremely low contact resistance. The deflection capability is quite large (see Table 10.1) and corner-shaped sections exist. Of course this type of joint by no means provide any climatic or pressure seal. So, if necessary, the finger row has to used with a weather gasket. Notice that this can increase the required pressure and stiffness of the covers and fastener or locking system.

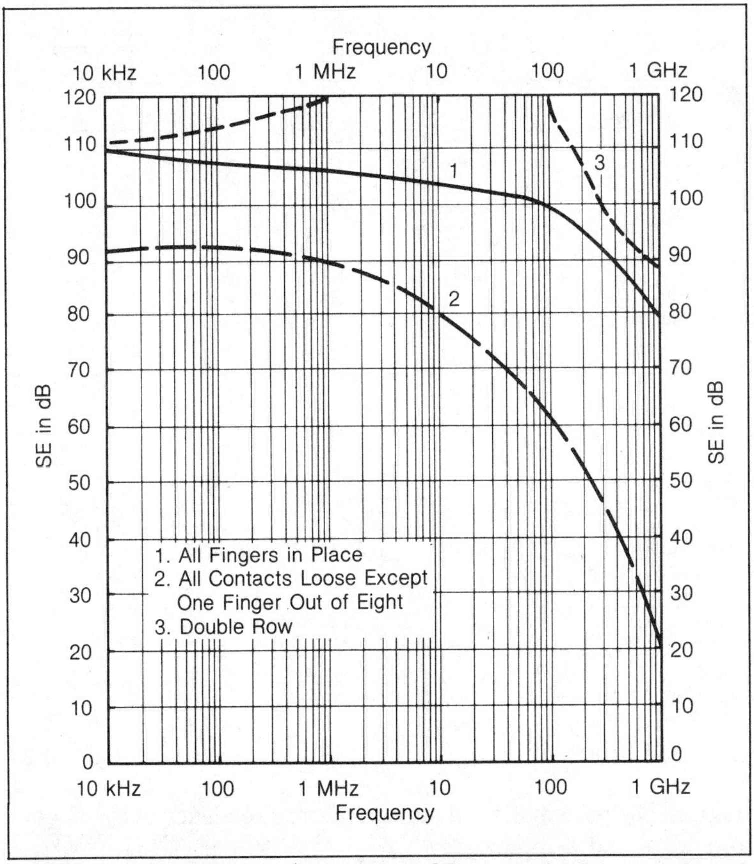

Figure 10.18—Measured Plane Wave SE, normalized to 1 m Length Gasket. Fingerstock copper berylium, Part #97-541. Fully compressed over clean, decreased steel surface (courtesy of Instrument Specialties).

Recent design improvements have made this type of gasket more competitive with other types. The new spring-finger contact strips now offer self-adhesive backing to replace older mechanical fastening methods. They are available in a wide variety of sizes and

Table 10.1—Comparison of Gasket Types & Materials

Comparison Factors \ Gasket Types		Knitted Wire Mesh	Oriented Immersed Wires	Conductive Plastics and Elastomers	Spring Finger Stock and Spiral Spring
Available Forms		Strips, Jointless Rings	Strips & Sheets, Jointless Rings, Die-cut Shapes	Strips & Sheets, Die-Cut, Molded, Extruded Shapes	Strips
Type of Seal	EMI only	Good to Excellent	Good	Also Seals Hermetically	Good to Excellent
	EMI plus Hermetic	NA	Fair to Excellent	Good to Excellent	NA
Conductive Material		Silver Plate, Monel, Aluminum, Steel Sn/Cu/Fe.	Monel, Aluminum	Many Tiny Silver Balls	Beryllium-Copper
Binder or Core Material		Rubber, Air Core, Neoprene, Silicone Sponge	Solid & Sponge Silicone	Silicone or Plastic	NA
Temperature Range		Limited to Core	(−70°F to 500°F) −56° to 260°C	(−100°F to +400°F) −73°C to 204°C	(−65°F to 100°F) −54°C to 38°C
Available Gasket Heights		1.6 to 12.7 mm	1.6 to 25.4 mm	0.5 to 4.1 mm	1.6 to 10.2 mm
Joint Unevenness Accomodations		0.5 to 4.1 mm	0.25 to 2.5 mm	0.08 to 0.8 mm	0.9 to 6.4 mm
Compression Height Range					7:1
Compression Pressure		0.35 to 7 kg/cm^2	1.4 to 7 kg/cm^2	1.4 to 7 kg/cm^2	
EMI Shielding Performance	10 kHz (H)	25 – 30 dB	> 45 dB	> 35 dB	> 40 dB
	10 MHz	> 100 dB	> 100 dB	> 100 dB	> 120 dB
	1 GHz	> 90 dB	> 90 dB	> 95 dB	> 100 dB
	10 GHz			> 70 dB	> 100 dB

shapes. The principal disadvantage lies in a tendency of the fingers to oxidize and to break off.

Another version of the spring-type gasket is the SPIRA® gasket. It consists of a tin-plated beryllium-copper strip wound in a spiral around a rubber core. It can operate with relatively low pressure, typically 100 to 200 g per cm length (Fig. 10.19). In Fig. 10.20, a few low-cost alternatives are seen where, instead of an uninterrupted seal, gasket pieces are used to provide scattered contact spots as for ESD hardening.[6]

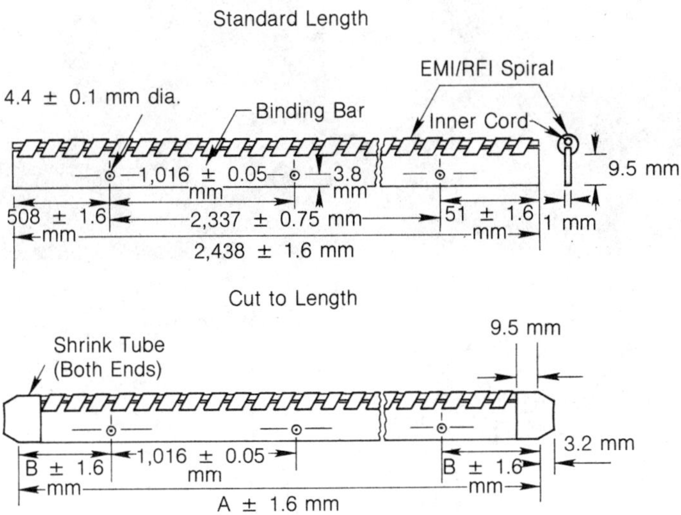

Specifications

EMI/RFI Bonding Spiral
Beryllium Copper per QQ-C-533. Tin Plated (90% Tin-10% Lead) per MIL-P-81728 (AS)

Stainless Steel per QQ-S-766

Binding Bar
Aluminum, 6061-T6 per QQ-A-250

Cord
PVC, Black per MIL-I-631D, Grd C. CL1, CAT 1, TYPE F, Form US, 80 Shore

Figure 10.19—Spiral Spring Gasket from SPIRA® Corp.

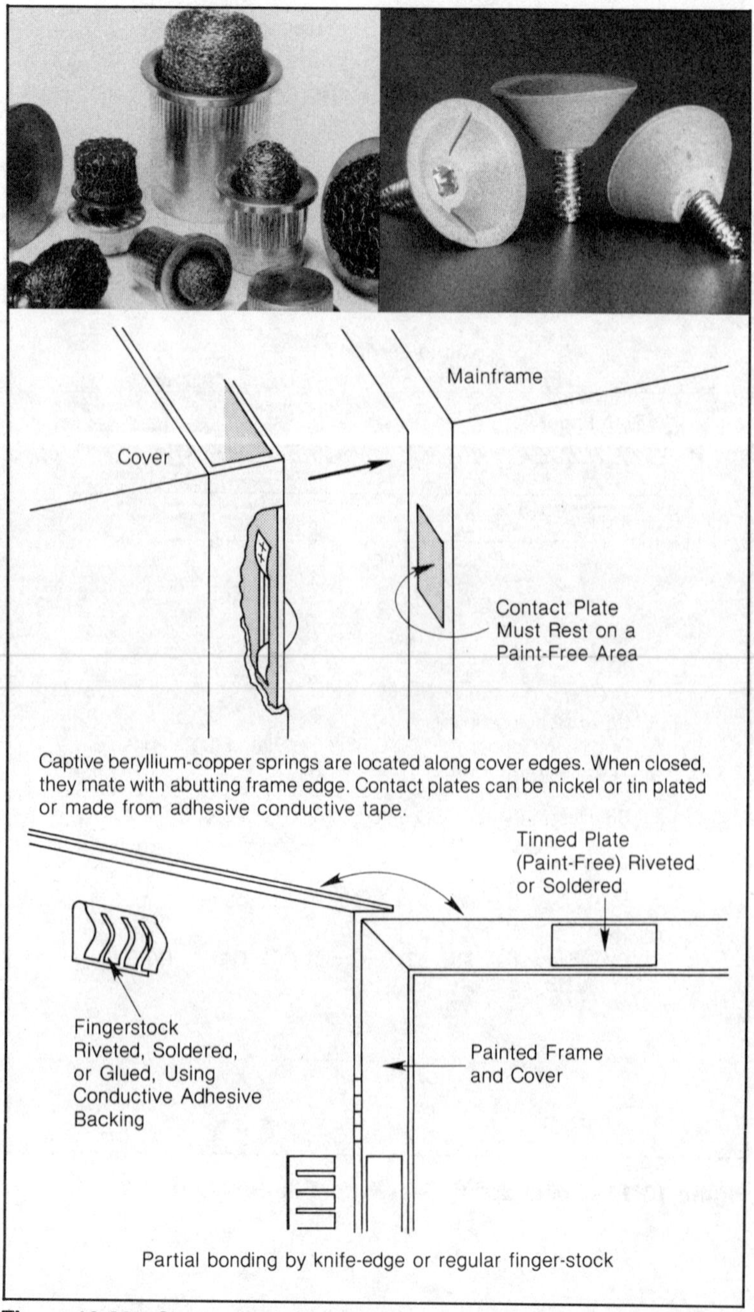

Captive beryllium-copper springs are located along cover edges. When closed, they mate with abutting frame edge. Contact plates can be nickel or tin plated or made from adhesive conductive tape.

Partial bonding by knife-edge or regular finger-stock

Figure 10.20—Cheaper Solutions for Cover Bonding (Spot Bonding) When Only a Modest SE is Necessary

10.2.3.1.6 Pressure-Sensitive, Foam-Backed Foil Gaskets

Another type of gasket is a beryllium-copper foil backed by a highly-compressible neoprene foam. The foam side, containing a synthetic rubber pressure-sensitive adhesive, is applied to cover plates. When placed over an electronics package containing shielded compartments, the foam-backed foil assumes the irregularities of the compartment heights including outside plates to result in a continuous EMI seal. This 1.6 mm gasket is available in sheet widths to 6" and lengths to 0.9 m or may be die cut. EMI shielding effectiveness of 90 db to electric fields is claimed over the 1 kHz to 10 GHz frequency spectrum. The price range is \$64.50 per m^2 (in 100 m^2 quantities) to \$129.00 per m^2 (in 0.5 m^2 quantities).

10.2.3.1.7 Comparison of Gasket Types and Materials

With the many different gasket types and materials available (over 1,000 variations), choosing the proper gasket may bc confusing to the design or specification engineer tasked with the responsibility of selecting the best candidates for a particular application. Table 10.1 provides a comparison of some of the principal characteristics of EMI gaskets.

No one type is best for all applications. For example, those gaskets having relatively low cost tend to have relatively high volume resistivity which results in a less impressive shielding effectiveness. Some gaskets are designed to operate down to cryogenic temperatures or up to 200°C, but not both. Since there exists several different methods of mounting, gaskets are available in sheets and strips, die cuts, molded shapes and extruded forms. At the risk of generalizing, conductive plastics and elastomers seem to offer the widest range of applications and price. To assist the reader in the further assessment of EMI gaskets, summary sheets offered by two leading gasket manufacturers are presented in Tables 10.2 and 10.3. No attempt has been made to standardize or to audit their data.

Table 10.2—Typical EMI Gasket Selection Chart for Eccoshield® Material (courtesy of Emerson & Cuming, Inc.)

	Eccoshield SV	Eccoshield SV-A	Eccoshield SV-M	Eccoshield SV-P	Eccoshield SV-SS	Eccoshield CLV	Eccoshield SV-NT	Eccoshield SV-MAS	Eccoshield SV-C
Claim	Original Plastic RF Gasket Spec C in Everywhere Avail. in Many Forms	Increases Temp. Capability to 250°C Low Durometer for Light Closing Pressures	Retains High Temp. of SV-R But Much Lower Cost Also Low Durometer	Lowest Cost Silver Silicone Large Gaskets and Large Cross Sections Economic	The Only Metal Fiber Filled Rubber Which Easily Compresses	High Performance For a Carbon System Low Cost, of Course	Continuous Lengths. Many Forms. Excellent Physicals for Doors, Cabinets	Simple Way to Hold Cover Or Door in Place Without Hinges/Latches RF Seal	Pour Your Own RF Gaskets in Place or in Simple Molds
Basic Composition	Silver + Vinyl Elastomer	Silver + Silicone Rubber	Silver + Silicone Rubber	Silver + Filler + Silicone Rubber	Stainless Steel + Silicone Rubber	Carbon + Vinyl Elastomer	Rubber Sheathed Silver-Filled Elastomer	Plastic Magnet Sheathed in Silver-Filled Elastomer	Silver Silicone Rubber
Forms Available	Flat Sheet & Gaskets Rod Tube O-Rings Extrusions	Flat Sheet & Gaskets	Flat Sheet & Gaskets	Flat Sheet & Gaskets O-Rings, Molded Gaskets	Flat Sheet & Gaskets	Large Flat Sheets & Gaskets	Rod, Tube Extrusions Molded Gaskets, Sponge Strip	Strips 1/2" × 1/4" × 2' + Foam-Backed Version	Castable Liquid
Color	Silver	Silver-Tan	Silver	Tan	Gray	Black	Silver	Silver	Silver-Gray
Density	2.6	3.5	1.1	1.2	1.3	1.4	Various	(Not Applicable)	1.8
Service Temp °C	56 to + 125	Cryogenic to + 250	Cryogenic to + 250	Cryogenic to + 250	Cryogenic to + 250	−66 to +126	−66 to +26	−66 to +126	Cryogenic to +250
Durometer Shore A	66	40	45	85	75	85	Various (Some Very Low)	--	75
Relative Compression Set	Good	Very Good	Very Good	Excellent	Excellent	Good	Excellent	--	Excellent
Volume Resistivity	2×10^{-3}	1×10^{-3}	$< 1 \times 10^{-2}$	$< 5 \times 10^{-3}$	$< 5 \times 10^{-3}$	5.0	--	--	$< 5 \times 10^{-3}$
Resistance	2×10^{-5}	1×10^{-5}	$< 1 \times 10^{-4}$	$< 5 \times 10^{-6}$	$< 5 \times 10^{-4}$	5×10^{-2}	--	--	$< 8 \times 10^{-6}$
Bond to Self	Eccoshield VSV	Eccoshield SV-C	Eccoshield SV-C	Eccoshield SV-C	Eccoshield SV-C	Eccoshield VSV	--	--	Eccoshield SV-C
Bond to Metals	Eccoshield BDM/VCA	Eccoshield RVS	Eccoshield RVS	Eccoshield RVS	Eccoshield RVS	Eccoshield VSM	Eccoshield VCA	Eccoshield VSM	Eccoshield RVS
Elongation at	>200	>60	>60	>40	>75	>160	Various (Most Very High)	--	>30
Tensile Strength At Rupture, pm	1,000	150	150	260	260	1300	Various (Most Very High)	(Not Applicable)	200
Thermal Conductivity	25	30	3	4	6	6	--	--	4
Relative Cost	Medium	High	Medium	Low	Medium	Very Low	Low	Medium	Very Low
Insertion Loss Improvement (Note 2)									
Magnetic Field, 200 kHz	Excellent	Excellent	Good	Good	Good	Not Recommended	Good	Good	Good
Electric Field 1 MHz	Excellent	Excellent	Excellent	Excellent	Good	Good	Excellent	Excellent	Excellent
Plane Wave, 400 MHz	Excellent	Excellent	Good	Good	Good	Good	Excellent	Excellent	Excellent

Table 10.3—Typical EMI Gasket Selection Chart (Chomerics Corp.)

Product Trade Name		Polastrip	Polasheet Polastick	Polaring	Pola-N
Schematic Cross Section					
Construction		Oriented wire in Matrix of Silicone Elastomer (Polastick has pressure sensitive adhesive)			Solid Metal Paths in Elastomer plus Special Contact Material
Available Forms Supplied by Metex		Strips, Gaskets Made by Joining Strips	Sheets Die Cut Gaskets	Jointless Rings	Strips, Gasket Made by Joining Strips
Type of Seal	EMI only	Good	Good	Good	Excellent
	EMI plus rain tight, dip proof contain ventilating air	Good	Good-Excellent	Good	Excellent
	EMI plus pressure to 30 psi	Excellent	Good	Excellent	Excellent
	EMI plus pressure over 30 psi	Fair	Fair	Good-Excellent	Fair
EMI Rating	14 KHz (H)	>46 dB	>35 dB	>46 dB	>56 dB
	18 MHz (E)	>102 dB	>102 dB	>102 dB	>102 dB
	1.0 GHz (P)	>93 dB	>93 dB	>93 dB	>93 dB
Max. Joint Un-evenness % of Gasket Height	Class A—Permanently Closed	20%	20%	20%	20%
	Class B—Open-Close in same position	17%	17%	17%	17%
	Class C—Completely Interchangeable	17%	17%	17%	17%
Minimum/Maximum Height		.062"/.625"	.055"/.250"	.602"/.625"	.125"/.625"
Minimum width (greater of actual dim. or portion of height)		.062"/ 1/2 H	.125	.062"/ 1/2 H	.125/ 1/2 H
Recommended Compression Pressure (psi)		20-100	20-100	20-100	20-100
Attachment or Positioning	In slot	Good	Possible	Excellent	Good
	Pressure Sensitive Adhesive	N/A	Excellent	N/A	N/A
	Bond non-EMI Gasket portion (4)	Combo Version Only	N/A	N/A	Combo Version Only
	Conductive Adhesive	Poor Because Conductive Adhesives Do Not Bond To Silicone			
	Bolts thru Bolt Holes	Excellent	Excellent	Good if Wide Enough	Excellent
Elastomer Temperature Range	Neoprene Version	N/A	N/A	N/A	N/A
	SiliconeVersion	−70°F to 500°F	−80°F to 400°F	−70°F to 500°F	−70°F to 380°F
Standard Metals Available in EMI Portion (Others also available)		Monel, Aluminum	Monel, Aluminum	Monel, Aluminum	Tin Plated Copper

(continued next page)

Table 10.3—Typical EMI Gasket Selection Chart (Chomerics Corp.) (Continued)

	Product Trade Name	EMI Strips (all metal)	Compressed EMI Gaskets	EMI Strips (elastomer core)
	Schematic Cross Section			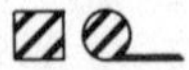
	Construction	Formed or Compressed Knitted Wire Mesh		Knitted Wire Mesh Over Elastomer Strips
	Available Forms Supplied by Metex	Strips, Gaskets Made by Joining Strips	Jointless Rings or Rectangular Gasekts	Strips, Gaskets Made by Joining Strips
Type of Seal	EMI only	Good-Excellent	Good-Excellent	Good
	EMI plus rain tight, dip proof contain ventilating air	No	No	Fair-Good
	EMI plus pressure to 30 psi	No	No	No
	EMI plus pressure over 30 psi	No	No	No
EMI Rating	14 KHz (H)	>20 × >30 dB	>25 × >30 dB	>25 × >35 dB
	18 MHz (E)	>102 dB	>102 dB	>102 dB
	1.0 GHz (P)	83 × >93 dB	>93 dB	>93 dB
Max. Joint Un-even-ness % of Gasket Height	Class A—Permanently Closed	30-40%	30%	30 × 50%
	Class B—Open-Close in same position	25-30%	25%	25 × 40%
	Class C—Completely Interchangeable	20-25%	20%	20 × 30%
	Minimum/Maximum Height	.062"/.500"	.040"/.375"	.125"/.625"
	Minimum width (greater of actual dim. or portion of height)	.062"/ 1/2 H	.062"/ 1/2 H	.125"/ 1/2 H
	Recommended Compression Pressure (psi)	5-100	5-100	5-100
Attach-ment or Posi-tioning	In slot 	Excellent	Excellent	Excellent
	Pressure Sensitive Adhesive 	N/A	N/A	N/A
	Bond non-EMI Gasket portion (4) 	Version with Fins Only	N/A	Versions with Fins Only
	Conductive Adhesive 	Poor to Good	Poor to Good	Poor to Good
	Bolts thru Bolt Holes 	Possible	Good If Wide Enough	Possible Fin Versions
Elas-tomer Temper-ature Range	Neoprene Version	N/A	N/A	− 30°F to 150°F
	SiliconeVersion	N/A	N/A	− 80°F to 400°F
	Standard Metals Available in EMI Portion (Others also available)	Monel, Ferrex™(1), Aluminum	Monel, Ferrex™(1) Aluminum	Monel, Ferrex™(1), Aluminum

(1) Ferrex™ is a trade name for tin plated, copper clad steel EMI gasketing.
(2) Two versions, and have fins especially designed for easy attachment.

(continued next page)

Table 10.3—Typical EMI Gasket Selection Chart (Chomerics Corp.) (Continued)

Product Trade Name		Combo Strips®	Combo Gaskets®	Metales®
Schematic Cross Section				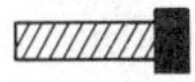
Construction		Formed Knitted Wire Strips in Parallel with Elastomer Strips or Die Cut Gaskets		Formed Knitted Wire Strips Clamped in Aluminum Extrusions
Available Forms Supplied by Metex		Strips, Gaskets Made by Joining Strips	Die Cut Elastomer with Joined EMI Strips	Strips, Fab Lengths Frames with Joined EMI Strips
Type of Seal	EMI only	Good-Excellent		Good-Excellent
	EMI plus rain tight, dip proof contain ventilating air	Good-Excellent		Excellent
	EMI plus pressure to 30 psi	Good-Excellent		Good
	EMI plus pressure over 30 psi	Fair	Excellent	Good-Fair
EMI Rating	14 KHz (H)	>30 dB		>30 dB
	18 MHz (E)	>102 dB		>102 dB
	1.0 GHz (P)	>83 dB × >93 dB		>93 dB
Max. Joint Uneven-ness % of Gasket Height	Class A—Permanently Closed	30%		30%
	Class B—Open-Close in same position	30%		25%
	Class C—Completely Interchangeable	25%		25%
Minimum/Maximum Height		.062"/.375"		.156"/.250"
Minimum width (greater of actual dim. or portion of height)		.125"/1 1/2 H		.500"
Recommended Compression Pressure (psi)		20-100		5-100
Attach-ment or Posi-tioning	In slot 	Excellent		No
	Pressure Sensitive Adhesive 	Excellent		N/A
	Bond non-EMI Gasket portion (4) 	Good-Excellent		Poor
	Conductive Adhesive 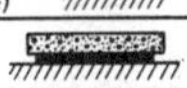	N/A		N/A
	Bolts thru Bolt Holes 	Excellent		Excellent
Elas-tomer Temper-ature Range	Neoprene Version	−30°F to 150°F		−30°F to 150°F
	SiliconeVersion	−80°F to 400°F		−80°F to 400°F
Standard Metals Available in EMI Portion (Others also available)		Monel, Ferrex™ (1), Aluminum		Monel, Ferrex™ (1)

(1) Ferrex™ is a trade name for tin plated, copper clad steel EMI gasketing.

(continued next page)

Table 10.3—Typical EMI Gasket Selection Chart (Chomerics Corp.) (Continued)

Product Trade Name		Porcuples Metalistic®	Metalistic	Xasco
Schematic Cross Section				
Construction		Expanded Metal in Elastomer	Woven Wire in Elastomer	Silver on Inert Substrate in Elastomer
Available Forms Supplied by Metex		Sheets, Die Cut Gaskets	Sheets Die Cut Gaskets	Sheets, Die Cut Gaskets, Molded Gasket, Strips
Type of Seal	EMI only	Good-Excellent	Fair-Good	Good-Excellent
	EMI plus rain tight, dip proof contain ventilating air	Good-Excellent	Fair-Good	Good-Excellent
	EMI plus pressure to 30 psi	Good	Fair	Excellent
	EMI plus pressure over 30 psi	Fair	Poor	Excellent
EMI Rating	14 KHz (H)	>35 dB	>35 dB	>35 dB
	18 MHz (E)	>102 dB	>102 dB	>102 dB
	1.0 GHz (P)	>45 dB	>40 dB	>93 dB
Max. Joint Uneven-ness % of Gasket Height	Class A—Permanently Closed	15%	10%	25
	Class B—Open-Close in same position	10%	7%	20
	Class C—Completely Interchangeable	10%	7%	20
Minimum/Maximum Height		.020"/.030"	.016"/.020"	.020"/.250"
Minimum width (greater of actual dim. or portion of height)		.140"	.125	.090"/H
Recommended Compression Pressure (psi)		20-100	20-100	20-100
Attach-ment or Posi-tioning	In slot	No	No	Good
	Pressure Sensitive Adhesive	N/A	N/A	N/A
	Bond non-EMI Gasket portion (4)	N/A	N/A	N/A
	Conductive Adhesive	No	No	Poor
	Bolts thru Bolt Holes	Excellent	Excellent	Excellent
Elas-tomer Temper-ature Range	Neoprene Version	N/A	−40°F to 225°F	N/A
	SiliconeVersion	−80°F to 400°F	−65°F to 500°F	−80°F to 400°F
Standard Metals Available in EMI Portion (Others also available)		Monel, Aluminum	Aluminum Only	Silver or Inert Substrate

(1) Ferrex[TM] is a trade name for tin plated, copper clad steel EMI gasketing.
(2) Two versions, and have fins especially designed for easy attachment.
(3) The aluminum extrusion is intended as a convenient attachment method.
(4) Most products for which this method is suitable are available from Chomerics with "dry back" (solvent activated) adhesives already applied.

10.2.3.2 Gasket Selection

EMI gasket selection involves making compromises between (1) available EMI gasket materials and their characteristics and (2) performance requirements of equipment and design constraints of mating surfaces. Gasket mounting (and hence selection) involves a number of alternatives.

In selecting one or more suitable EMI gaskets electrical characteristics, application requirements, application constraints and price factors are the major considerations. These topics are summarized as follows:

Application Requirements

These are usually dictated by equipment performance specifications. They include: amount of shielding, pressure sealing and environmental exposure (e.g., temperature, salt spray, ambient pressure, and corrosive material). Also critical are the vibration requirements since gasket SE typically degrades under vibration.

Application Constraints

These are usually governed by the equipment housing design. They include space available, compression force, joint unevenness, contact surface characteristics and attachment possibilities and limitations.

The important compromises between application requirements and constraints on one hand and gasket characteristics and price on the other are:

1. Gasket height and compressibility must be large enough to compensate for joint unevenness under the available force.
2. The gasket must be capable of providing the required EMI sealing and hermetic sealing (when applicable) when compressed by the available force.
3. There must be sufficient space for the gasket within the design limitations of the application.
4. The gasket must be attached or positioned by a means that fits in with the joint design.
5. If there is a metal portion in the EMI gasket, it must be

sufficiently corrosion resistant and compatible with the mating surfaces relative to the equipment environment.

6. The EMI gasket must meet the temperature and other environmental requirements of the equipment specifications.

Gasket manufacturers and suppliers provide design guide tables to assist the user to select the gasket most nearly meeting the application requirements and constraints.

Price Factors

On the topic of price, EMI gaskets cover a wide range depending upon type, size, quantity and performance. However, to give the reader some idea of price range, the following remarks are offered:

For silver-silicone elastomers having sheet thicknesses of 0.25 mm to 1.5 mm, prices may range from $170 to $1,300 per square meter in small quantities. While unit price comes down to about 30 or 40 percent of this in large quantities, a tooling die charge from about $100 to $500 would be spread over the cost of the entire lot. Knitted-wire mesh gaskets will range from about $1 to about $12 per meter depending upon uncompressed diameter, elastic core if any, and methods of adhesion.

Typical 170 × 20 cm custom gaskets of all types will cost about $4 to $16 each, and about 40 percent of this in production. Several suppliers will provide a free sample of their gaskets. Alternatively, some supply a kit of different sizes and shapes for about $10 to $25.

10.2.3.3 Joint Unevenness

The degree of misalignment or misfit of the mating surfaces is commonly called **joint unevenness** and is designated ΔH in Fig. 10.21a. It is the **maximum** separation between the two surfaces when they are just touching, and in the limit becomes the sum of the peak irregularities of both surfaces. If the surfaces are not rigid, the joint unevenness also includes any additional separation between the two surfaces due to bending when pressure is applied.

Figure 10.21b shows the same joint with a gasket installed. The dashed lines indicate the gasket height, H_g, before compression.

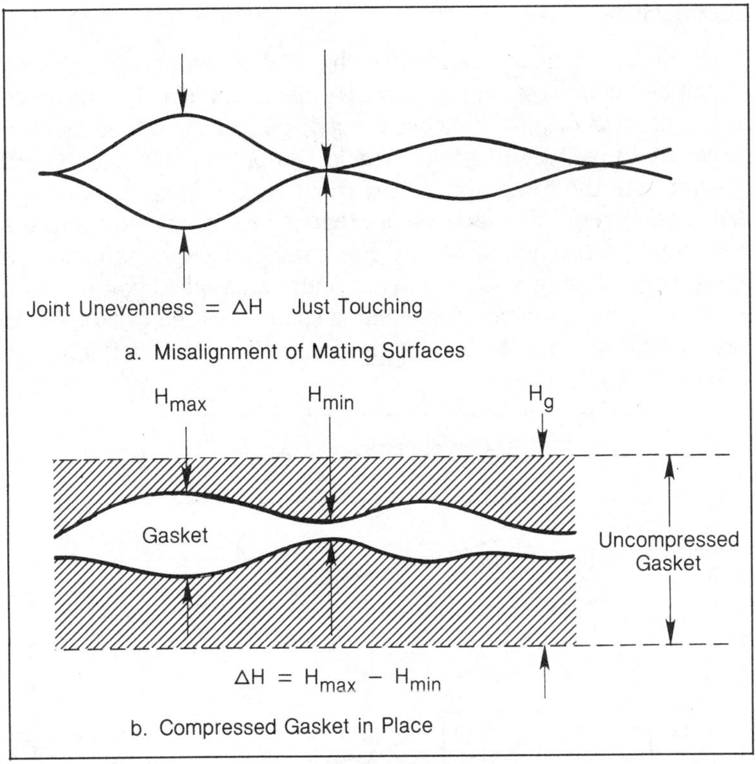

Figure 10.21—Description of Joint Unevenness

The minimum compressed gasket height H_{min} occurs at the point where the height H_{max} is at maximum joint separation. Thus, joint unevenness of the mating surface is:

$$\text{Joint unevenness} = \Delta H = H_{max} - H_{min} \quad (10.1)$$

10.2.3.4 Required Compression Pressure

Three factors determine the required compression pressure on a gasket, viz, its resiliency, the minimum pressure required for a seal and the total joint unevenness. They are discussed below:

Resiliency

Resiliency is the amount by which a gasket compresses per unit of applied compression pressure. Resiliency is usually expressed in percent of original (uncompressed) gasket height divided by pressure in psi. A soft gasket would compress more than a hard gasket with the same applied pressure. Stated in another way, a soft gasket requires less pressure than a hard gasket to compress the same percentage of height. For example, a sponge neoprene gasket might compress 10 percent under an applied compression pressure of 0.42 kb/cm^2 (6 psi), but a solid neoprene gasket would require 2.8 kg/cm2 (40 psi) for the same 10 percent deflection as shown in Fig. 10.22.

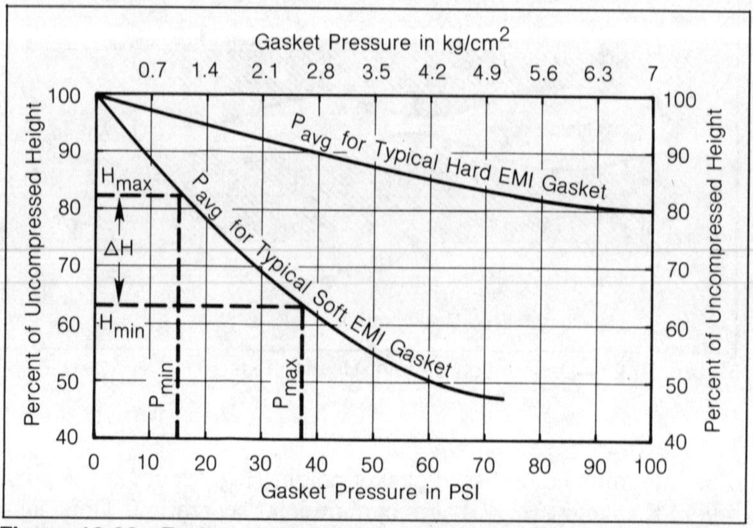

Figure 10.22—Typical Hard and Soft EMI Gasket Height vs. Pressure Relations (Shown for Mesh Strips)

Minimum Pressure for Seal

A gasket must at least make contact at the point of maximum separation between mating surfaces, i.e., $H_{max} \leqslant H_g$ in Fig. 10.21. Actually, the pressure at this point must be stated as a minimum amount in order to assure an EMI seal.

This is easy to understand by making a comparison with a high-pressure lubricating system. If there is not some required minimum pressure at the point of H_{max}, oil will blow out between the flanges and the gasketing material. Thus, the pressure at the H_{max} point must be high enough to prevent blowout.

For EMI gaskets, this minimum pressure, P_{min}, is determined by the pressure required to break through corrosion films and to make a suitable low-resistance contact. P_{min} is typically about 1.4 kg/cm^2 (20 psi), but can be as small as 0.35 kg/cm^2 (5 psi).

Average Pressure

The average pressure applied to the gasket must also be large enough to compress the overall gasket so that the difference between the minimum height and the maximum gasket height (determined by P_{min} from the previous paragraph) is equal to the joint unevenness, i.e., $\Delta H = H_{max} - H_{min}$, in Eq. (10.1). In general, the average pressure should equal or exceed that corresponding to the average compressed gasket height, H_{avg}:

$$H_{avg} = (H_{max} + H_{min})/2 \qquad (10.2)$$

and,

$$P_{avg} = (P_{min} + P_{max})/2 \qquad (10.3)$$

The required compression force F may be calculated from P_{avg} by determining the area "A" of the gasket to be sandwiched between the mating members:

$$F = P_{avg} \times A \qquad (10.4)$$

for P and A in consistent units,

where,

A = gasket area in in^2, or cm^2

10.2.3.5 Required Gasket Height

To obtain the required EMI seal from a gasketed joint, the gasket height must meet these criteria:

The pressure at the point of maximum joint separation (H_{max}) must correspond to the minimum pressure to obtain the required

EMI seal. The pressure P_{max}, at minimum gap H_{min} corresponds to a not-to-exceed value beyond which the gasket could loose its elasticity and remain permanently flattened.

The difference between maximum and minimum compressed heights of the gasket must equal the joint unevenness of the mating surfaces. If the average pressure, available to compress the gasket is P_{avg}, the maximum pressure, P_{max}, is obtained from Eq. (10.3):

$$P_{max} = 2P_{avg} - P_{min} \tag{10.5}$$

where, P_{min} is a selected value at least as much as that required to obtain a gasket seal.

The percentage of uncompressed height corresponding to P_{min} and P_{max} in Fig. 10.22 are H_{max} and H_{min}, respectively. To calculate the required uncompressed gasket height H_g as a dimension:

$$Hg = \frac{\Delta H \text{ inches or mm}}{\Delta H \text{ decimal}} \tag{10.6}$$

Thus, the required height is the actual joint unevenness in specified units divided by the joint unevenness expressed in decimal equivalent of percent gasket compression.

Illustrative Example 10.1

Suppose that for a specified soft EMI gasket, the minimum required pressure for a seal is 0.7 kg/cm^2 (10 psi). Picking a value somewhat greater than this, P_{min} is selected at 1.05 kg/cm^2 (15 psi) as shown in Fig. 10.22. This corresponds to a maximum uncompressed gasket height of 81 percent. If the average pressure available to compress the gasket is 1.82 kg/cm^2 (26 psi), the maximum pressure from Eq. (10.5) is:

$$P_{max} = 2 \times 1.82 \text{ kg/cm}^2 - 1.05 \text{ kg/cm}^2 = 2.59 \text{ kg/cm}^2$$

or,

$$P_{max} = 2 \times 26 \text{ psi} - 15 \text{ psi} = 37 \text{ psi}$$

The corresponding minimum uncompressed gasket height from Fig. 10.22 is 65 percent. Thus:

$$\Delta H \text{ decimal} = 81\% - 65\% = 16\% = 0.16$$

The required gasket height for a joint unevenness of 0.813 mm (0.032") is obtained from Eq. (10.6):

$$H_g = \frac{0.813}{0.16} = 5.08 \text{ mm}$$

or,

$$H_g = \frac{0.032}{0.16} = 0.20''$$

10.2.3.6 Compression Set

Some gaskets do not return to their original uncompressed height after release of compression. This is called compression set or **fluence**. It may be visualized by assuming that the lower curve shown in Fig. 10.22 applies for a particular soft gasket. When pressure is removed, the gasket returns to a lesser height whose properties might look somewhat like the upper curve in Fig. 10.22 (this is exaggerated for illustrative purposes). The importance of compression set depends upon how the gasket is to be used; three classes of use are now defined:

Class A: Permanently Closed

Compression set is unimportant since the gasketed component in all probability will never be removed.

Class B: Repeated Identical Open-Close Cycles

Examples include hinged doors and symmetrical covers. In class B applications, compression set problems are marginal; further examination of details, however, is indicated.

Class C: Completely Interchangeable

This classification implies complete freedom to reposition gasket on repeat cycles, e.g., round gasket in waveguide. Since the compression-set height at a point of maximum compression may end up being less than minimum compressed height, no contact at all would result between gasket and mating surfaces at this point. For class C applications, use a soft gasket with a wide compliance (i.e., a large value of ΔH) Otherwise, do not re-use gaskets with compression set limits; instead, use a new gasket.

10.2.3.7 Gasket-to-Metal Contact Resistance

The gasket is the electrical media between the two mating members, so the calculations of Chapters 7 and 8 on box leakages lead to the following statements:

1. If the seam between two fasteners (screws, bolts, etc.) has a naked attenuation higher than the gasket (meaning the impedance of the basic seam is already lower than the gasket resistance), the seam SE will dominate. (In other words, the thin aperture would attenuate as well if the gasket was not there.)
2. If the naked seam SE is less than the gasket SE, the latter dominates (i.e., the lower gasket resistance is "shunting" the seam impedance).
3. If there are no conductive fasteners, the gasket SE should be used at all frequencies.

At the risk of oversimplifying, we can say that the metal density in the gasket core is generally too low to provide the several skin depths required for a significant absorption. Therefore, a typical gasket will work mainly by reflection and the SE can be grossly predicated from the gasket impedance augmented by the two gasket-to-metal contact resistance. So, there is a need for *both* a low contact resistance R_c and a low gasket resistance R_g. This latter depends on the gasket volume resistivity and gasket size:

$$R_{TOT} = 2\,R_c + R_g \qquad (10.7)$$

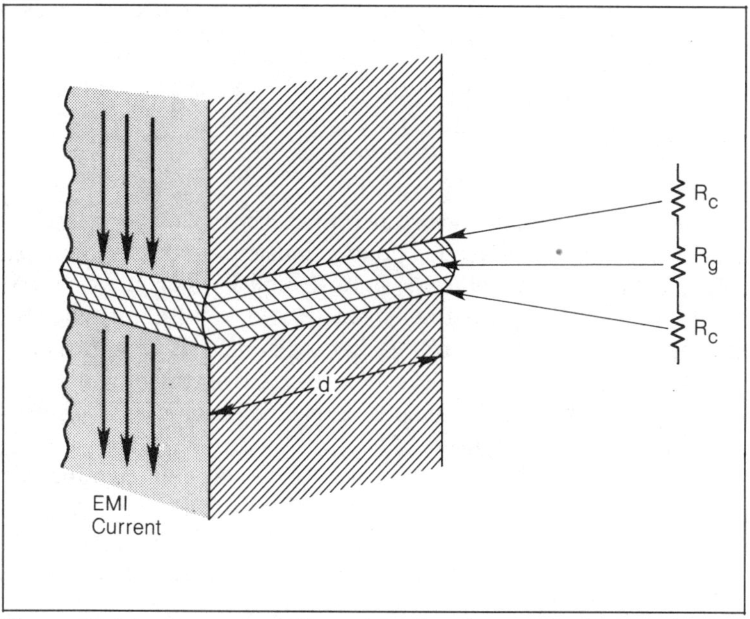

Figure 10.23—Conceptual View of Gasket Contact Resistance (Inductance and Capacitance Not Shown)

R_g can be characterized, for a given gasket resistivity, by the gasket resistance per unit length in the direction of the current flow for worst-case field polarization. In general, the thicker the gasket, the higher the resistance will be, and the wider the gasket (dimension d in Fig. 10.23), the lower the resistance.

Illustrative Example 10.2

A certain conductive gasket with a resistance of 3 mΩ/cm is mounted between two flanges with 5 cm spacing between two consecutive screws. Under adequate pressure, a **uniform** contact resistance of 1 mΩ is ensured at each interface (typical default value for a good bond). What is the SE for this gasket?

From Eq. (10.7):

$$R_{TOT} = 2 \times 1 + 5 \times 3 = 17 \text{ m}\Omega$$

From the reflection equation (for a barrier thickness less than skin depth and plane wave conditions):

$$SE = 20 \log \frac{377}{4 \times 17 \times 10^{-3}} = 75 \text{ dB}$$

The above example shows once more that, even for a gasket, a proper methodical selection process must consider the combined effect of the two contributors, here R_g and R_c. For instance, trying to improve the contact resistance would be wasteful, since SE is mainly determined by the gasket resistance. By contrast, had the gasket resistance been 0.1 mΩ/cm (a value obtainable with some wide wide-mesh gaskets), some effort should be directed at decreasing the contact resistance.

In all cases, the above discussion serves to demonstrate that because the SE of a gasket depends on its length, any SE data given without specifying the length is incomplete. When length is not specified, the designer might assume by default that the MIL-STD-285 method, with a 30 cm × 30 cm panel and approximately 2.5 cm screw spacing, has been used. Unfortunately, this may not be the case, and errors in SE calculations may result.

For accuracy, such measurements should use nylon or other nonconductive fasteners, otherwise **the low leakage of the naked seam may mask the actual SE of the gasket.** For instance, with a stiff flange, the 2.5 cm screw spacing indicated above already provides >90 dB SE at 1 MHz, i.e., the naked seam is better than many gaskets! This precision is seldom provided in manufacturers' literature.

10.2.4 Design and Dressing of Mating Faces for Gasket Mounting

Several methods are available to position a gasket on a metal mating surface:

1. Slot mounting
2. The use of pressure-sensitive nonconductive adhesive
3. Bonding the non-EMI portion of the gasket
4. The use of conductive adhesives

5. Attachment via bolt or screw holes
6. Special attachment methods

Each of these methods is summarized later in this section.

The one guideline which is common to all mounting styles is that the fasteners or cover locking devices must be capable of applying the pressure to within +10 percent of the average closure pressure; a gasket should be neither overloaded under fasteners nor underloaded between them. Generally, pressures ranging from 0.3 to 5 kg per cm of gasket length may be required, and even more if the EMI gasket is paralleled with a weather seal. For instance, a large cover with a 1 m perimeter requires quite a hefty closure force of several hundred kilograms. Spring-latch or quarter-turn fasteners may not be easy to handle at such compression levels, so other types of fasteners or preloaders may be required.

As a general rule, flat, wide gaskets (e.g., width = 2 × height) and round gaskets handle better than square ones (no rollover). To avoid overcompression and permanent gasket damage, compression stops can be used in the form of regular studs, washers, etc. When a slot-mounted gasket or a gasket with integrated fins is used, these profiles provide a natural compression stop.

Slot Mounting

This method is recommended if the slot can be provided at a relatively low cost, such as in a die casting. All solid elastomer materials, which constitute the gasket material, are essentially incompressible. These products appear to compress because the material flows while it maintains a constant volume. Therefore, when these products are used in a slot, extra cross-sectional area must be allowed for the material to flow axially. At least 10 percent extra space (more if possible) is recommended as shown in Fig. 10.24a. Determination of gasket and groove tolerances must be coordinated to avoid an underfilled or overfilled groove.

Pressure-Sensitive Adhesive

This method of mounting is often the least expensive for attaching EMI gasket materials. Installation costs are substantially lower than with other mounting methods, with only a slight increase

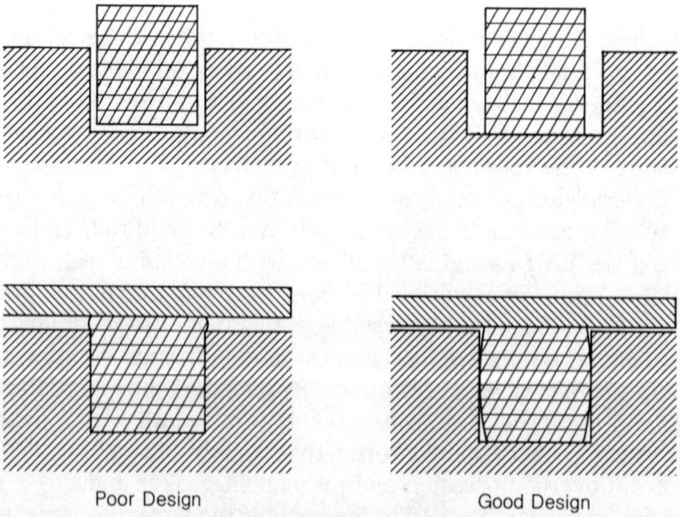

a. Making Allowance for Solid Elastomer Gasket Flow

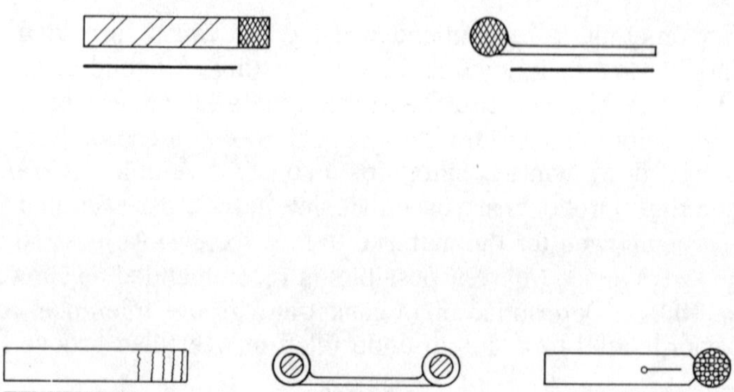

b. Areas where Nonconductive or Dry-Back Adhesive Can Be Used

Figure 10.24—Different Methods of Mounting Gaskets (continued next page)

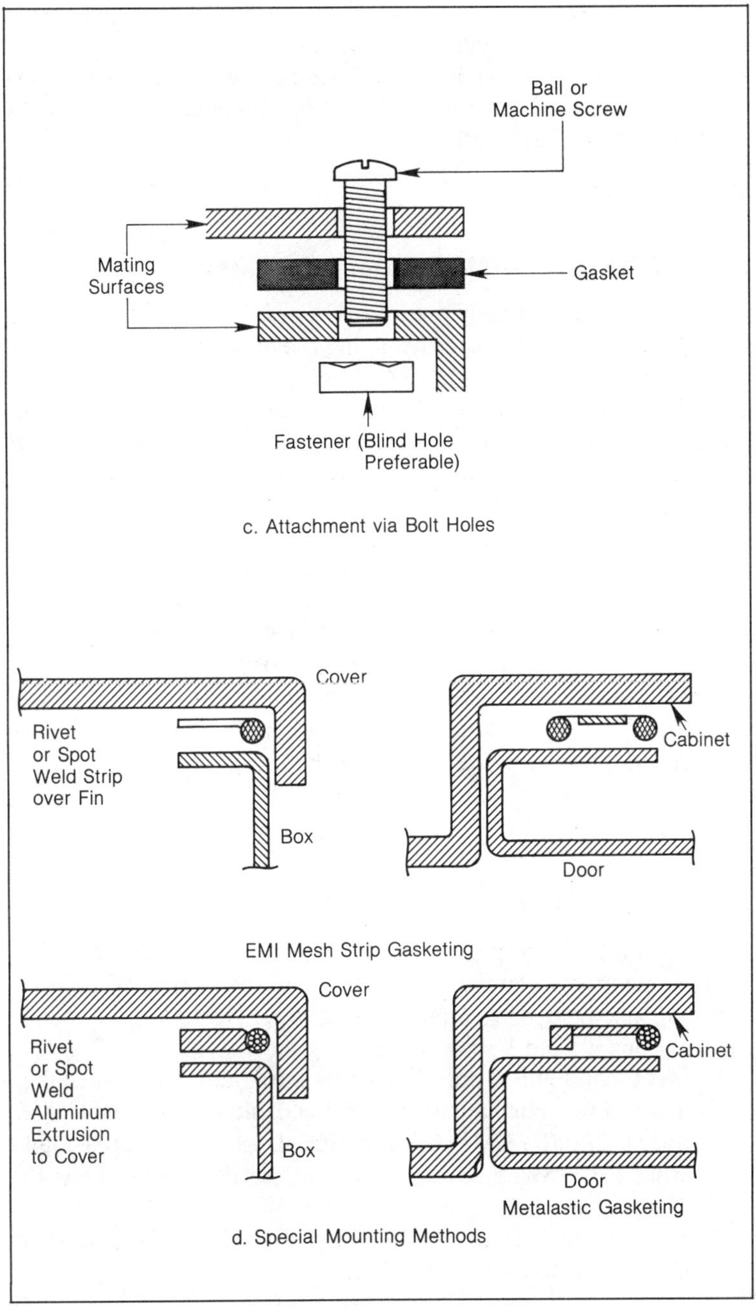

Figure 10.24—(continued)

in gasket cost versus a material without adhesive backing. Most sponge-elastomer materials are used for applications which do not require any hermetic sealing, and the adhesive-backed rubber portion of this material serves only as an inexpensive attachment method for the EMI portion.

Bonding the Non-EMI Portion of a Gasket

Many good nonconductive adhesives are now available to bond an EMI gasket in position by applying the adhesive to its non-EMI portion. As shown in Fig. 10.24b, this can be insulated from the mating surfaces by a nonconductive material and is often a good application method. The designer specifying nonconductive adhesive attachment must include adequate warnings in applicable drawings and standard procedures for production personnel. These cautions should state that adhesive is to be applied only to the portion of the gasket material not involved with the EMI gasketing function. Experience indicates that installation workers, either through carelessness or a misguided desire to do a better job, will apply the nonconductive adhesive to the **entire** gasket including the EMI gasket portion. It is not uncommon to hear "This gasket would hold better if I glued all of it rather than half of it." This occurrence completely degrades the EMI performance.

Conductive Adhesives

Since good conductive adhesives can provide an adequate electrical contact between the EMI gasket and the mounting surfaces, they can also be used to mount the gaskets. However, the following cautions should be observed:

1. Most conductive adhesives are hard and incompressible. Thus, if too much adhesive is applied and it is allowed to soak too far into the EMI gasket material, gasket compressibility will be destroyed. Irregularly applied adhesive also has the effect of increasing joint unevenness.
2. The volume resistivity of the adhesive should be 0.01 Ω cm or less, preferably 0.001 Ωcm.

3. Most conductive adhesives do not bond well to either neoprene or silicone. This is why all products that have conductive paths in elastomer are rated **poor** for conductive adhesive bonding by the manufacturers.
4. Applying a 3 to 6 mm (1/8" to 1/4") diameter spot of conductive adhesive every 25 to 50 mm (1" to 2") is preferable to a continuous bead.
5. Conductive epoxies will attach the gasket permanently. Thus, removal of the EMI gasket without destroying it is almost impossible.

Attachment via Bolt or Screw Holes

This is a common and inexpensive way to hold gaskets in position, as shown in Fig. 4.24c. For most products, providing bolt holes involves only a small initial tooling charge; there generally is no extra cost for bolt holes in the piece price of the gasket. Bolt holes can be provided in the fin portion of EMI strips or in rectangular cross-sectional EMI strips if they are wide enough, e.g. >10 mm (>3/8").

Special Attachment Methods

The knitted-mesh fins provided on some versions of EMI gaskets and the aluminum extrusions in aluminum gasketing were designed to attach these products as shown in Fig. 10.24d. The mesh fins could be clamped under a strip of metal which is held down by riveting or spot welding, or the mesh fins can be bonded with an adhesive or epoxy. The aluminum extrusions of aluminum gasketing can also be held in position by rivets or bolts.

Except with finger stock, EMI gaskets should be positioned so they experience little or no sliding motion when being compressed. This is illustrated in Fig. 10.25. The EMI gasket shown in Fig. 10.24a is subject to sliding motion when the door is closed. This may cause it to tear loose or to wear out quickly. In Fig. 10.24b, the gasket is subject to almost pure compression-only forces. This is the preferred position. Finally, Fig. 10.26 provides a checklist for gasket selection and design.

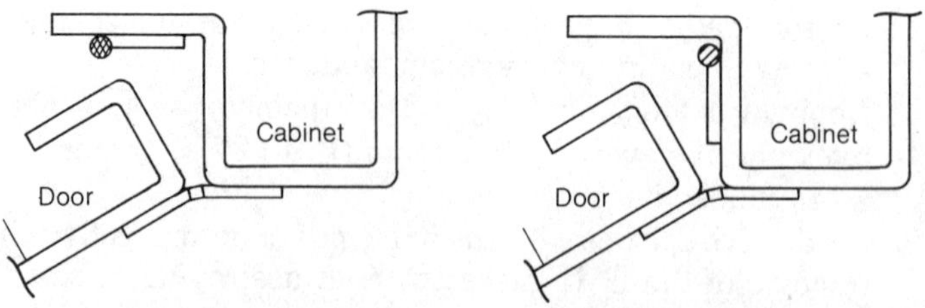

a. Poor Design: Door Slides on EMI Gasket

b. Good Design: Door Compresses on EMI Gasket

Figure 10.25—Proper Method of Mounting Gasket in Cabinet Door Well

Checklist for Gasket Selection

[ ] 1. Determine the environmental constraints:

____

____

[ ] 2. Identify the shape and material of mating members:

____

____

____

____

____

[ ] 3. Select the gasket type:
- [ ] 3.1 Check material against (1) and (2).
- [ ] 3.2 Select shape with respect to (2).
- [ ] 3.3 Any "combination" gasket needed?

____

____

[ ] 4. Calculate minimum and maximum pressure, compare with resistance and SE objectives.

[ ] 5. Define mating face machining tolerances, determine surface treatment and painting precautions.

Figure 10.26—Checklist for Gasket Selection

10.2.5 The Role of Cover Gasketing in Internal Shielding

Previous sections have emphasized the role of gasketing as part of the barrier between the inside of a box and the surrounding space. However, there are cases where electromagnetic isolation is also needed between compartments of the same housing, i.e., to prevent an equipment from self-jamming. One example of this need is shown in Fig. 10.27, where there exists a potential

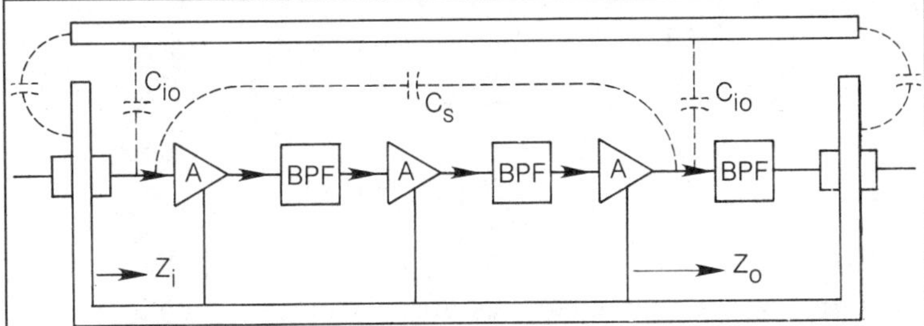

a. Box Shield for High-Gain Amplifier Showing Two Paths of Output-Input Parasitic Feedback

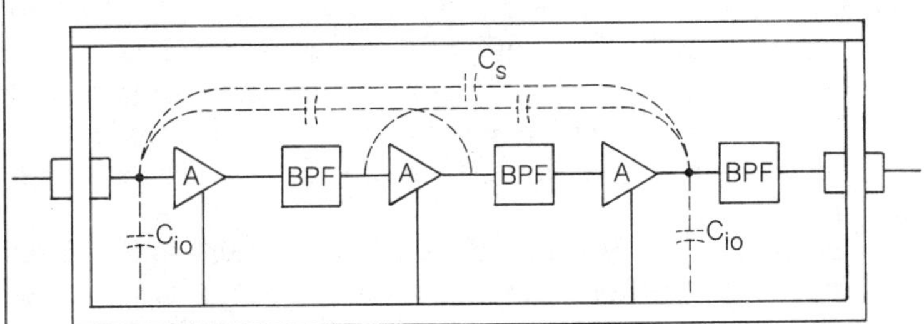

b. Box Shield Cover Plate in Position to Eliminate Larger Feedback Path

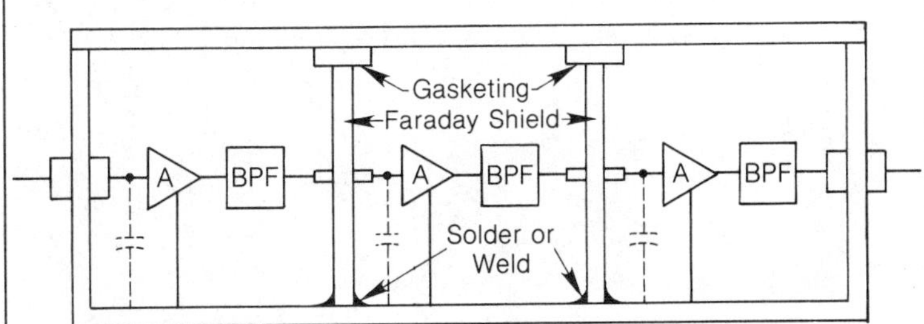

c. Box Shield With Compartment Septum Shields to Ground Only Interstage Feedback

Figure 10.27—The Need for Gasketing with Compartmented Shielding

capacitive coupling between several high-gain amplifier stages.

In Fig. 10.27a, the cover plate is floating or poorly grounded (paint, oxide, etc.). Thus, a portion of the high-level output voltage V_o is capacitively coupled back to the amplifier input via the conductive cover, which acts as a coupling bus. It is remarkable that in this case the floated cover makes things even worse than if there were no cover at all; without the cover, C_{io} would be smaller (shown as C_s in Fig. 10.27a).

The feedback voltage to the input stage, for instance, is:

$$V_i \approx V_o \frac{Z_i}{2/\omega C_{io}} = 0.5\ V_o\ Z_i\omega\ C_{io}$$

If V_i is on the order of and in phase with the normal low-level input signal of the input stage, instability can result due to positive feedback, and the amplifier will oscillate.

When the cover plate is bonded to the box, the capacitive feedback current is shunted to ground and does no harm (Fig. 10.27b). However, the residual capacitance C_s between the first and last stage (or between intermediate stages) can still cause parasitic feedback. Thus, internal compartments are added which must be bonded to the top plate by an integral or die-cut gasket as shown in Figs. 10.27c and 10.28.

In Fig. 10.28, note that the left-hand option results in no wasted material but requires an expensive custom design. The right-hand option leaves the gasket foil uncut. The foil needs only to have one side conductive, with the rubber layer acting as a pressure medium between the actual mechanical cover and the box edges. For this design to work, it is imperative that the screw holes be blind.

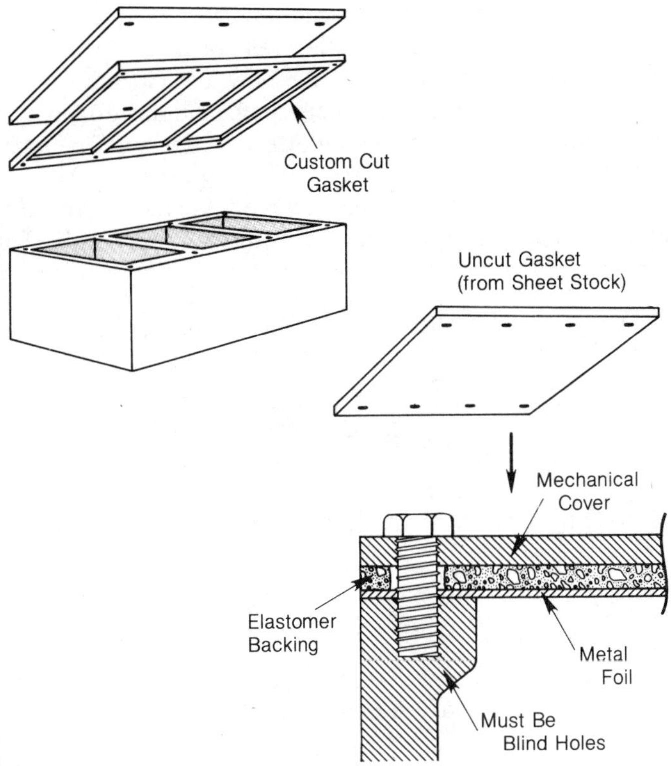

Figure 10.28—Two Different Ways of Having a Gasket to Fit Compartmented Boxes

10.3 Ventilation Openings

Most shielded housings or enclosures require either convection or forced-air cooling. Since ventilation openings will compromise the integrity of the basic shield material, a suitable electromagnetic mask must be installed to provide substantial RF attenuation while not significantly impeding the flow of air. Two approaches are possible: screened covers and honeycomb aperture covers. Screens are inexpensive solutions to this problem but are limited in shielding effectiveness and tend to block the flow of air (increase in aerodynamic impedance). Thus, a honeycomb material is often used because it provides higher shielding effectiveness and maintains a streamlined flow of air.

The SE performance of wire-mesh air filters can be predicted from Fig. 10.29 against plane-wave or near magnetic fields. The

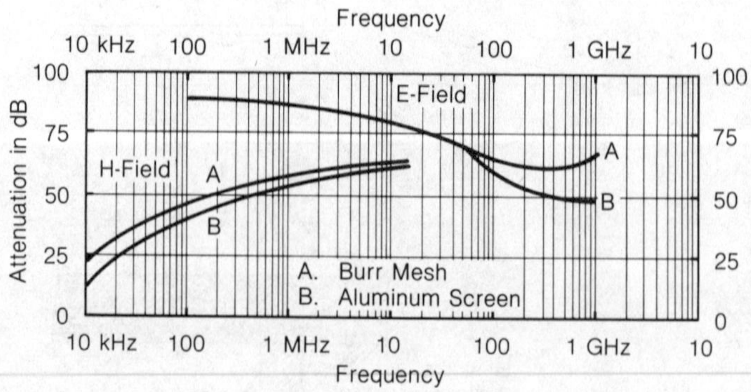

Figure 10.29—Typical Shielding Effectiveness of Shield Screen Mesh Vent Covers

advantage of screen meshes, besides their low cost, is their relative thinness and ability to combine dust filtering and EMI attenuation, usually via several layers of woven mesh captive between two retaining grids (see Fig. 10.30). Wet or viscous filters are available.

In many cases, however, mesh-type air filters introduce too much air resistance. Also, their SE performance approaches 0 dB when the mesh apertures sizes reach $\lambda_{EMI}/2$. When more than 80 dB of SE is needed up to 1 GHz, honeycomb structure becomes necessary (see Fig. 10.31). Honeycomb filters are usually made of hexagonal cells constructed either from extruded aluminum, corrugated aluminum, brass or steel foils bonded together or metalized plastic molded parts.[7] For types made of corrugated foils held together, the conductivity may be marginal, especially with aging. Better quality filters are obtained when preformed steel or brass foils are welded, brazed or fused together then plated. A further increase in performance can be obtained by stacking two honeycomb arrays, 90° polarized (Fig. 10.32), or by creating cancellation slots

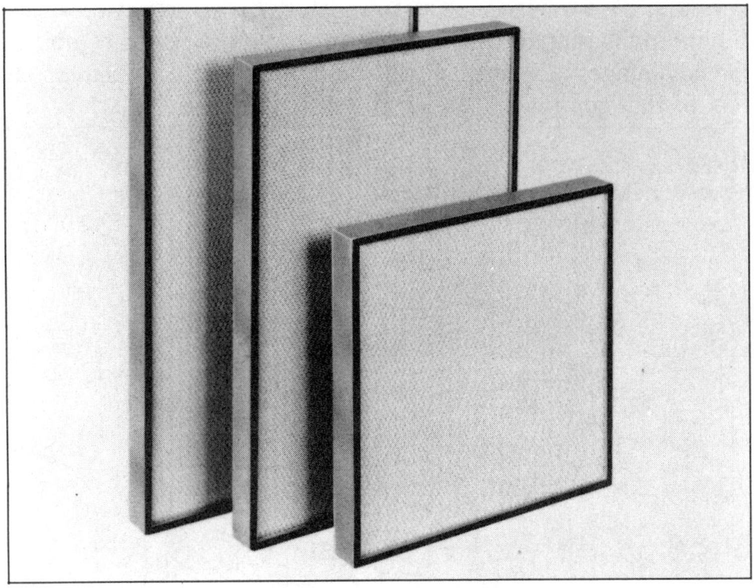

Figure 10.30—Shield Screen Mesh Ventilation Permitting Dust Removal by Oil Impregnation

Figure 10.31—Representative Honeycomb Configurations

to obtain the same benefit as the adjacent holes in a wire mesh. Figure 10.33 illustrates an interesting structure which combines the advantages of a deep aperture and the mutual cancellation effect of thin contiguous openings.

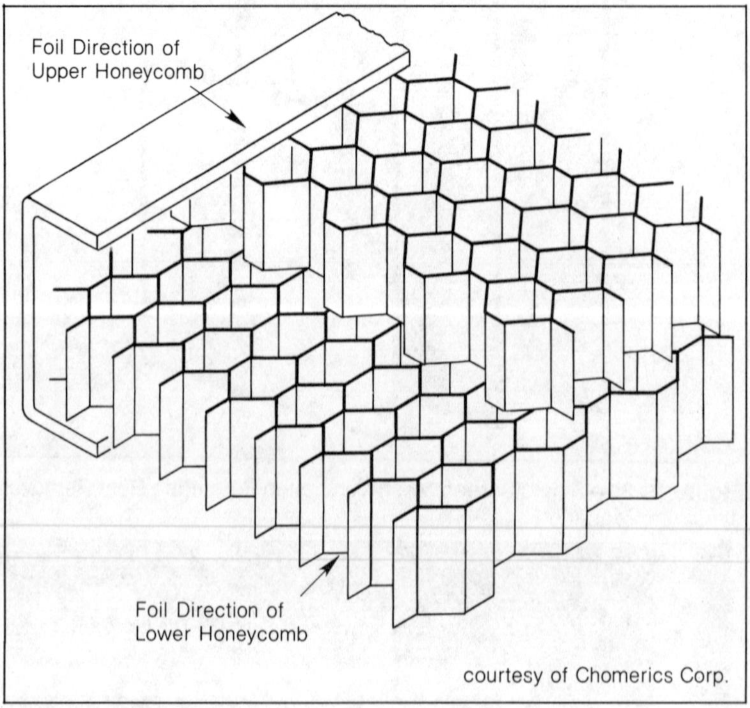

Figure 10.32—Typical Honeycomb Construction

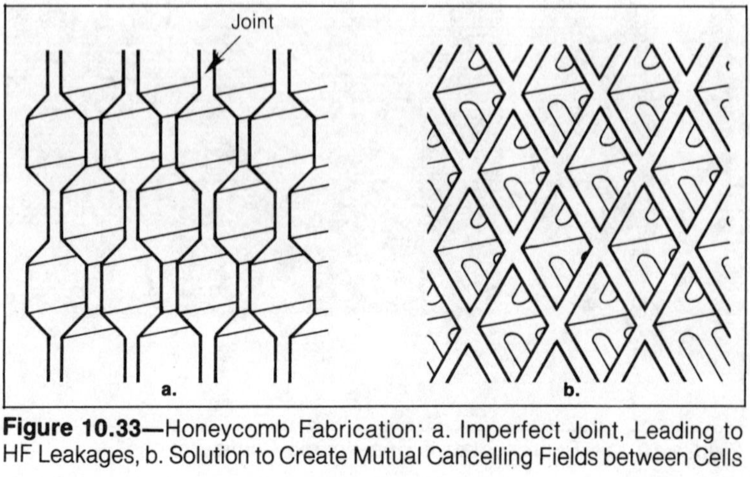

Figure 10.33—Honeycomb Fabrication: a. Imperfect Joint, Leading to HF Leakages, b. Solution to Create Mutual Cancelling Fields between Cells

The principal features of honeycomb vent panels are:

1. Higher attenuation
2. Minimum air pressure drop (see Fig. 10.34)
3. Resistance to damage
4. Higher cost
5. More depth required versus mesh, especially if a dust filter has to be added
6. Difficult to mount "flush"

Typical performance curves of honeycomb filters are shown in Fig. 10.35. At frequencies below 30 to 100 MHz, there is evidence that the dynamic range of the measuring instrumentation becomes

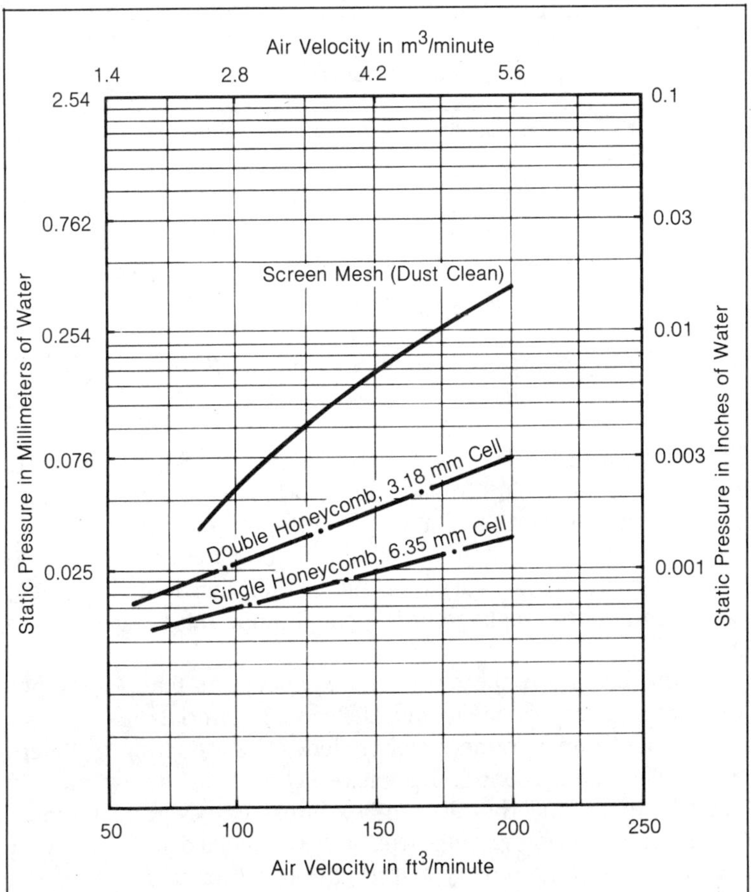

Figure 10.34—Aerodynamic Pressure Drop of Several EMI Vent Panels

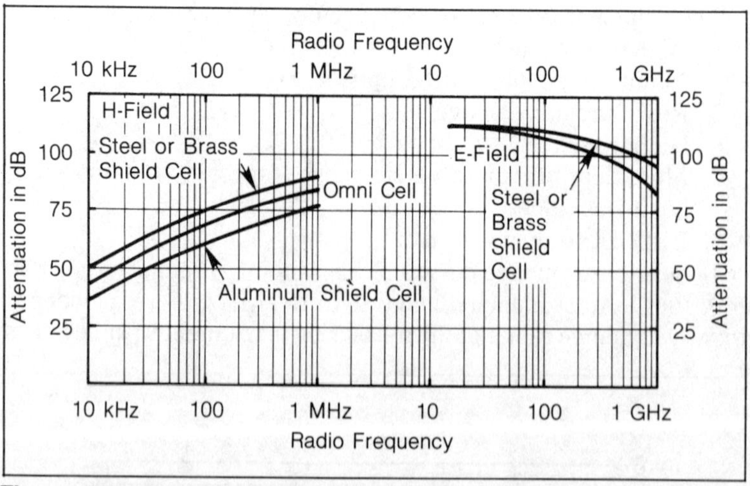

Figure 10.35—Typical Shielding Effectiveness of Honeycomb Vent Covers

the limiting factor.[8] Using the model described in Section 7.2.5, one sees, for instance, that a honeycomb panel having 3 mm wide, 12 mm deep cells arranged in a 100 mm × 100 mm array (about 1,000 holes) gives a plane-wave attenuation of:

$$\begin{aligned} SE_{dB} &= 100 - 20\log(3) - 20\log(f) + 30 \times 12/3 - 20\log(1{,}000) \\ &= 150 - 20\log(f) \end{aligned}$$

Therefore, at 10 MHz, the theoretical SE is 130 dB, which is greater than the dynamic measurement range of the vast majority of SE evaluation setups. Often, SE measurement results of vent panels report a constant SE of 100 to 120 dB below 100 MHz, while what they actually measured was either their own dynamic range, background noise, leakage of the panel mounting flanges or box skin SE.

To fully benefit from the ventilation panel performance, whether it is screen mesh or honeycomb, the screen material must be kept rigid by a stiffener frame which includes the mounting fasteners and a good quality conductive gasket. Otherwise, a new leakage could be created at the filter-to-box interface. Even in the most carefully performed measurements of EMI-shielded air filters, what is often seen is the quality of the gasket rather than that of the screen itself. Therefore, the gasket must have, if possible, an intrinsic SE equal to or better than that of the ventilation panel honeycomb or mesh; the mounting frame must be plated, or have a chromate or other noncorrodible finish and the mating area on the box must be free of paint, grease, etc. If required, a sealant can be applied to the joint after tightening the panel to avoid moisture penetration. Mechanical mounting of a ventilation panel is illustrated in Fig. 10.36.

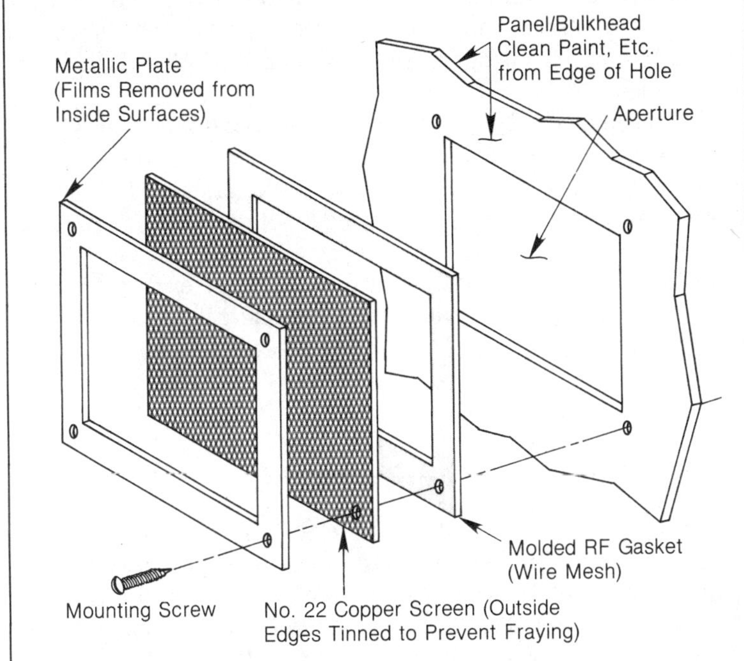

Figure 10.36—Mechanical Mounting of EMI Ventilation Panels

10.4 Viewing Apertures

Another compromise in the integrity of the basic shield material is the need for video displays, panel meters, digital displays, CRTs and other types of status monitors or readouts contained inside the shielded housing or enclosures. Such viewing apertures can also be windows in buildings (see Chapter 13), windows in hospital shielded enclosures, canopies in aircraft, transparent domes in missiles containing cameras, etc.) For instance, the leakage created by a CRT tube face is so large that, if not considered in the initial design, it may create difficult problems when the product is tested for compliance with EMC standards and regulations.

Display openings can be shielded by incorporating laminated-screen windows, transparent, conductive substrates or backside-mounted (doghouse) shields. This section discusses the display apertures purely from a leakage point-of-view. The display itself (especially CRT) can be a source or victim of radiated EMI for which its emission and susceptibility profile needs to be characterized. This is addressed in Volume 5 of this EMC handbook series.

EMI shielding can be combined (sometimes advantageously) with antireflection and contrast enhancement. To be effective (make the glare unnoticeable) the reflection of the ambient light at the air-to-glass coating interface must be <1 percent. This is accomplished by optical filters absorbing all optical wavelengths except the narrow-wavelength light from the tube or display itself. Also, one-way light-scattering layers and optical polarizers affecting only the light coming to the tube facade can be beneficial.

10.4.1 Screen Windows

The screen materials used in shielded windows are made of either a thin mesh of knitted wires (up to 12 openings per cm) or a high-density woven screen (thin meshes with 40 or even 60 wires per cm are obtainable). This mesh is incorporated in or laid over the transparent glass or acrylic sheet (see Fig. 10.37). The open area represents 60 percent (for super-dense screen) to 95 percent of total area (Fig. 10.38). The SE performance tracks the value given in Section 7.2.3; therefore, it decreases when frequency increases as shown in Fig. 10.39. The denser mesh offers more shielding since the aperture size is a minute fraction of the EMI wavelength, but this is at the expense of visual transmission. For instance, to compensate for the light losses, a CRT would have to be driven harder, reducing the tube life. The mesh must be plated to avoid corrosion, and a durable electrical contact must be guaranteed at each wire crossing. Heavy plating is not recommended and may be counterproductive since it affects the percentage of optical transmission.

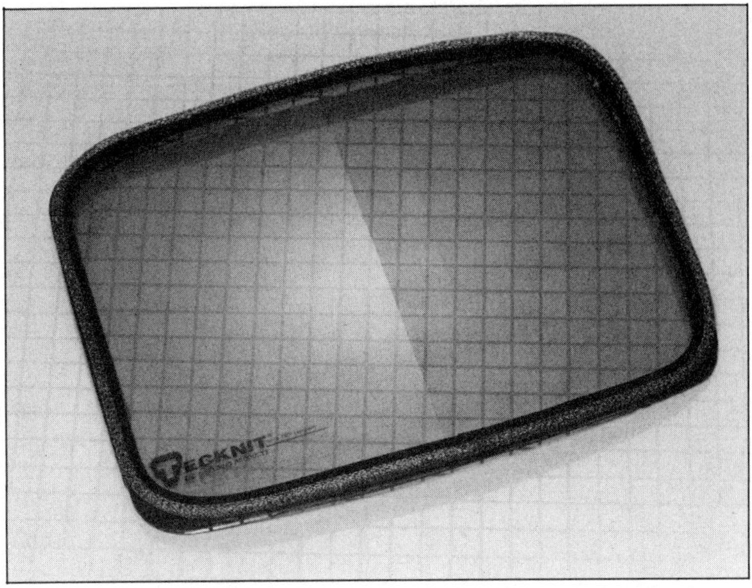

Figure 10.37—Representative Shield Screen Windows for Viewing

With knitted mesh, the non-square shape of the cell (Fig. 10.40) minimizes the risk of regular, periodic obscuring patterns, i.e., diffraction interference to alphanumeric characters. With woven type screens, the higher density and perfectly square pattern causes a greater risk of optical interference since the wire threads occlude portions of the pixels. This is aggravated with the shadow-mask CRT types if the center-to-center dimension of the mask is an integer multiple of the mesh size. To decrease this effect and avoid discomfort under certain viewing angles, the mesh should be put on the inner side of the glass sheet, as close as possible to the shadow mask itself.

Another similar concern exists with plasma and electroluminescent displays. In a 512 × 256 matrix panel (25 cm × 15 cm) for instance, the fine resolution of 0.5 mm would degrade rapidly if a metal mesh were put on top. A conductive transparent film as presented in the next paragraph would sacrifice luminosity, which

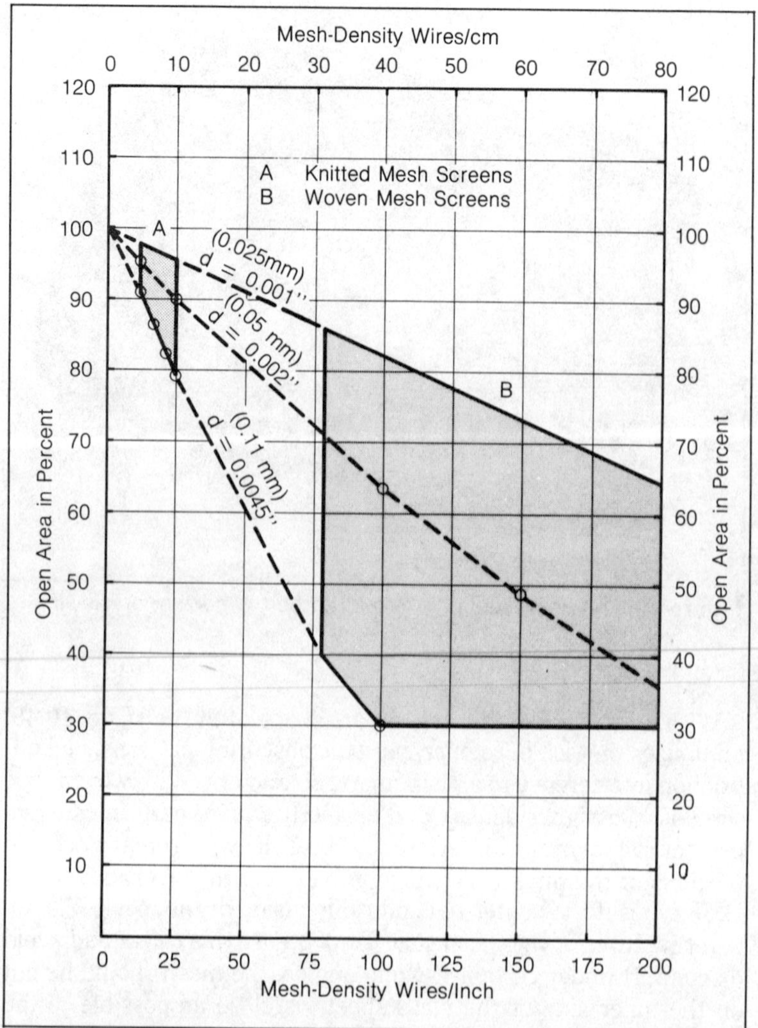

Figure 10.38—Percentage of Open Area of Window Mesh vs. Type of Mesh and Wire Diameter d (Source = Tecknit® Design Guide)

is already marginal with these low-power displays. Until conductive films with better light transmittance are found, the solution is either to use a fine-line mesh whose grid size fits between the matrix columns and rows or to equip the display with a backside shield as explained in Section 10.4.3.

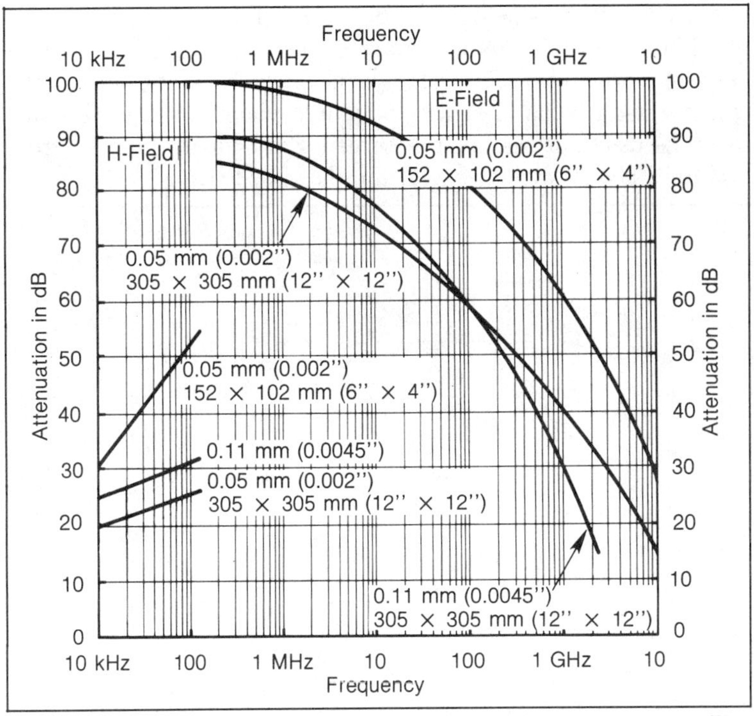

Figure 10.39—Shielding Effectiveness of Shield Screen Windows (Top Figures = Wire Size, Lower Figures = Aperture Size)

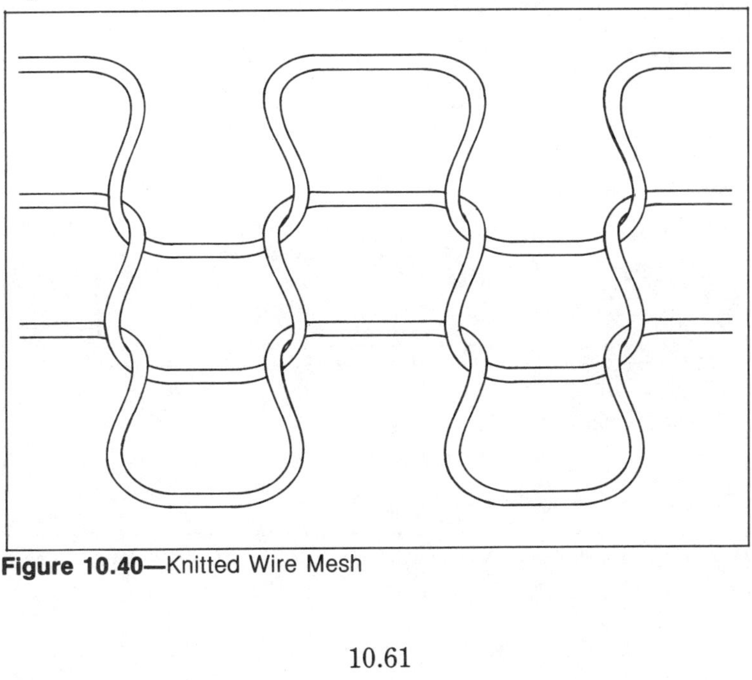

Figure 10.40—Knitted Wire Mesh

The edge of the shielding window must provide for a tight electrical bond to the metallic surface of the equipment box. This is generally done by an interface gasket, which can be part of the window itself. This interface, of course, must have an SE equal to or better than that of the screen glass itself. The different methods, shown in Fig. 10.41, are discussed below.

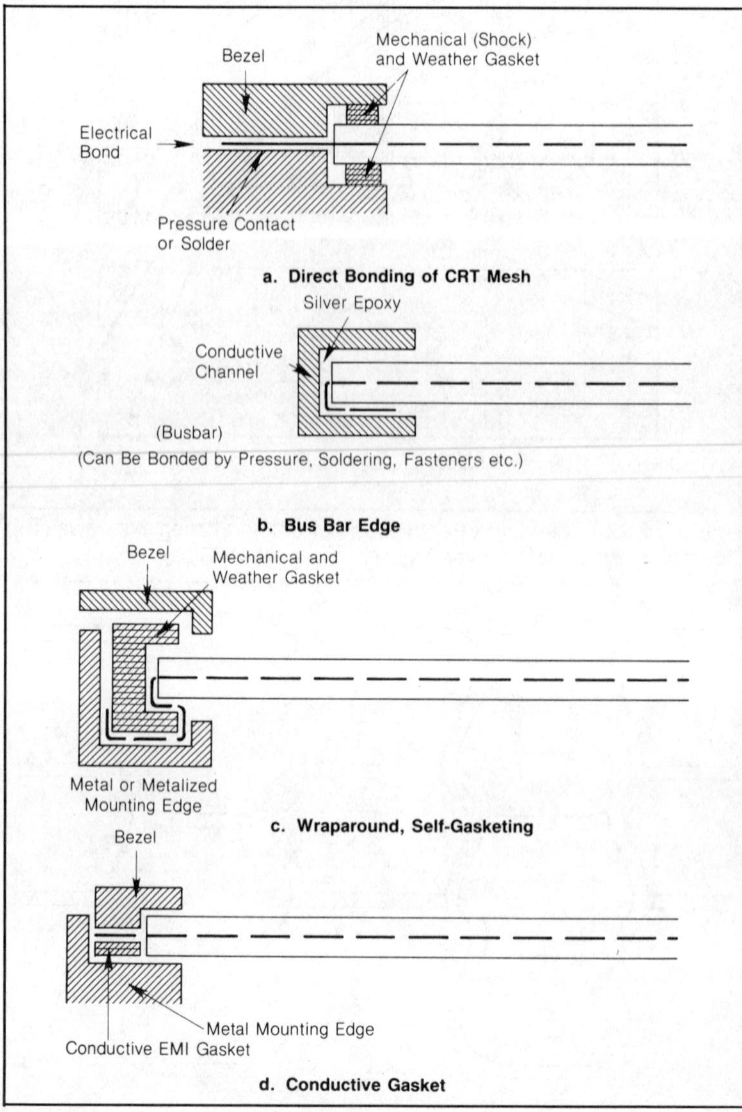

Figure 10.41—Various Bonding Methods for Mesh-Shielded Windows and CRT Shields

Direct Bonding

The shield screen can be more or less permanently bonded (by welding or soldering) to the perimeter of the box opening (Fig. 10.41a). Although probably the best method, it is seldom used because of the cost and the need to have a significant amount of the screen exposed around the glass perimeter. Also, since the window may need an environmental gasket anyway, other methods are generally preferred.

Bus Bar Edge

This method (Fig. 10.41b) is similar to the previous one except that the edges are already shaped by a metallic rim which provides an easy fastening medium.

Wraparound, Self-Gasketing

This style is shown in Fig. 10.41c.

Conductive Gasket

This is illustrated by Fig. 10.41d.

10.4.2 Conductive Optical Substrate Windows

The use of conductive windows is another approach for providing shielding of apertures through which either optical viewing or light transmission is also necessary. In this technique, a thin film of metal (gold) or indium oxide is vacuum deposited on a transparent substrate. These conductive window designs, such as shown in Fig.

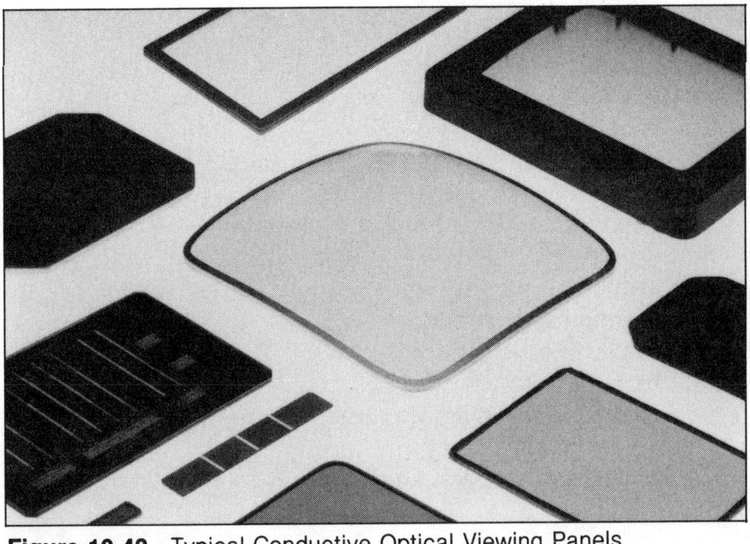

Figure 10.42—Typical Conductive Optical Viewing Panels

10.42, are determined by establishing some or all of the six basic design parameters, as applicable:

1. Window material
2. Reticle requirements
3. Conductive coatings
4. EMI gasketing
5. Optical coatings and finishes
6. Framing and mounting

Most plastic and glass materials are suitable for use as substrates for the application of conductive coating. The standard materials are glass, acrylic, polycarbonate and fluorocarbon plastics. The substrates may be clear or colored as required by the application. There are no restrictions on substrate thickness. Curved or three-dimensional parts generally can be coated. Most thermosetting and thermoplastic substrates have minute surface scratches produced in their normal manufacture. Application of the coating will inherently make these more apparent, although actual user experience indicates no functional problem will arise. The following list illustrates a sample of the large selection of common substrate materials suitable for conductive coating application:

1. Glass
    - plate
    - single-strength
    - float
    - tempered
    - laminated
    - PVB-film
    - quartz
2. Crystals
    - ruby
    - quartz
3. Vycor[1]
4. Pyrex[1]
5. Lexan[2]
6. Plexiglas[3]
    - thermoplastic acrylic
    - transparent, colorless
    - frosted, colorless
    - colored (yellow, amber, grey, bronze, green, blue)
    - self-extinguishing
7. Homalite[4] thermosetting plastic

8. Kapton[5]
9. Mylar[5]
10. Abcite[5] coated acrylic
11. Polycarbonate
12. Fluorocarbons

Trademarks of: 1. Corning, 2. General Electric Co., 3. Rohm & Hass, 4. Homalite and 5. Dupont

In the plastic substrate group, the most scratch-resistant materials are Abcite followed by Homalite. Polarized filter laminate finishes are available for contrast improvement. Coatings are unaffected by the application of laminated circular polarizers.

Translucent or frosted finishes, rough in surface nature, are available. They are best employed on the side opposite the conductive face. They can only be used for rear projection displays or where the object is extremely close to the window surface. Antireflective, vacuum-deposited coatings may be applied to windows before coating. Figure 10.43 illustrates typical SE versus frequency for different film coating thicknesses on glass measured in surface resistance units of ohms per square.[9]

Data reported by manufacturers may be somewhat misleading if the user is not aware that they are generally measured per MIL-STD-285 at 30 cm distance (see Section 3.1). Therefore, up to about

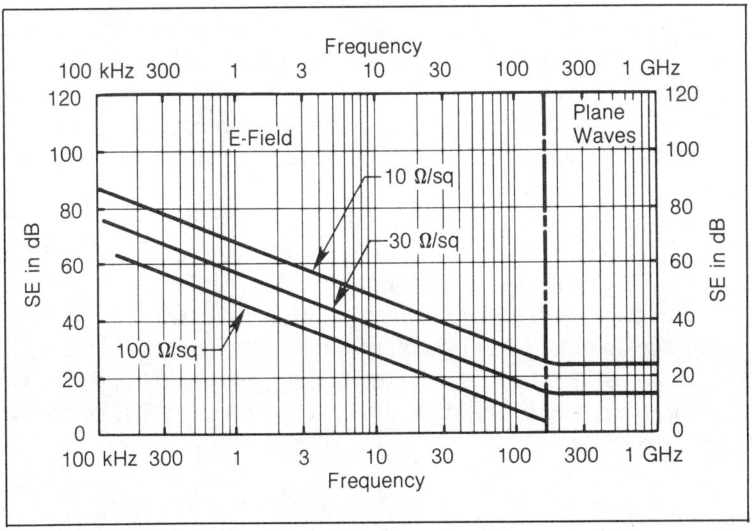

Figure 10.43—Shielding Effectiveness of Conductive Glass for a 30 cm Source Distance

200 MHz, the attenuation will follow a decreasing slope simply because the test setup creates near E-field conditions. Sections 2.1 and 6.1.2 show that when the thickness t of the shield is much less than the skin depth δ, the only shielding is due to reflection **and is constant with frequency** for plane-wave conditions. For conductive glass, the film thickness is much less than a micron. Gold, for instance, is often used because it offers the best compromise between corrodibility and low resistance. The thickness of gold required to stay within a tolerable range of opacity is 10 to 100 Å (one Å = 10^{-10} m or 10^{-4} microns). An 80 Å film will correspond to 75 to 80 percent light transmittance in the mid-visible spectrum (500 to 700 nm).

Very thin metal films tend to have a higher resistance than what one would simply calculate from bulk metal resistivity. This is due to electron scattering when the film thickness is equal to or less than the electron **mean free path**, which is the average distance an electron moves without collisions.

For gold, at room temperature, this mean free path is 570 Å. Except for near-field condition and in contrast to wire meshes, the SE of conductive coated windows remains constant with frequency up to several hundred gigahertz. At that point, the film thickness would approach one skin depth, and absorption occurs.

Because of their $\epsilon_r > 1$, the intrinsic impedance of the glass or acrylic substrate is often in the 150 to 200 Ω range, causing a slight additional impedance mismatch to the EM wave. Thicker films can result in lower surface resistance, with the minimum acceptable light transmission being 60 to 65 percent, corresponding to a surface resistance of 5 to 8 Ω/sq depending the type of metal, substrate type and thickness.

The light transmission is generally given for the narrow gap of the visible spectrum. Eighty percent transparency (for visible light) may very well correspond to only 10 or 30 percent of the total light transmittance from upper infrared to low UV boundaries. Depending on the dominant display color (orange, green, grey, blue, etc.), different coatings may be selected to optimize the visibility of the display figures and attenuate unwanted wavelengths, reducing glare, moire, etc.

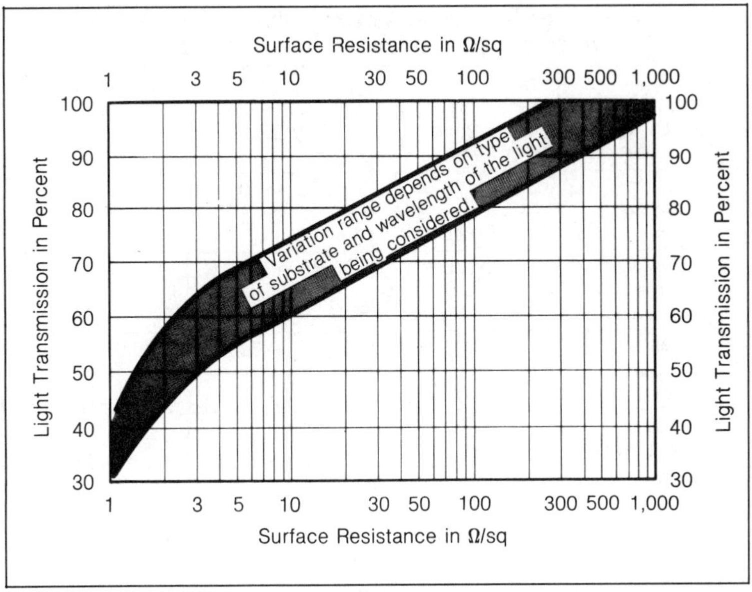

Figure 10.44—Range of Light Transmission of Conductive Windows vs. Film Resistance. Film resistances down to about 3 Ω/sq are achievable, but below this level, transparency falls off abruptly.

Recognizing that some variance exists (depending on the type of metal, type of substrate and so forth), light transmission versus surface resistance for the above conductive glass is shown in Fig. 10.44. Transmission values of 60 to 80 percent correspond to resistances of about 10 to 100 Ω/sq. Thus, these values may now be compared with the attenuation data of the shield screen depicted in Fig. 10.39 to determine relative performance for comparable area size specimens.

The screen mesh window is everywhere superior to the film-coated type, as shown in Table 10.4 in which the difference would only vanish above 30 GHz. Thus, it is concluded that if significant VHF and UHF attenuation is required for viewing apertures, shield screen windows should be used. If the esthetics or other considerations do not permit this, conductive glass cannot be relied upon to provide RF attenuation in excess of 20 dB, except for near E-fields.

Table 10.4—Comparison of Shielding Effectiveness of Mesh vs. Conductive Glass Windows

Frequency & Field Condition		Thin Mesh Screen: (15.5 wires/cm) (100 wires/in)	Conductive Glass: 10 Ω/sq	Superiority Mesh-Conductive Film
1 MHz (30 cm distance)	E	>100	65 – 70	>30
	H	40	0	>40
10 MHz (30 cm distance)	E	>100	45 – 50	>50
	H	60-70	0	>60
100 MHz (30 cm distance)	E	80	25 – 30	50 – 55
	H	60-70	15	55
1 GHz Plane Waves		55	20 – 25	30 – 35
10 GHz Plane Waves		35	20 – 25	10 – 15

Finally, another decisive parameter for determining the applicability of conductive transparent coatings is the possible requirements for large power handling, in watts/cm^2. Examples are de-icing/defrosting and exposure to powerful, narrow-beam ambient fields in the millimeter wavelength where absorption may show up and cause heating of the window.

10.4.3 Shielding the Display from the Rear Side

This solution does not really belong to the transparent window family since it consists of reconstructing a shield behind the display itself. However, since it is part of the general subject of preventing leakages from the large display openings, the designer must be aware of it.

We have seen that neither conductive glass nor fine mesh can provide efficient shielding against VLF H-fields such as power-mains magnetic fields, nearby power converters, etc. Therefore, a variety of cone-shaped magnetic shields are available which can be fitted to all current CRT sizes. They are fabricated from flat, unannealed Mumetall®, Conetic® or other highly permeable foil.

After stamping, forming and welding, the shield is annealed. From then on, it should not be bent or cold worked lest the annealing be destroyed. Figure 10.45 shows a few aspects of this possible option.

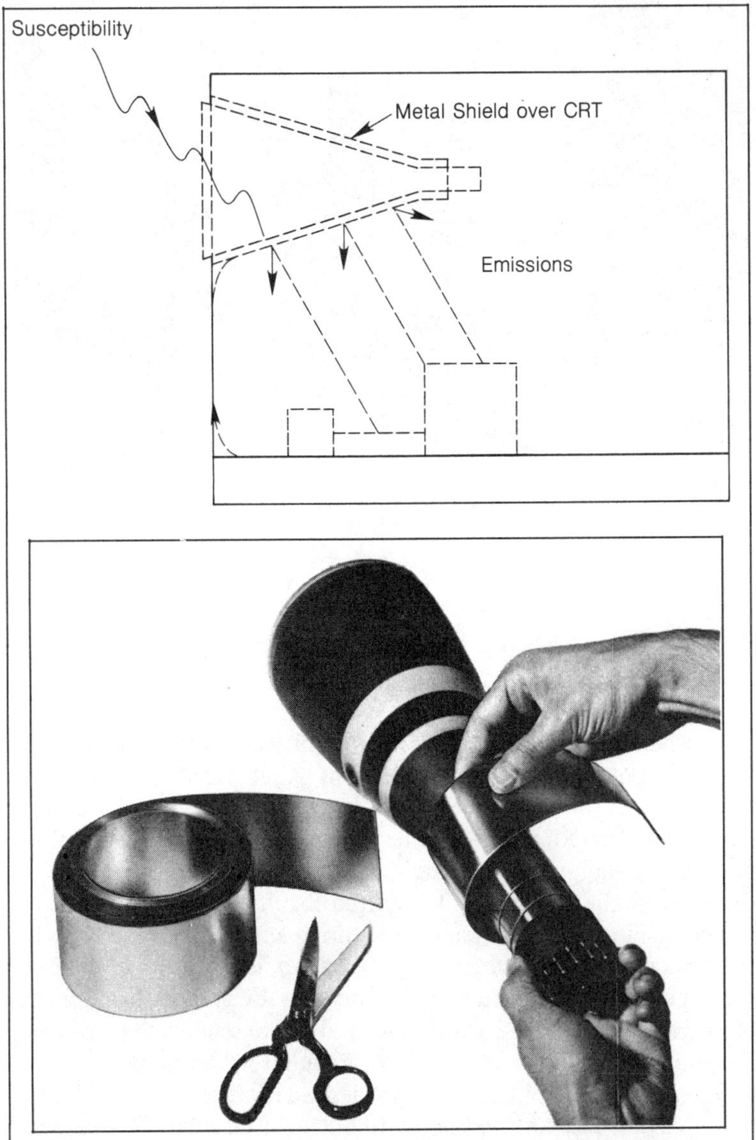

Figure 10.45—Other Options Available for Shielding a CRT. Instead of blocking the radiated EMI at the front of the tube, a shield, generally magnetic, is added behind it. (continued next page)

Figure 10.45—(continued)

10.5 Switches and Control Shaft Openings

Another aperture type which compromises the shielding integrity of an equipment housing or instrument panel is that resulting from the shafts of potentiometers, tuning dials and control devices. Generally, an external metallic front panel or housing is either drilled or punched with enough clearing tolerance through which the control shaft extends. This often results in a leaky aperture.

The inside wall of the panel hole forms an outer conductor, with the coaxially situated internal control shaft acting as inner conductor. In other words, potential EMI can enter or exit through this effective short-length coaxial line, and the extended shaft beyond the panel acts either as a pickup or radiating antenna.

One method of minimizing this degradation of SE is to design a supporting bushing extender to act as a circular WGBCO attenuator (see Chapter 7). For 100 dB attenuation in a circular waveguide, the length of the waveguide must be somewhat more than three times its diameter. In Eq. (7.12), this would be expressed as $d/g > 3$. Figure 10.46 shows an acceptable use of a metal tube bonded to the wall containing the clearance aperture for control shafts. The tube then makes contact with the metal control shaft which seals the hole. In Figs. 10.46a and 10.46b, the shaft must be nonconductive or make electrical contact with the shield wall via an iris-type spring or conductive guard.

If the preceding solution were implemented without regard to the control shaft properties within the added metal tube, little improvement would result for metal shafts. This situation corresponds to a low-impedance coaxial line in which an intervening dielectric is present due to contaminants such as oil films or oxides. To

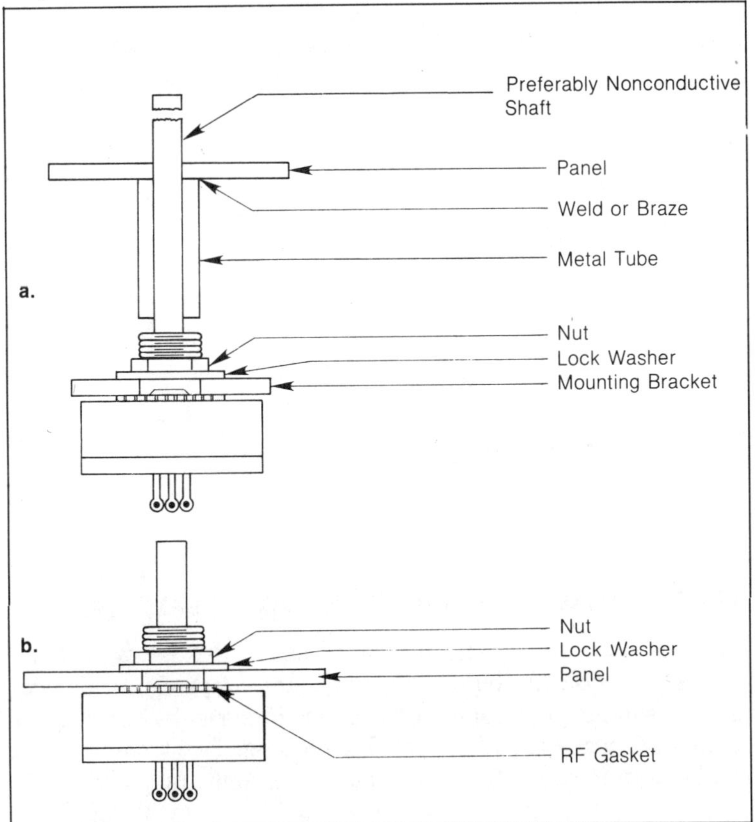

Figure 10.46—Use of Circular Waveguide in a Permanent Aperture for Shaft EMI Leakage Control (From USAFSC DH 1-4) (continued next page)

preclude this, one of two techniques is followed:

1. Replace the metallic control shaft with a nonconductive shaft as shown in Fig. 10.46c.
2. Use a cylindrical-shim EMI gasket between the shaft and tube. This method does not require modification of existing control shafts.

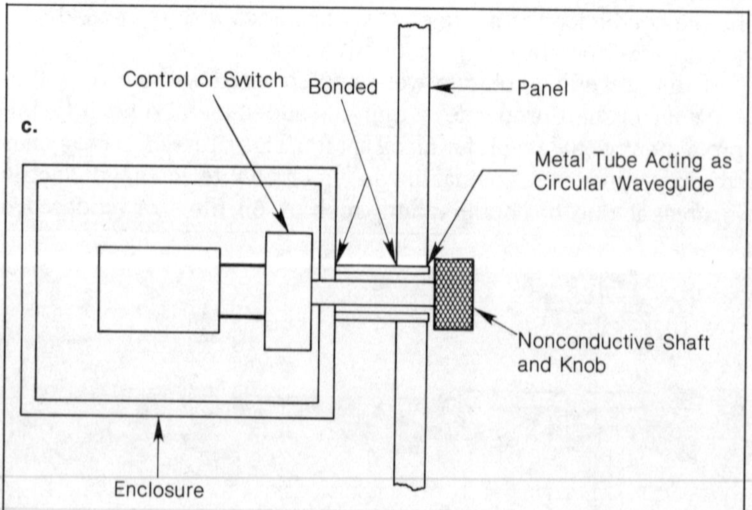

(from USAFSC DH 1-4)

Figure 10.46—(continued)

10.6 Indicator Buttons and Lamps

Instruments and equipments often require the use of pushbuttons, status indicator buttons and indicator lamps. These devices provide another compromise of shielding integrity by virtue of the required apertures in a front panel or housing. Two techniques are available to mitigate the EMI leakage through such devices:

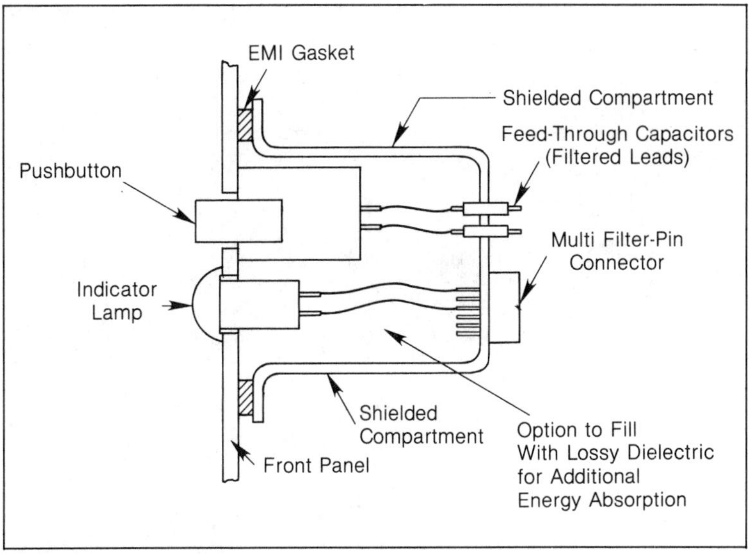

Figure 10.47—Shielded and Filtered Compartment Technique to Restore Shielding Integrity of Button and Lamp Ampertures

1. Encase them in a shielded compartment behind the front panel when they are mounted as shown in Fig. 10.47. Feed-through capacitors or filter-pin connectors are used for hard wiring from outside the compartment to the buttons or indicator lamps since conducted EMI could exist on either side of the barrier.
2. Use special EMC-designed hardware where such devices are mounted directly to a front panel. Examples of this include wire-mesh indicator lamps (which look like miniature photoflash bulbs). This mesh serves to reflect and absorb entering or exiting EMI energy. Some of these devices are shown in Figs. 10.48 and 10.49.

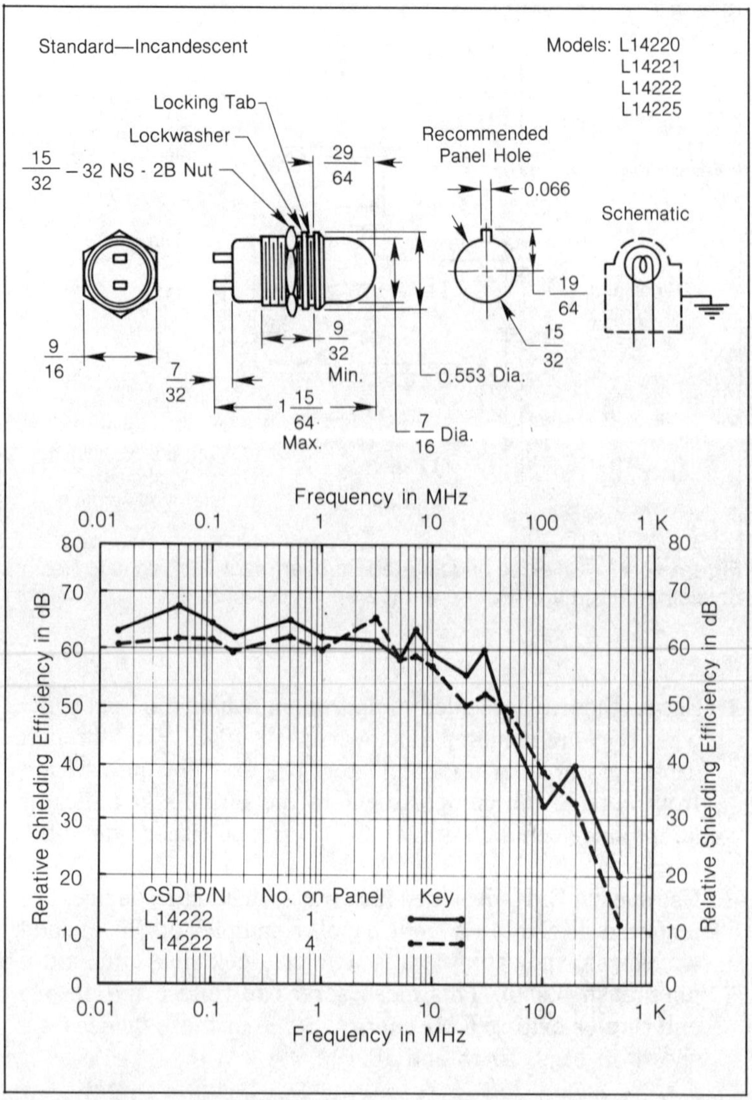

Figure 10.48—EMI-Shielded Incandescent Indicators (courtesy of Cutler-Hammer, Aerospace/Commercial Controls) (continued next page)

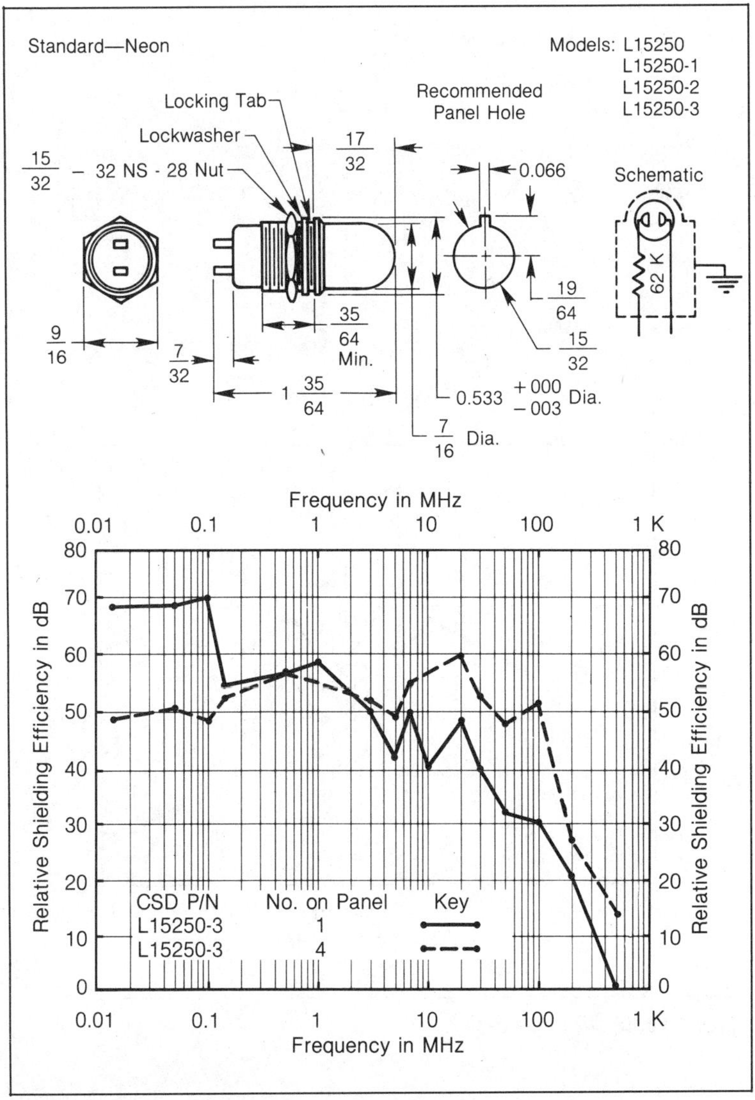

Figure 10.48—(continued)

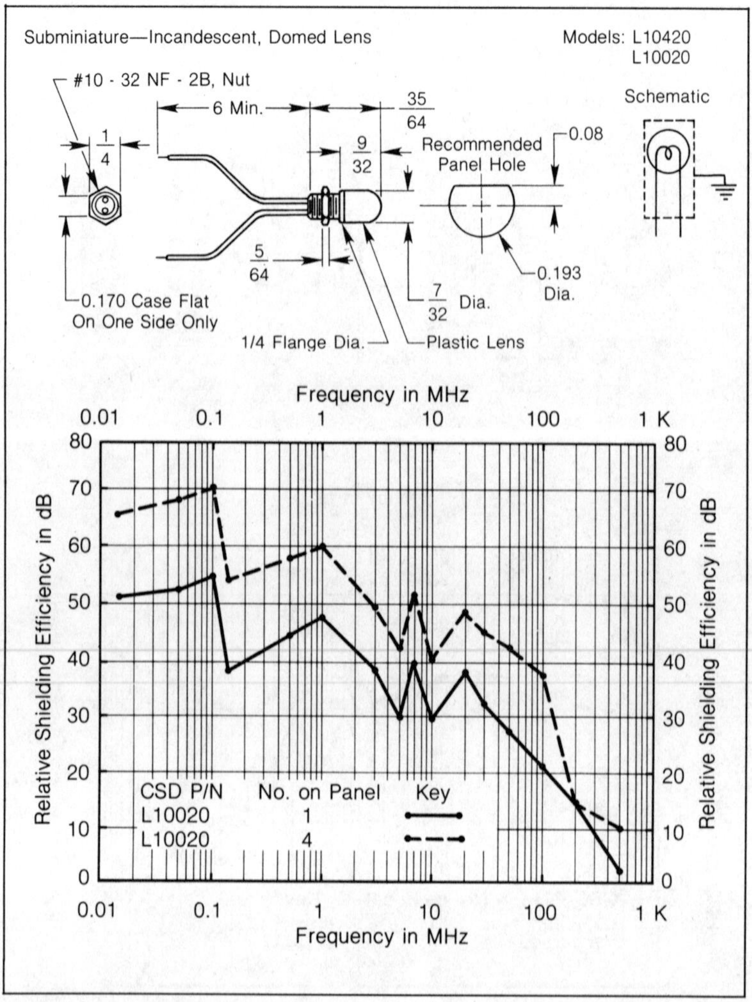

Figure 10.48—(continued)

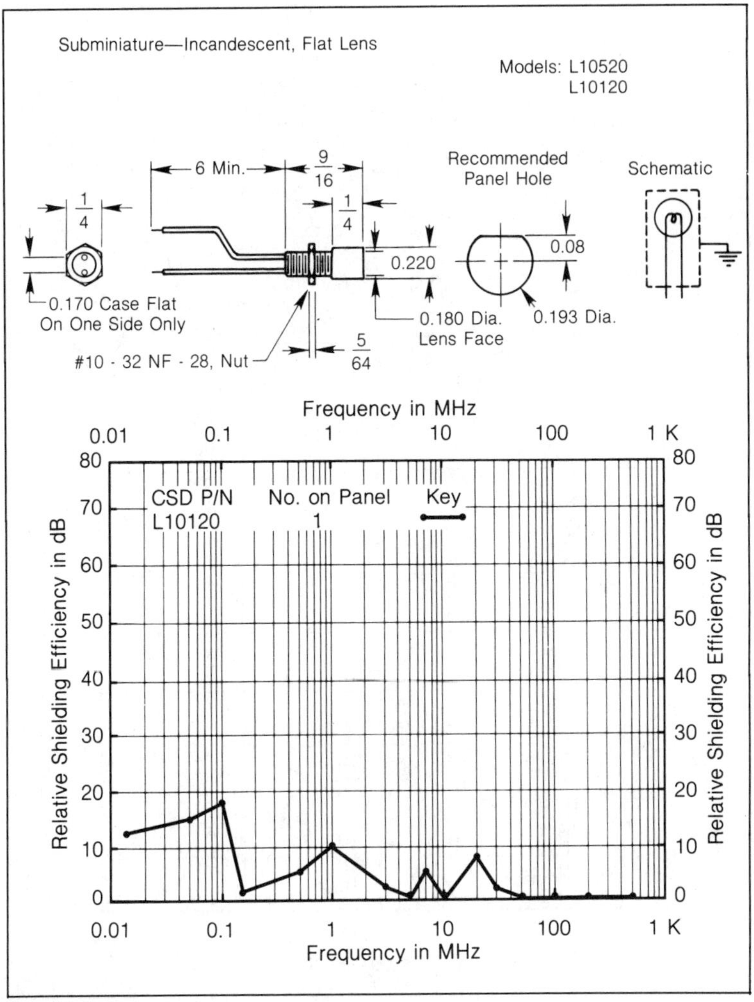

Figure 10.48—(continued)

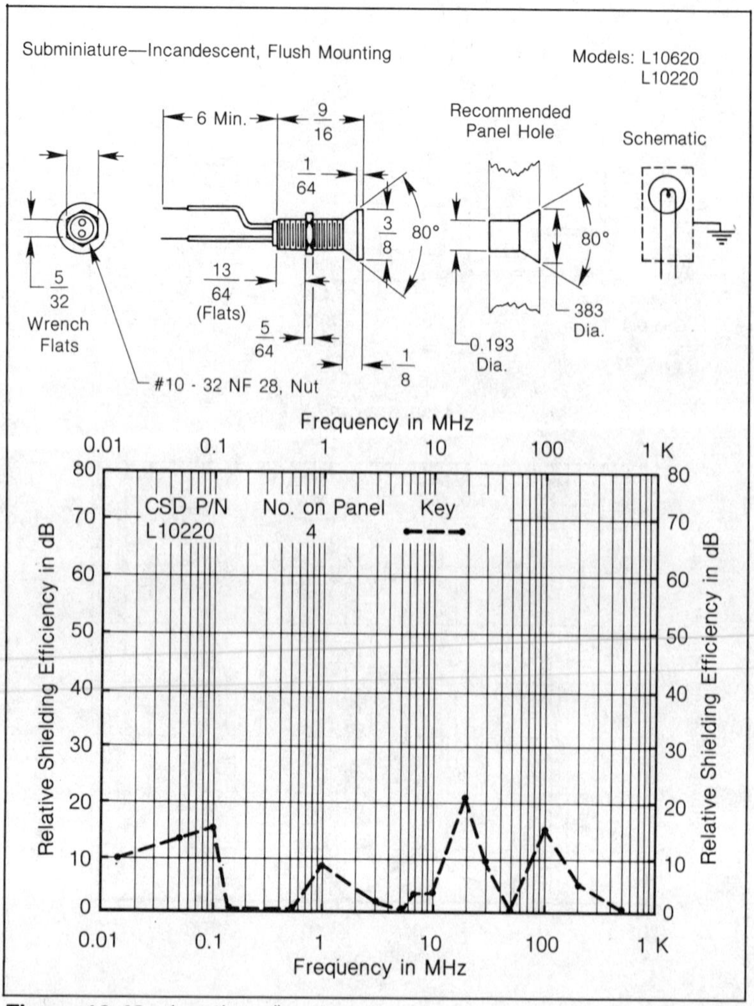

Figure 10.48—(continued)

10.7 Fuse Openings

Fusesholders, of both the indicating and nonindicating type, create another housing leakage which is aggravated by the fact that, for obvious shock-protection reasons, the holder and cap are normally made of Bakelite® or another insulating material. The simple existence of a 12 to 20 mm hole would be bad enough already,

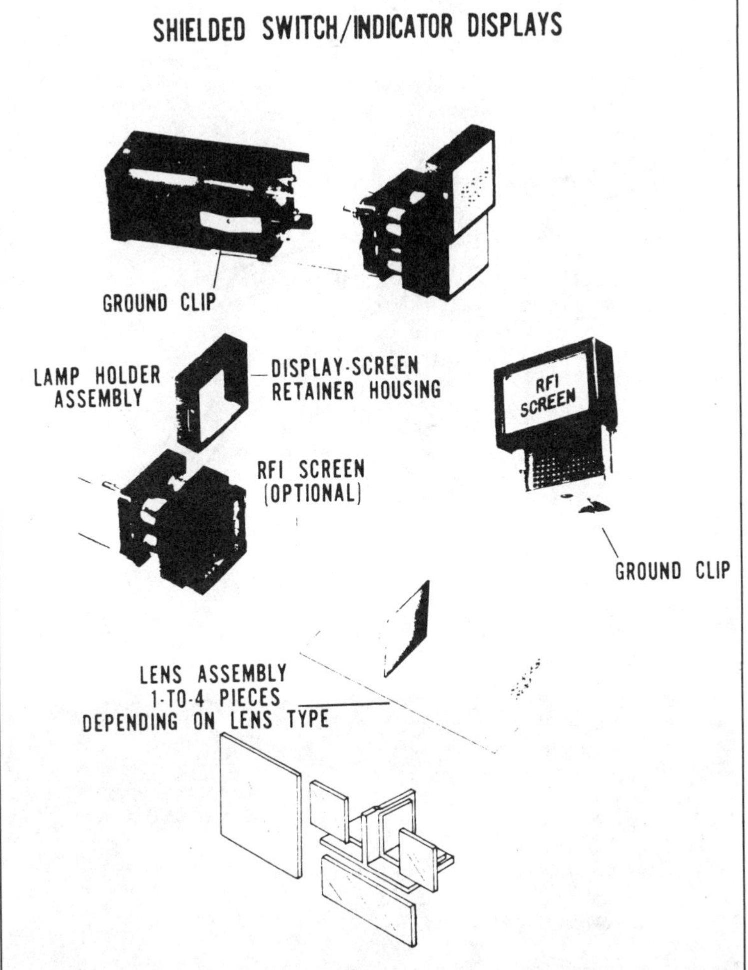

Figure 10.49—Shielded Switch/Indicator Displays (courtesy of Master Specialties Company)

but in the case of a fuse, the fuse cartridge itself acts as a pickup (or radiating) antenna which is located right in the middle of a leaky aperture (see Fig. 10.50).

The result is a stronger coupling than if the conductor were deep inside the box. For instance, a 2.5 cm fuse cartridge is a tuned

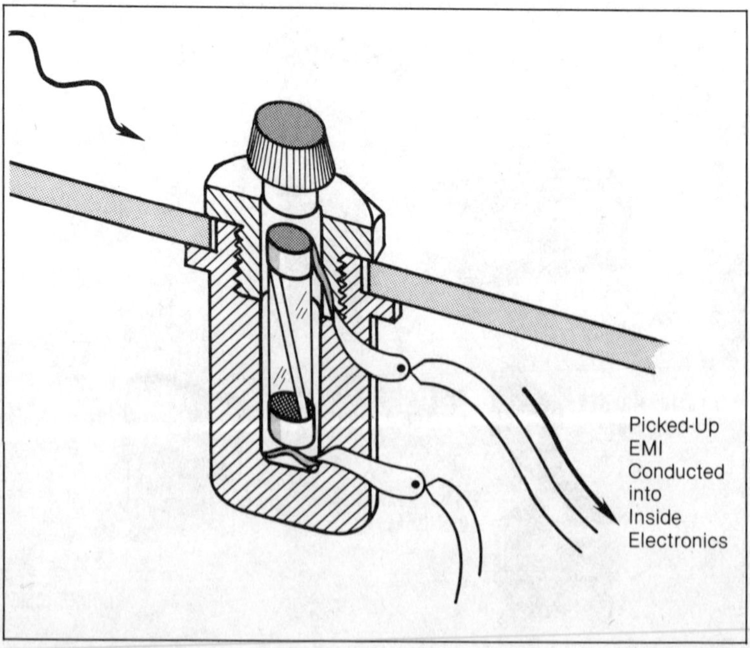

Figure 10.50—The Problem of Radiated Susceptibility or Emission Coupling through a Fuseholder

Figure 10.51—Shielded Fuseholder (courtesy of FIC Corp.)

antenna at 3 GHz, and it is still a rather effective aerial down to the 100 MHz region. It can behave in either of two ways:

1. It can intercept internal signals and radiate them outside of the box.
2. It can pick up external signals and conduct or reradiate them to the inside of the box.

Shielded fuseholders are available. The insulating body and cap are canned in a metal envelope, providing a SE better than 90 dB from 25 MHz to 10 GHz (see Fig. 10.51).

10.8 Cable Entries, Connectors and Nonconductive Inlets

This last type of penetration corresponds to electrical wiring entering the box directly or through a connector. It also includes the nonconductive type of entry ports associated with pressure inlets, optical fibers, fluids, etc. The design process follows the decision chart shown in Fig. 10.52.

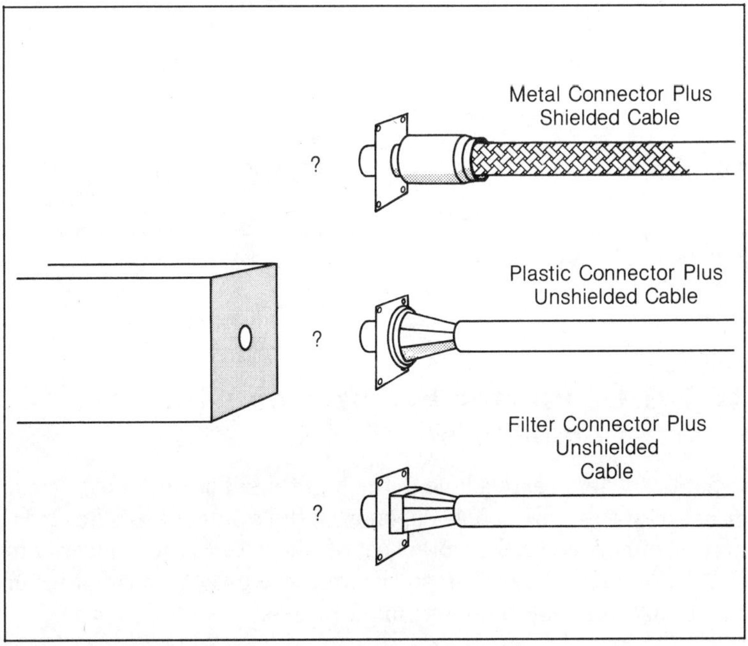

Figure 10.52—Decision Chart for Shielded Cables and Connectors vs. SE Objectives of the Whole System (continued next page)

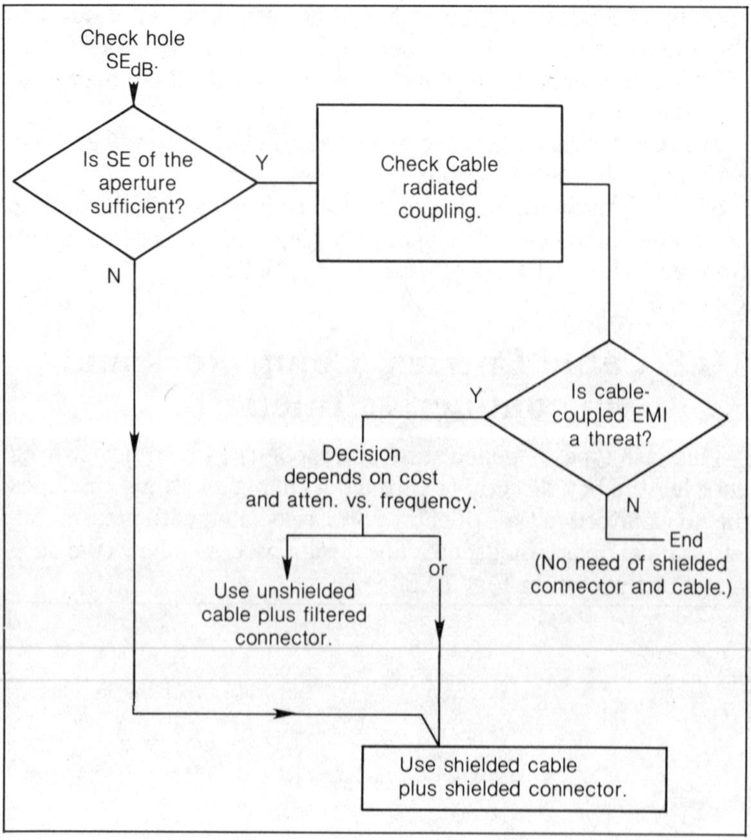

Figure 10.52—(continued)

10.8.1 Conductor Penetration without Connectors

Cable entries create a hole in the box skin. The decision to treat or not treat this type of hole depends on the result of the SE calculations summarized on the worksheet of Chapters 7 and 9, under item B.4, "Component Apertures." From these calculations, either of two situations may be determined to exist:

1. **The calculations show that the leakage of the cable entry hole can be tolerated**. However, even if the box is safe, the exterior cable can behave as a transmitting or receiving antenna. Therefore, if the cables are shielded, the cable shield must properly terminate at the box interface via serrated clamps, fittings, etc. as explained in Chapter 11. This is illustrated in Fig. 10.53. If the cables are not shielded, their

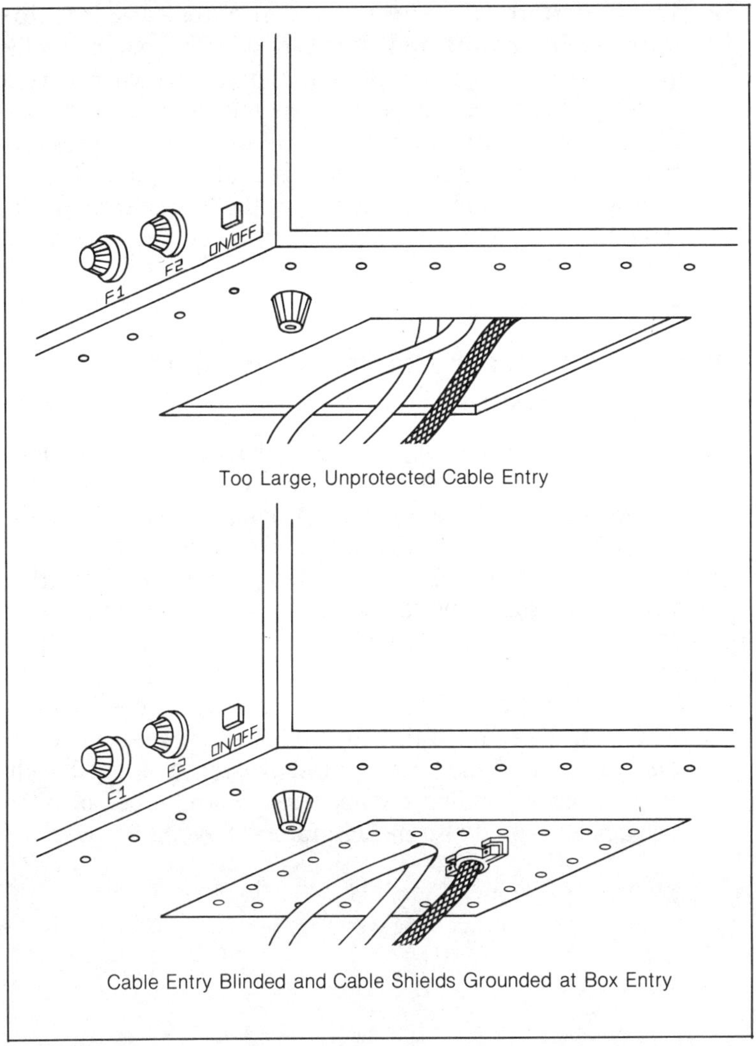

Figure 10.53—Maintain Shield Integrity at Cable Entries

role as antennas must be at least minimized by **filtering** each wire with a capacitive or pi filter at the box wall. The amount of filtering required depends on how critical cable pickup is. This can be found by running a simulated system test or, better still, by performing a prediction of cable radiation coupling by the methodology described in Volume 8 of this handbook series.

2. **The calculations show that the leakage of this entry hole cannot be tolerated.** Merely creating a tube in an attempt to get WGBCO attenuation would not work because we have a conductor in the middle: instead of having a nontransmitting waveguide, we would have a very efficient coaxial configuration. The only remaining solution is to have a shielded cable terminating at the box interface or installed with an appropriate metallic connector as explained next.

10.8.2 Conductor Penetrations with Connectors

Here again, the question arises as to whether or not the naked connector hole is a threat to the combined SE objective (line B.4 of the SE worksheet). If the naked entry hole can be tolerated, the amount of EMI picked up by the cable and conducted through the box must be evaluated. If this contribution is tolerable, no further evaluation is necessary and the connector can be an inexpensive plastic type. If the cable contribution is not tolerable, two options exist:

1. Use a shielded cable, which implies a shielded connector as detailed in Figs. 10.54 through 10.56. A shielded connector must have a metallic (or metalized plastic) shell, good shell-to-cable shield bonding and a low receptacle-to-box bonding resistance. It may also include filters. Several authors and

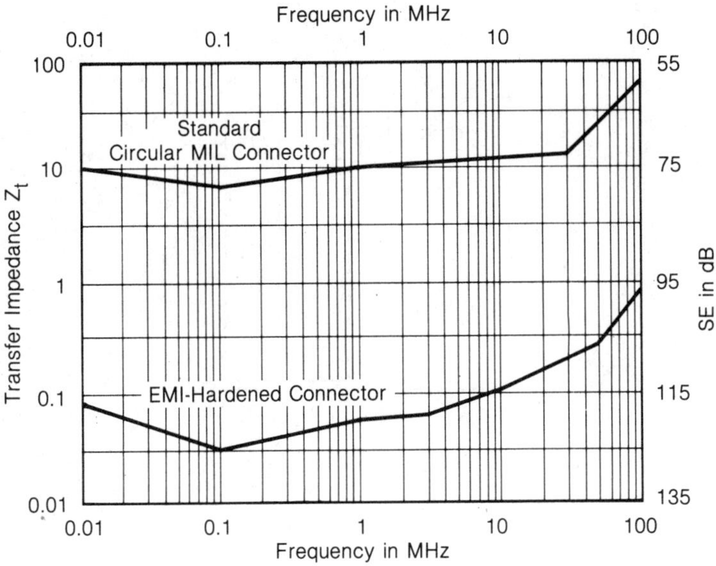

Figure 10.54—Shielding Performance of Shielded Connector

manufacturers identify these connector features either as transfer impedance (see Chapter 11) or as connector SE. The low transfer impedance is sometimes obtained via a rim of miniature fingers inside the connector shell which mate with the receptacle when installed. A low-impedance (high-SE) bond at the receptacle-to-box interface is obtained in the same manner as for cover seam bonding, described in Section 10.2. This includes the use of flat gaskets cut to size and wall surface preparation (paint-free, degreased and attached with conductive epoxy). This option is the one generally adopted,

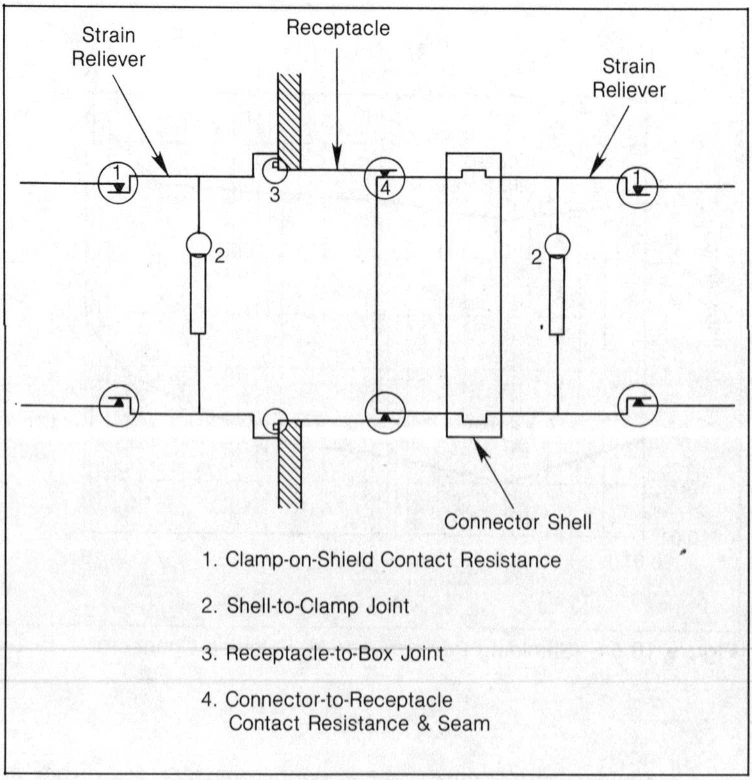

Figure 10.55—Shielded Connector Details: Typical SE Weak Spots

1. Mounting Surface Nickel or Silver-plated- Or Receptacle Made of Passivated Stainless Steel
2. Stronger Locking Nut- Tighter Tolerances
3. Conductive Elastomer Ring Seal
4. Receptacle-to-Box EMI Gasket

5. & 6. Gold Plating Inside

7. & 8. Brazed Spring Contacts

Figure 10.56—Shielded Connector Details: Treatment of Improved Connector (continued next page)

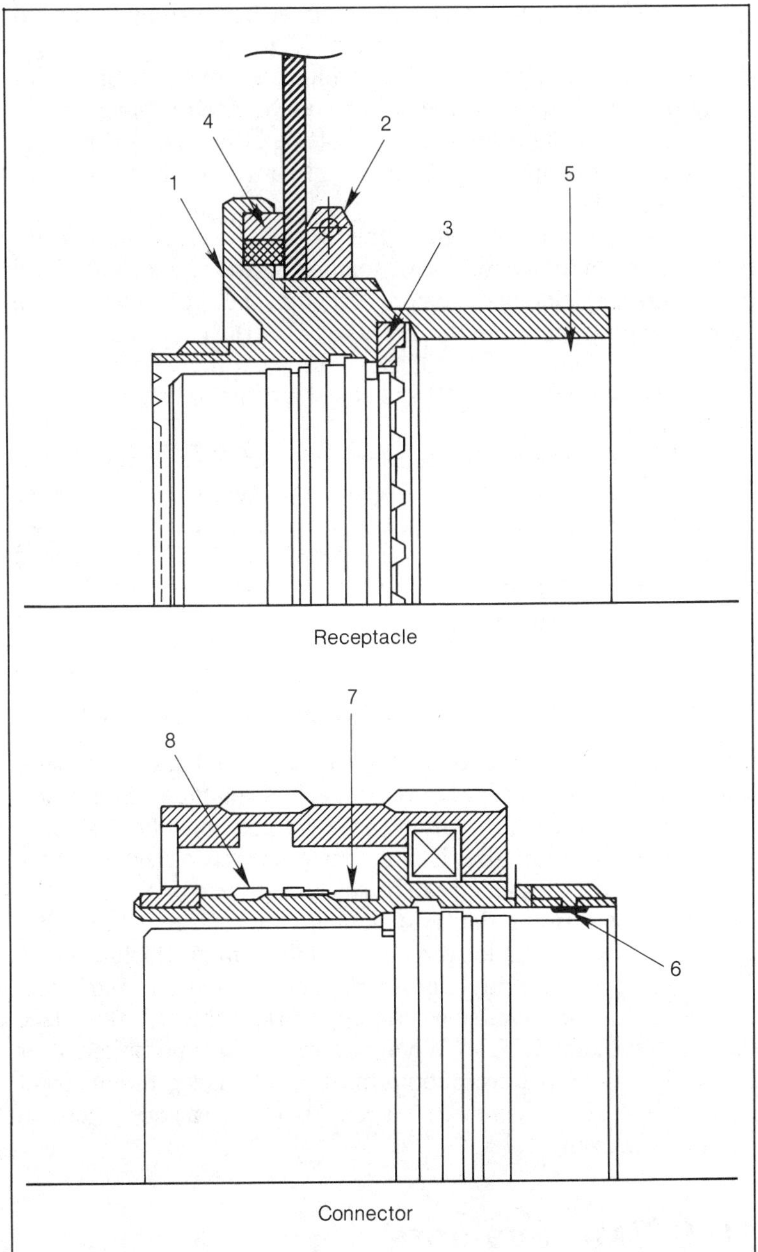

Figure 10.56—(continued)

though more often by tradition or recipes than by analytical approach.

2. Use an unshielded cable and a filtered connector. Since this option assumes that the naked hole by itself is not a threat, we can let the cable pick up EMI on either side of the box. However, it must be cleaned up by a filter when it crosses the barrier.

The last case would be where the SE of the naked hole is insufficient: this requires adoption of solution 1, above, i.e., a shielded cable **and** a shielded connector. Although there is no one-to-one correspondence between the two, the SE of the cable and its connector must be on the order of that of the box. It would be unwise to aim at a box SE of 100 dB and select a cable or connector with 20 dB of SE.

Although the above is a series of fairly simple statements, the decision process involved in specifying cable and connector shielding can easily turn into a can of worms. The designer may be aided by the decision chart presented in Fig. 10.52. Cable shield terminations and their contribution to the overall system SE are discussed in Chapter 11.

10.8.3 Entry Ports for Nonconductive Lines

Entry ports may exist for such nonconductive lines as pressure sensor tubes, fluid lines, fiber optics, etc. If the tube, shaft, fiber, etc. is nonconductive, the SE of the entry hole must be evaluated using Eq. (7.4). If the SE_{dB} is found insufficient for the desired objective, this type of opening can be easily improved by the WGBCO technique, selecting a depth-to-diameter ratio larger than unity.

A welded or brazed tube can be added if the metal thickness cannot be built up sufficiently. For fiber optics, transmitters and receptors, packages are available with appropriate metal fittings (see Illustrative Example 7.5). When such a metal tube is bolted or screwed in, all proper precautions must be taken (degreasing, paint removal and the use conductive epoxy) to ensure a good electrical bond and prevent leaks.

10.9 EMC Sealants

This section discusses another form of EMI shielding integrity protection in the form of conductive epoxies and caulking.

10.9.1 Conductive Epoxies

Conductive epoxies are used to join, bond and seal two or more metallic mating surfaces. The silver-epoxy resins replace soldering and other bonding techniques and cure at room temperatures. The conductive epoxy adhesive material family is used to make electrical connections in the following applications:

1. Heat-sensitive components
2. Capacitor slugs
3. Ferrites
4. Integrated circuits
5. Electroluminescent panel connections
6. Forming bus bars or strips on conductive glass
7. Bonding flanges to waveguides
8. Bonding waveguide sections
9. With bolt holes and fasteners on electronic enclosures
10. Joining dissimilar metals
11. Sealing IC packages against moisture and EMI
12. PCB repair
13. Interconnecting conductive metal gaskets
14. Field repairs to circuits
15. Permanent seam shielding
16. Sealing EMI shields

10.9.1.1 Preparation and Curing

Conductive epoxies are easily mixed by volume, although some are prepared by weight. They are formulated with mixed viscosities that produce a light, creamy paste; this makes application with standard dispensing equipment reasonably easy and foolproof. Typical cure times are one day at room temperature or 30 min at 93°C (200°F).

10.9.1.2 Typical Properties

Depending upon the type of silver-epoxy resin used, typical volume resistivity will range from 0.001 to 0.02 Ωcm. The operating temperature range is about −63 to +120°C. Shear strength is about 85 kg/cm^2, and tensile strength varies with type but averages about 180 kg/cm^2. As a rule, higher metal content produces better conductivity but poorer adhesion. The best tradeoff seems to be 60 to 70 percent by weight of conductive content. Conductive epoxies exhibit excellent moisture resistance. The cured specific gravity is about 2.0, suggesting its relative light weight for many limited-payload applications.

10.9.2 Conductive Caulking

Conductive caulking is used to EMI shield and seal two or more metallic mating members mechanically held by other means. Silver particles are suspended in resin to provide conductive sealing. Conductive caulking is used in the following applications:

1. Caulking EMI-shielded shelter panels
2. Caulking EMI-tight cabinets and enclosures
3. Improving joint and seam integrity of electronic enclosures
4. Protecting mating members of shielded conduits
5. EMI sealing and grounding bulkhead panel fittings
6. Moisture sealing of mating members
7. Attaching metal-foil tape to shielded room joints
8. Repairing damaged conductive gaskets
9. Sealing ends on wire-mesh gaskets
10. Caulking fasteners, panels and handles

10.9.2.1 Preparation and Use

The conductive caulking compounds, as in any EMI sealant and bond, requires that the surfaces be thoroughly degreased and cleaned of oxide coatings. The caulking may be applied with conventional caulking guns and dispensing equipment such as small bead-orifice syringes. Hand application with a spatula or putty knife is also acceptable. The caulking is free of any corrosive binders. It is used at room temperatures, and most caulking will not cure (permanently nonsetting). This feature permits easy disassembly of caulked parts for movement or maintenance.

10.9.2.2 Typical Properties

Depending upon the type of silver resin used, typical volume resistivity will range from 0.005 to 0.02 Ωcm. The operating temperature is −62 to 200°C (−80 to +400°F). Moisture resistance is excellent. The final specific gravity is about 1.8, suggesting its relative light weight for many limited-payload applications.

10.10 Conductive Grease

Conductive grease is not a member of the EMI gasket and sealant families discussed in this chapter. However, it is related in that one of its functions is to provide a low-resistivity contact to mating members. Here, mating members may engage and disengage more often than in most EMI gasketing applications, with the exception of finger stock used in shielded enclosures.

Conductive grease is a low-resistivity, silver-silicone grease which contains no carbon or graphite fillers. The material will maintain its electrical and lubricating properties over a broad environmental range. Environmental conditions include high and low temperatures, moisture and humidity, chemicals, ozone and radiation. Most conductive greases are viscous pastes which can be applied at elevated operating temperatures to vertical or overhead surfaces without dripping or running.

Conductive grease is used on power substation switches and in suspension insulators to reduce EMI noise. It also reduces make-break arcing and pitting of the sliding metal contact surfaces of switches and fills in pitted areas with silver/silicone. In addition, normally-closed switches are prevented from sticking due to corrosion or icing. The grease is effective in maintaining a continuous electrical path between contact surfaces which must be free to move under extended environmental conditions. These include ball and socket connections of power insulators which, if allowed to arc, can generate EMI.

Conductive grease is designed to maintain low-resistance electrical contact and thereby maintain equipment operating over extended environmental conditions, helping to deliver continuous electrical service. Conductive grease is used on the contacting surfaces of circuit breakers and knife blade switches. It reduces localized overheating or **hot spots**, in turn maintaining the blades' spring properties and the current rating of the switch or breaker at original equipment levels. Conductive lubricants prevent freeze-up in operating equipment and permit restoration of marginal or discarded breakers to rated capacity.

Typical volume resistivity is about 0.02 Ωcm. The operating temperature range is −54 to 232°C (−65° to +450°F). Conductive grease provides excellent moisture resistance and has no corrosive effect on metals. Its pot life is unlimited, and unused portions are returned to its container.

Users of conductive lubricants must be quite careful if the equipment contains PCBs, connectors, bare contacts, etc.; even a minuscule film of grease can create a short between copper traces, pins, lands of edge connectors, etc.

10.11 References

1. Kunkel, G., "Corrosion Effects on Field Penetration through Apertures," *1978 IEEE EMC Symposium Record* (New York: IEEE), Atlanta.
2. McElvein, D., Chomerics Co., "Common Misconceptions in the Use of EMI Gaskets," 1980.
3. "Corrosion in EMI Gaskets," 1984 *ITEM*, p. 190.
4. Instrument Specialties, Delaware Water Gap, PA, general catalog.
5. Brewer, R., (Instrument Specialties), Unpublished Test Data on Fingerstock SE, 1984.
6. Mardiguian, M., *Electrostatic Discharge: Understand, Simulate and Fix ESD Problems*, (Gainesville, VA: Interference Control Technologies, Inc., 1986).
7. Bereuter, W.A. and Chang, D., "SE of Metallic Honeycombs," *IEEE Transactions on EMC*, February 1982.
8. Madle, J.P., "Transfer Impedance and Admittance of Gasketed Panels and Honeycombs," *1976 IEEE EMC Symposium Record* (New York: IEEE).
9. Lassiter, H.A., "Low Frequency Shielding Effectiveness of Conductive Glass," *IEEE Transactions on EMC*, Vol. EMC-6, July 1964, pp. 17-30.

10.12 Bibliography

Booth, A., "Conductive Elastomers for EMI/RFI Shielding Based on Silver-Plated Aluminum Filler," *1982 IEEE EMC Symposium Record* (New York: IEEE), Santa Clara.

Boothman, G., "Designing Cabinets for Optimal EMI Shielding," *1982 IEEE EMC Symposium Record* (New York: IEEE), Santa Clara.

Faught, A., "An Introduction to Shield Joint Evaluation Using EMI Gasket Transfer Impedance Data," *1982 IEEE EMC Symposium Record* (New York: IEEE).

Faught, N., "Shielding Material Insertion Loss Using a Dual TEM Cell," *1983 IEEE EMC Symposium Record* (New York: IEEE), Washington, D.C.

Kunkel, G.M., "Shielding Characteristics of Honeycomb Filters," *1986 IEEE International EMC Symposium Record* (New York: IEEE).

MIL Handbook 419, Chapter 5.

Schreiber, O.P., "Some Useful Analogies for RF Shielding and Gasketing," IEEE Third National Symposium on Radio Frequency Interference, Washington, D.C., June 1961.

Schreiber, O.P., "Designing and Applying RFI Shields and Gaskets," *Proceedings of Second Conference on Radio Interference Reduction,* Armour Research Foundation, Chicago, Ill., March 1956, pp. 343-359.

Schulz, R.B., "Feasibility Study of Shielding Techiques," Presentation to Interference Reduction Branch, Engineering Support Services Department, U.S. Army Electronics Command, Forth Monmouth, New Jersey, May 1964, pp. 4-5.

Chapter 11

Cable Shields

Although one could easily devote an entire book to cables alone, a cable shield is not an independent entity but a part of the whole shielding of the system. Overlooking the cable contribution to the system EMC performance can ruin the effect of the best box shielding. In fact, cable shielding very often is the weakest link in the overall system shielding. The performance of a shielded cable depends on the shield material, its termination, the geometry of the installation and the nature of the field.

This chapter focuses on the first aspect and will briefly describe the other parameters. The reader who needs a deeper study of the shield termination methods will find them described in the Volume 2 of this EMC handbook series. For the calculation of the coupling of an ambient EMI field into shielded cables or the emission from these cables, consult Volume 8 of this series.

11.1 Generalities Regarding Cable Shields

For a given frequency, the key contributors to the overall performance of a shield placed upon a cable are:

1. The shield material (foil, braid, tube)
2. Its termination method (clamp, solder, pigtail, shielded connector)
3. The geometry of the installation in the system (cable length,

cable height, orientation versus the field pointing vector, type of conductive ground)

In the final analysis, each of these contributors behaves differently, depending on the nature of the field: near E-field, near H-field or far field.

11.1.1 Shield Material and Thickness vs. the Type of EMI

A cable shield can be made of a thin aluminum film flashed over a paper or mylar substrate, a copper braid, a thinly wrapped metal foil, a corrugated pipe or a conduit. As explained in Chapter 1, all these materials have basic shielding properties which depend on two mechanisms: **reflection** and **absorption**.

Reflection is a radiated phenomenon and is the result of the mismatch between the impinging wave impedance (the ratio $Z_w = E/H$ of the incident field) and the shield barrier impedance. Therefore, for a metal of given resistivity, the reflection will be largely different against predominantly E-fields from that of predominantly H-fields. Even a mediocre conductor will provide a decent E-field reflection, but it takes extremely good conductivity to provide some H-field reflection under near-field conditions.

Absorption depends on the skin effect (see Chapter 1). Above a certain frequency, the current does not penetrate the whole shield thickness and circulates primarily on the skin of the shield facing the source. In other words, the outer skin if the EMI source is external, or the inner skin if the EMI source is internal.

It is clear that at low frequencies typical cable shields, with a thickness of 0.2 mm or less, cannot exhibit absorption properties because the skin effect does not yet occur unless the shield material has a high permeability. So the only low-frequency performance of an ordinary shield would be via reflection, which requires high conductivity.

At frequencies high enough for the skin effect to occur (e.g., above 1 MHz for a 0.1 mm thick copper or aluminum shield) absorption becomes significant. If the shield is a solid tube, this absorption will increase exponentially and soon reach values which are not even measurable. If the shield is a braid, all the minuscule rhombic apertures due to the weave of copper wires will spoil the absorption effect by making the shield more and more "transparent" as frequency increases. Depending on the density of the braid (termed **optical coverage**), this effect can show up as low as 1 to 10 MHz, ruining the SE of the braid. Figure 11.1 is a comparison of the SE of foil versus braid.

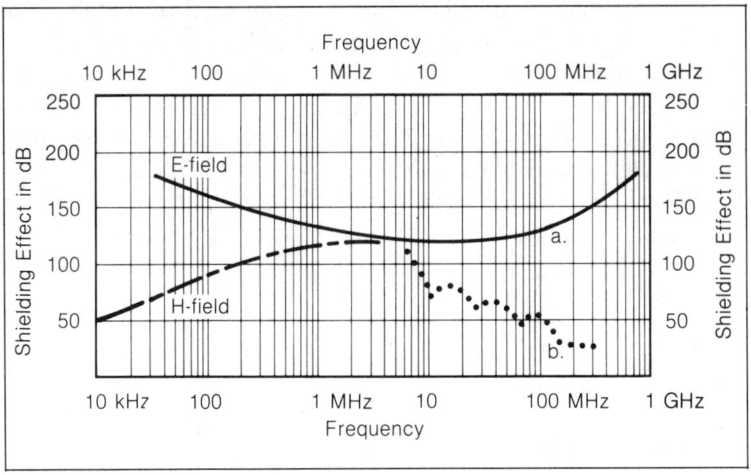

Figure 11.1—Combined Reflection and Absorption Shielding for a 25 μm Thick Shield Wall at One Meter Distance for (a) Solid Foil and (b) Single Braid

11.1.2 Shield Termination

Where and how the shield is terminated can radically affect the performance of a shield. Shielded cable manufacturers generally provide performance data based on ideal installation of the shield, i.e. corresponding to the intrinsic performance of the cable shield wall. Once a cable shield is mounted, its termination may be the weakest link in the chain, especially at higher frequencies.

As a first approximation, the quality of a shield can be associated to its dc sheath resistance, although this statement becomes invalid above the HF region. Since it is very difficult to have the shield terminated by a connector clamp or pigtail whose impedance is much less than the shield material impedance, the termination hardware is **always** the limiting factor of the in-situ performance. That is, the bonding or grounding method of the shield always degrades its basic performance or, at best, gives an overall performance that is almost equal to the shield alone; it cannot improve performance.

Let us consider, for instance, a very tight braid, with a braid resistance of 3 mΩ/m. Assume a 0.75 m piece of this shielded cable interconnecting two metallic racks and terminated via metallic connectors. Unless special surface treatment, degreasing and tightening techniques are used, a typical bonding resistance of the braid to the serrated clamp is 0.5 mΩ. An ordinary hand-tightened connector shell exhibits a contact resistance of about 3 mΩ to the receptacle. Finally, the contact resistance of the connector socket flange to the rack wall, assuming it is normally tool-tightened using four screws or a nut-and-washer fitting, is again 0.5 mΩ.

Since there are two ends to this cable, the total termination resistance is:

$$R_{term} = 2(0.5 + 3 + 0.5) = 8 \text{ m}\Omega$$

Comparing this to the braid shield resistance:

$$R_{sh} = 0.75 \text{ m} \times 3 \text{ m}\Omega/\text{m} = 2.25 \text{ m}\Omega$$

We see that the termination contribution to the whole shield resistance will be more than three times that of the braid alone. It is possible to improve this situation: a careful selection of connector and tightening torque can bring the total termination resistance down to 2 mΩ.

Not only can an improper shield termination spoil the cable shield efficiency, it can be the source of secondary mechanisms which can cause more EMI than if there were no shield at all. Figure 11.2 shows an example where a pigtail can act either as a pickup element (susceptibility problem) or a radiating element (emission problem). The key for preventing or solving such problems is always the same: identify any existing EMI current paths. Recognizing where the shield current originates, where it goes and how it returns to its source (including all the "sneaky" paths) is the most important step to the solution. Often, people end up interrupting the shield, while in fact it was the current flow into sensitive circuits which needed interruption.

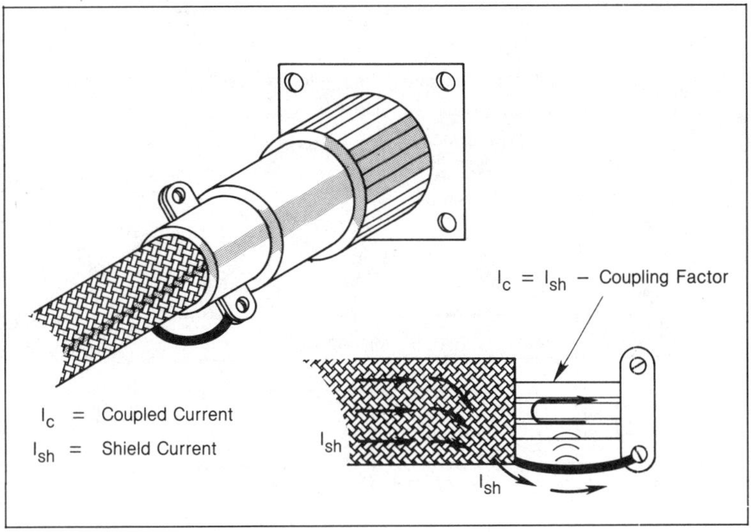

Figure 11.2—Possible Emission or Susceptibility Problem with a Pigtail

11.1.3 Overall Geometry of the Installed Cable

A shielded cable can be regarded as a conjunction of two transmission lines, with the shield being the transfer medium between the two. The two circuits are (see Fig. 11.3):

1. The intended circuit (2) has well controlled parameters as far as source and load impedances Z_g and Z_L, and characteristic impedance Z_{02}. These parameters set the values of voltage and current at every point in this circuit, including reflection and standing-wave effects.
2. The exterior circuit (1) has very poorly controlled parameters since each of its terminating impedances Z_{g1} and Z_{g2} can vary from zero to infinity depending on the grounding conditions, and its characteristic impedance Z_{01} depends on the ratio h/d of the height above ground plane versus cable diameter.

The effectiveness of a shield is in fact the measure of the percentage of energy transferred from circuit 2 into circuit 1 in the case of EMI emission, or transferred from circuit 1 into circuit 2 in the

Z_{g1} = Source Impedance
Z_{g2} = Common-Mode Impedance, Source to Ground
Z_{01} = Characteristic Impedance, Line 1
Z_{02} = Characteristic Impedance, Line 2
P_L = Power Input Load
P_{out} = Power Outside Shield
I_s = Shield Current
Z_L = Load Impedance

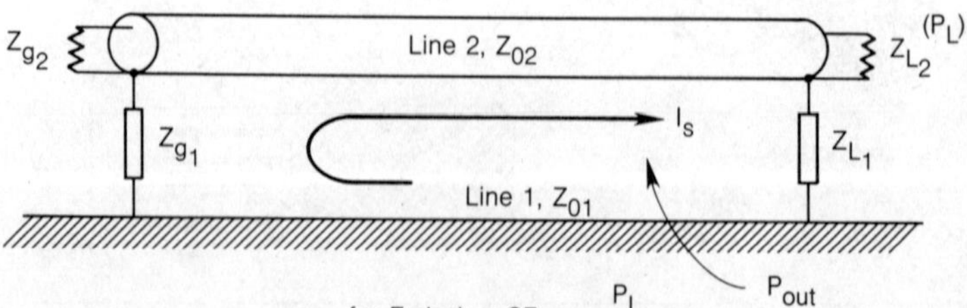

$$\text{for Emission: SE} = \frac{P_L}{P_{out}}$$

$$\text{or Shielding Attenuation} = \frac{P_{out}}{P_L}$$

$$\text{for Susceptibility: SE} = \frac{P_{out}}{P_L}$$

Figure 11.3—A Shielded Cable as a Conjunction of Two Transmission Lines, with the Shield Being the Transfer Media between the Two

case of susceptibility. The characteristic impedance of circuit 1 has a large effect on this transfer of energy, especially at integer multiples of λ/4. Therefore, a given shield with given terminations will behave differently when installed, depending on its length and its height above ground.

11.1.4 What Is Expected from the Shield?

A cable shield may be installed for emission suppression; that is, to prevent the EMI in the cable to leak outside. Or a cable shield may be installed for susceptibility problems; that is, to protect signal conductors from external EMI. In many cases, a cable shield will be installed for both purposes. The above sections have illustrated that the same cable shield with given installation will behave differently against emission and susceptibility, due to the very different conditions of E-fields and H-fields in each case.

11.1.5 Unbalanced vs. Balanced Shielded Cables

An unbalanced shielded cable (e.g., coaxial or triaxial) is a transmission line where the return conductor (the shield) surrounds the signal conductor (see Fig. 11.4a). The "shield," therefore, is an **active conductor**, carrying load current back to the source. A balanced shielded cable (e.g., shielded pair, twisted shielded pair, shielded ribbon or multipair cable) is a line where the currents circulate back and forth in a signal/return pair while the shield performs an essentially **protective** role (Fig. 11.4b) while also providing some impedance balancing of the pair wires to ground.

The preceding distinctions may sound trivial, but it is fundamental to identify a cable type in terms of what is expected from it. A common misunderstanding is illustrated in Fig. 11.5, where someone tries to reduce radiated EMI by shielding a signal wire. However, the return current path is via some common 0 V wire which is external to the shield. Installing such a shield will be for naught since the radiating loop is practically unaffected. To produce some beneficial effects, the shield has to **surround both wires** (i.e., the cable should be a shielded pair). Otherwise, the shield itself has to be the return conductor (i.e., using a coaxial cable) with the signal ground attached to both ends of the cable.

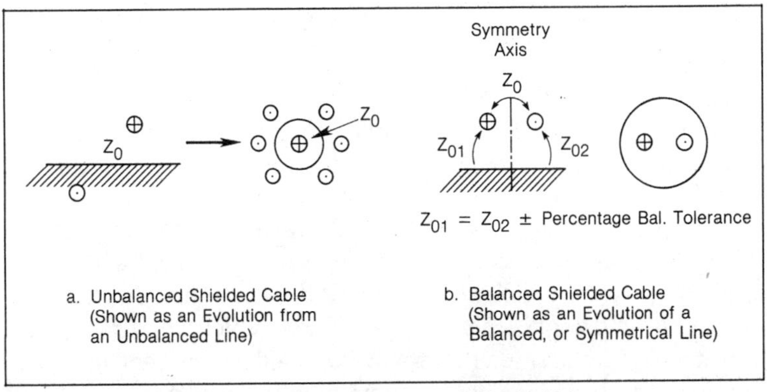

Figure 11.4—Balanced and Unbalanced Shielded Lines

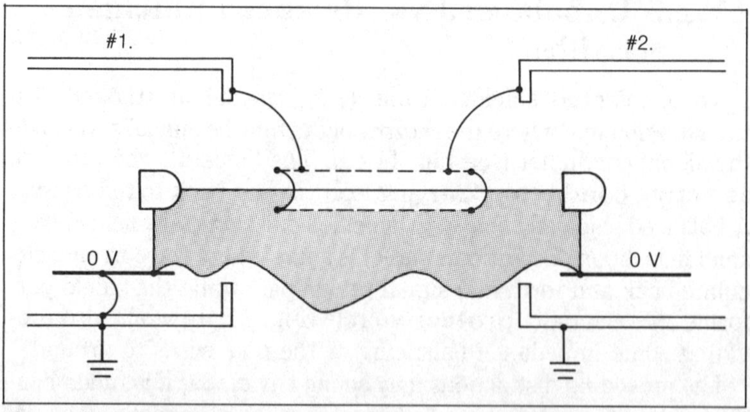

Figure 11.5—A Common Misconception of the Shield's Role

11.2 Definition of Shielding Effectiveness for Cable Shields

As it is defined in Chapter 1 and other preceding chapters, the SE of a shield is a rather simple concept: the ratio of the field received at a given location before and after the shield has been installed, that is:

$$SE_{dB} = 20 \log \frac{E\ \text{before}}{E\ \text{after}}$$

Although this definition could be applied to a cable shield, an immediate difficulty arises as to how to accurately measure the field inside the conductive tube of a cable shield. Miniature field probes could be used, but this would limit the measurement to highly specialized, laboratory-type instrumentation which is totally impractical for simple, recurrent EMI evaluations. Fortunately, an in-situ field probe exists in a shielded cable: the wire or wire pair itself. If we assume that the voltage induced in this fortuitous antenna directly relates to the impinging field, the basic definition of SE becomes:

$$SE_{dB} = 20 \log \frac{V\ \text{induced without shield}}{V\ \text{induced with shield}} \equiv \frac{E\ \text{or}\ H\ \text{before}}{E\ \text{or}\ H\ \text{after}}$$

This SE definition for shielded cables has been (and is still) used widely. Unfortunately, it suffers from some basic weaknesses. The voltage measured at either end of the line depends on the type of termination, the degree of impedance mismatch at the ends of the line, and the line losses. These very parameters are affected by the presence of the shield, introducing some inaccuracy into the comparison of the two voltages with and without the shield. Also, the comparison requires that the reference unshielded line and its shielded version have exactly the same geometry, pair twisting (for balanced lines), etc. Furthermore, the shield termination method will strongly influence the result and can even totally obscure the parameter being measured, i.e., the quality of the shield alone.

Because of all these problems, SE data given for a shielded cable may apply only to the exact type of circuit and mounting hardware of that particular setup and could be misleading for any other application. Consequently, another measure of a shield quality, the shield **transfer impedance Zt** is often preferred, as explained in the next section.

11.3 Transfer Impedance of Coaxial Cable Shields

Transfer impedance Z_t (sometimes called **surface transfer impedance**) is an especially convenient way of characterizing the quality of a cable shield. For any shield, it provides a figure of merit which is an absolute parameter, instead of being a relative term like SE. Also, the measurement of Z_t is relatively easy to perform, and the test setup provides far better accuracy and repeatability than conventional SE measurements. This is because transfer impedance measurement is basically a conducted measurement and is less sensitive to environmental factors than a radiated measurement.

11.3.1 Definition of Shield Transfer Impedance

The transfer impedance concept, initially conceived by Schelkunoff in the late 1930s, relates the current flowing on the shield surface to the voltage it develops on the other side of this surface. Figures 11.6 and 11.7 show that by diffusion the transfer impedance converts the cable surface current I_s into an inner longitudinal voltage V_i, which appears in **series** in the inner (intentional) circuit. As a result, a voltage V_o will appear at each termination of the coax, equal to:

$$V_o = \frac{V_i}{2} = \frac{I_s \times Z_t}{2} \tag{11.1}$$

if both ends are matched.

Figure 11.8 shows the Z_t values for few typical coaxial cables. Z_t is expressed in ohms and normalized to a 1 m shield length. As seen from Eq. (11.1) and Fig. 11.7, the lower the Z_t, the better the shield quality for EMI reduction.

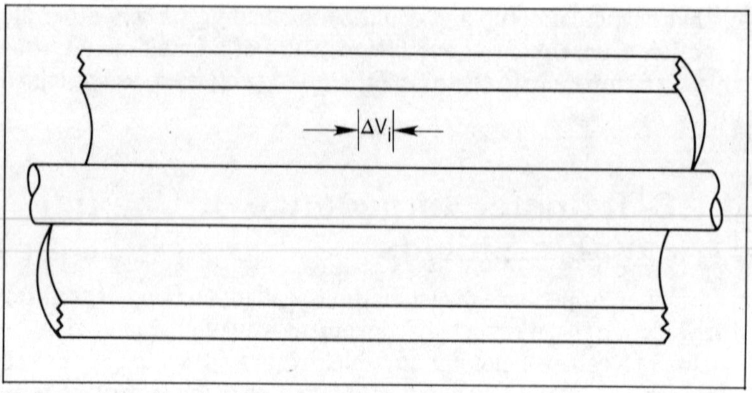

Figure 11.6—Coaxial Transfer Impedance

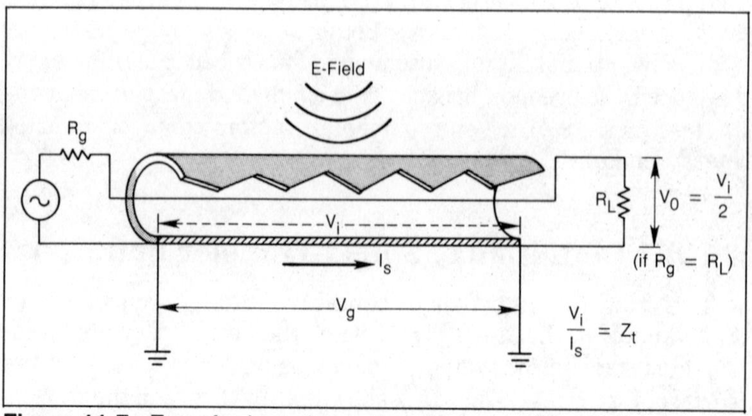

Figure 11.7—Transfer Impedance Coupling: I_s = Sheath Current Due to Ambient Field, or Ground Potential Difference Vg.

Below about 100 kHz (this is called the **Ohm's law or diffusion region**), Z_t is practically equal to the shield dc resistance. Above about 10 MHz, Z_t is proportional to the leakage inductance between the shield and the inner conductor, because of the multiplicity of minuscule rhombic loops of the braid (see Fig. 11.8). From then on, Z_t is mostly due to inductance, which is expressed in nanohenries per meter. Above several megahertz for loose braids, a capacitive coupling between the shield and the inner conductor becomes significant (see Fig. 11.9). A simplified model of shield-to-inner-conductor coupling is shown in Fig. 11.10.

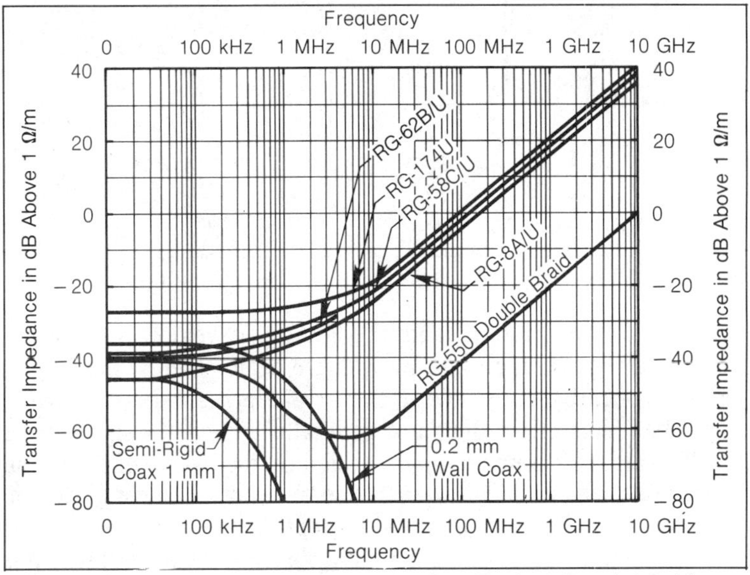

Figure 11.8—Transfer Impedance of a Few Typical Coaxial Cables

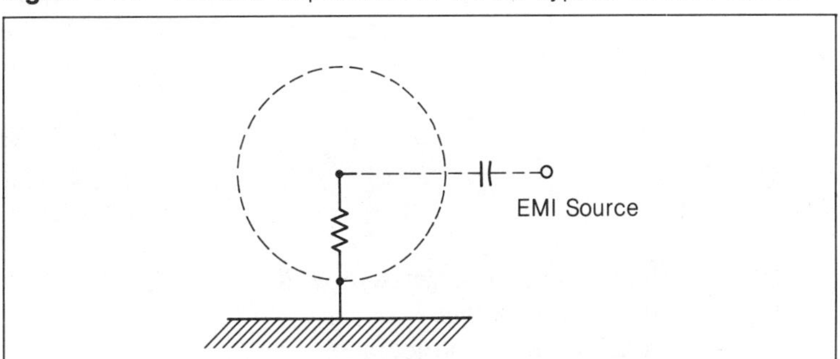

Figure 11.9—Electrostatic Coupling through Cable Shield Apertures

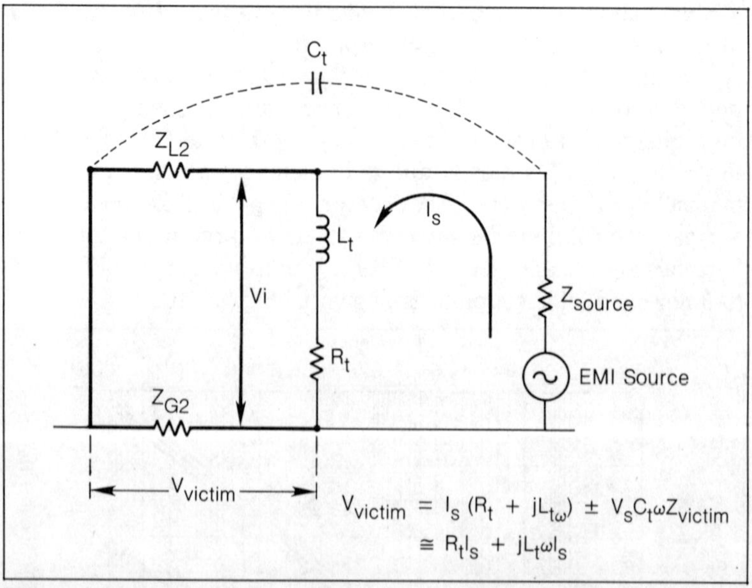

Figure 11.10—Simplified Equivalent Circuit of EMI Coupling via the Shield

Between the quasi-dc and the inductive regions, Z_t passes through a transition region. For double braid shields, this region shows a negative slope due to the skin depth effect as shown in Fig. 11.8. This effect does not generally appear for single-braid shields since it is immediately nullified by the weave of the braid. Some optimized braids, however, due to a careful knitting pattern, may show the beginning of a skin depth effect.

Neglecting the capacitive coupling through the shield, which is generally a second order term (except for very poor, leaky braids) we can say that the transfer impedance of a braided shield has the general form of:

$$Z_t = R_t + j\omega L_t \tag{11.2}$$

where,

Rt	=	shield transfer resistance = $R_s/ 1 - e^{-\delta/t}$ for $\delta \leqslant t$
R_s	=	shield dc resistance
L_t	=	shield transfer inductance
δ	=	skin depth
t	=	shield wall thickness
ω	=	radian frequency

11.3.1.1 Shield Transfer Resistance R_t, for $\delta > t$

At low frequencies, where the shield thickness represents less than one skin depth, the shield transfer resistance R_t is equal to R_s, and the voltage appearing inside the shield is merely equal to $R_s \times I_{sh}$. There is no difference between the longitudinal voltages over the inner and outer surface of the shield. Incidentally, at these low frequencies, the $L_t\omega$ term is small and approximately:

$$Z_t \approx R_s$$

11.3.1.2 Shield Transfer Resistance R_t, for $\delta \leqslant t$

As the frequency of the EMI current flowing on the shield increases, the skin depth may become smaller than the shield thickness, and absorption takes place (see Chapter 1 for a complete discussion of skin depth). Therefore, the diffusion current appearing on the other side of the wall decreases and so does the resulting voltage in the victim circuit. This is that phenomenon which is translated as an equivalent **transfer resistance** R_t decreasing with frequency. If the shield is a homogeneous metal sleeve or tube, Z_t will simply follow the value of R_t, as seen in Fig. 11.8 for the semi-rigid or corrugated coaxial cables. Since the shield more typically is made of some braid or wrap, the many openings allow some magnetic field lines surrounding the shield current to find a closed path through the shield as explained next. Also, the fact that each wire in a strand alternately runs above and below the other strands (called **porpoising** effect) partially defeats the advantage of skin effect.

11.3.1.3 Behavior of Transfer Inductance L_t

Figure 11.11 shows that the magnetic leakage inside the shield links the area between the conductor and the shield; therefore, a voltage is induced in the victim circuit proportional to the derivative of the flux over the time. Relating this voltage to the rate of change of the shield current, dI_s/d_t, defines the transfer inductance L_t as:

$$V_i = L_t \frac{dI_s}{dt} = L\omega I_{sh} \text{ for sinusoidal current}$$

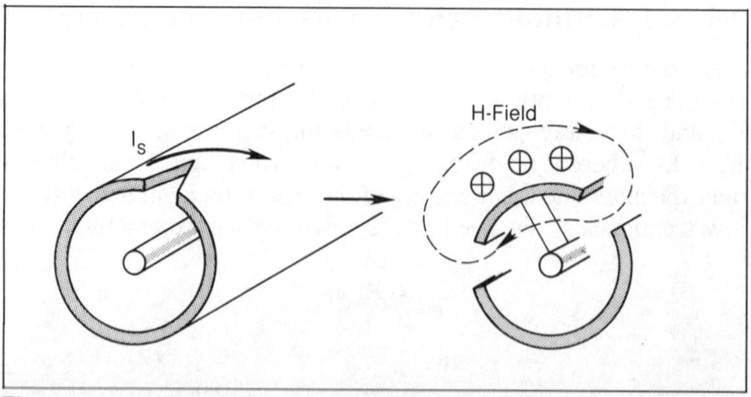

Figure 11.11—Magnetic Flux Linkage into the Inner Circuit of a Braid-Shielded Conductor through Shield Apertures

For braided shields, the transfer inductance can be minimized by a good, tight-weave electrical contact between each strand and an optimal braiding angle. The tightness of a braid is often termed as **optical coverage** in units of percent.

The fabrication of a good braided jacket is an art in itself, and the simple characterization of the braid by its dc resistance, although giving some indication of the shield quality, may not be sufficient. Comparisons between two manufacturers' braids with the same dc resistance and even the same percentage of optical coverage can show different values for L_t and, therefore, different shielding performances. Table 11.1 gives detailed data on R_s and L_t of the most popular coaxial cables.

Because Z_t of a coaxial shield is normalized to a 1 m length, all applications require that the nominal Z_t of the shield in ohms per meter be multiplied by the physical length of the cable, l in meters.

Illustrative Example 11.1

Calculate the transfer impedance of a 4 m piece of RG-58C/U coaxial cable at 5 MHz. Derive the voltage at either one of the matched ends for 100 mA (−20 dBA) of 5 MHz EMI current flowing on the shield.

The curve in Fig. 11.8 for RG 58C/U indicates −26 dBΩ/m. Thus, for the total cable length:

$$Z_T = -26\ \text{dB}/\Omega\text{m} + 20 \log (4\ \text{meters}) = -\ 14\ \text{dB}\Omega \text{ or } 0.2\ \Omega$$

Noise voltage at one end:

$$V_o = \frac{V_i}{2} = \frac{Z_T I_s}{2} = 14\ \text{dB}\Omega + (-20\ \text{dBA}) - 6\ \text{dB} = -40\ \text{dBV} = 10\ \text{mV}$$

Table 11.1—Transfer Impedance Parameters for Several Popular Coaxial Cable Shields

$Z_t = \sqrt{R_s^2 + L_t^2\omega^2}\ \Omega/\text{m}$

R_s = Shield Resistance in Ω/m (Total Parallel R_s if 2 Braids)

L_t = Shield Transfer Inductance

L_s = Shield External Self-Inductance (for Shield about 5 cm above Ground)

Coaxial Type	Characteristic Impedance in Ω	Number of Braids	R_s in Ω/m	L_t in nH/m	L_s in μH/m
RG8	50	1	0.0044	1	0.65
RG-11	75	1	0.004	0.25	0.65
RG 58C	50	1	0.014	0.93	0.85
RG 59A	75	1	0.008	0.32	0.79
RG 63	125	1	0.004	0.34	0.65
RG 114	185	1	0.005	0.7	0.65
RG 122	50	1	0.015	0.20	0.89
RG 174	50	1	0.036	2.3	0.96
RG 188	50	1	0.025		0.98
RG 213	50	1	0.004	0.25	0.65
RG 6-A	75	2	0.006	0.0087	0.75
RG 55A	50	2	0.008	0.016	0.85
RG 142	50	2	0.011	0.015	0.85
RG 156	50	2	0.005	0.0015	0.64
RG 181	125	2	0.002	0.00018	0.61
RG 189	50	2	0.0011	0.00018	0.48
RG 214	50	2	0.0055	0.0058	0.64
RG 223	50	2	0.013	0.035	0.85
RG 226	50	2	0.006	0.004	0.59

11.3.2 Calculating the Induced Voltage, Given the Incident EMI Field

This calculation is given a complete treatment in Volume 8 of this handbook series. Here we will give a brief review of the mathematical model for field coupling into coaxial cable shields.

An impinging field illuminating a shielded cable creates circulating shield currents by two mechanisms:

1. A shield displacement current is due to the intrinsic capacitance of the length of shield in the air (about 10 pF/m), which can be roughly modeled as an elementary dipole. This displacement is independent of the grounding or floating of the shield ends. This capacitance is the **free-space** capacitance of the shield. This first current is approximately equal to:

$$I_{s1} = 2\pi f l^2 \epsilon E/\sqrt{2} \qquad (11.3)$$

for an average (diagonal) polarization and $l \leqslant 0.5\ \lambda$,

where,

l = cable length in meters
E = incident field in volts per meter
ϵ = capacitance of free space = 8.8 pF/m
f = EMI frequency in hertz

Substituting all the constants gives:

$$I_{s1} \approx 3.9 \times 10^{-5}\, l^2 f_{MHZ}\, E \qquad (11.4)$$

2. A ground-loop current is due to the fact that the shield, the boxes and circuits it terminates into and the ground reference constitute a loop antenna. This antenna picks up an induced voltage equal to:

$$V_i \approx E\sqrt{2}\, l \sin(0.0074\, h\, f_{MHz})$$

$$\approx E\sqrt{2}\, l \times 0.0074\, h\, f_{MHz}) \text{ for } hf < 100 \quad (11.5)$$

where,

h = cable height above ground

Then this open circuit voltage forces a current to circulate in the loop. If the shield is grounded at both ends, this current in the shield is:

$$I_{s2} = \frac{V_i}{Z_{loop}} = \frac{V_i}{R_s + jL_s\omega} \quad (11.6)$$

with,

L_s = shielded-cable-to-ground inductance = 10^{-6} H/m for typical cables a few centimeters to 1 meter above ground

At least up to $l = \lambda/4$, this current depends strongly upon the way the boxes or circuit references (0 V returns) are floated or grounded to the common ground plane. Above $l = \lambda/4$, this consideration has little importance and the loop impedance is represented by the characteristic impedance of the cable above ground.

Therefore, to summarize:

1. The victim-induced voltage due to the coupling mechanism (1) called **direct differential coupling** is:

$$V_{o1} = 1/2\ I_{s1} \times Z_t \times l$$

with,

$$Z_t = \text{transfer impedance of the shield in } \Omega/\text{m}$$

2. The victim-induced voltage due to the coupling mechanism (2) called **common-mode coupling** is:

$$V_{o2} = 1/2\ I_{s2} \times Z_t \times l$$

Since these two voltages are manifestations of the same EMI field, the final result is:

$$V_{vict(tot)} = V_{o1} + V_{o2}$$

Generally, one of the two voltages predominates.

Illustrative Example 11.2

Calculate the voltage at the termination of a 4 m piece of RG 58 coax, 30 cm above ground, when illuminated by a 10 V/m, 5 MHz field. The shield is grounded at both ends.

From Illustrative Example 11.1, the 4 m length of RG58 represents a total transfer impedance Z_T of 0.2 Ω. From Eq. (11.4), I_s due to differential coupling equals 0.03 A or 30 mA.

From Eq. (11.5), V_i = 0.6 V induced in the shield-to-ground loop.

$$\begin{aligned} Z_{loop} &= R_s + jL_s \times 2\pi \times 5 \times 10^6 \\ &= 4 \times (10\ m\Omega/m) + j(10^{-6}\ H/m \times 4 \times 2\pi \times 5 \times 10^6) \\ &= 40\ m\Omega + j \times 125\ \Omega \approx 125\ \Omega \end{aligned}$$

Due to loop voltage,

$$I_{s2} = \frac{0.6\ V}{125\ \Omega} = 4.8\ mA$$

At the termination,

$$V_o = 1/2\ Z_t\ (30\ mA + 4.8\ mA) = 3.5\ mV$$

11.3.3 Grounding Coaxial Shields

In a coaxial cable, **the shield is the designated current return path.** Therefore, both ends of the shield must at least be connected to the signal reference. Supported by the transfer impedance concept, the following rules and rationale can be established.

11.3.3.1 Equipment with Floated Signal Reference

If the 0 V reference (signal ground) has been intentionally floated from the chassis, the standard mount of the coaxial connector to the chassis cannot be used because it automatically forces a 0 V-to-chassis connection. This nullifies the purpose of the floating. As an illustration, Fig. 11.12 shows the ground loop coupling mechanism (GLC) which is detailed in Volume 8.

At frequencies below a few kHz, the automatic grounding to chassis caused by the connector results in a common-mode rejection of only 6 dB. This rejection improves with frequency due to the increasing loop impedance as Z_t stays constant. If, instead, a special isolated coaxial connector is used (isolated BNC or other type), or if the connector is PCB-mounted instead of chassis-mounted, this restores the signal-to-chassis isolation, giving an enormous improvement at low frequency.

This improvement vanishes as frequency increases and becomes null in the 1 to 10 MHz region. Interestingly (and this is not a coincidence) this is where the single-point or star ground systems fail and multipoint grounding should be substituted. From this we can derive the first rule of coaxial shield grounding:

Rule 1: With floated signal references (single-point ground systems), coaxial cables should be connected via insulated connectors if low-frequency ground loops problems are anticipated. Above a few megahertz, this precaution becomes useless, as does as the whole concept of floated reference, and ordinary coaxial connectors, bonded to the chassis, are preferred (see Fig. 11.12).

From this rule, we can derive several corollaries:

1. At low frequencies, the designer must realize that for the sake of opening the ground loop, he weakens the overall system shielding. However, if the EMI problem is mainly a low-frequency structural noise or earth gradient, the solution is valid.
2. If the system is susceptible to lightning, the floated coaxial end is a real danger, and a "transient" ground should be provided via a surge protector.

Curve	h/d	$Z_0 \Omega$
A	0.6 – 1	50
B	1 - 5	120
C	5 – 100	250

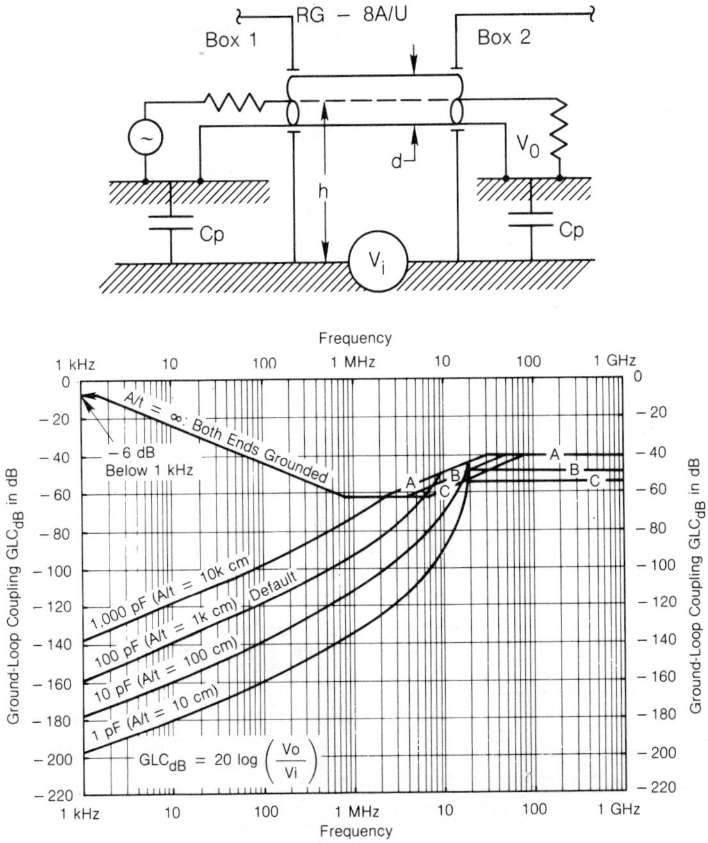

Figure 11.12—Ground-Loop Coupling (GLC) for an RG-8A/U Coaxial Cable, Showing the Effect of Insulating the Coaxial Shield from the Chassis

3. Because of these complex tradeoffs, a coax is not the best medium for low-frequency situations; balanced twisted shielded pairs provide a better opportunity to apply common-mode rejection techniques.
4. Triaxial or quadraxial cables reconcile the need for a floated signal ground and shield continuity to chassis (see Section 11.3.4 and Fig. 11.14).

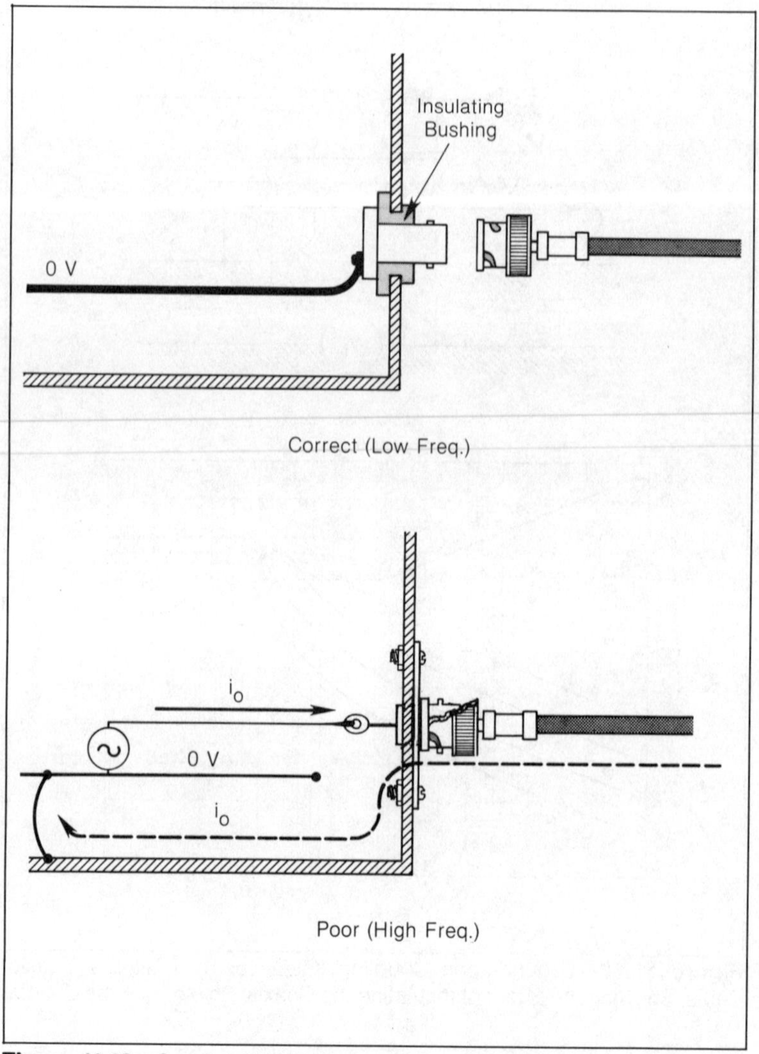

Figure 11.13—Coaxial Shield Connection (continued next page)

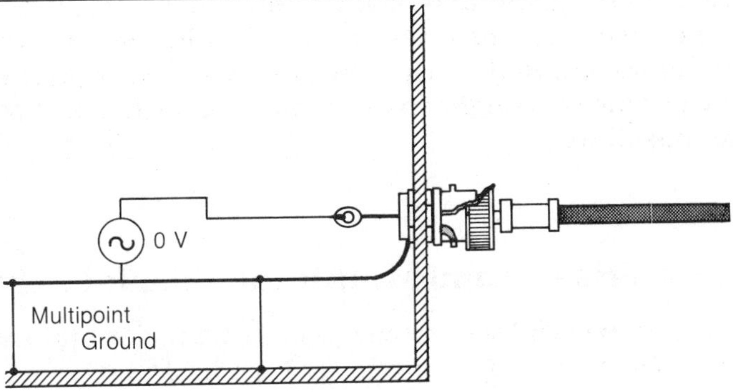

Better (High Freq.)

Figure 11.13—(continued)

11.3.3.2 Equipment with Grounded Signal Reference

If the 0 V reference (signal ground) is purposely connected to the chassis, as in a multipoint ground, there is no need to try isolating the shield from the chassis, and this case indeed reflects the various corollaries of 11.3.3.1. However, two things must be pointed out:

1. Even though the signal ground is connected to the chassis ground, if this connection is physically remote from the coaxial connector, a path, as direct as possible, must be established between the signal reference and the connector backshell. This is to prevent the coaxial return current from flowing throughout the chassis (most of the current will seek the lowest impedance).
2. Coaxial cables with high-quality braid tend to isolate the EMI currents which flow on the outer skin from the circuit current which flows on the inner skin (at least between 100 kHz to 10 MHz). The mounting method recommended in no. 1, above, will enhance this segregation. Thus, a second rule of coaxial cable grounding can be stated:

Rule 2: For grounded signal references (multipoint ground systems), coaxial cables should use standard connectors bonded to the chassis, with the signal 0 V plane or traces brought to the connector socket as directly as possible.

11.3.4 Triaxial and Multibraid Coaxial Cables

The word **triaxial** is sometimes confused with **double-braid**, but they are not exactly synonymous. One should distinguish between several types:

1. **Double-braided** coaxials are cables with one braid wrapped directly over the other. Because the two braids are not electrically insulated, the low-frequency benefit is merely that of a reduced dc braid resistance, typically on the order of 2 or 3 mΩ/m. However as frequency increases, the isolation due to skin effect will appear between the two braids, as described in 11.3.3.2b but will be more pronounced than for a single braid. Rules 1 and 2, described above, are applicable as for ordinary coaxial cables.
2. **True triaxial cable** is constructed with the two braids mutually isolated. This construction permits connection of the outer braid to the chassis and the inner braid to the signal return at both ends while maintaining a floated electrical ground at one end. To preserve this isolation at low frequencies, a special connector (special isolated "N" connector for instance) (Fig. 11.14) is necessary where the outer and inner sockets are insulated from each other. At high frequencies, the two braids still provide the 30 dB improvement shown in Fig. 11.8 due to the decoupling between the two braids' leakage inductances (low interbraid coupling).
3. **Quadraxial cables** are constructed with three conductive layers. Taking the concept one step further, the triax can be covered with a third conductive layer in several variations. For example, one of the two outer layers can be made of a ferrous wrap or high permeability braid to enhance the cable's magnetic shielding effectiveness.

These multiple-shielded cables offer more efficiency, the penalty being increased weight, stiffness and fitting complexity. The connection of the shields should obey the same rule as triaxial cable, that is:

Rule 3: True triaxial cables must have the inner braid connected to the signal ground at both ends and the outer braid to the chassis at both ends. To benefit fully from the triaxial concepts, special "N" connectors must be used which provide two insulated sockets.

11.3.5 Emission Aspects: Reciprocity of the Transfer Impedance

So far, the concept of Z_t has been applied only to susceptibility situations. However, the parameter is perfectly reversible. A cable with a low Z_t will be immune to ambient EMI threats but will also radiate less if properly connected. The process by which the reciprocity of Z_t can be seen is the following:

1. The intentional signal current I_o returning by the shield induces an EMI voltage $V_i = Z_t \times I_o$ that appears in series on the outer side of the shield.
2. This voltage V_i generates both a displacement current (dipole mechanism) and a ground-loop current, causing the shield to become a radiating antenna.

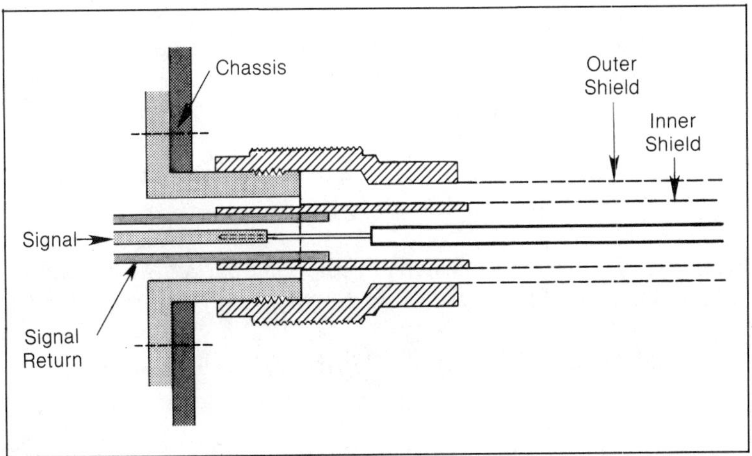

Figure 11.14—Triaxial Termination via an Insulated "N" Connector

Illustrative Example 11.3

A 2 m piece of RG58 coax connects two equipment boxes. The average cable height above ground is 30 cm. The carried signal is a 16 MHz video carrier with 12 V peak amplitude. The link has a 75 Ω matching impedance at each end. What is the radiation at 3 m of the 16 MHz signal, assuming good quality BNC connectors at each end and both equipment boxes connected to the common ground?

First calculate the internal shield return current I_O:

$$I_o = \frac{12\text{ V}}{75\ \Omega} = 0.16\text{ A}$$

$$= -16\text{ dBA}$$

Then calculate the shield external longitudinal voltage:

$$V_i = Z_t \times I_o$$

From Fig. 11.8, Z_t for RG58 is −18 dBΩ/m.

$$V_i = -18\text{ dB}\Omega/\text{m} + 6\text{ dB (two meters)} - 16\text{ dBA} = -28\text{ dBV}$$

This voltage is driving a loop antenna representing a loop area of:

$$2.0\text{ m} \times 0.30\text{ m} = 0.6\text{ m}^2 \text{ or } 6{,}000\text{ cm}^2$$

The loop inductance can be estimated as equal to 2 μH, giving an impedance of about 200 Ω at 16 MHz.

From Fig. 9.5 in the "Emissions Prediction" part of this book, the field at 3 m from this coaxial cable due to ground-loop radiation is:

$$E\text{ dB}\mu\text{V/m} = -4\text{dB}\mu\text{V/m} - 28\text{ dBV} + 20\log(6{,}000\text{ cm}^2)$$

$$= 44\text{ dB}\mu\text{V/m}$$

Coaxial Shield Grounding with Regard to EMI Emissions

To minimize EMI emission, the grounding requirements for coaxial cables are similar to those described so far. For instance, at very low frequency, with both ends of the shield connected to the signal grounds and those signal grounds connected to the structure, some of the return currents will flow in the chassis and structural grounds instead of the shield. This results in two effects:

1. They will contribute to the general ground noise.
2. They will cause the whole shield-to-ground loop to radiate.

Let us consider a 10 m cable, 1 m above ground. Assume for instance that current returning by the unintended path is only 1 percent of the total circuit current. In other words, 99 percent of the current going to the coaxial load returns to the signal source via the shield (the intended path) and 1 percent by the chassis, frame, safety grounds, etc. (the unintended path). If we compare the radiation from this loop to that of an unshielded, balanced link carrying the current, I, via a parallel wire pair with 1 mm wire separation, we can establish a relative comparison based on the respective magnetic moments such as:

$$\frac{\text{Radiation with coax}}{\text{Radiation with ideal (balanced) pair}} =$$

$$(0.01\ \text{I} \times 10\ \text{m}^2)\ /\ (\text{I} \times 10\ \text{m} \times 10^{-3}) = 10$$

Therefore, in relative terms, the low-frequency radiation of the coaxial link due to ground loop will be 10 times higher than the radiation from a typical untwisted pair with 1 mm wire separation. In this case, there will be no benefit in having the coax.

When frequency exceeds a few kilohertz, an interesting phenomenon occurs which is somewhat reciprocal to the upper left part of the GLC curve. Due to a mutual inductance between the center conductor and the surrounding shield, the return current is "invited" to flow in the shield rather than in the increasing impedance of the ground loop.

The value of the shield current I_s versus the total current I_c is expressed as:[1]

$$I_S = I_C\left(\frac{j\omega}{j\omega + R_S/L_S}\right) \tag{11.7}$$

where R_S is the shield resistance and L_S the external shield-to-ground self-inductance.

For typical single-braid coax, R_S is in the 5 to 10 mΩ/m range, while L_S is 0.5 to 1 μH/m, depending on the shield diameter and its height above ground. Therefore a typical value for R_S/L_S is 10^4. The ratio R_S/L_S is the reciprocal of a time constant and is termed ω_C, the shield cutoff pulsation in radians per second (rad/s), or in hertz:

$$F_C = \frac{\omega_C}{2\pi} = \frac{R_S}{2\pi L_S} \tag{11.8}$$

For typical shields, ω_C being in the 10^4 rad/s range, F_C is in the 1.5 kHz range.

The following table gives the approximate proportion of the total current which "elects" to return by the shield in a coaxial cable grounded at both ends to a metallic structure as a function of frequency:

Frequency	I_S	$I_C - I_S$ = I Ground Current
1.5 kHz	0.7 I_C	0.3 I_C
7.0 kHz	0.98 I_C	0.02 I_C (−34 dB)
15.0 kHz	0.995 I_C	0.005 I_C (−46 dB)
150.0 kHz	0.9995 I_C	0.0005 I_C (−66 dB)

11.3.6 Use of Z_t when Shields Become Electrically Long ($l > \lambda/4$)

The Z_t curves of Fig. 11.8 are normalized to a 1 m length of shielded cable. The process of multiplying Z_t in ohms per meter by the actual length of the cable, then multiplying by the shield current I_s to obtain the induced voltage, implies that the current is uniform along the cable length. This assumption is valid as long as the cable is an electrically short line, i.e. $l << \lambda_{EMI}$. It becomes invalid when the line is electrically long, because the distribution of current along the shield is sinusoidal with alternate peaks and nulls every $\lambda_{EMI}/4$. Therefore, it becomes illegitimate to continue to multiply Z_t by the physical length, since each half-wave length segment of the shield will see an induced voltage V_i of opposite sign. Without an excessive error, a significant simplification can be made by "clamping" the length of the cable l to $\lambda_{EMI}/2$ when doing the transfer impedance length adjustment. To summarize this rule:

$$\text{for } l < \lambda_{EMI}/2\text{: use } Z_t \text{ (cable)} = Z_t\Omega/\text{m} \times l$$

$$\text{for } l > \lambda_{EMI}/2\text{: use } Z_t \text{ (cable)} = Z_t\Omega/\text{m} \times \lambda/2$$

Illustrative Example 11.4

Using the same cable as in Illustrative Example 11.2, calculate the induced voltage for an incident EMI field of 10 V/m at 100 MHz.

The wavelength at 100 MHz is 3 m, so all cable lengths for field coupling and Z_t calculations will be replaced by $\lambda/2$ or 1.50 m. The characteristic impedance of this cable 0.3 m above ground is about 200 Ω. The Z_t of an RG58 cable at 100 MHz is −2 dBΩ/m or 0.8 Ω/m., so $Z_T = 0.8\ \Omega/\text{m} \times 1.50\ \text{m} = 1.2\ \Omega$. I_{s1} induced by displacement current from Eq. (11.4) is:

$$3.9 \times 10^{-5} \times (1.5)^2 \times f \times E = 0.09\ \text{A}$$

From Eq. (11.5), voltage V induced by a ground loop is 4.4 V; therefore the shield current due to this mechanism is:

$$I_{s2} = \frac{4.4\ \text{V}}{200\ \Omega} = 0.022\ \text{A}$$

Therefore, to compute the induced voltage V:

$$V = 1/2\ Z_T\ (I_{s1} + I_{s2}) = 0.067\ \text{V} = 67\ \text{mV}$$

11.3.7 Calculating Time-Domain Coupling via Transfer Impedance

Another interesting aspect of Z_t is that it allows simple calculation of cable shield coupling when the EMI threat is characterized in the time domain, without necessarily performing a Fourier transform of the EMI field or current pulse. In this case, the voltage induced longitudinally in the shield for a skin current I_s varying in a given time increment:

$$V_i = R_t \Delta I_s + L_t \frac{\Delta I_s}{\Delta t} \tag{11.9}$$

where R_t and L_t are the transfer resistance and inductance as defined with Eq. (11.2) and listed in Table 11.1.

Illustrative Example 11.5

Calculate the voltage induced in a 10 m length of RG62, 1 m above ground, when illuminated by the magnetic field of a 30,000 A lighting stroke 30 m away. At the location of the cable, this corresponds to a flux density of 2×10^{-4} Tesla, varying in 1 μs.

The voltage induced in the external cable loop (ground lŏop) is:

$$V = d\text{Ø}/dt = -\ dB/dt \times \text{Area in m}^2$$

$$= -\ \frac{2 \times 10^{-4} t}{10^6} \times 10\ \text{m} \times 1\ \text{m} = 2{,}000\ \text{V}$$

This voltage forces a shield current to flow:

$$I_s = \frac{V}{Z_s}$$

where Z_s is the external impedance of the cable shield above ground. This impedance is equal to the total shield resistance R_{sh} plus the shield-to-ground external reactance L_s/dt.

The first term, representing 10 mΩ/m × 10 m = 100 mΩ is negligible compared to the second one, which is equal to:

$$\frac{10^{-6}\ \text{H/m} \times 10\ \text{m}}{10^{-6}} = 10\ \Omega$$

So,

$$I_s = \frac{2{,}000\ \text{V}}{10\ \Omega} = 200\ \text{A}$$

The voltage induced internally in the coax will be:

$$V_1 = R_t \times 200\ \text{A} + L_t \times \frac{200\ \text{A}}{10^{-6}\text{s}}$$

In the above expression, it is assumed for simplification that the rise time of the current in the shield is equal to that of the lightning pulse. This may not always be true if the current rise is limited by the line parameters (time constant, capacitance, etc.).

The thickness of a one-layer braid represents less than one skin depth in this frequency range (< 300 kHz), therefore:

$$R_t = R_{sh} = 10\ \text{m}\Omega/\text{m} \times 10\ \text{m} = 100\ \text{m}\Omega$$

From Table 11.1, $L_t \approx 10^{-9}$ H/m $= 10^{-8}$ H for the total length.

Therefore:

$$V_i = 200\ \text{A} \times 0.1\ \Omega + 10^{-8} \times \frac{200}{10^{-6}} = 22\ \text{V}$$

This voltage divides evenly between the two ends since they are matched, and we can approximate a uniform voltage and current distribution along the 10 m cable. The propagation velocity in an overhead cable is about 250 m/μs, therefore the 10 m cable has a propagation delay equal to 1 μs × 10/250 = 0.04 μs. This is much shorter than the 1 μs duration of the pulse. Therefore, at either end of the cable, V_0 = 22 V/2 = 11 V.

The previous abbreviated calculation assumed, for simplification, that only the inductive term of Z_t is into play and that both ends were sharing equally the shield noise. For more accuracy in the time domain, DeGaucque (Ref. 2) has developed a set of equations to predict accurately the maximum voltage at each end for a pulsed excitation. This takes into account not only the inductive leakage of the braided shield (leakage or transfer inductance L_t) but also the capacitive coupling through small apertures in the shield (referred to as **shield leakage capacitance** or sometimes **admittance**).

If, from Fig 11.3, we call **near-end** and **far-end** the sides of the coaxial cables which are near to and far from the disturbing source side, respectively the peak voltage at these ends will be:

$$V_{near} = -\frac{V_0}{4\ Z_{01}}\left[R_t + \frac{M'}{\tau_r}\right] \times l \qquad (11.10)$$

$$V_{far} = -\frac{V_0}{4\ Z_{01}}\left[R_t + \frac{M''}{\tau_r}\right] \times l \qquad (11.11)$$

where,

V_0 = Voltage into the disturbing circuit (no. 1)

Z_{01} = characteristic impedance of the disturbing line, to be replaced by actual shield-to-ground impedance if shield is open-ended or shorted to ground

R_t = shield transfer resistance (see Section 12.3.1) in ohms/meter

$M' = L_t + Z_{01}\ Z_{02}\ C_{12}$ (near-end coupling term per meter length)

$M'' = L_t - Z_{01}\ Z_{02}\ C_{12}$ (far-end coupling term per meter length)

τ_r = disturbing pulse rise time or transition time

Z_{02} = coaxial cable characteristic impedance

L_t = shield transfer inductance, in Henry per meter

C_{12} = shield leakage capacitance, in F per meter (range from 0.1×10^{-12} F/m for excellent, thick braid to 10.10^{-12} F/m for leaky braid with poor optical coverage)

l = length of the coaxial cable, in meter

Illustrative Example 11.6

A 2 m length of typical 50 Ω coaxial cable has the following characteristics:

$$R_t = 10 \text{ m}\Omega/\text{m (in the skin depth region)}$$

$$L_t = 1 \text{ nH/m}$$

$$C_{12} = 0.2 \text{ pF/m}$$

The coax is subjected to a common-ground impedance pulse from a neighboring circuit. The shield is grounded at both ends with a typical shield-to-chassis inductance of 0.8 μH/m. The disturbing pulse shown in Fig. 11.15 has the following characteristics:

$$\text{Amplitude} = 1 \text{ V}$$

$$\tau_r = 10 \text{ ns}$$

$$\text{Duration } \tau = 50 \text{ ns}$$

The shield-to-ground impedance is mainly inductive and, in this case, will be used as Z_{01}:

$$Z_{01} \cong \frac{L}{dt} \cong \frac{0.8 \times 10^{-6} \text{ H/m}}{10^{-8}} = 80 \ \Omega$$

Applying Eq. (11.10) and Eq. (11.11):

$$V_{near} = -\frac{1}{4 \times 80}\left[10^{-2} + \frac{M'}{\tau_r}\right] \times 2 \text{ m}$$

$$V_{far} = -\frac{1}{4 \times 80}\left[10^{-2} + \frac{M''}{\tau_r}\right] \times 2\ \text{m}$$

$$M' = 10^{-9} + 80 \times 50 \times (0.2 \times 10^{-12}) = 1.8 \times 10^{-9}$$

$$M'' = 10^{-9} - 80 \times 50 \times (0.2 \times 10^{-12}) = 0.2 \times 10^{-9}$$

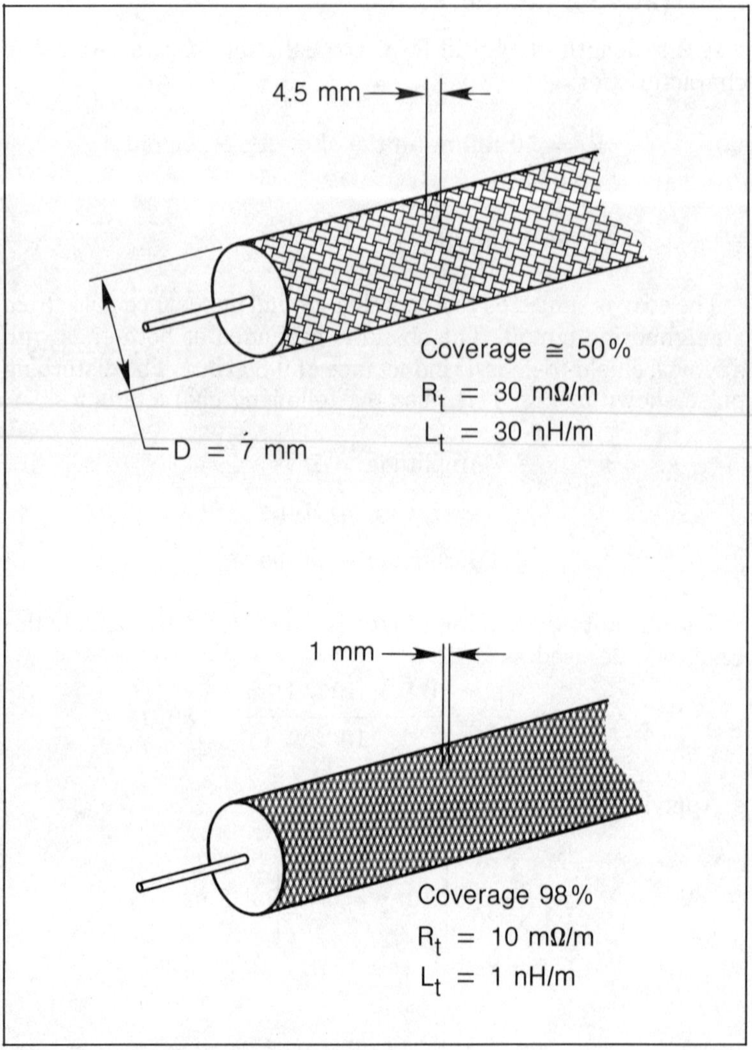

Figure 11.15—Shape of Induced Noise

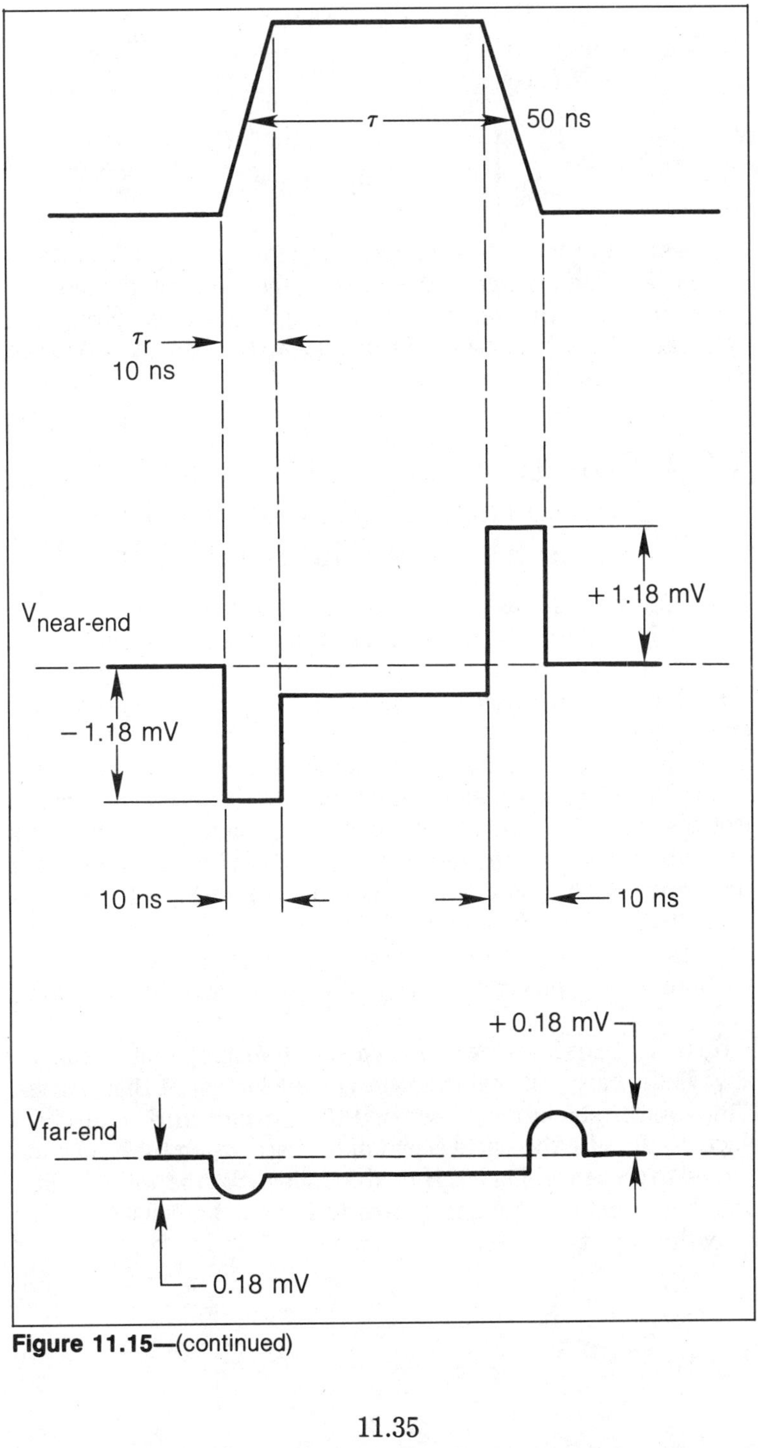

Figure 11.15—(continued)

$$V_{near} - \frac{1}{320}\left[10^{-2} + \frac{1.8 \times 10^{-9}}{10 \times 10^{-9}}\right] \times 2 = -1.18 \text{ mV}$$

$$V_{far} = -\frac{1}{320}\left[10^{-2} + \frac{0.2 \times 10^{-9}}{10 \times 10^{-9}}\right] \times 2 = -0.18 \text{ mV}$$

The shape of both induced noises is shown in fig. 11.15. Since there is a change in sign of the dV/dt at the fall-time, the induced noise appears positive. A smaller coupling exists during the "plateau" of the disturbance, mainly due to the transfer resistance coupling times the shield current.

11.4 Effectiveness and Transfer Impedance of Multiconductor Cable Shields (Balanced Lines)

As previously discussed, in a shielded balanced line (parallel wire pair, twisted pair, ribbon cable, etc.) the shield acts purely as a barrier and does not carry the intentional current. Therefore, it seems that the designer has complete latitude to ground or float one or both ends.

Re-examining the fundamentals of interference control, we see in Fig. 11.16 (parts a and b) that, ideally, if a system must be isolated from a coupling to or from its environment, **it must be completely surrounded by a conductive barrier that prevents both space waves (radiation) and guided waves (conduction) from passing through it.** However, systems need to be referenced to a common ground by one or more connections (safety, power return, lightning, static sink, etc.) as shown in Fig. 11.16c.

If such a barrier is deemed necessary (meaning that this is not always the case), the cable shield is one element of that barrier. This tubular shield must surround the forward **and** return conductors. Because the shield is part of the barrier, it must be uninterrupted to prevent E-field or H-field coupling with the wiring. Thus, whether or not the shield is **grounded**, it must at least be closed at both cabinets.

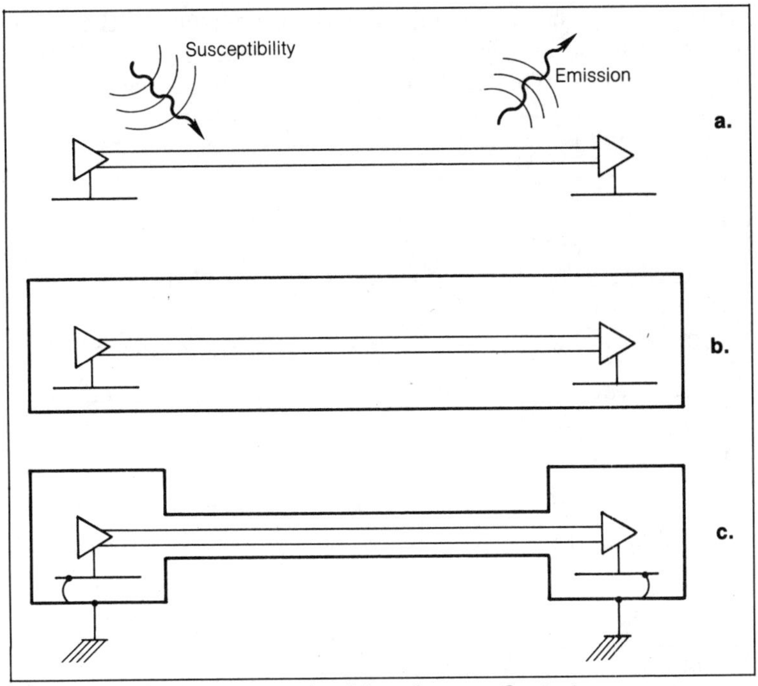

Figure 11.16—Fundamentals of Interference Control

The pro and con arguments about grounding a cable shield are an often illegitimate extension of audio-frequency technology. What is important is to interrupt the current flowing in the shield because shield excitation invariably means a certain amount of energy coupling to the interior conductors. The idea of interrupting the shield current sometimes has degenerated into concepts where it is the shield which becomes interrupted.

Figure 11.17 shows examples of connecting the shield to both boxes (11.17a) and connecting the shield at one end only (11.17b). The first approach will preserve the overall system shielding concept. In the case of noisy grounds (earth or chassis V_{cm} or induced loop voltage (H-fields), however, a large shield current may circulate. This can aggravate EMI into the inner circuit. Connecting the shield only on the end where the 0 V reference is grounded will avoid the ground loop contribution by the shield but will create a noisy end if AB approaches $\lambda/4$.

Interrupting the shield current without interrupting the shield implies that the loop is interrupted elsewhere in path ABCDEF of Fig. 11.17b. This in turn may pose installation or safety

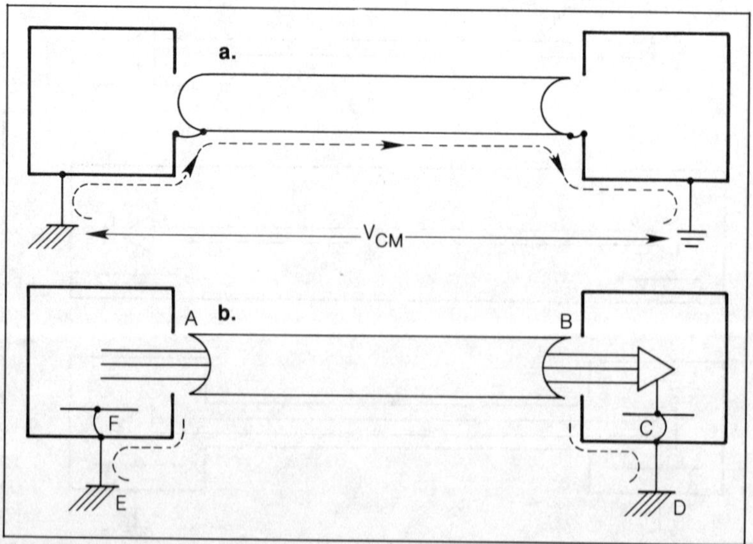

Figure 11.17—Connecting Shield to (a) Both Boxes and (b) the End where the 0 V Reference Is Grounded

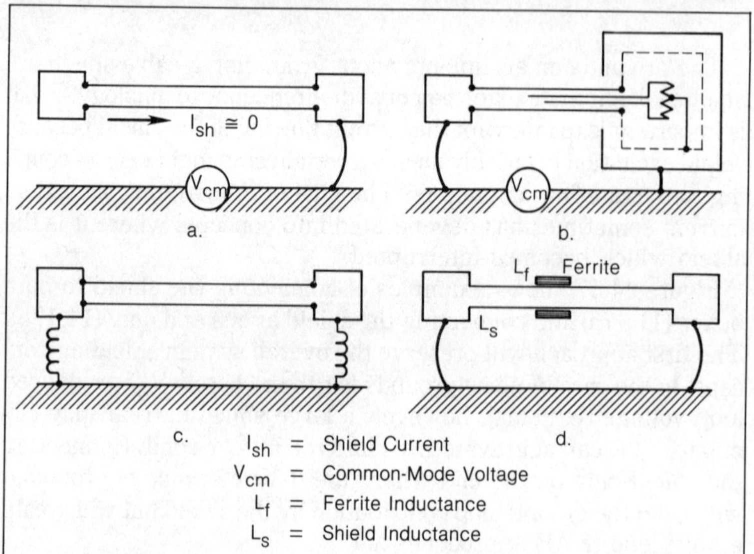

Figure 11.18—Several Ways to Reduce Cable Shield Current without Interrupting the Shield

problems. However, as shown in Fig. 11.18, there are several ways to reduce the shield current without opening the shield. The next sections will analyze these different facets and try to reconcile them in an orderly method.

11.4.1 Common-Mode vs. Differential-Mode Shield Performance

The purely differential action of a shield around a balanced line, infinitely far from any ground, is depicted in Fig. 11.19. This is often assumed in cable shield catalogs where the SE of the "tube" (due to its type of metal and thickness) is supposed to attenuate radiation to and from the wire pair. The mechanism of this DM coupling can be broken into the following steps:

1. The time-varying field induces a shield current or a shield longitudinal voltage since its barrier impedance is not zero. Note that the shield current (or displacement current) is due to the free-space capacitance of the shield element, acting as a short dipole. This is similar to the case of a coaxial shield.
2. This current is more or less attenuated through the shield material, and the residual current and voltages appear inside. This is the transfer impedance coupling of the shield. This too is similar to the case of a coaxial cable shield.

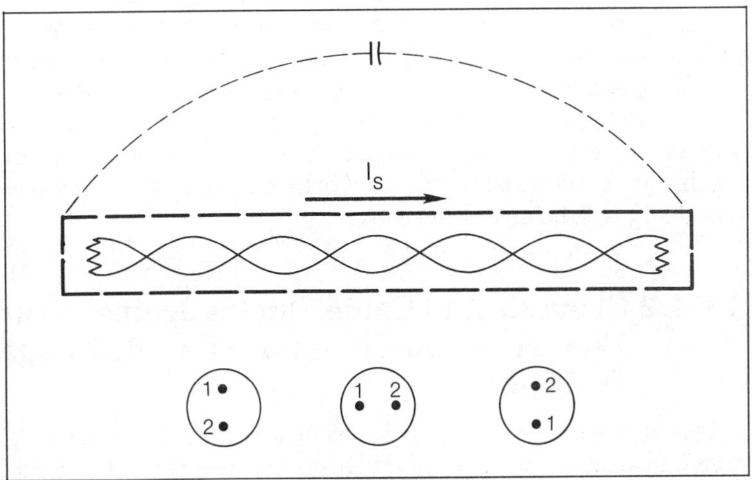

Figure 11.19—Purely Differential Coupling into a Balanced Shielded Pair

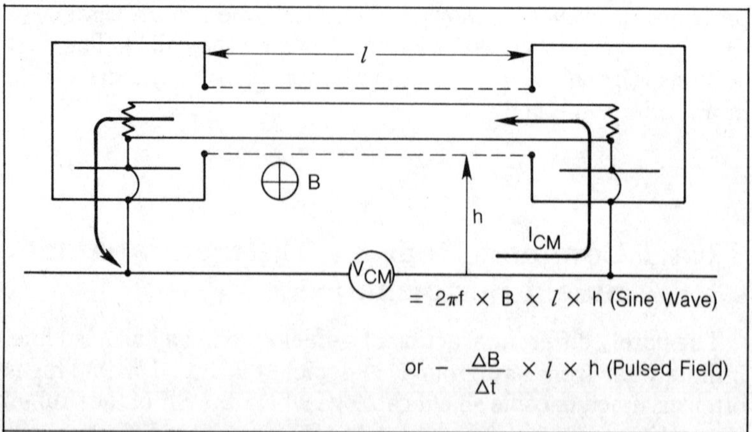

Figure 11.20—A More Realistic Configuration of Electric and Magnetic Coupling into a Shielded Cable between Two Equipment Boxes

3. The net magnetic or capacitive coupling to the wire pair inside would be null if the balancing and twisting of the pair were perfect. Since it is not, some voltage appears across the pair, proportional to the percentage of unbalance of the pair.

However, this configuration is purely academic; except in a laboratory experiment, no one would use a shielded pair that way. A more realistic installation is that of Fig. 11.20 where an actual ground surface is present and the cable terminates into two boxes which are strongly coupled. Most often, they are connected to this ground, forming a ground loop. For worst-case orientations of this ground loop versus field polarization, a low-impedance voltage source will appear in series in the loop, due to magnetic induction, and a high-impedance current source will appear across the loop due to electric excitation. Depending on which one of these mechanism predominates, or even if both of them create a nuisance, the shield will behave differently.

11.4.2 Grounding of Cable Shields against Near Electric-Field Crosstalk: Faraday Cage Isolation

The case of a near electric-field coupling, as from low-frequency, high-impedance sources or capacitive crosstalk, is illustrated in Fig. 11.21. In the case of an unbalanced load (but still a balanced line), the electrical coupling can be depicted as a set of coupling

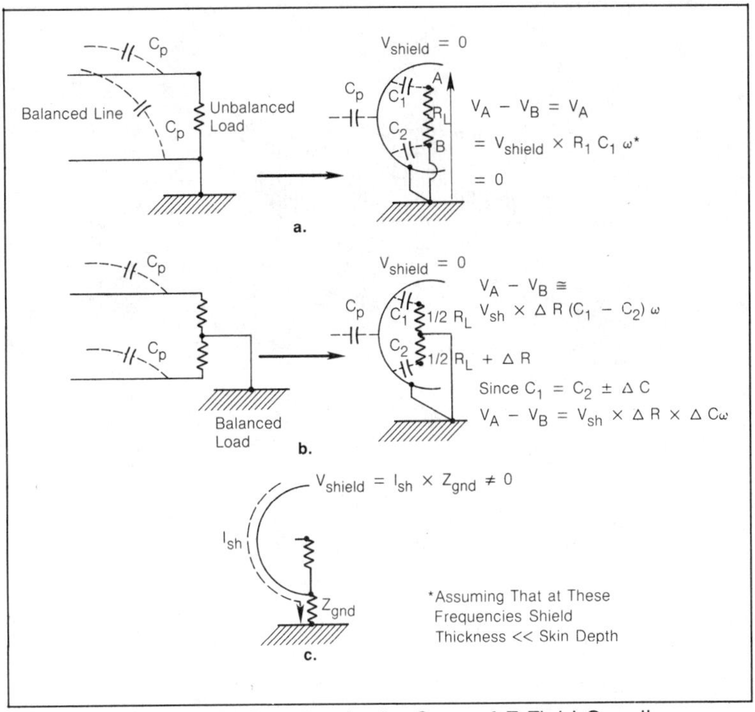

Figure 11.21—Shield Coupling in the Case of E-Field Coupling

capacitances Cp between the circuit of concern and the external EMI source (Fig. 11.21a). When a shield is added and grounded to the signal reference, the new coupling consists of a set of capacitances in series: the exterior-to-shield capacitance (approximately identical to the previous C_p), plus the shield-to-pair capacitances C_1 and C_2. C_1 and C_2 are several orders of magnitude larger than C_p.

For instance, C_p may be fractions of picofarads per meter of cable while C_1 and C_2 are in the range of 30 to 100 pF/m. The voltage appearing across R_L is:

$$V_A - V_B \approx V_A = V\ \text{shield} \times R_L\, C_1 \omega \qquad (11.10)$$

However, if the shield is perfectly grounded, $V_{shield} = 0$ and $V_A = 0$.

In the case of a balanced load (Fig. 11.21b), the two leakage currents ideally would create equal and opposite voltages across R_L.

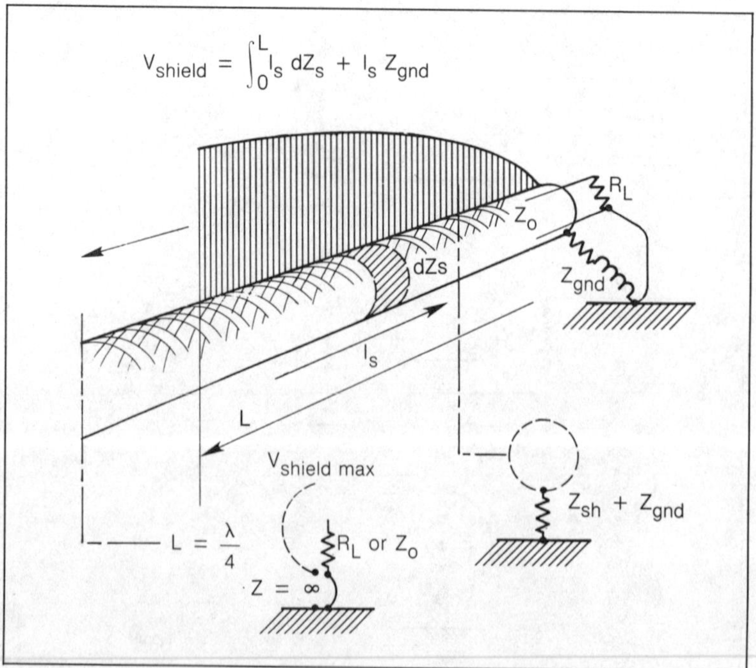

Figure 11.22—General Model of E-Field Shielding

Since in fact neither the load nor the pair-to-shield capacitance is perfectly balanced, Fig. 11.21b shows that a residual voltage due to the shield voltage coupling into ΔR_L, through ΔC, will exist. However, here again, because $V_{shield} = 0$, $V_A - V_B = 0$.

Now, as shown in Fig. 11.21c, if a fortuitous or deliberate impedance Z_{gnd} exists between the shield and the ground, the shield voltage is no longer zero. The circuit becomes surrounded with a noisy shield which is strongly coupled via C_1 (or via ΔC for balanced loads).

A more general model of this E-field screening is shown in Fig. 11.22 where the impedance between the shield and the ground is in fact made of the shield termination (pigtail, solid wire, grounding pin, etc.) plus the shield impedance Z_s. Since a current generally will flow on the shield, the product of this current and the total impedance $(Z_s + Z_{gnd})$ will cause the shield to become increasingly noisy as we move away from its grounding point.

As long as the shield and its ground connection are still short with respect to the EMI wavelength, the amplitude of V_{sh} is generally small and tolerable. But if the worst case happens, i.e., the shield reaching $\lambda_{EMI}/4$ or an integer multiple thereof, the shield impedance to ground becomes infinite. In this case, the wiring is surrounded by a floating shield. Because C_1 and C_2 are so large, the whole shield voltage is strongly coupled into the inner circuit. So, oversimplifying the model we may say that when $l = \lambda/4$, a thin shield whose thickness is less than a skin depth has nearly zero effect; the shielding is equal to 0 dB. Decreasing monotonically with frequency as 1/F, we may induce that for $l = \lambda/40$, the same shield will offer appreciable EMI decoupling of 1/10, or −20 dB. When $l = \lambda/400$, we have 40 dB of shielding, and so on.

Extending this idea, it becomes obvious that grounding the other end, or grounding at the middle or even at multiple points, will parallel the grounding impedance and reduce the shield voltages. Indeed, if instead of focusing on the term "grounding," we simply said that the other end of the shield is also connected to the box frame, the shield noise would decrease, and we would in addition achieve the overall shield integrity advocated in Fig. 11.16. Unfortunately, doing so may create another EMI problem, as in the case where the two boxes are remotely grounded and a large ground potential difference exists. This mechanism will be explained in the next section, through the use of differential transfer impedance, a concept introduced by the author[2] and also by DeGaucque.[3]

11.4.3 Quantifying the Ground-Loop Contribution of the Shield Differential Transfer Impedance Z_{td}

In the absence of a shield, the existing ground loop causes some EMI induction and common-to-differential-mode conversion effects. The ground loop coupling coefficient (GLC) is defined as the ratio of V_o at the victim's input to the common-mode voltage V_{cm} induced in the cable-to-ground loop. This concept is thoroughly addressed in Volume 8 of this handbook series. When a shield is added to the balanced cable, how does it interact with GLC? Does the shield render the two boxes equipotential by shorting out the longitudinal CM voltage or does it instead aggravate the ground-loop coupling?

Looking at Fig. 11.23, we see that when a shield is added, an additional path is present for the common-mode current to flow. Since the ground plane source impedance Z_g is small compared

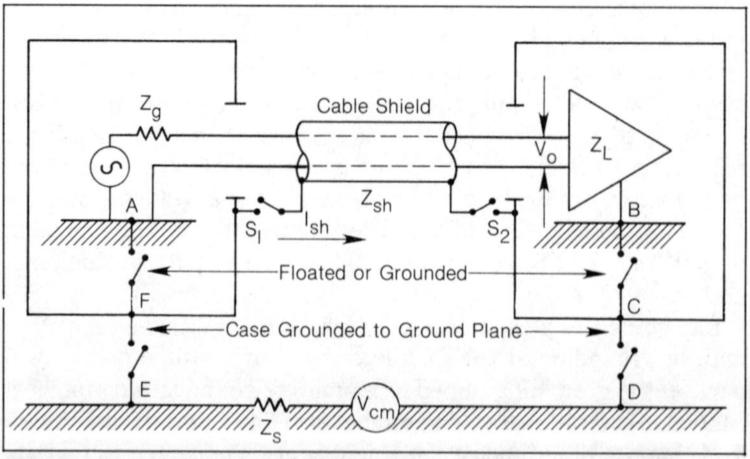

Figure 11.23—Cable Shield Interaction and Grounding Conditions.

to the external shield impedance Z_s, it appears that the flow of I_s does not influence the GLC factor which existed before the introduction of the shield. I_s is simply a new branch, draining a new current from the common mode voltage V_{cm}, without reducing the currents already drained in the wire pair via the circuit loop ABCDEF. If either switch S1 or S_2 is open, we are in the configuration of previous Section 11.4.3, i.e., a single-ended Faraday shield, and V_1 can create virtually no shield current I_s, but in return this shield is only efficient against near E-fields (see previous Section 11.4.2) and $l << \lambda/4$.

However, when S_1 and S_2 are both closed, the shield transfer impedance Z_t (see Section 11.3) causes an inner voltage V_1 to appear inside the shield. This voltage V_1 is equal to V_{cm} if all boxes are grounded and the shield is a thin layer, with no skin effect benefit. It is less than V_{cm} if the boxes are floated or if the boxes' skins and the shield skin are thick. If the cable is coaxial, this noise V_1 would appear directly in the signal return circuit. This mechanism already was addressed in Section 11.3 and Fig. 11.11. But since the circuit is balanced, an additional reduction is caused by the wire pair balance; only a fraction of V_1, proportional to the degree of pair symmetry, actually will show up as an undesired differential voltage V_o across the load. So, a more complex parameter, differential transfer impedance, Z_{td}, is used. This is defined as:

$$Z_{td} = \frac{V_0}{I^s} \tag{11.11}$$

One sees that Z_{td} includes the cascading effects of Z_t, a characteristic of the shield material, and the balance percentage of the pair. Using typical values and measured results for both parameters, and since $I_s = V_{cm}/Z_s$, we can derive a shield coupling factor SC:

$$SC = \frac{V_o}{V_{CM}}$$

This will give a figure of merit of the shield contribution to the overall CM coupling. The smaller this coupling, the better.

Typical values of SC are shown in Fig. 11.24. For a better understanding of the overall mechanism, SC has been overlaid on the GLC (mode conversion) curve, an effect which exists regardless of whether a shield is there. At any frequency, the highest coupling level predominates. We see that at frequencies below a few megahertz (e.g., for a 10 m cable) the fact that the shield is connected to ground at both ends compromises the excellent isolation (100 dB or more) offered by floating schemes. (i.e., the PCB floated

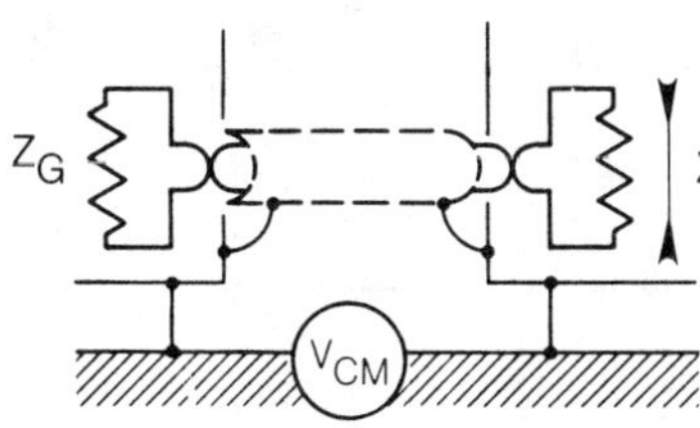

$$SC = 20 \log \frac{V_o}{V_{CM}} = 20 \log \left| \frac{Z_t C_b Z_L}{Z_{sh}(Z_L + Z_G)} \right|$$

Z_t = Surface Transfer Impedance of the Shield

Z_{sh} = Shield External Impedance ($R + JL\omega$)

A = Ordinary Shielded Pair: R Shield = 30 mΩ/m,
Pair Balance C_b = 10%

B = RG22U Good Quality Sh. Pair R Shield = 5 mΩ/m
Pair Balance C_b = 3%

C = RG22B/U Super Shielded Pair: 2 Layers of Shield Touching Together. R Shield = 3mΩ/m,
Pair Balance C_b = 3%

Figure 11.24—Ground Loop Coupling Due to Double-Grounded Braided Shield (SC Factor) (continued next page)

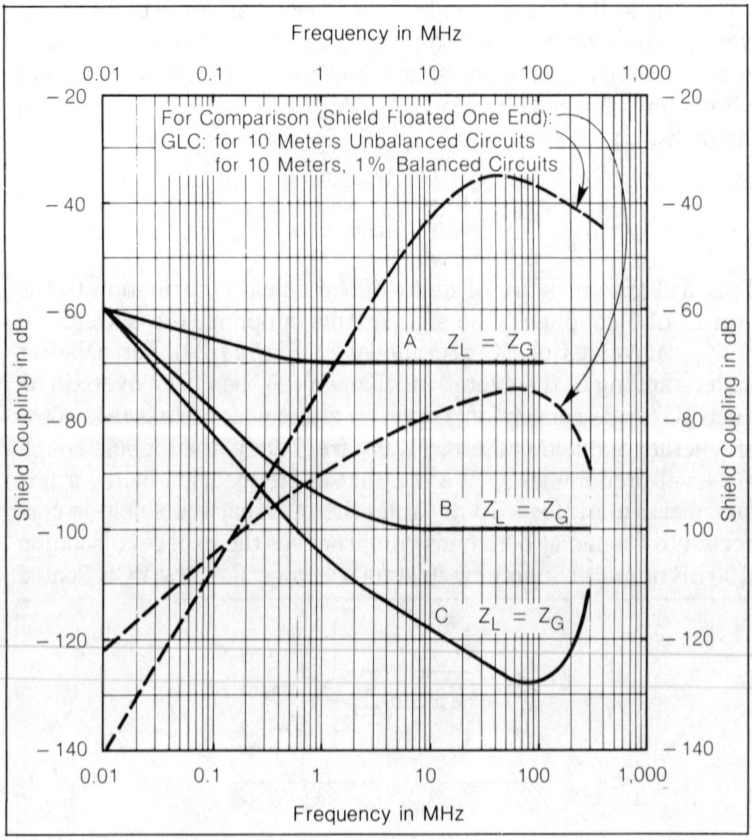

Figure 11.24—(continued)

at one or both ends)

At higher frequencies, the shield contribution is overridden by the predominant GLC, and grounding the shield at both ends does not create any aggravation of coupling, while it contributes to the overall system shielding integrity per Fig. 11.23. This model validates, *a posteriori*, the old precept which recommended single-ended shield grounding for low frequency EMI. The "low frequency" frontier, depending on the source of the tradition, defined as λ/4, or λ/10, or λ/16, etc.

If we look at Fig. 11.24, observing that 1 MHz is approximately the crossover frequency below which SC predominates over GLC. For a 10 m cable, this corresponds to λ/30. This number tracks with the most stringent empirical findings but reveals that the most optimistic ones (λ/4) were definitely too lax.

Volume 2 of this EMC handbook series gives abundant details and alternatives concerning the optimal ways of grounding multiconductor shielded cables. Table 11.2 recapitulates the most essential cable shield grounding rules.

Table 11.2—Cable Shield Grounding Decision (For Other than Coaxial Lines)

Options Available:

- Shield Grounding: Neither End, Load End, Source End, Both ends
- Shield Termination: Use pigtail (inexpensive, practical and generally poor) or 360° bulkhead (expensive).
- Shield Material Choice: High Z_t (inexpensive, flexible, light) or low Z_t (costly, rigid, heavy)

Purpose of Shield	l/λ EMI $\leqslant 0.1$	l/λ EMI > 0.1
To Reduce Susceptibility (EMI is Outside the Shield)	**Electrically Short Shield** System should be a single point ground zero, with pereferably 0 volt-to-chassis connection at receiver side. Ground cable shield at this point. Ordinary braid and short pigtail are acceptable.	**Electrically Long Shield** Ground both ends of shield to chassis. Use low Z_t shield and integral clamp. No Pigtails
To Reduce Emission? (EMI Source is inside the Shield)	**Electrically Short Shield** System should be a single point ground zero, with preferably 0 volt-to-chassis node at transmitting side. Ground cable shield at this point. Ordinary braid and short pigtail are acceptable.	**Electrically Long Shield** Ground both ends of shield to chassis. Use low Z_t shield and integral clamp. No Pigtails

11.5 Importance of Shield Termination Hardware for Preserving Cable Shield Effectiveness

With the exception of coaxial shields (discussed in Section 11.3) where the shield must be fitted in the coaxial connector shell as a part of the transmission line, several methods exist for balanced shield termination:

1. Terminating jumpers or drain wires and lugs (Fig. 11.25)

2. Terminating jumpers or drain wires and use of some connector pins as a shield/ground carrier (Fig. 11.26)
3. Forming a shield in a pigtail and soldered to ground pin or tab (Fig. 11.27 and 11.28)
4. Using a halo ring or interlacing strap (Fig. 11.29)
5. Using a 360° crimped clamp and flat strap (Fig. 11.30)
6. Using integral clamps at connector strain reliever (Fig. 11.31)
7. Using a 360° clamp at box penetration (Fig. 11.31)

Methods 1 through 3 imply the use of some length of discrete wire whose self-inductance seriously impairs the shield efficiency at high frequencies as explained in Sections 11.1.2 and 11.4.2. Also this piece of discrete wire can strongly couple with the signal wire pairs which are locally stripped from the shield protection. And finally, this piece of wire, when it becomes an appreciable fraction of EMI wavelength (e.g., $\lambda/12$ or less) acts as a radiating or pickup element.

Good practices, learned from experience and tradition, dictate that

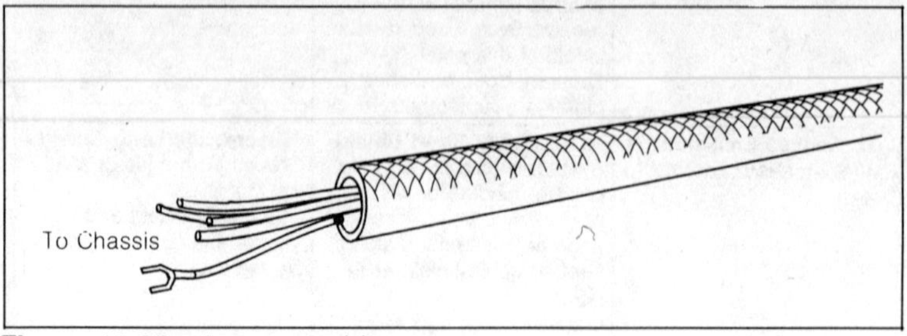

Figure 11.25—Jumper and Lug

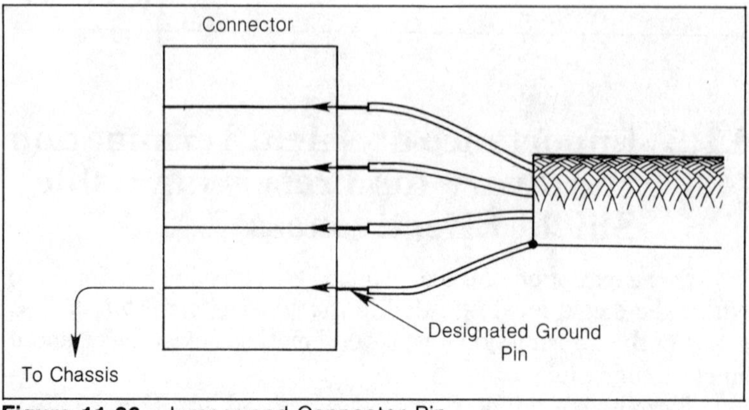

Figure 11.26—Jumper and Connector Pin

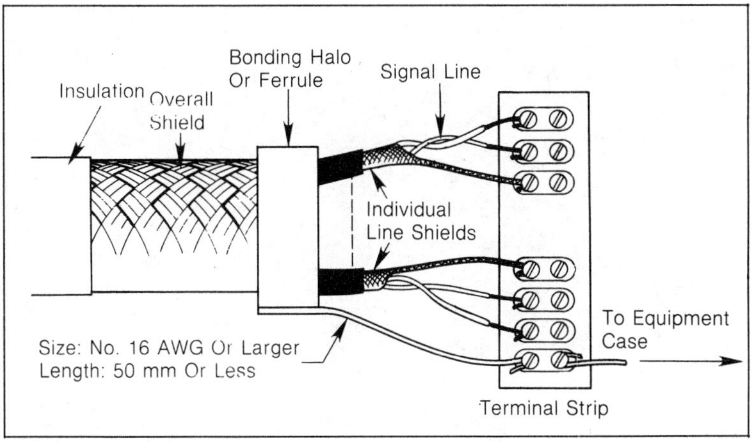

Figure 11.27—Pigtails

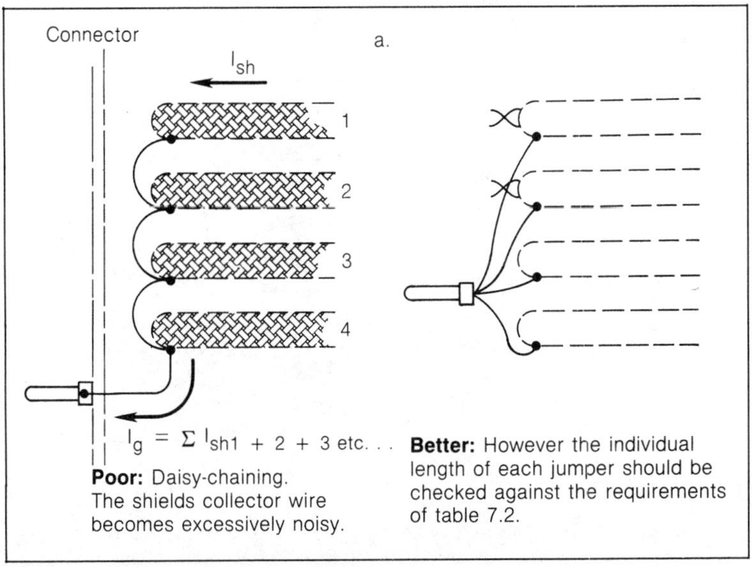

Figure 11.28—Pigtail Daisy-Chaining

pigtails and jumpers are adequate at low frequencies but should not be used at high frequencies. However, the borderline between "low" and "high" is often ambiguously defined, if at all. For instance, the λ/12 criterion may be far too lenient for a shield termination as it would allow a shield to be terminated by a 2.5 m wire against a 10 MHz EMI. Therefore a better rationale is needed, as described in the following paragraphs.

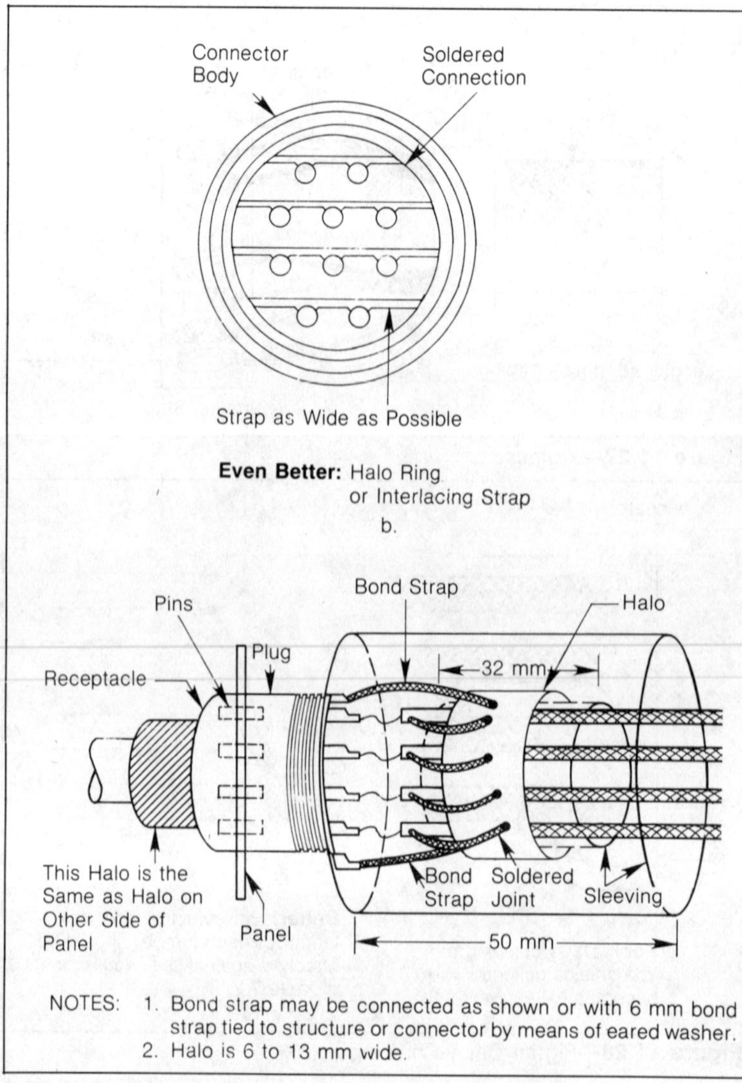

Figure 11.29—Halo Ring

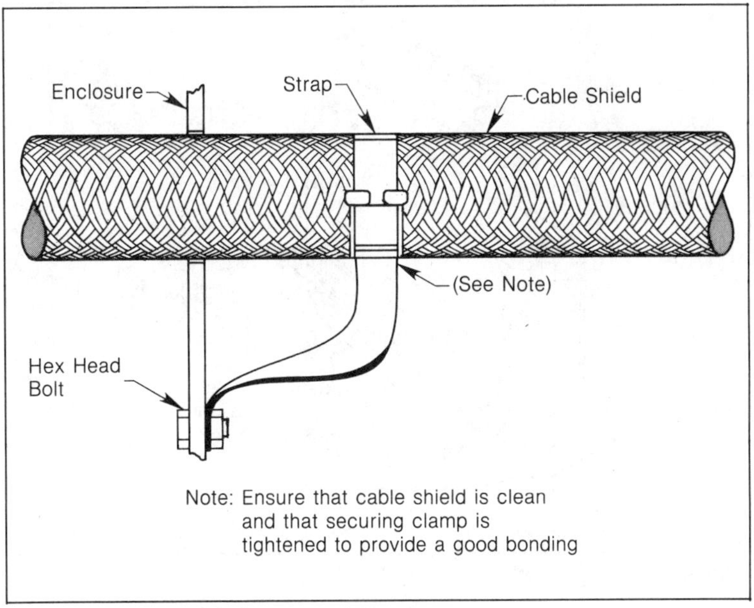

Figure 11.30—360° Clamp and Flat Strap

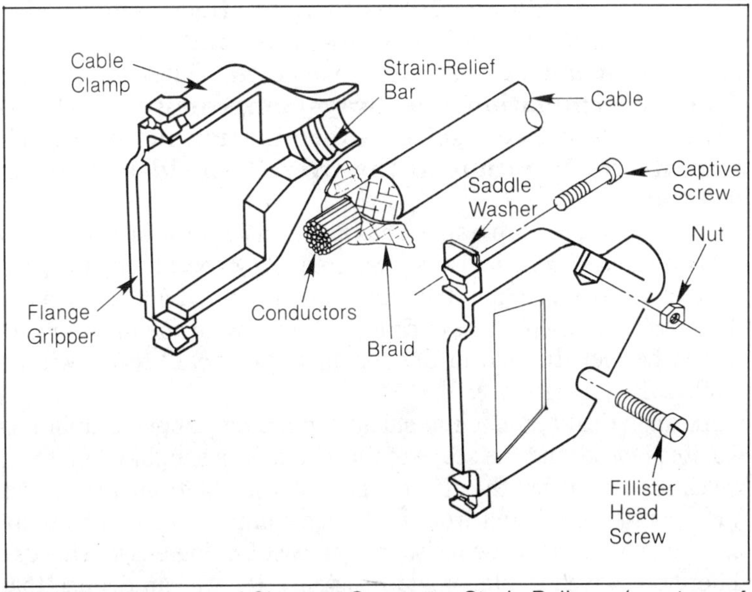

Figure 11.31—Integral Clamp at Connector Strain Reliever (courtesy of AMP and Alpha Wire) (continued next page)

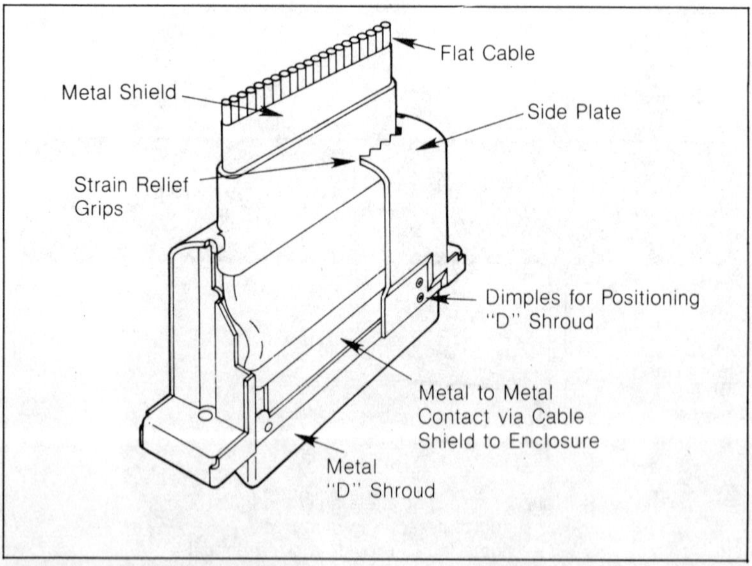

Figure 11.31—(continued)

Given a certain objective of shield effectiveness (SE), select a cable which provides shielding performance equal to or better than the desired SE. Then, knowing that the termination will always compromise the shield SE to some degree, select a termination method such that its contribution to the overall shield leakage is tolerable.

For instance, if a cable shield has an intrinsic performance of −40 dB (1 percent of coupling) at a given frequency, considering the principle of superimposition, the jumper alone should not cause more than −60 dB (0.1 percent of coupling) if we want the initial objective to be met. In fact, in this example the overall result will be −39.1 dB (1.1 percent).

An analytical approach to shield grounding jumper coupling is feasible but beyond the scope of this book. As a simplified method, we can consider that the effectiveness of a shield, from dc to upper VHF, can be predicted from its transfer impedance Z_t which approximately tracks its dc resistance (at least for single-layer shields). Therefore, given a total shield resistance R_s, we can deduce that if in the whole Ohm's law region of the Z_t curves (that is, up to

100 kHz) the grounding jumper impedance Z_{gnd} does not exceed 0.1 times R_s, the shield performance will be maintained. Beyond this, both Z_{gnd} and Z_t will vary in frequency with the same slope. Since $Z_{gnd} = R + j\omega L$, the simple procedure would be to:

1. Identify the transfer impedance Z_t in ohms per meter of the shield being used, and multiply by the cable length to obtain the total shield transfer impedance, $Z_T = Z_t \times l$.
2. Identify the highest frequency F_x where EMI needs to be suppressed by the given amount.
3. Find the jumper length whose impedance at frequency F_x does not exceed $0.1 \times Z_T$. This can be found from wire impedance nomograms or calculated quickly by assuming a jumper inductance of 1 μH/m.

To help in quickly establishing the maximum permissible jumper length, Table 11.3 has been devised. For instance, if one uses 10 m of a mediocre shielded cable (e.g., microphone cable) against a 100 kHz EMI threat, up to 5 cm of jumper length can be tolerated without performance degradation. But if excellent shielding is needed and a cable with a shield resistance of only 5 mΩ/m is used, only a 0.75 cm pigtail is permitted at 100 kHz. Above this frequency, no jumper termination at all can be used since the short length requirement would be impractical. Integral clamping or bonding should be used instead.

Note that as the length increases, jumper length requirements are relaxed since the SE of the cable decreases. For termination methods using a metallic connector or fitting, the effectiveness of the method can be graded by the transfer impedance of the connector assembly.

Remarkably, the concept of transfer impedance can be perfectly extended to the connecting hardware. In this case, Z_t for a threaded, bayonet, etc. connector **assembly** is defined as the ratio of the voltage appearing in the inner circuit to the current circulating on the outer surface of the connector body, once the two mating parts have been assembled. Figure 11.32 shows some typical data on transfer impedance of some connectors. Contrary to that of a shielded cable, Z_t for connectors does not need to be expressed per meter length; therefore, it is simply expressed in ohms or dB ohm.

Here again, the transfer impedance of a metallic connector is a

Table 11.3—Length for a Shield Grounding Jumper vs. EMI Frequency Range and Type/Length of Shielded Cable (AWG 16, or 1.3 mm Dia. Jumper)

Cable Shield dc Resistance		Total Cable Length	SE Range Defined At < 100 kHz	Max. Length of Tolerable Pigtail at Highest EMI Frequency of:				
				1 kHz	10 kHz	100 kHz	1 MHz	10 MHz
Poor	50 mΩ/m	1 m	60 dB	15 cm	7.5 cm	0.75 cm		
		10 m	40 dB	1.50 m	75 cm	7.5 cm	0.75 cm	
		100 m	20 dB	no limit	7.50 m	75 cm	7.5 cm	0.75 cm
Good	16 mΩ/m	1 m	70 dB	5 cm	2.5 cm			
		10 m	50 dB	50 cm	25 cm	2.5 cm		
		100 m	30 dB	no limit	2.50 m	25 cm	2.5 cm	
Excellent	5 mΩ/m	1 m	80 dB	1.5 cm	0.75 cm			
		10 m	60 dB	15 cm	7.5 cm	0.75 cm		
		100 m	40 dB	1.50 m	75 cm	7.5 cm	0.75 cm	

more reliable measure of its EMI immunity than the traditional SE approach. Measuring the SE of a connector would require a radiated field susceptibility test setup, where a shielded cable is installed with a portion of the shield stripped, corresponding to the connector length. Then a second measurement is made with the connector installed. The comparison of the two voltages induced in the cable with and without the connector gives the SE of the connector.

Using Z_t of a connector allows a systematic approach to cable shielding by ensuring that the connector Z_t is not worse than the cable Z_t. It would be wasteful, for instance, to specify a shielded interconnection with 2 m of cable having a nominal Z_t of 3 mΩ/m at frequency f_x, resulting in a Z_t of 6 mΩ for the cable while using

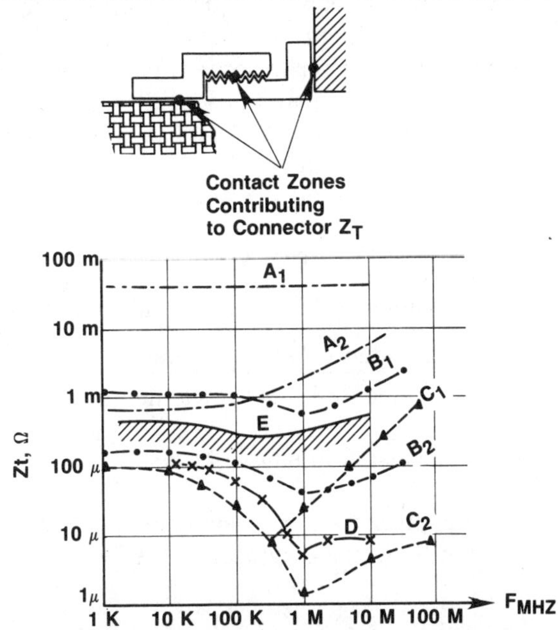

A_1 Worst Case Multi-Pin Connector (Bayonet or Other Non-Thread Style)
A_2 Best Case Multi-Pin Connector (Threaded Shell)
B_1 Worst Case BNC (Sideway Traction)
B_2 Best Case BNC
C_1, C_2 N Connector (Worst Case & Best Case)
D SMA Connector
E Range of EMP-Hardened Connectors (Single Shell)

Figure 11.32—Measured Transfer Impedance Z_t of Standard Coaxial Connectors, Mating Face Only (Bulkhead and Cable Clamp Not Included)

a connector which, for the same frequency f_x, has a Z_t of 10 mΩ or more.

11.6 Relationship between SE and Z_t

Since there is a large data base of shield performances still given in terms of SE_{dB}, it is interesting to have a way of correlating SE with Z_t so that when either one is known, we can derive the other without a need for repeating many measurements. Another reason for wanting a correlation is that the shielding performance of a box is always expressed by its SE. For instance, when doing a system EMC analysis, it is a great advantage to be able to rate cable immunity in the same terms as the individual box immunity. In doing so, any weak points and discrepancies between cable and box performances are immediately apparent. Although some authors argue that such correlation is not dependable, it has been given more and more attention in recent years.

11.6.1 Simplified Theory

If we accept the statement that the SE of the cable is the ratio of:

$$\frac{\text{V induced before shielding}}{\text{V induced after shielding}}$$

this ratio can be calculated in simple terms.

For instance, in the presence of a given magnetic field, expressed in amps per meter, the voltage induced in the loop $l \times h$ without shield is:

$$V_1 = 4\pi \times 10^{-7} \times H_{A/m} \times 2\pi f_{Hz} \times l_m \times h_m$$

for, l and $h < \lambda/2$

When a shield is added, this same voltage V_1 is now impressed on the shield. We know, from the transfer impedance concept explained in Section 11.3, that only a fraction of this voltage will appear inside the shield. This voltage is:

$$V_i = Z_T I_S$$

where Z_T is the **total** transfer impedance of the shield.
If the impedance of the large loop is termed Z_{01} we have:

$$I_s = \frac{V_i}{Z_{01}}$$

therefore,

$$V_i = Z_T \times \frac{V_1}{Z_{01}}$$

and,

$$SE = \frac{V_1}{V_i} = \frac{V_1}{Z_T \times V_1/Z_{01}} = \frac{Z_{01}}{Z_T}$$

$$SE_{dB} = 20 \log Z_{01} - 20 \log Z_T \qquad (11.2)$$

If the cable of length l has a transfer impedance equal to $Z_{tx}\Omega/m$ at frequency f_x,

$$SE_{dB} = 20 \log Z_{01} - 20 \log(Z_{tx}\ \Omega/m \times l_m) \qquad (11.13)$$

Predictably, this indicates that SE and Z_t evolve in opposite ways: the lower the Z_t, the better the SE.

Illustrative Example 11.6

A 3 m shielded cable is installed 5 cm above a conductive ground. The characteristic impedance Z_{01} of this line above ground is approximately 100 Ω. The transfer impedance of this type of shield is −16 dBΩ/m at 30 MHz. Find the equivalent SE:

$$SE_{dB} = 20 \log(100) - (-16 \text{ dB} + 20 \log 3)$$

$$= 40 + 16 - 10 = 46 \text{ dB}$$

Among the criticisms directed at the possibility of correlating SE with Z_t is the fact that the SE seems to depend on cable length, while in the conventional definition of SE (see Chapters 1 and 2), only the material type and thickness should be relevant. However, it must be considered that the length l is only entered to calculate the effective Z_T of a particular piece of shield from the reference length of 1 m, i.e. to integrate the combined effect of all the shield apertures.

11.6.2 A More Complete Model for the SE-Z_t Relationship

Several authors have approached the relation between SE and Z_t in a very rigorous manner. The transfer impedance, as has been shown in Fig. 11.3, can be related to SE in the following way.

Let us treat the outside line in Fig. 11.3 as transmission Line no. 1, and the inside of the shield (the intentional path) as transmission line no. 2. The power into line 1 is $P_{out} = I_s^2 Z_{01}$, where Z_{01} is the characteristic impedance of line 1. The power at the end of line 2 is $P_L = V^2_L/Z_{02}$, where Z_{02} is the characteristic impedance of line 2.

Assuming that the cable length is shorter than λ, and neglecting the capacitive leakages through the shield, it has been established that:[2,3]

$$Z_t = \frac{\sqrt{Z_{01} Z_{02}}}{l(1+K)} \sqrt{\frac{P_L}{P_{out}}} \times \left[\frac{4(1 - \rho_{12}\rho_{22})(1 - \rho_{11}\rho_{21})}{(1+\rho_{22})(1-\rho_{11})(1-\rho_{12})(1-\rho_{21})} \right] \quad (11.14)$$

where,

$K = Z_g/Z_{01}$
Z_g = impedance of the source driving line 1
l = cable length

ρ_{11} and ρ_{12} = near-end and far-end reflection coefficients in line 1

ρ_{21} and ρ_{22} = near-end and far-end reflection coefficients in line 2

If we now define the last term of the equation as a determinant D, and if we recognize that the actual expression for SE is:

$$SE = \frac{P_{out}}{P_L}$$

we see that Eq. (11.14) can be rewritten as:

$$Z_t = \sqrt{Z_{01} Z_{02}} \times \sqrt{\frac{1}{SE}} \left(\frac{D}{l}\right) \tag{11.15}$$

or

$$SE = \frac{Z_{01} Z_{02}}{Z_t^2} \left(\frac{D}{l}\right)^2 \tag{11.16}$$

That is, in dB:

$$SE_{dB} = 10 \log \frac{P_{out}}{P_L} = 10 \log Z_{01} + 10 \log Z_{02}$$

$$+ 20 \log D - 20 \log l - 20 \log Z_t \tag{11.17}$$

The first term depends on the height of the cable above the surrounding ground. The second term depends on the internal construction of the cable. The third term characterizes the dependence of SE on the way the shield and the conductors are terminated. The two last terms in fact correspond to the actual transfer impedance Z_t in ohms per meter × l in meters.

A typical range for Z_{01} is 70 to 300Ω (corresponding to a height-to-diameter ratio of 1 to 50). An average value for h/d = 3 is 120 Ω. Z_{02} falls in line with the most standard values for coaxial cables, i.e., 50 Ω, 75 Ω, 90 Ω and so on. Therefore, the term 10 log (Z_{01} Z_{02}) can be assigned a typical range of 36 to 42 dB with an average (default) value of 38 dB. The term 20 log D depends on how well transmission lines 1 and 2 are matched at their respective ends. In general, line no. 2 will be correctly matched at source and load ends for maximum power transfer, so ρ_{11} and ρ_{12} typically will be small.

In contrast, line no. 1 will be almost always mismatched (shorted or open ended) at both ends, ending in ρ_{21} and ρ_{22} being rather large, like ± 0.8 or 0.9. Even though they vary for each installation, a reasonable worst case for 20 log D is 4 dB.

Entering these approximate values into Eq. (11.17) results in:

$$SE_{dB} = 38 + 4 - 20 \log Z_{t\ \Omega/m}\ l_m$$

$$= 42 - 20 \log Z_t\ l \qquad \text{for } L \leqslant \lambda/2 \qquad (11.18)$$

$$\approx -2 - 20 \log Z_t + 20 \log f_{MHz} \qquad \text{for } L > \lambda/2 \qquad (11.19)$$

The last expression results of the multiplication of Z_t by 150/f when $l > \lambda/2$. Obviously, if the designer knows more exactly the value of Z_{01}, Z_{02} and the reflection coefficients in his application, he should use them for a more exact prediction. Incidentally, one sees that if it were not for the mismatch term 20 log D of 4 dB, this expression would be about the same as our simplified model of Eq. (11.13) for an external line characteristic impedance of 100 Ω.

11.6.3 Influence of Aging on Cable Shields

The degradation in service of the effectiveness of a cable shield is due essentially to repeated flexions or mechanical abuse and, to some degree, to the corrosion of the braid where exposed to atmospheric agents. Both flexion and corrosion cause the contact resistance between the many strands of the braid to increase and the leakage (or transfer) inductance L_T to increase because some apertures get larger.

A limited data base exists on these aging effects. Figures 11.33 and 11.34, for instance, show the degradation of Z_T over 10 years of intensive service for several single braid, foils and double braid coaxial cable combinations. The double braided cables, being naturally more rigid, seem to stabilize better.

Overall, the degradation over 10 years ranges from a 5 to 30 times increase in Z_T, with an average value of 10 times; i.e. a 25 percent increase (or 2 dB) per year. Therefore, it seems prudent to anticipate a 2 dB/year loss in SE for any cable shield which is not mechanically held in place and has no insulating top jacket.

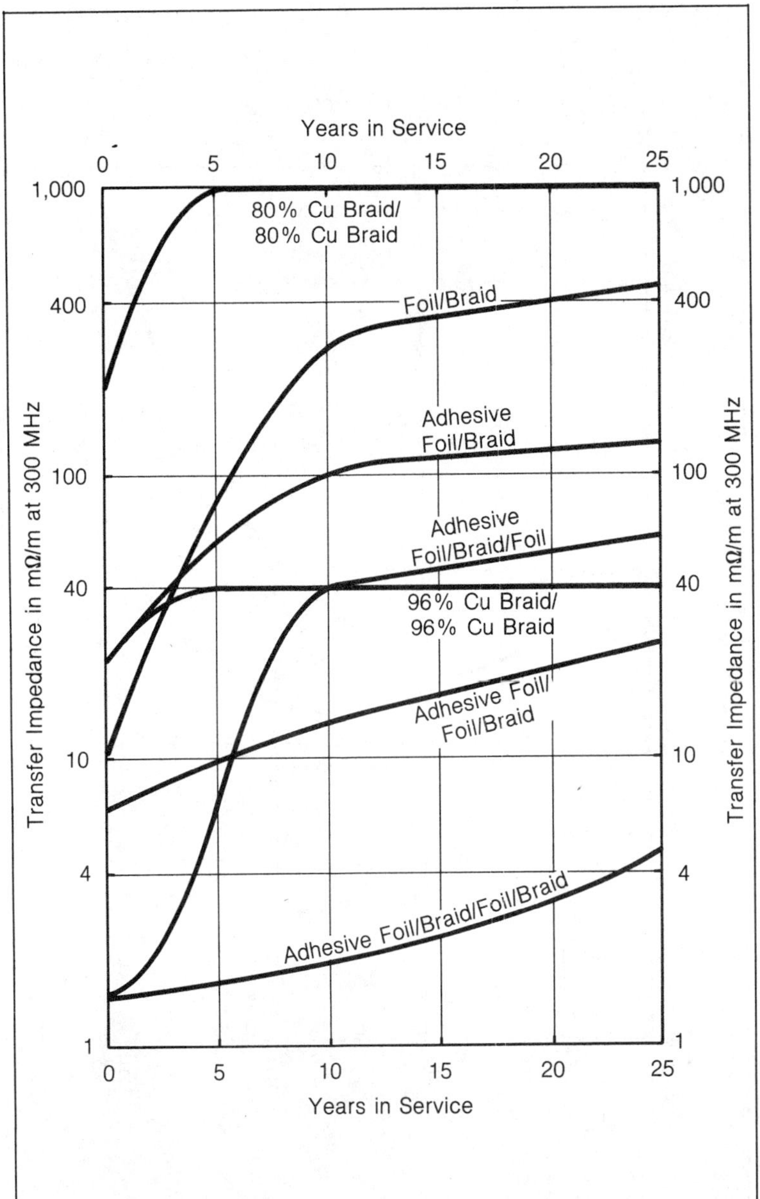

Figure 11.33—Aluminum Foil/Braid Transfer Impedance versus Frequency (courtesy of *EMC Technology*)

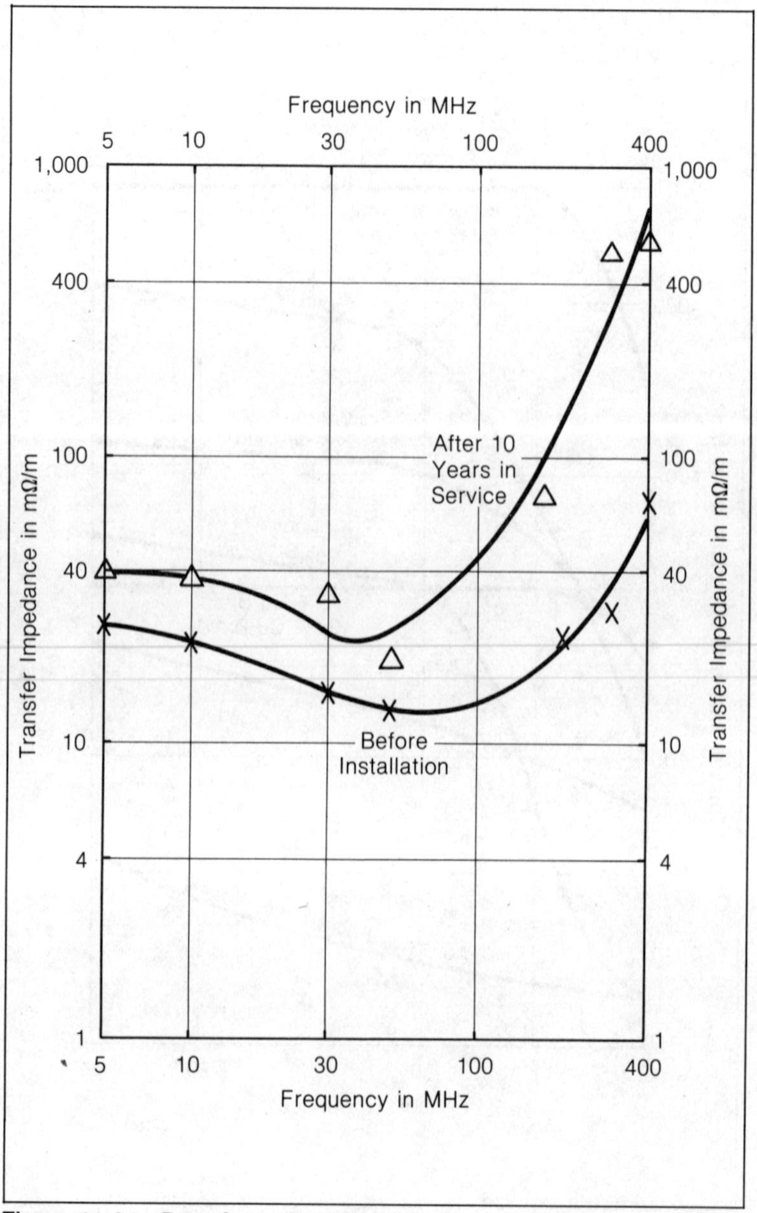

Figure 11.34—Drop Cable Predicted Transfer Impedance versus Years in Service (courtesy of *EMC Technology*)

11.7 References

1. Ott, H.W., Noise Reduction Techniques in Electronic Systems (New York: John Wiley & Sons, Inc., 1976)
2. DeGauque, P., "Transient Response of Braided-Wire Shields," *1979 IEEE EMC Symposium Record* (New York: IEEE).
3. Mardiguian, M., "Transfer Impedances of Balanced Shielded Cables," *EMC Technology,* July 1982.

11.8 Bibliography

Bridges, J.E. and Zalewski, R.A., "Magnetic Field Pickup by Flexible Braid Coaxial Cables," *IEEE Transactions on EMC*, Vol. EMC-10, no. 1, March 1968, p. 130.

DeGaùque, P. and Demoulin, B., "Tensions Perturbatrices Induites sur un Cable Bifilaire Blinde," 1983 French EMC Symposium, Tregastel, France.

DeMilt, M.D.; Loh, Y.P.; and Showers, R.M., "Resonance Properties of the Shield of a Coaxial Cable over a Ground Plane" (Abstract), *IEEE Transactions on EMC*," Vol. EMC-10, no. 1, March 1968, p. 135.

Ervin, H.W., "Shield Termination Prediction Method," *Proceedings of the Eighth Tri-Service Conference on Electromagnetic Compatibility,* Armour Research Foundation, Chicago, October 1962.

Farhat, N.H.; Loh, Y.P.; and Showers, R.M., "Effects of Partial Shields on Transmission Lines at Low Frequencies," *IEEE Transactions on EMC,* Vol. EMC-10, no. 1, March 1968, pp. 42-51.

Fisher, F.A., "Changing-H Field on Shielded Conductors," *1977 IEEE EMC Symposium Record* (New York: IEEE).

Fowler, "Screening Measurement in Time and Frequency Domains," *IERE Journal UK*, April 1985.

Hager, J.J. and Van Brunt, M.K., "Advanced Cable Shield Termination Process for Broad-Band EMI Suppression," *1986 IEEE EMC Symposium Record* (New York: IEEE), p. 63.

Hoeft, L., "Transfer Impedance of Backshells," *1982 IEEE EMC Symposium Record* (New York: IEEE), Santa Clara.

Homann, K., "Screened Cables with Optimum Braids," *Nachrichtentechnische Zeitschrift*, 1968, no. 3.

Liaw, Y.S., "Calculation of Shielding Effectiveness of Cylindrical Cable at Low Frequencies," M.S. Thesis, Moore School of Electrical Engineering, University of Pennsylvania, Philadelphia, 1967.

Loh, Y.P., "Shielding Theory of Coaxial Cylindrical Structures," *IEEE Transactions on EMC*, Vol. EMC-10, no. 1, March 1968, pp. 42-51.

Martin, A., "Introduction to Surface Transfer Impedance," *EMC Technology*, July 1982.

"Measured Degradation of Shielding Effectiveness by Pigtailing," *IEEE Transactions on EMC*, February 1986, p. 40.

Pearlston, C.B., "Case and Cable Shielding, Bonding and Grounding Considerations in Electromagnetic Interference," *IRE Transactions on Radio Frequency Interference*, Vol. RFI-4, no. 3, October 1962, pp. 1-6.

Peel, R., "Relation Between ZT and SE," *1982 IEEE EMC Symposium Record* (New York: IEEE), Santa Clara.

Schelkunoff, S.A., "The Electro Magnetic Theory of Coaxial Transmission Lines and Cylindrical Shields," *Bell Systems Technical Journal,* 1934, no. 4, p. 532-579.

Schor, F.W., "Measurement of RF Leakage in Multipin Electrical Connectors," *IEEE Transactions on EMC*, Vol. 10, no. 1, March 1968, pp. 135-141.

Smith, K., "Drop Cable RF Leakage throughout 20 Years of Service," *EMC Technology*, July 1982.

Vance, E.F., "Coupling to Shielded Cables," (New York: John Wiley and Sons, 1979).

Walters, G.E.; Van Brunt, M.K.; and Hager, J.J., "Shielded Electrical Connectors: Performance and Testing," *EMC Expo 86 Proceedings,* p. E05.1.

Chapter 12

Cabinets, Shielded Rooms and Large Enclosures

12.1 Shielded Cabinets

This section discusses the repercussions of increasing the physical size of shields from the chassis or box-size equipment of the previous chapter to console-size units. Shielded cabinets presented in this section include the 19" rack, single or double-bay consoles and boxes occupying several cubic feet.

An example of a typical shielded cabinet is shown in Fig. 12.1. It is noteworthy that the cabinet appears to be similar to most other racks or consoles which have not been EMI hardened. Both provide some degree of shielding, but an EMI-shielded cabinet distinguishes itself from other cabinets, typically in the following ways:

1. The rear access door uses a continuous piano hinge for better door-to-frame bonding by equalizing the pressure along the hinge side.
2. The rear door is EMI gasketed all around to provide shielding integrity when closed.
3. The cabinet frame enclosure is usually seam-welded to provide a continuous homogenous bond.
4. The cabinet is often made of light-gauge steel to provide im-

Figure 12.1—Example of Commercially Available Shielded Cabinets (courtesy of Equipto Electronics Corp.)

proved magnetic shielding at low frequencies in contrast to that offered by aluminum.

5. Separate heavy power and signal grounding busses are run vertically up the rear side walls.
6. Large strike-plate holes on the backs of drawers and a dagger-pin arrangement on the back of cabinet grounding busses result in lower common-mode impedance problems. In lieu of this, jumper wires may be installed between the insulated grounding bus and drawers.

Finally, shielding effectiveness (SE) performance of three differently configured racks is shown in Fig. 12.2. Note the substantial improvement in SE produced by proper gasketing and seam treatment.

12.2 Shielded Rooms and Enclosures

This section continues with shielded housings but at the increased size level of the shielded-room configuration. Here sizes may range from a 2.5 × 2.5 × 2.5 m small room to a large aircraft hangar. The **shielded room** or **enclosure** has been in use for many years for performing electronic measurements where a low electromagnetic ambient is required or where potentially damaging emissions must be contained. (The term **screen room** is still used in some

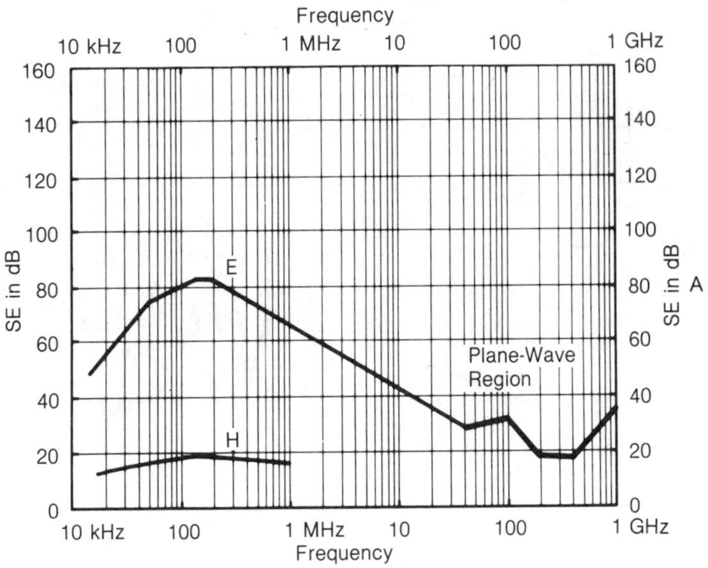

a. Rigid, Reinforced Cabinet, No Gasketing

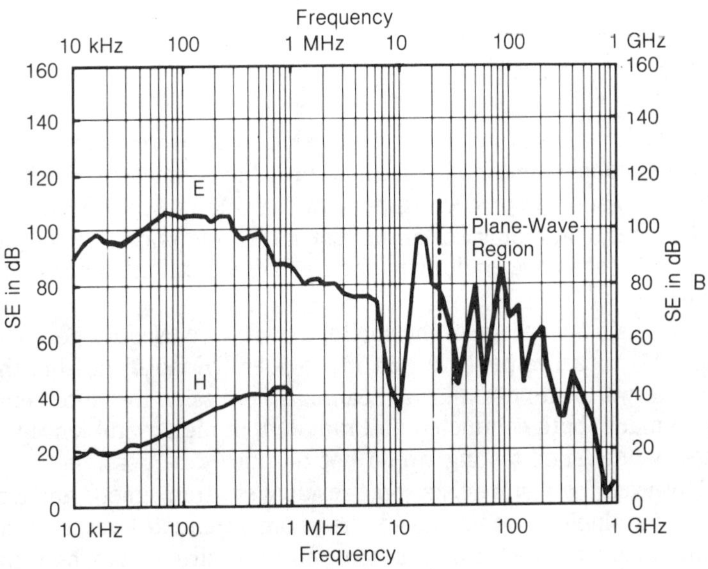

b. Stainless Steel Cabinet With 3 cm Fingerstock Strips Every 17 cm On All Joints

Figure 12.2—Measured SE of 19" × 70" (0.48 m × 1.75 m) Steel Cabinets per MIL-STD-285, 30 cm Distance (continued next page) (courtesy of Equipto Electronics Corp.)

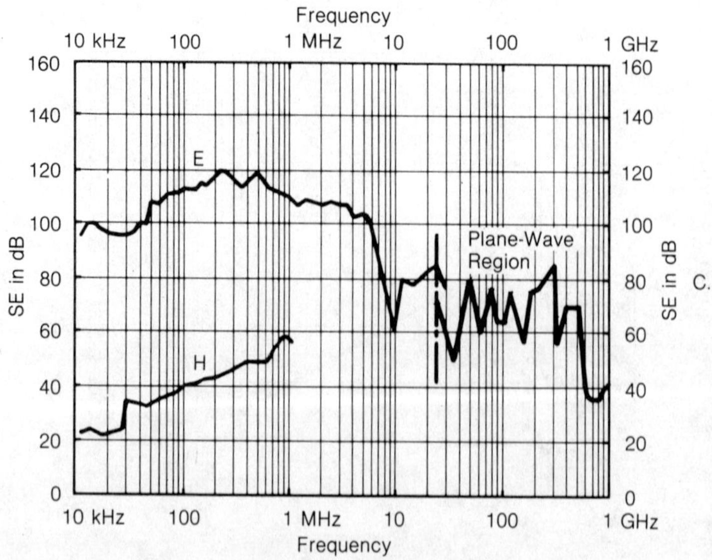

c. Same as B, but 100% Fingerstocks and Honeycomb Air Filters

Figure 12.2—(continued)

circles. It comes from the early shielded enclosures which were made of copper screen.) Its use has spread to nonmeasurement applications such as protecting personnel working near high-power radar sites, containing certain industrial RF emission sources and protecting sensitive equipment such as biomedical instruments and computers.

The main advantage of the shielded enclosure used for performing EMI measurements is that it provides RF isolation from the outside world. Its use allows meaningful emission measurements to be made, both conducted and radiated, in higher ambient locations where such testing would not ordinarily be possible.

However, there still exists some residual electromagnetic ambient inside a shielded room, since the room attenuates rather than eliminates outside-world emissions. In most situations where the outside RF environment is not abnormally high, a modern shielded room provides sufficient attenuation to reduce all outside emissions to levels below the sensitivity of typical receivers connected to their normal antennas. Magnetic-field shielding at and below ELF is the exception. For example, the enclosure walls offer little attenuation to 60 Hz fields, and input power lines allow easy emission entry. To reduce 60 Hz magnetic fields and their first few harmonics, one must use **extended-frequency-range shielded enclosures.** (This is a term which has been adopted by the EMI community for shielded enclosures to imply significant attenuation to magnetic fields at power-line frequencies.)

12.2.1 Enclosure Types and Sizes

Some distributed electronic systems are physically too large to employ shielded enclosures for testing. However, shielded enclosures have been built to enclose aircraft hangars and other large structures. One example is the entire computer facility that was built for early Atlas ICBMs, located in a large shielded enclosure such as shown in Fig. 12.3. Some structures are lined with absorbent material to form an anechoic chamber as shown in Fig. 12.4. RF anechoic chambers are specifically addressed in Volume 6, "EMI Test Methodology and Procedures," of this handbook series.

Figure 12.3—Example of a Large Shielded Enclosure

Figure 12.4—A Large Shielded Anechoic Chamber

Mobile shielded enclosures are also available which can be used as roving electronic laboratories; i.e., shielded enclosures constructed in a trailer or van. This subject is discussed in Section 12.3.

Either steel or copper is usually employed as the basic material in the shielded enclosure. Steel is generally used in the form of galvanized sheet as shown for the enclosure in Figs. 12.5 and 12.6, while copper is used in either sheet form or as fine mesh screen. The best material in a given application depends on weight restrictions, shielding desired, cost and other variables. Steel generally provides comparable performance to copper at radio frequencies down to 150 kHz. Below this, the permeability of steel begins to provide better magnetic-field shielding than nonpermeable copper.

The first of the modern shielded-enclosure construction methods was developed a number of years ago at the Naval Air Development Center. This room was made with two layers of copper screen, separated by a 25.4 mm (1") frame. The room was constructed of several panel edges, called cells, each measuring 1.22 × 2.44 m (4' x 8'). Individual panel edges were butted together and were bolted through the wood frame that provided the shape and strength for each panel.

One feature of this room was that it could be disassembled, moved and reassembled at another location without major modifications.

Figure 12.5—Typical Double-Shielded Room

Figure 12.6—Solid, Double-Wall Enclosure

The door was well braced, framed with wood and also covered with copper screen. Two sets of spring finger stock were attached around the periphery of the door; one to provide contact with the inner edge of the doorjamb, and the other set to push against the outer edge of the door frame. The second set of spring fingers actually overlapped the door frame opening.

A similar enclosure construction is the cell type which employs only a single layer of either screening or sheet metal. This enclosure is not widely used since its shielding performance is less satisfactory and it is almost as costly as the double-layer, cell-type. Figure 12.7a illustrates the three more common types of demountable shielded enclosure construction.

There appears to be a controversy over the purported advantages of completely separate layers in the double-shield, isolated-type room as opposed to the cell-type room. Installations show little inherent advantage in either type of construction where the same total thickness of the same metal is used. Where bolted seams are used in both, the double-wall isolation construction may be more effective due to the second opportunity to clamp seams closed.

Properly-made welded seams (see Fig. 10.2) are favored in single-shield construction because electromagnetic leaks are virtually eliminated and there is no possibility of intra-wall resonances. The trick is to seam weld without developing voids which may not be seen by the eye. This type of construction uses a single sheet of steel on a metal framework. The sheet is under some tension because of the manner in which it is welded to the frame, and the frames or panels are welded together. Seam-welded enclosures cannot be moved without destroying them in the process. Thus, those ordering an enclosure of the seam-welded type should be sure that there will be no future requirement for demounting it and erecting it elsewhere (see Fig. 12.7b).

For double-layer modular enclosures (i.e., the double-shielded room), a "sandwich panel" is made with two steel sheets bonded to a 19 mm (0.75") plywood core. The panels are not butted together all the way, but are clamped on each edge by special continuous channels and strapping. The method of joining panels both along the sides of a room and at the corners is shown in Fig. 12.8. Machine screws pull the channel and strap together every few inches.

A different approach to the above sandwich-type enclosure is a single solid-wall type in which the shield material is 3.2 mm (0.125")

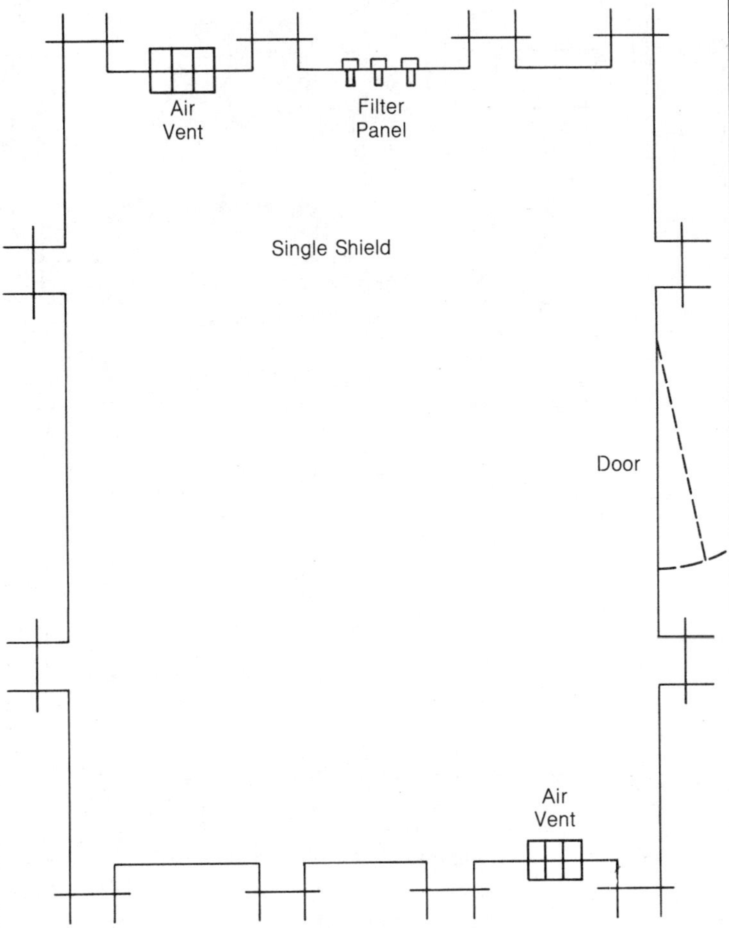

a. Bolted or Clamped-Together Enclosures
"Example 1"

Figure 12.7—Types of Demountable and Seam-Welded Shielded Enclosures (continued next page)

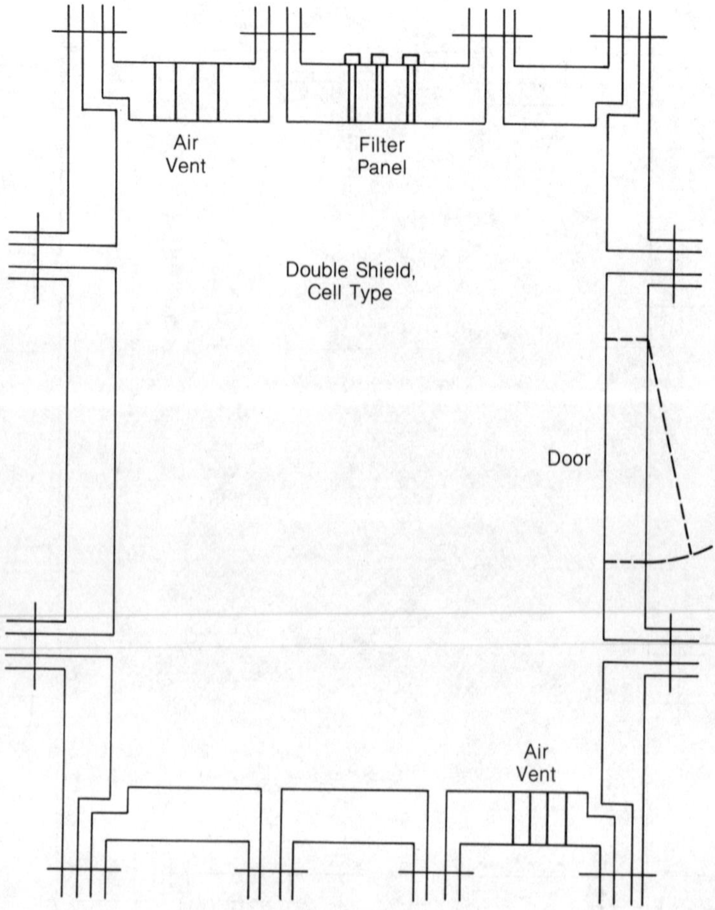

a. Bolted or Clamped-Together Enclosures
"Example 2"

Figure 12.7—(continued)

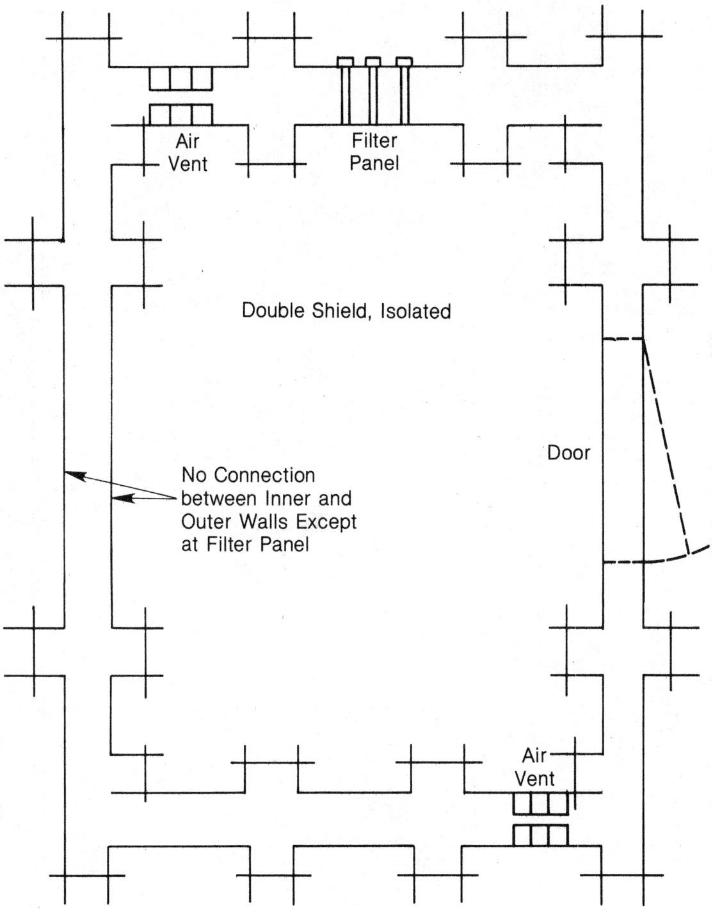

a. Bolted or Clamped-Together Enclosures
"Example 3"

Figure 12.7—(continued)

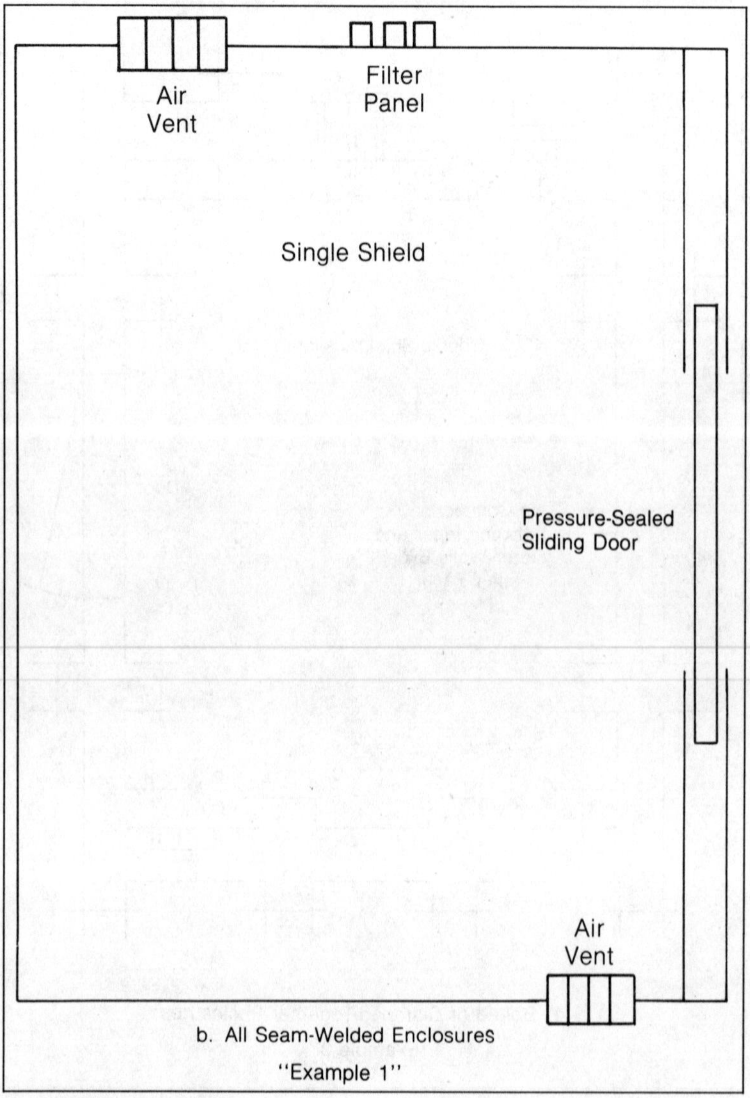

Figure 12.7—(continued)

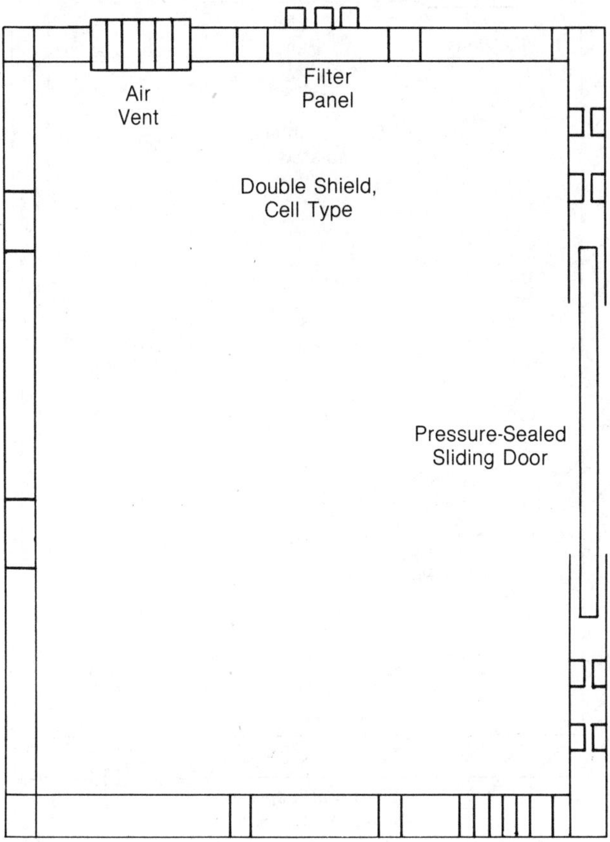

b. All Seam-Welded Enclosures
"Example 2"

Figure 12.7—(continued)

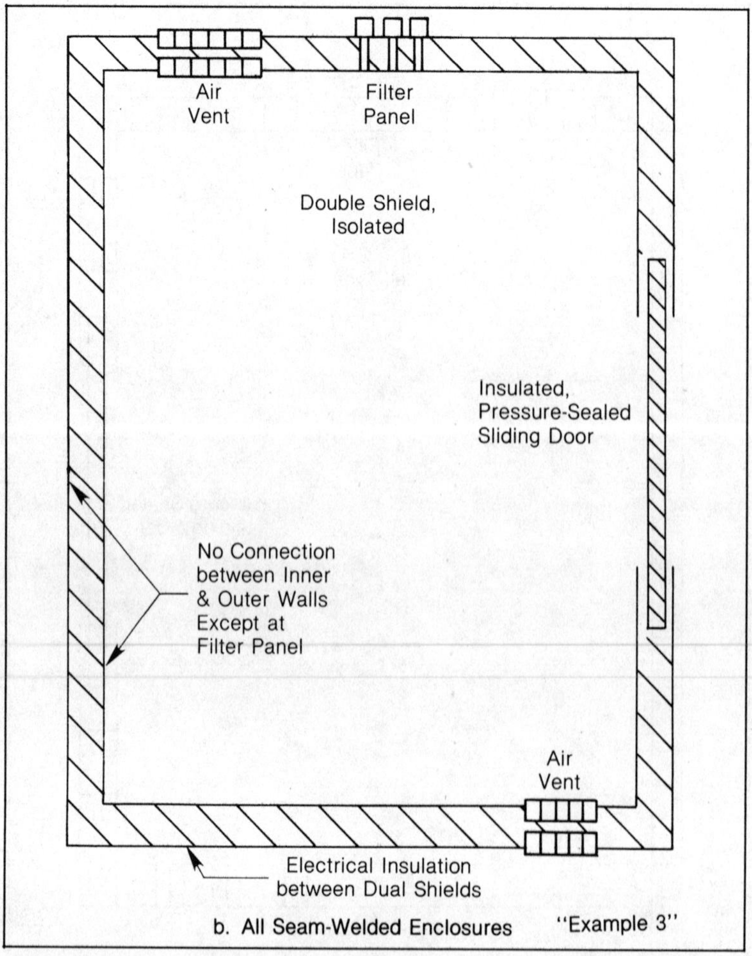

Figure 12.7—(continued)

thick-rolled steel which includes "U"-channels and "U"-tensioners as shown in Fig. 12.9. All fittings, bolts, screws and spline nuts used to fasten the framing members together are plated steel.

Variations are used by several shielded-enclosure manufacturers. Some newer developments in clamp design include a preassembled, welded and interlocking three-way corner which eliminates leakage problems. Also incorporated into the design is the use of a closed thread insert to eliminate RF leakage and penetration at each of

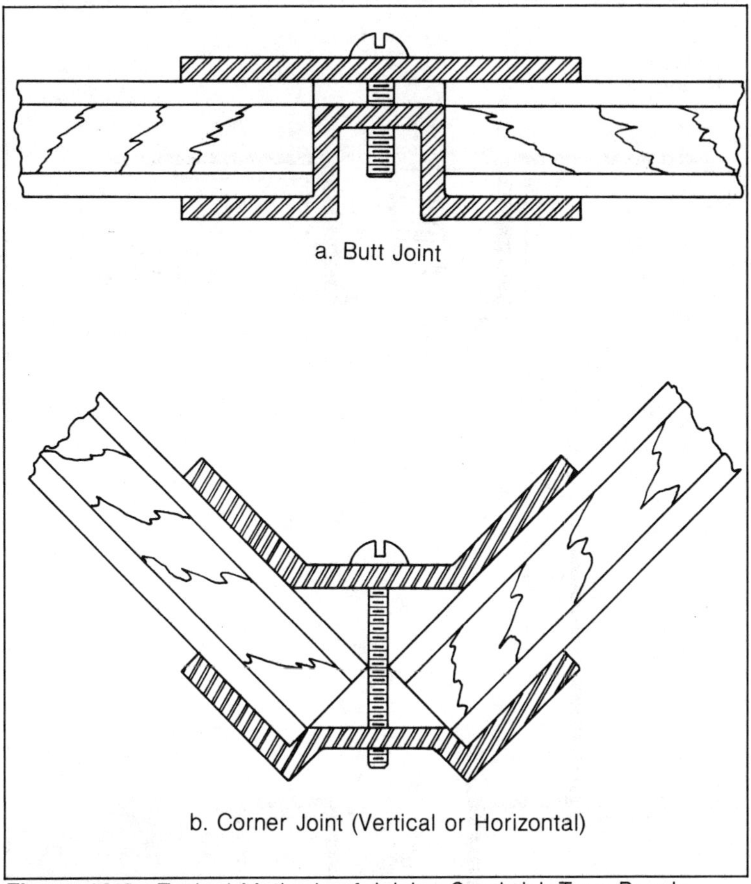

Figure 12.8—Typical Methods of Joining Sandwich-Type Panels

the clamping bolts. Typical features of this design are shown in Fig. 12.10.

Shielded enclosures require periodic maintenance if they are to retain their designed attenuation. The vulnerable areas are the joints and seams of bolted structures, the air vent shields and the door. Fasteners between panels must be kept tight. The enclosure manufacturer usually gives a torque rating on the fasteners. Exclusive of doors, no such maintenance is required on all-welded enclosures since they use no bolts for fastening.

Finger stock gasketing along the edge of the door must be kept in good condition. If fingers are damaged or broken off, a replace-

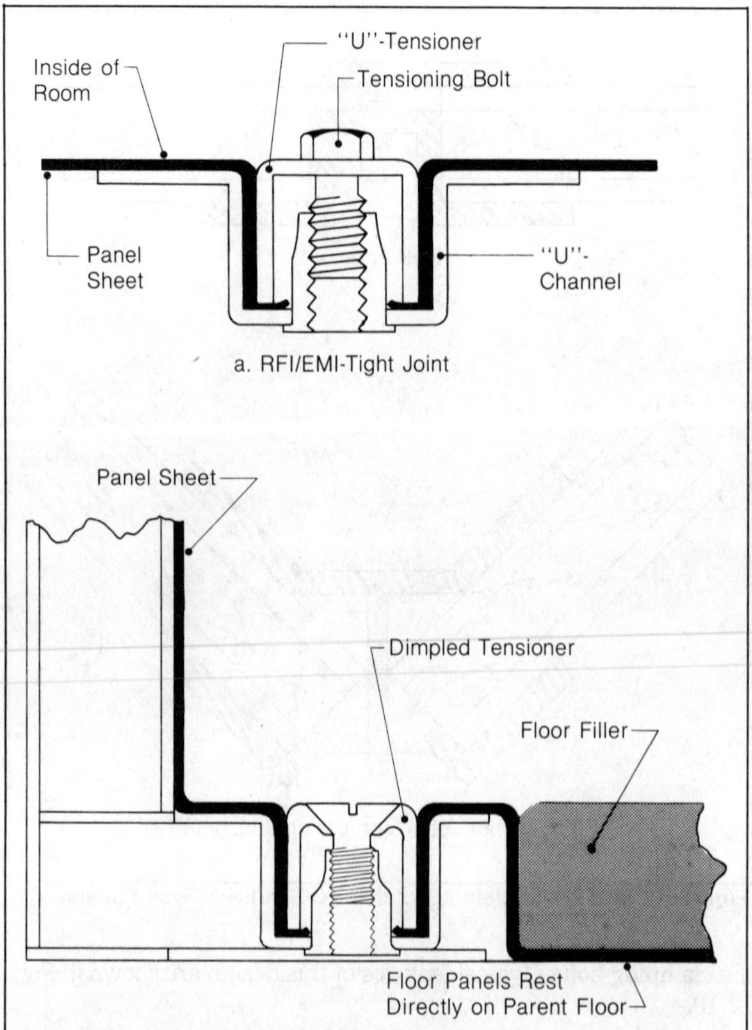

Figure 12.9—Joiners for Solid Sheet-Stressed Metal Enclosures

ment section of fingers may be soldered on or bonded in position (see Section 10.2). As discussed in Chapter 10, these fingers provide a good bond between the enclosure and the door via a wiping motion over a short distance along the door frame (see Fig 12.11). To maintain good gasketing action, the door frame must be kept

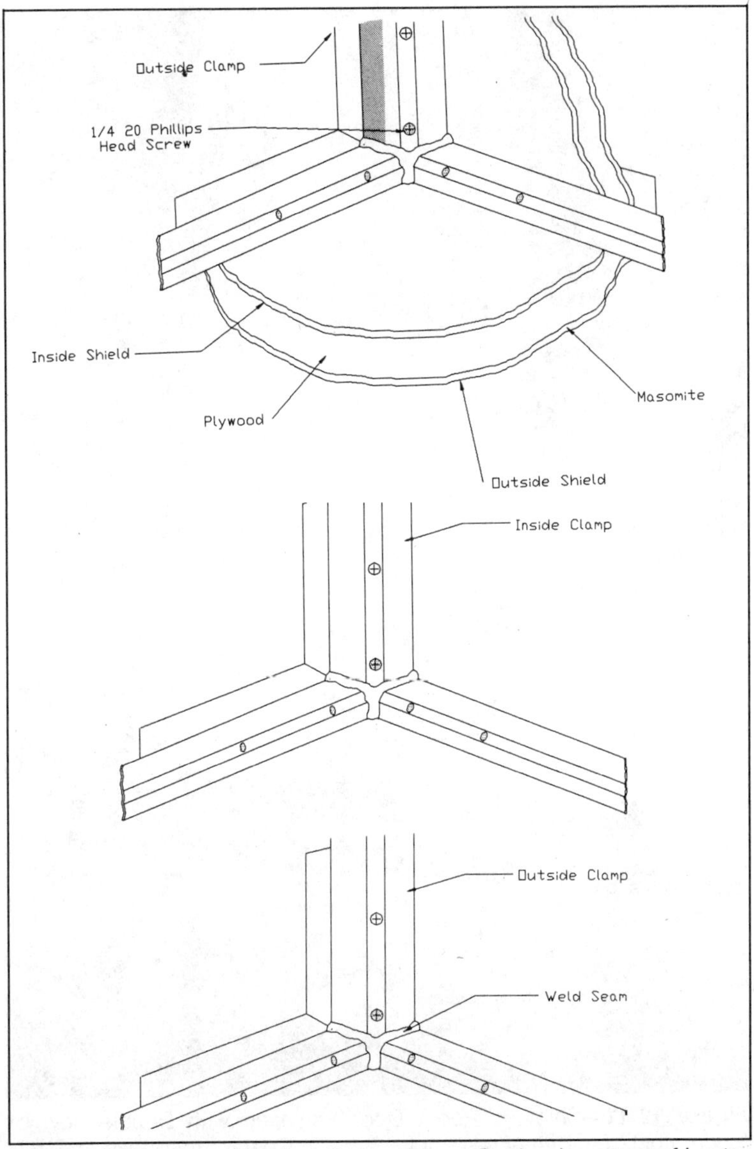

Figure 12.10—Seam-Welded, Corner-Clamp Design (courtesy of Lectro Magnetics, Inc.)

smooth and clean. Some doors use finger stock on the door frame; here the door must be kept clean of oxides and grease. An exception is the pneumatic door, which does not use finger stock.

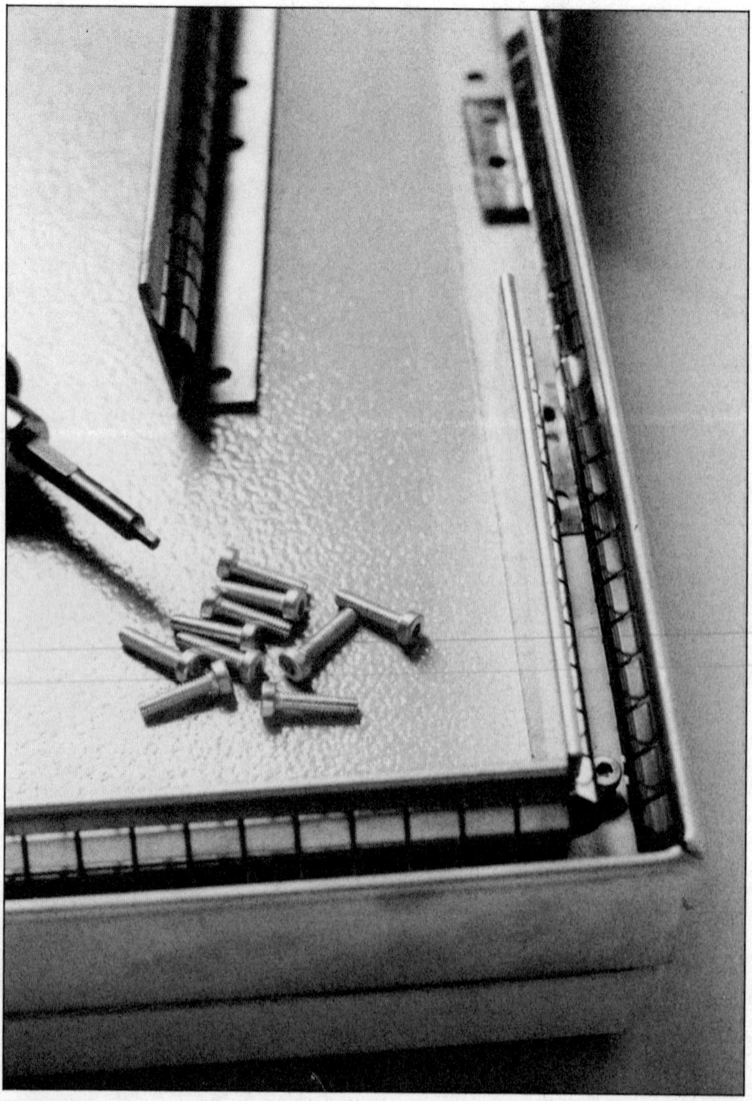

Figure 12.11—Shielded Room Door Equipped with Double Row of Fingerstock, for Knife-and-Groove Type of Contact (courtesy of SIDT, France)

Other types of shielded enclosure doors are considerably different in concept. They are stronger, with stronger frames to provide better attenuation to RF emissions around the door. They still use two rows of metallic finger stock or hidden rows to effect a good seal between the door and frame. An example is shown in Fig. 12.12. There are also improvements in the latching arrangements, so that the door may be opened and closed easier. Doors for all-welded

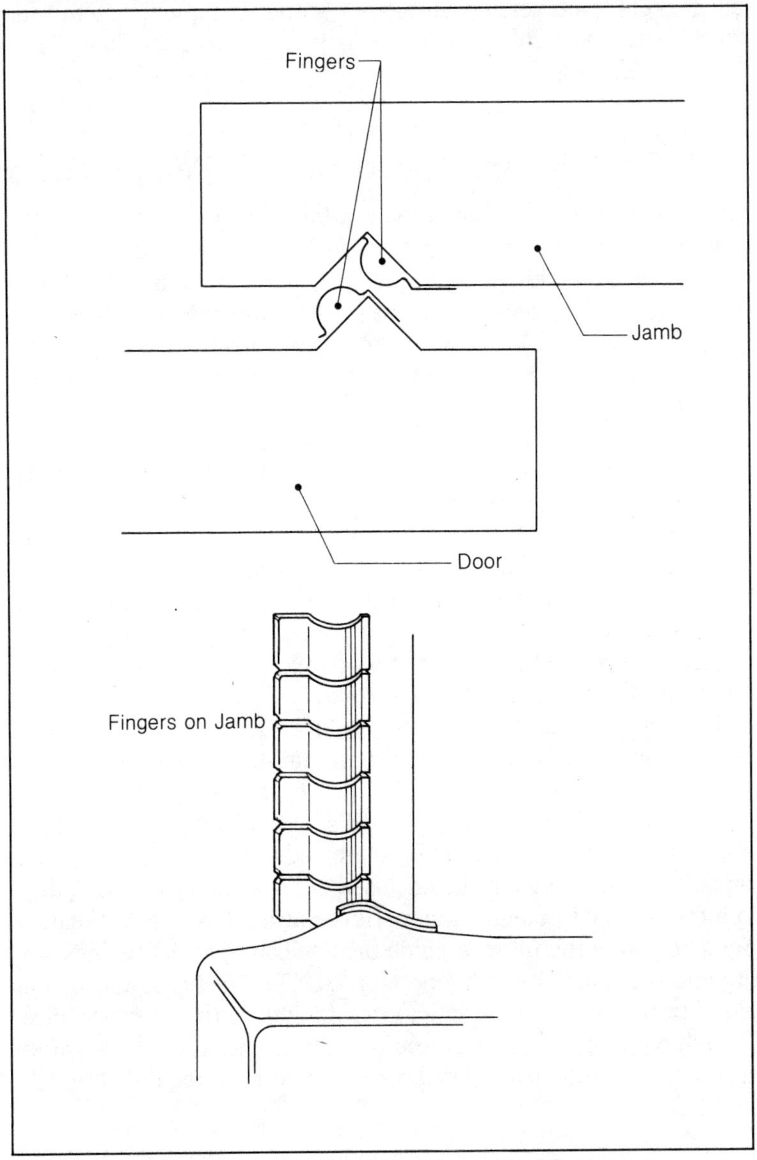

Figure 12.12—Contact Fingers of Enclosure Door and Jamb

shielded enclosures require uniform clamping or sealing pressures which surpass those achieved by the standard approach.

A very effective and reliable high-performance shielded door is the pneumatic-sealed door. This door uses neither finger stock nor RF gaskets. One version employs a rather complex system that makes use of a set of pressurized air bags to force the door edges against a mating flange.

12.2.2 Enclosure Apertures and Penetrations

Shielded enclosure attenuation is attained by the structure itself, including the wall sections, seams and door. These are of little value, no matter how well designed, if the shielding integrity of the enclosure is compromised by one or more apertures such as suggested in Fig. 12.13. Many of these aperture considerations are similar in principle to the box integrity factors covered in Chapters 7 through 10.

Other than the access door, the most prominent aperture involves the ac power service entrance and associated filtering to remove RF noise. The basic requirement for shielded enclosure filters is that they provide a certain minimum attenuation over the useful frequency range of the room, typically from approximately 14 kHz to 10 GHz. One problem is to determine how much attenuation the filters must offer to complement a particular room design or application, since the filter attenuation requirements are only roughly related to the attenuation of the room.

One approach to the problem is to provide filters with attenuation performance somewhat less than that of the room. For example, if the room offers attenuation of 120 dB over most of the frequency range, filters offering 100 dB should prove adequate. If the required enclosure performance exceeds the power line filter capability, the filters may be enclosed in a shielded electrical panel to increase low-frequency magnetic isolation. Normal installation practice places the filters outside the enclosure, with the line coming into the room through pipe nipples. The mechanical design of high-attenuation filters is an important factor in their performance. To reduce coupling between the input and output of the separate sections, well designed isolated compartments are required (see Fig. 12.14).

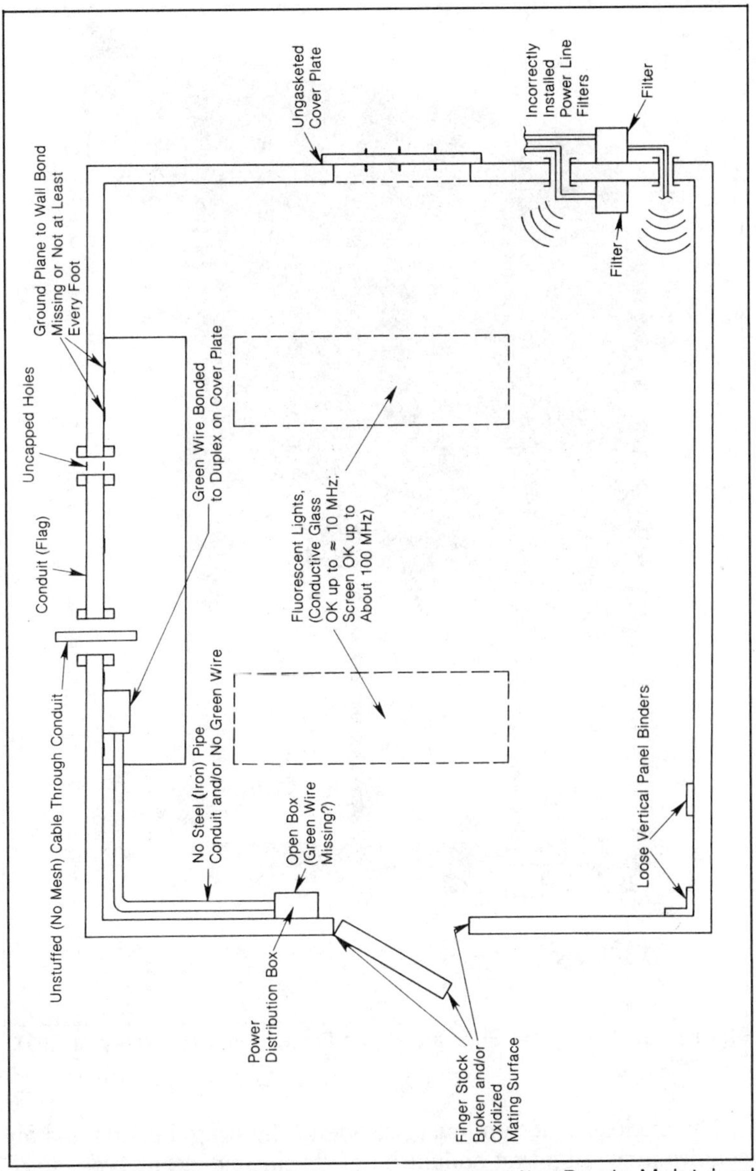

Figure 12.13—Example of Poorly Installed and/or Poorly Maintained Shielded Enclosure

Figure 12.14—Typical Shielded Room Power Filter (courtesy of SIDT, France)

The enclosure design must provide for lighting, heating and air conditioning. Lighting should be of the incandescent type, since other types usually involve ionization which produces substantial RF noise. Special fluorescent fixtures with built-in filters, shields, and conductive cover windows are available. Although this type of lighting is effective in lowering the electromagnetic ambient within buildings and laboratories, it is not particularly suitable for installation in shielded enclosures.

For heating and air conditioning, it is not generally necessary to provide additional facilities if they are available in the immediate area where the shielded enclosure is to be located. Consequently, forced-air duct work is extended from existing ducts to enter the enclosure at the top or sides.

To assure adequate forced ventilation, however, a return system is needed. This generally consists of an exhaust fan which forces interior air to the outside ambient through a honeycomb vent which protects the shielding integrity of the enclosure (see Section 10.3). Additional penetrations of the enclosure walls are often necessary to provide other services. Gas, water, compressed air, fire detector and Halon fire extinguisher lines may be furnished through steel or copper piping which acts as a WBCO. If pipe is joined to the enclosure wall in a clean, tight bond, SE of the room should not be compromised.

The same method is used to bring coaxial lines through the enclosure wall. Special fittings are available that are similar to a threaded pipe nipple, except that suitable coaxial fittings are used. Coaxial cables feeding the fittings from outside the enclosure can reduce the SE of an enclosure by providing a path of entry for EMI. This may develop when energy penetrates the coaxial cable outer conductors and is conducted into the enclosure. For this reason, triaxial or quadraxial cables (see Section 11.3) should be used. No cable will stop EMI from entering the enclosure if it exists in the driving or loading equipment located outside the room.

12.2.3 Enclosure Performance and Checkout

Well designed and properly installed shielded rooms of the modular clamp-together type (as discussed in a previous section) conform to well known shielding requirements of MIL-STD-285, USAF Class I shielding, or NSA Spec. 65-6. The SE of some modular rooms using two sheets of 24-gauge steel sandwiched on 19 mm (0.75") plywood is shown in Fig. 12.15. The performance of all-welded rooms is also shown in the figure for different classes of shielded rooms based on the thickness of the shielding steel used. The performance of all-welded rooms tends to follow the predictable magnetic field attenuation more closely, while the lesser slope

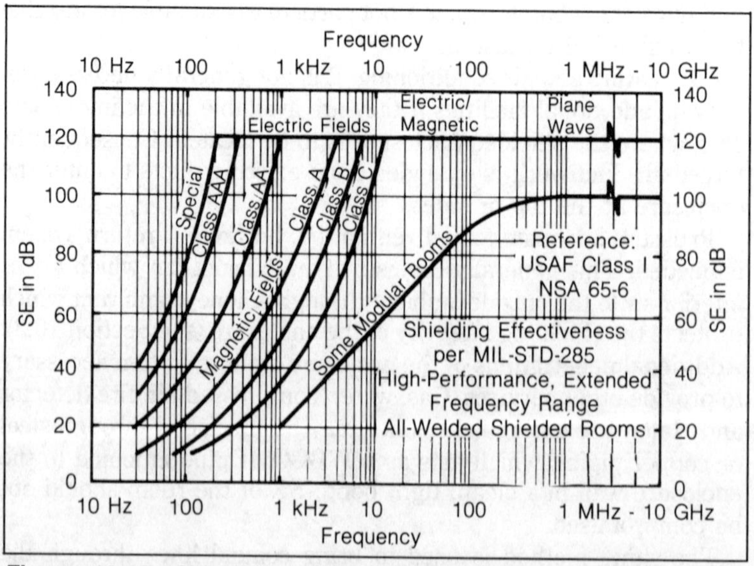

Figure 12.15—Shielding Effectiveness of Seam-Welded and Some Modular Rooms (courtesy of Lectro Magnetics, Inc.)

of magnetic shielding for the modular room is due to the magnetic seam impedance of the clamping arrangement.

To avoid creating the impression that all cell-type shielded enclosures are inferior to seam-welded designs in SE, Fig. 12.16 illustrates the performance of a double, electrically-isolated enclosure. Here the SE is everywhere superior to that captioned **some modular rooms** shown in Fig. 12.15. The magnetic-field attenuation of the double 24 oz. copper-steel design in Fig. 12.16 exceeds that of Classes A, B and C of Fig. 12.15 up to 1 kHz. It is difficult to make such comparisons since the parameters are often different and measurement details may not be the same. One attempt to present enclosure attenuation on a comparative basis is shown in Table 12.1.

The attenuation of a newly installed shielded enclosure is usually measured to determine if it performs to some specification, typically MIL-STD-285. This standard prescribes test frequencies and equipment, as well as antenna separation distances. Below UHF, the testing of an enclosure is performed in the near-field, yielding results which may vary widely as a function of distance, antenna type and frequency. Thus, it is important that the methods

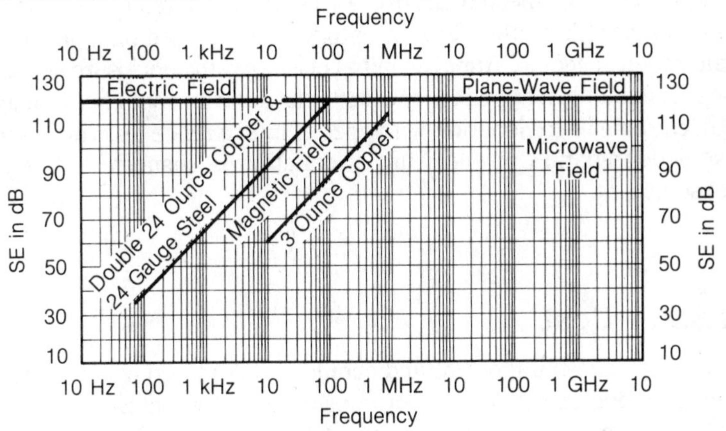

Figure 12.16—Typical Shielding Effectiveness of a Double, Electrically-Isolated Shielded Enclosure (courtesy of Lindgren & Associates, Inc.)

Table 12.1—Summary of Typical Shielded Enclosure Performance

Enclosure Type	Wall Type	Material	Magnetic Fields 60 Hz	15 kHz	Electric & Plane Waves 1 GHz	10 GHz
Double Electrically Isolated	Screen	Copper		68 dB	120 dB	
		Bronze		40 dB	110 dB	
		Galvanized		50 dB	50 dB	
	Solid	24 Ga. Steel	15 dB	82 dB	120 dB	90 dB
		Cu & Steel	18 dB	93 dB	120 dB	105 dB
Cell Type	Screen	Copper		48 dB	90 dB	
		Bronze				
	Solid	24 Ga. Steel		68 dB	100 dB	
		Copper				
Single Shield	Screen	24 Ga. Steel	6 dB		60 dB	
		Bronze				
	Solid	Copper				
		24 Ga. Steel		48 dB	90 dB	

indicated in MIL-STD-285, if this is the test specification to be met, be followed as closely as possible so that repeatable results can be obtained (see details on shielding test methods in Chapter 2).

Testing of an enclosure should be performed periodically to verify that its attenuation continues to meet the original specification. In this respect, the shielded enclosure is often considered to be an item of equipment in a laboratory inventory, and is placed on a periodic calibration schedule. After the initial checkout, the enclosure should be retested at least once every other year. Interim spot checks may be desirable in conjunction with special interference tests or in the event degradation of enclosure attenuation is suspected for any reason.

12.2.4 Cost

Cost is always important and should be addressed early when a new shielded enclosure is being considered. Cost and SE are interrelated. If extended performance is not required, the modular room is less expensive than the all-welded room in its initial cost for enclosures up to approximately 8 × 6 m. The operational environments and maintenance must be considered to determine the end cost over a number of years.

If performance or environmental factors are paramount, the cost of an all-welded room becomes secondary. One exception to the above statement involves structures which are significantly larger than 7.4 × 9.1 m (24' × 30'), and in particular when such structures must meet building codes and other environmental factors. Here the all-welded enclosure is generally less expensive than the modular type regardless of the degree of SE required. In such applications, modular rooms are not self-supporting and require considerable framing, whereas all-welded room construction is in accordance with the Uniform Building Code and is self-supporting. Some cost data are supplied in Table 12.2 and Fig. 12.17. Figures 12.18 and 12.19 show a large shielded room at beginning and later steps of construction.

Finally, when budgets are tight, the purchaser should not overlook the used market as a source of modular shielded enclosures. The EMI instrument and rental houses are often aware of such sources. Used modular enclosures generally can be purchased at 20 to 50 percent of the original cost. Prospective purchasers must remember the added cost to disassemble, physically move, reassemble and verify the SE of such enclosures.

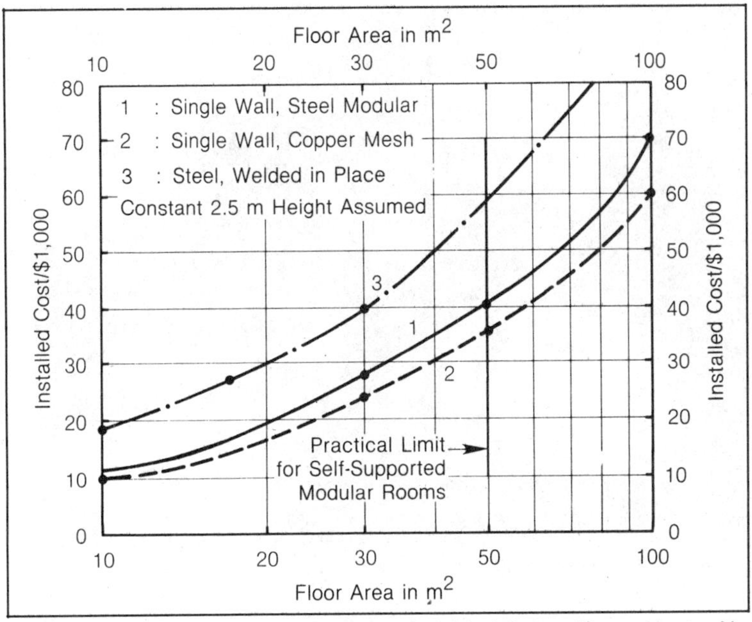

Figure 12.17—1986 Cost of Installed Shielded Room (Door, Vents, No Filters)

Figure 12.18—Construction of a Large Shielded Room, Beginning with the Draping of Galvanized Steel Plates.

Figure 12.19—Later Stage of the Erection of Shielded Walls for the Shielded Room of Fig. 12.19

Table 12.2—Cost of Installed Shielded Room

	Joining System	Base Price per m² (ft²)*		Advantages	Disadvantages
Copper Mesh	Nailed or Taped	$220	($20)	Semi-Transparent Lightweight	Low performance Not reliable for long term
Steel Modular	Clamped	$275	($25)	Good Performance Good Value Over 20,000 Installed	Can degrade if not installed properly
Copper Modular	Clamped	$333	($30)	Non-ferrous for MRI Market	More expensive steel
Steel-14 gauge	Welded	$333	($30)	Reliable	Expensive
6.4 mm (1/4")	Welded	$359	($35)	Good magnetic field performance	Labor intensive Unsuitable for construction in existing offices

*Base room cost per square foot of shield (all six sides) with a door, air vents but no filters.
These are installed costs and are June 1986 estimates.

12.3 Shielded Huts and Transportable Shelters

Shielded huts and vehicles represent a special case of shielded enclosures discussed in the previous section. They are characterized by their transportability, which subjects them to a substantial amount of flexure when in transit. Consequently, modular construction is seldom used and seam-welding is the most popular design.

Techniques similar to those used in stationary shielded rooms are used to maintain the shielding integrity of transportable enclosure apertures and penetrations. However, due to structure flexure, the SE of the access door usually becomes the limiting factor in attenuation performance. Consequently, shielded huts and vehicles typically exhibit 20 to 30 dB less SE than their stationary counterparts.

Figure 12.20 shows one form of shielded hut. The air vent flaps are typically backed by gasketed honeycomb. The power supply is generally a remotely located, engine-driven generator. Some versions provide a 28 Vdc battery bank for quiet operation. The hut may be transported by either a flat-bed trailer or by helicopter. When shielded vehicles are used, an auxiliary power supply, heating and air conditioning are furnished external to the driver's cab. To minimize enclosure flexure, an air-bag suspension system is installed on the vehicle.

Figure 12.20—A Typical Transportable Shielded Hut

12.4 Temporary, Low-Cost Enclosures

There are many cases in EMI testing, or EMI protection, where what is needed is simply an easy to erect, inexpensive, shielded volume that provides at least 60 dB SE at 30 MHz and up to 30 dB at 1 GHZ or so.

Such enclosures can be:

1. "Chicken wire" or window screen cages above ground
2. "Chicken wire" or window screen cages below ground
3. Metalized air-inflatable structures

The purpose of such structures is to permit acceptable EMI (or RF) measurements by reducing the ambient field to a tolerable level.

The screen cage is illustrated Fig. 12.21. The vertical members are 5 × 10 cm or 10 × 10 cm wooden poles, ranging in height from 2 to 6 m or more, as required. They are set only a few centimeters into the ground or bolted to earth-driven (or concrete-buried) metal bars, then secured with guy wires like for a tent. Then 5 cm mesh (chicken wire) or window screen (about 1 mm mesh) is draped in position and stapled to the wood studs. The same material is used for the ceiling and floor, using generous overlaps at corners and

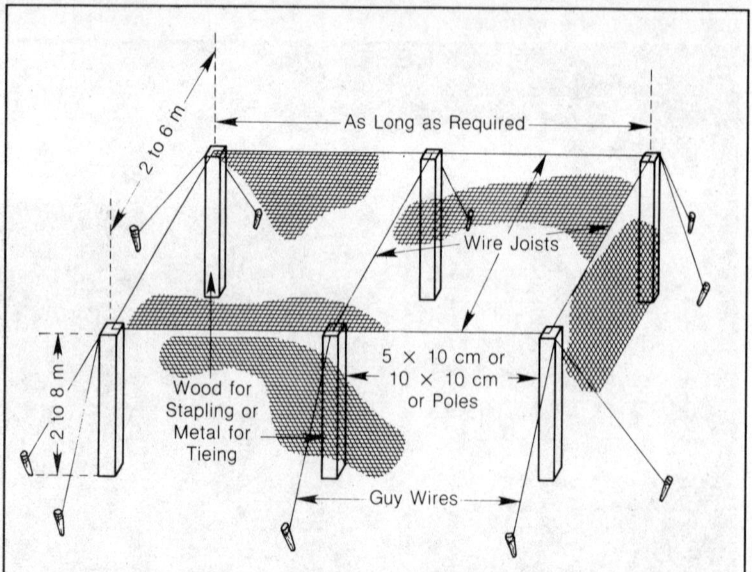

Figure 12.21—Low Cost, Temporary Shielded Hut and Measured Performances. The measured cage had a ≅ 1 cm galvanized steel mesh.

angles with the overlaps folded tight and stapled every 5 cm or pressed with conductive copper tape. Such a structure, measuring 5 × 10 × 5 m, has been assembled in 3 man-days at a total cost (labor and materials) of about $1,800 (1986 cost).

The choice between chicken wire and window screen is based on desired performances. Chicken wire (5 cm mesh) provides about 40 dB SE at 30 MHz, dropping to 10 dB at 1 GHz. Window screen (about 1 to 1.5 mm mesh) provides at least 30 dB improvement. On the other hand, chicken wire is easier and faster to install.

Concerning the necessary access gate, since no sophistication is sought, at 1 m × 1 m flap can be cut in the mesh, equipped with a wood frame and secured with many wing-nuts or other fasteners. Alternatively, a "storm window" screen door can be purchased in any hardware store and installed with EMI mesh gasket along the edges.

The power can be provided through a 60 dB line filter, mounted through a metal plate which is bolted on one of the wooden poles, over the screen mesh. If the enclosure is in an outdoor site, a light, portable, engine-driven generator which has been EMI suppressed can be located inside the enclosure. Because of the mesh, no ventilation and no special lighting is necessary.

Another approach to the temporary site is to dig, with a bulldozer, a few meter deep excavation to form a rectangular pit. One end has an approach ramp to facilitate access of test vehicles and equipment. If the earth is sufficiently wet and conductive, it is only necessary to provide the mesh to cover the hole, with the four sides extending about 1 m outside the perimeter and buried vertically. A definite advantage of this enclosure is its low Q, provided by the absorptive earth all around. The main drawback is that water collects in the bottom after a rain. Therefore, a water drain pipe needs to be provided.

Finally, air-inflatable structures are constructed of vinyl fabric and are able to cover large areas without resorting to rigid support. Their installed cost (as of 1987) ranges from $100 to $200 per square meter of floor. The vinyl fabric is coated with aluminum or other metal film (see Chapter 2). The weak points are the access door and the need to have a compressed air supply to build up and maintain the required pressure.

12.5 Bibliography

Albin, A.L., "Optimum Shielding of Equipment Enclosures," *Electronic Design,* Vol. 8, no. 3, February 3, 1960.

Butterfield, B., "Polyform—A New Approach to Electromagnetic Shielding Enclosures," *Electrical Design News*, August 1962.

Cohen, D., "A Shielded Facility for Low-Level Magnetic Measurements", *Journal of Applied Physics*, Vol. 38, no. 3, 1967, pp. 1295-6.

Dolle, W.C., Van Steenberg, G.N., and Jouffray, O.L., "Effects of Shielded Enclosure Resonances on Measurement Accuracy," *1970 IEEE International EMC Symposium Record* (New York: IEEE), Vol. 70C28-EMC, July 14-16, 1970.

Free, W.R., et al., "Electromagnetic Interference Measurement Methods-Shielded Enclosure," Final Report on Contract no. DA 28-043 AMC-02381 (E), EES, Georgia Institute of Technology, 1967.

Free, W.R., "Radiated EMI Measurements in Shielded Enclosures," *1967 IEEE EMC Symposium Record* (New York: IEEE), Vol. 27-C80, July 18-20, 1967, pp. 43-53.

Genistron, Inc., Final Report, "Proposed Specification for Electromagnetic Shielding of Enclosures and Buildings," U.S. Naval Civil Engineering Laboratory Contract NBY-33220.

Hollway, D.L.,"Screen Rooms and Enclosures," *Proceedings of the IREA*, Vol. 21, no. 10, October 1960, pp. 660-8.

Intrator, A.M., "Using Sheet Steel in the Construction of Shielded Rooms", *Electrical Engineering*, September 1953.

Jorgenson, C.M., "Shielding in Modern Computer Design," *Data Control*, December 1958.

Kanellakos, D.P., and Schulz, R.B., "New Techniques for Evaluating the Performance of Shielded Enclosures," Fifth Conference on Radio Interference Reduction and Electronic Compatibility, Armour Research Foundation, Oct. 1959.

Kenny, H.W., and Conard, B.L., "A practical Approach to RF Shielded Enclosure Design" ACE Engineering and Machine Co., Inc. 1962, (also appears as "RFI Shielded Enclosures" in *Electrical Design News*, September 1962).

Klouda, J.C., "Practical Aspects of Evaluating Shielded Rooms," Electro-Technology, June 1961.

Kozakoff, D.J.; Bolt, A.T.; and Howard, F.D., "Shielding Effectiveness of Various Shaped Geometrical Enclosures in Terms of Normalized Parameters," *1970 IEEE Regional EMC Sym-*

posium Record (New York: IEEE), Vol. 70C64-REGMC, October 6-8, 1970, pp. V-A-11.

Lindgren, E.A., "Contemporary RF Enclosures", Lindgren and Associates, 1967.

Lockwood, R.O., "New Technique for the Determination of the Integrity of Shielded Enclosure by the Measurement of the Perpendicular Magnetic Field," 1967 IEEE EMC Symposium Record (New York: IEEE), Vol. 27C80, July 18-20, pp. 61-9.

Mendez, H.A., "A new Approach to Electromagnetic Field-Strength Measurements in Shielded Enclosures," *IEEE EMC Symposium Record* (New York: IEEE), Vol. 69C3-EMC, June 17-19, 1969, p. 137.

MIL-STD-285, "Method of Attenuation Measurements for Enclosure Electromagnetic Shielding, for Electronic Test Purposes."

Osburn, J.D.M. and Morris, F.J., "Problems Encountered During the Design and Fabrication of an ELF-VLF-Shielding Enclosure," *1970 IEEE International EMC Symposium Record* (New York: IEEE), Vol. 70C28-EMC, July 14-16, 1970, pp. 472-8.

Patton, B.J., "Room-Size Enclosure for Geomagnetic Shielding," *1970 IEEE International EMC Symposium Record* (New York: IEEE), Vol. 70C28-EMC, July 14-16, 1970, pp. 89-96.

Patton, B.J., and Fitch, J.J., "Design of a Room-Size Magnetic Shield," *Journal of Geophysics Research*, Vol. 67, 1962, pp. 1117-21.

Pearlston, C.B., "Enclosure Shielding in Radio Interference," IEEE Third National Symposium on RFI, June 1961.

Schulz, R.B., and Kanellakos, D.P., "Shielding Enclosure Performance Utilizing New Techniques," Armour Research Foundation of Illinois Institute of Technology.

Stuckey, C.W.; Free, W.R.; and Robertson, D.W., "Preliminary Interpretation of Near-Field Effects on Measurement Accuracy in Shielded Enclosures," *1969 IEEE EMC Symposium Record* (New York: IEEE), Vol. 69C3-EMC.

Toler, J.C., and Evans, R., "Shielded Enclosure Specification in Perspective," *1970 Regional Electromagnetic Compatibility Symposium Record*, Vol. 70C64-REGEMC, Oct. 6-8, 1970, pp. III-A01 to III-A-3.

U.S. Naval Civil Engineering Laboratory, "Proposed Specification for Electromagnetic Shielding of Enclosures and Buildings,"

Final Project Report, 31 July 1963, Contract NBy-32220, pp. 97-113.

Van Steenberg, G.N., and Willman, J.F., "Shielded Enclosure Measurements Can be Accurate", *21st Annual SWIEECO Record* (New York: IEEE), IEEE Catalog no. 69C16-SWIEECO, April, 1969.

Vasaka, C.S., "Problems in Shielding Electronic Equipment," *Proceedings of the Conference on Radio Interference Reduction*, Armour Research Foundation, pp. 86-103, December, 1954.

Vasaka, C.S., "Theory, Design and Engineering Evaluation of Radio Frequency Shielded Rooms," Report no. NADC-EL-54129, August 1956.

Chapter 13

Application to Buildings and Vehicles

In a growing number of cases, some electromagnetic shielding is needed at the building level because of either the severity of the outside threat or the criticality of some system inside. Some examples are:

1. Buildings where EMI testing or other RF measurements are performed
2. Hospitals using magnetic resonance imaging (MRI) scanners
3. Buildings hosting some laboratories using electron beam microscopes, etc.
4. Buildings housing ordinary electronic equipment, computers etc. and located near a powerful broadcast transmitter, radar or other noise source, since ordinary office computers can exhibit errors when exposed to field strengths exceeding 3 to 5 V/m particularly in the VHF band
5. Embassies, headquarters and government facilities handling confidential information which must be protected against eavesdropping

This chapter shows that most buildings offer a fair (30 to 50 dB) attenuation to HF electric fields and very little attenuation beyond (see Fig. 13.1). If one wants to provide reliable shielding against strong ambient electromagnetic environments, it is necessary to

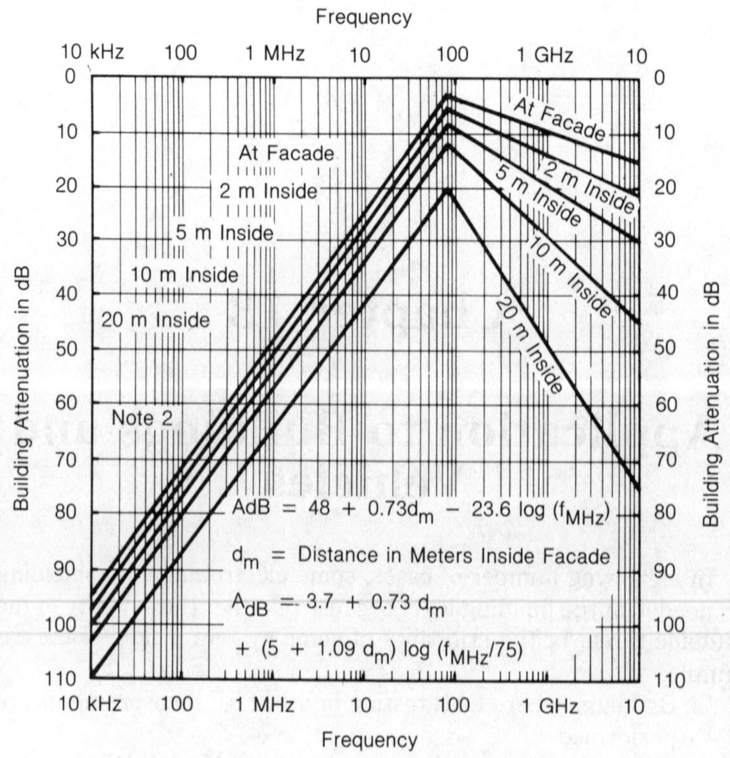

Notes: 1. Buildings surveyed are business-type, with metal girders.
2. Suggested extrapolation for E-Field Attenuation—H-Field attenuation decreases significantly below 1 MHz, to approach 0 dB around 10 kHz.

Figure 13.1—Building Attenuation to Outside Radiations vs. Frequency and Distance Inside Facade

use one form or another of screening on the exterior and, eventually, interior walls. Depending on the frequency range of concern and the amount of attenuation required, this may go as far as shielding and gasketing windows, doors, ventilation grilles, attics, etc. In a sense, a building is nothing more than a giant version of our shielded box of Chapters 4 through 10.

Very little reported data exists regarding the mean attenuation of vehicles, aircraft and so on. However, this chapter presents some actual measured results as well as an analytical approach to determine SE.

13.1 Typical Building Attenuation

Two types of building attenuation are of interest:

1. SE versus frequency of exterior rooms located just inside the building facade
2. SE versus frequency of deep interior rooms which are protected by at least 30 m of inner walls, corridors, etc.

Several measurement surveys were conducted by Mir & White and A.A. Smith.[1,2] The surveys addressed ordinary unshielded buildings. The latter survey, incidentally, has been used as one of the data bases of CBEMA study leading to the FCC-15J and CISPR-22 radiated EMI limits for computers.

Although a large variance appeared among the buildings surveyed, one can make several general observations:

1. Predictably, brick, cinder and stone buildings with little or no metal armature offer practically no low-frequency H-field attenuation.
2. Steel-frame, multistory buildings exhibit a significant attenuation (50 dB average) in the AM range (a few hundred kilohertz).
3. Both E-field and H-field attenuations are at a null between approximately 50 and 200 MHz, where typical distance between girders is of the order of $\lambda/2$.
4. There is a significant increase in attenuation in the center core of the building, where mean SE of more than 60 dB exists at 1 MHz.
5. Some of the measurements are inaccurate because power and telephone lines entering the building and reradiating can mask the actual building SE. For example, in a steel-frame building, you can pick up AM stations with a pocket radio, simply because the building attenuation is shunted by the power wiring reradiation.
6. Most of the measurements were performed using remote external sources, i.e. the myriad of broadcast fields as reference. No information seems to be available concerning the reverse, i.e., the SE of a building for sources located inside. It can be assumed that this attenuation will be poorer

since sources will create near H-field situations (lesser reflection SE) and also will be closer to the building apertures.

Fig. 13.1 is a plot of the fitted mean values of building SE. Compared to the large data base available, there is a standard deviation of +10 to 15 dB (depending on frequency) around these mean values.

13.2 Exterior Building Materials

The exterior of most buildings is either cinder block, concrete or brick. Less frequently, exteriors are of wood, glass or stone. Studies conducted by the National Bureau of Standards (NBS) showed that the median SE for a single-layer brick wall is only 2.5 dB between 500 and 1,000 MHz. Above 1 GHz, absorption loss becomes significant because even a poorly conductive clay or concrete will have a skin depth of about 10 cm at 1 GHz.

An interesting material is coke, which offers a typical dc resistivity of 0.1 ohm-centimeters (Ωcm) at dc. Therefore, a thick layer of coke chunks can achieve some reflection and absorption above HF. Typical SE of a 30 cm layer of coke chunks about 30 cm (12") in diameter is 60 dB at 100 kHz (plane waves) to 90 dB at 1 GHz. These numbers are for untreated coke. Treated coke, where the sulphur content has been extracted, still shows 50 dB at 100 kHz to 70 dB at 1 GHz per 30 cm thickness.

One approach to using coke as a building shielding material is to pour coke dust or granules between two layers of brick, concrete slabs, or behind a wooden siding. However, typical architectural practices do not allow a 30 cm waste in the thickness of each wall, and only 5 or 10 cm can be reasonably accommodated. This provides only 10 to 20 dB of SE—hardly an impressive figure. Another approach would be to manufacture bricks, cinder blocks, etc., with a specified percentage of coke dust, such that structural strength is not significantly affected. Here again, this would probably just give about 10 to 20 dB per 30 cm thickness.

One obvious solution to improve the RF attenuation of the facade is to use a conductive top coat such as aluminum foil, conductive textile (see Chapter 2) or conductive paint. For example, a minimum SE of about 50 dB can be achieved by using aluminum foil. To do this, the wall must be dressed to provide a flat surface for the application of the foil. This can be done by using furring strips overlaid with 1 cm plywood, particle board or gypsum boards. Aluminum foil, 60 μm thick and in rolls about 1.20 m wide, can be installed like wallpaper and overlapped about 10 cm on the edges. It is then affixed with staples on 10 cm centers. A strip of ordinary nonconductive adhesive tape can be rolled over the edges for mechanical protection. The underlayer must be thick enough to prevent the staples from touching any metal in the building which may be connected to an earthing conductor.

A second foil is laid over the first layer and stapled. This layer is offset by half a layer width so that the edges of the second layer fall in the middle of the first one. All the inside corners of the rooms must be covered first so that there are no seams at the corners. A single 1 cm crack can compromise the whole building SE at as much as 50 dB at 500 MHz.

The next problem is matching the foil to the door stills and the room floor and ceiling. Ideally, these should be made of sheet metal, too, or installed with an underlayer of wire screen below the floor and above the ceiling boards.

A better installation can be made using copper foil backed by a material similar to roofing paper and glued to the underlayer boards. Conductive copper tape can be used to seal the seams. Copper tape strips can extend down at regular intervals to connect to the floor. Finally, a horizontal strip is criss crossed with the vertical strips. The same technique can be used to bond the conductive foil to the door and window frames.

Another material recently developed by **Fortifiber Corp.** of Attleboro, MA. is a copper armored sisalkraft constructed of a 3 oz (0.1 mm thick) copper sheet laminated to a heavy-duty fiberglass reinforced kraft paper. The fiber mesh gives the material the physical strength to withstand construction job abuse. The kraft backing facilitates gluing to the walls. The seams are bonded with copper tape which is soldered to the 3 oz copper sheets.

Looking at the performance of wire meshes and textiles (see Chapter 2), we see that a thin mesh (e.g., 22 wires/inch, approximately corresponding to 1 mm openings) provides more than 60 dB up to 100 MHz. These screens or fabrics can be impregnated in plaster or concrete during construction or laid over the walls for post-construction shielding. The installation procedure is similar to that for wallpaper and follows the guidelines described previously for aluminum foil. Ceilings and floors can be covered with mesh, which is overlapped slightly (10 cm) at the walls and corners. Then the ceiling and floor can be covered as usual. Installers must take proper precautions so that lighting fixtures do not pierce or deteriorate the ceiling mesh.

While expensive when compared to normal construction, such "Faradization" of a building section is considerably cheaper than installing even the least sophisticated shielded room. While conductive paints would be an interesting alternative, and much easier to apply, they are not widely used. One possible reason is the relative fragility of such a thin coating over an interior wall: it can be easily scratched, peeled and abused in many ways. Most people would not suspect the consequences of a scar a few centimeters in width.

13.3 Treatment of Doors, Windows and Other Openings

Above a few megahertz, the SE performance of a building will be dominated by the penetrations. All openings such as windows and doors, though surrounded with conductive materials or a steel structure, will be RF transparent (see the slot leakage model in Chapter 7) when their larger dimension reaches $\lambda/2$, i.e., at approximately 75 to 100 MHz. Hollow-core steel or aluminum doors can be used. Finger-stock gasketing should be used to seal the door periphery if better than 60 dB SE is required up to 1 GHz. Otherwise, cheaper conductive gaskets can be used. Windows should be equipped with screen-type glass panels with a conductive strip around the entire edge to make contact with the window frame (see Chapter 10). When SE has to be improvised on the spot, screen mesh or metalized textile can be laid over the window and folded over at the edges to make contact with the frame.

Other penetrations, such as HVAC duct work, can be equipped with screen vents or stretched metal grids. If better than 60 dB is required, or if the purpose of the shielding is to contain confidential information within government offices, embassies, etc., screen vents are insufficient. Here, honeycomb air filters should be used as described in Chapter 10.

Finally, other penetrations include water, sewage and other utility lines. Where nonconductive pipes are used, the WGBCO technique is applied by welding or fitting a heavy galvanized steel or copper pipe through a flange mounted over the surface of a conductive panel. Because the WGBCO cannot work when a conductor is driven inside, a power line filter must be used on incoming ac wiring such that all lines including the neutral are filtered. The same applies to telephone or other data lines (local area networks, CATV, CCTV, etc.).

Single-point grounding of each of these improvised shielded rooms is maintained by isolating grounded plumbing and pipes. This is accomplished by using dielectric sections of approximately 10 cm long pipe which is inserted into the plumbing and other metal pipes. Then the wall shielding is connected to the room power panel ground bus (safety earth) via a single green-yellow (solid green in the United States) well-identified insulated wire. Abundant details on the grounding of shielded rooms are provided in Volume 2 of this handbook series.

When more attenuation is required, like 80 or 100 dB, it is almost mandatory to contract out either the construction of a custom made shielded room, or to install a prefabricated shielded room inside the building. Both are described in Chapter 12.

13.4 SE of Buried Facilities

A question often arises as to how much shielding is provided by earth. This topic may be expanded to underwater facilities and submarines as well. Earth and sea water offer natural attenuation via absorption through skin effect. The attenuation can therefore be calculated by the absorption equation given in Chapter 1. Taking into account absorption only (earth is a very mediocre conductor, so reflection loss on typical earth is poor) produces:

$$A_{dB} \approx 131\, t \sqrt{f_{MHz}\, \mu_r\, \sigma_r}$$

where,

t = thickness of earth in mm
f = frequency in MHz
μ_r = relative permeability

This attenuation can be calculated for various types of soils and for sea water. The results are shown in Table 13.1, which shows the depth below the surface that corresponds to one skin depth (8.7 dB absorption) and 11.5 skin depths (100 dB absorption).

Table 13.1—Natural Absorption Loss of Water and Land

	Desert σ = 0.03 mho/m		Average Land σ = 0.03 mho/m		Marsh Land σ = 0.111mho/m		Sea Water σ = 5 mho/m	
Frequency	8.7 dB	100 dB	8.7 dB	100 dB	8.7 dB	100 dB	8.7 dB	100 dB
50-60 Hz	647 m	7.5 km	392 m	4.5 m	205 m	2.4 km	30 m	349 m
400 Hz	239 m	2.7 km	145 m	1.7 km	76 m	874 m	11 m	130 m
1 kHz	152 m	1.7 km	92 m	1.1 km	48 m	552 m	7.1 m	82 m
100 kHz	15 m	175 m	9 m	106 m	4.8 m	55 m	71 cm	8.2 m
10 MHz	1.5 m	17.5 m	92 cm	11 m	48 cm	5.5 m	7.1 cm	82 cm
1 GHZ	15 cm	1.75 m	9 cm	1.1 m	5 cm	55 cm	7.1 m	8 cm

13.5 Actual SE of Typical Vehicles and Aerospace Structures

Although the subject is of great interest, very few published data are available on field attenuation through automobiles, armored vehicles, ships, aircraft, etc. The following is a compilation of the few scattered data appearing in open literature, plus some of the author's experiences and calculations.

13.5.1 Vehicles

Ordinary vehicles such as civilian cars and trucks offer a decent plane wave attenuation below 30 MHz. The largest opening in steel frame bodies being the windshield and windows, the few measured data correspond quite well with the model of a 1.50 to 1.80 m aperture leakage. The trunk and engine hoods correspond to long slots with about the same length. Around 80 to 100 MHz, a typical vehicle is therefore completely transparent (see Table 13.2). Some random weaknesses in SE can appear below this frequency and are attributable to imperfect or corroded welded joints in the car body.

Table 13.2—SE of Ordinary Vehicles (Default Values)

Frequency	**SE_{dB}** H-Field, 1 m	**SE_{dB}** Plane Waves
100 kHz	6-10	60-70
10 MHz	6-10	20-30
100 MHz and Beyond	NA	0

Combat vehicles such as tanks have much smaller apertures. The largest apertures are access hatches, axles and transmission-shaft through holes, gun apertures and other artillery clearance holes. Furthermore, for other EMI reasons (e.g., minimizing intermodulation by random contacts), hatches and maintenance panels are often equipped with bonding straps. Therefore, the widest dimension of a tank aperture is typically in the 50 to 75 cm range. One particular leak appears at the turret joint, which typically represents a diameter of 2 or 3 m. This corresponds to a developed slot length of 6 to 10 m. Although this joint is generally equipped with brushes to provide a sliding contact, the resultant SE is rather poor.

Based on a limited data base, the following default data are suggested for tanks and armored vehicles:

Frequency	**SE (Plane Waves)**
100 kHz	> 69 dB
1 MHz	43 dB
10 MHz	23 dB
50 MHz and beyond	15-18 dB

13.5.2 Aircraft and Helicopters

Knowledge of the SE for a given aircraft is of prime importance for both military and civilian applications. Besides the tactical aspects, the immunity to lightning-induced effects when 100 kA or higher levels are flowing on the fuselage and susceptibility to NEMP are serious concerns.

An aircraft fuselage behaves electrically as a long conductive tube with several holes (cockpit, windows, landing gear apertures, etc.). Several complex models have been devised to predict the penetration of such structures by EM fields. The most elaborate ones use Harrington's method of moments (MOM) by breaking the structure into hundreds of small filaments and integrating the contribution of each one of these minuscule antenna elements. Another method breaks the structure into a finite number of elementary surfaces.

A simpler approach for rough estimation is to use the surface transfer impedance concept which was already described for shielded cables. Approached this way, the SE of an aircraft can be viewed as the coupling of the EMI-induced skin current through the metal or composite at low frequencies (below skin depth) and through the apertures and joints at higher frequencies. Some limited data exist where the attenuation of several transporters (such as the KC 135 tanker aircraft) was rigorously measured by using H-field loops inside and outside the airplane.[3]

One particular problem that exists with a metallic aircraft is that the hollow metallic tube of the fuselage is extremely prone to resonance. When the length of the fuselage reaches $\lambda/2$, a first resonance can be found. This resonance causes the whole fuselage (for an E-field parallel to the roll axis) to become a resonant dipole, i.e., the voltage maximum exists at nose and tail and the current maximum in the middle. These reinforcements of V or I (vs. the uniform distribution if $L << \lambda/2$) cause more current to be coupled through the skin (diffusion) or through the apertures (leakages), giving a poorer effective SE than expected. When the diameter, in turn, reaches $\lambda/2$, a very sharp resonance is usually found since the cavity can reach a Q of 20 or more. For instance, in a large metallic aircraft with a fuselage diameter of about 3 m, this resonance occurs at 50 MHz. Even if some attenuation is normally expected at this frequency because the total aperture leakage level has not been reached, some regions inside the cabin may exhibit a peak in the received field.

Table 13.3 shows some default SE values for two types of aircraft. For helicopters, SE depends strongly on the design. Closed-

Table 13.3—Approximate SE_{dB} of Various Aircraft

Frequency	Fighter Length ≅ 11 m Fuselage Dia = 1.80 m 2 Seat		Transporter/Bomber Length = 25 m Fuselage Dia = 3 m
	> 50% Carbon Composite (8-Ply, Nonmetalized)	Aluminum	Aluminum
10 kHz	0 (H-Field)	50 (H-Field)	50 ± 30 (H-Field, Measured)
100 kHz	10-15 (H-Field)	70 (H-Field)	50 ± 30 (H-Field, Measured)
1 MHz	20-25 (H-Field)	50	45 ± 25 (H-Field, Measured)
10 MHz	40-45 (Plane Wave)	30	40 ± 30 (H-Field, Measured)
30 MHz	20-25 (Plane Wave)	20	35 ± 12 (Measured Plane Wave)
	No Sharp Resonance Due to Lesser Q of Cavity	*83 MHz Transverse Resonance SE ≡ 0 dB	*50 MHz = Transverse Resonance SE ≡ 0 dB
100 MHz and Beyond	10-15 dB	10-20 dB	10-20 dB

fuselage type helicopters can be expected to have SE behavior similar to that of an aircraft. Many civilian helicopters have virtually no metal in the fuselage; the cabin is a large Plexiglass[R] dome, and the tail is often an open structure. Unless the Plexiglass has a conductive coating (see the discussion of transparent conductive films in Chapter 10), the SE is virtually nil.

13.6 References

1. Mir, S. and White, D.R.J., "Building Attenuation, Impact on Product Susceptibility," IEEE EMC Symposium, 1974.
2. Smith, Jr., A.A., "Attenuation of Electric and Magnetic Fields by Buildings," (IEEE Transactions on EMC, August 1978).
3. Birken, J.A., "Advanced Composite Aircraft Design," 1982 IEEE EMC Symposium.

13.7 Bibliography

Air Force Handbook AFSC DH 1-4, Design Note 5 F8, "Enclosures."

Gierhart, G.D.; Hause, L.G.; and Farrow, J.E., "Insertion Loss of a Brick Wall at UHF" NBS Report 7272, June 1962.

Helene, F., "Architectural Shielding," *ITEM* 1986.

Jecko, B. and Castenot, A., "Attenuation of Reinforced Concrete Buildings," 1983 French National EMC Symposium, Tregastel.

Sauvain, H. and Chung, Chi Lin, "Attenuation des Ondes Electro-Magnetiques par l'acier des Armatures en Beton," *Revue Ingenieurs et Architectes Suisses*, no. 9, Mai 1980.

Snider, J.B., "A Statistical Approach to RF Attenuation Measurements of Buildings," NBS Report 8863, July 1965.

Appendix A

Decibel Terms and Definitions

It may seem very basic and academic to review decibels in a book of this type. Yet, it is observed that not all readers are fluent in the use of decibels.

The Bel is defined as the logarithm of the ratio of two powers:

$$\text{Bel} = \text{B} = \log\left(\frac{P_2}{P_1}\right) \tag{A.1}$$

where,

P_2 = After (a change) or second measurement in watts, mW, pW, etc.

P_1 = Before, first or second reference measurement in watts, mW, pW, etc., but same unit as P_2

Since the Bel results in a coarse measurement, a finer gradation is achieved through the decibel:

$$\text{dB} = 10\log\left(\frac{P_2}{P_1}\right) \tag{A.2}$$

$$\text{Since } P_1 = V_1^2/Z_1 \text{ and } P_2 = V_2^2/Z_2 \tag{A.3}$$

where,

V_1 = First or reference voltage in volts

Z_1 = Impedance of first voltage in ohms

V_2 = Second or measured voltage in volts

Z_2 = Impedance of second voltage in ohms

P_1 and P_2 are both in watts.

Substituting Eq. (A.3) into Eq. (A.2) produces:

$$dB = 10 \log\left(\frac{V_2^2}{Z_2} \Big/ \frac{V_1^2}{Z_1}\right) = 10 \log\left[\left(\frac{V_2}{V_1}\right)^2 \times \left(\frac{Z_1}{Z_2}\right)\right] \tag{A.4}$$

$$= 10 \log\left(\frac{V_2}{V_1}\right)^2 + 10 \log\left(\frac{Z_1}{Z_2}\right) \tag{A.5}$$

$$= 20 \log\left(\frac{V_2}{V_1}\right) + 10 \log\left(\frac{Z_1}{Z_2}\right) \tag{A.6}$$

When $Z_1 = Z_2$ (which is considered to apply most of the time):

$$dB = 20 \log (V_2/V_1) \text{ for } Z_1 = Z_2 \tag{A.7}$$

If current, I, were used in defining Eq. (A.3) instead of voltage,

$$P_1 = I_1^2 Z_1 \text{ and } P_2 = I_2^2 Z_2,$$

Eqs. (A.6) and (A.7) would become:

$$dB = 20 \log\left(\frac{I_2}{I_1}\right) + 10 \log\left(\frac{Z_2}{Z_1}\right) \tag{A.8}$$

$$= 20 \log (I_2/I_1) \text{ for } Z_2 = Z_1 \quad \text{(A.9)}$$

Also, if field strength or flux density is used in Eq. (A.3), when Eq. (A.2) is defined in terms of power density, there results:

$$\text{dB} = 20 \log \left(\frac{E_2}{E_1}\right) + 10 \log \left(\frac{Z_{w1}}{Z_{w2}}\right) \quad \text{(A.10)}$$

$$= 20 \log (E_2/E_1) \text{ for } Z_{w1} = Z_{w2} \quad \text{(A.11)}$$

$$\text{dB} = 20 \log \left(\frac{H_2}{H_1}\right) + 10 \log \left(\frac{Z_{w2}}{Z_{w1}}\right) \quad \text{(A.12)}$$

$$= 20 \log (H_2/H_1) \text{ for } Z_{w2} = Z_{w1} \quad \text{(A.13)}$$

$$= 20 \log (B_2/B_1) \text{ for } Z_{w2} = Z_{w1} \quad \text{(A.14)}$$

where,

E_1 = First or reference electric-field strength in V/m (or μV/m)
Z_{w1} = Wave impedance of E_1 in ohms
E_2 = Second or measured electric-field strength in V/m (or μV/m)
Z_{w2} = Wave impedance of E_2 in ohms
H_1 = First or reference magnetic-field strength in A/m (or μA/m)
Z_{w1} = Wave impedance of H_1 in ohms
H_2 = Second or measured magnetic-field strength in A/m (or μA/m)
Z_{w2} = Wave impedance of H_2 in ohms
B_1 = First or reference magnetic flux density in Tesla (gauss or gamma)
B_2 = Second or measured magnetic flux density in Tesla (gauss or gamma)

When Eqs. (A.2), (A.7) and (A.9) are computed for various values, Tables A.1 and A.2 result. For power density, field strength and magnetic flux density, Eqs. (A.10) - (A.14) result in Tables A.3 and A.4.

Table A.1—Conversion of Voltage, Current and Power Ratio to dB

Voltage or Current Ratio	Power Ratio	dB	Voltage or Current Ratio	Power Ratio	dB
1.00	1.00	0	1.00	1.00	0
1.02	1.04	.2	.98	.96	−.2
1.04	1.08	.3	.96	.92	−.4
1.07	1.14	.6	.93	.86	−.6
1.10	1.21	.8	.90	.81	−.9
1.15	1.32	1.2	.87	.76	−1.2
1.20	1.44	1.6	.85	.72	−1.4
1.30	1.69	2.3	.80	.64	−1.9
1.40	1.96	2.9	.75	.56	−2.5
1.50	2.25	3.5	.70	.49	−3.1
1.60	2.56	4.1	.65	.42	−3.7
1.70	2.89	4.6	.60	.36	−4.4
1.80	3.24	5.1	.55	.30	−5.2
1.90	3.61	5.6	.50	.25	−6.0
2.00	4.00	6.0	.48	.23	−6.4
2.20	4.84	6.8	.45	.20	−6.9
2.40	5.76	7.6	.43	.18	−7.3
2.50	6.25	8.0	.40	.16	−8.0
2.60	6.76	8.3	.38	.14	−8.4
2.80	7.84	8.9	.35	.12	−9.1
3.00	9.00	9.5	.33	.11	−9.6
3.25	10.6	10.2	.30	.090	−10.5
3.50	12.3	10.9	.28	.078	−11.1
3.75	14.1	11.5	.26	.068	−11.7
4.00	16.0	12.0	.24	.058	−12.4
4.50	20.3	13.1	.22	.048	−13.2
5.00	25.0	14.0	.20	.040	−14.0
5.50	30.0	14.8	.18	.032	−14.9
6.00	36.0	15.6	.17	.029	−15.4
6.50	42.3	16.3	.15	.023	−16.5
7.00	49.0	16.9	.14	.020	−17.1
7.50	56.3	17.5	.13	.017	−17.7
8.00	64.0	18.1	.12	.014	−18.4
9.00	81.0	19.1	.11	.012	−19.2
10	100	20.0	.10	.010	−20.0
30	900	29.5	.03	9×10^{-4}	−30.5
100	10^4	40	.01	10^{-4}	−40
300	9×10^4	49.5	.003	9×10^{-6}	−50.5
1000	10^6	60	.001	10^{-6}	−60
3×10^3	9×10^6	69.5	3×10^{-4}	9×10^{-8}	−70.5
10^4	10^8	80	10^{-4}	10^{-8}	−80
3×10^4	9×10^8	89.5	3×10^{-5}	9×10^{-10}	−90.5
10^5	10^{10}	100	10^{-5}	10^{-10}	−100
10^6	10^{12}	120	10^{-6}	10^{-12}	−120
10^7	10^{14}	140	10^{-7}	10^{-14}	−140
10^8	10^{16}	160	10^{-8}	10^{-16}	−160
10^9	10^{18}	180	10^{-9}	10^{-18}	−180
10^{10}	10^{20}	200	10^{-10}	10^{-20}	−200

Table A.2—Conversion of dB to Voltage, Current and Power Ratio

dB	Voltage or Current Ratio	Power Ratio	dB	Voltage or Current Ratio	Power Ratio
0	1.00	1.00	0	1.00	1.00
.5	1.06	1.12	−.5	.94	.89
1	1.12	1.26	−1	.89	.79
2	1.26	1.58	−2	.79	.63
3	1.41	2.00	−3	.71	.50
4	1.58	2.51	−4	.63	.40
5	1.78	3.16	−5	.56	.32
6	2.00	4.00	−6	.50	.25
7	2.24	5.01	−7	.45	.20
8	2.51	6.31	−8	.40	.16
9	2.82	7.94	−9	.35	.13
10	3.16	10.0	−10	.32	.10
11	3.55	12.6	−11	.28	.079
12	4.00	16.0	−12	.25	.063
13	4.47	20.0	−13	.22	.050
14	5.01	25.1	−14	.20	.040
15	5.62	31.6	−15	.18	.032
16	6.31	39.8	−16	.16	.025
17	7.08	50.1	−17	.14	.020
18	7.94	63.1	−18	.13	.016
19	8.91	79.4	−19	.11	.013
20	10.0	100	−20	.10	.010
25	17.8	316	−25	.056	.0032
30	31.6	1000	−30	.032	.0010
35	56.2	3.16×10^{3}	−35	.018	.00032
40	100	10^{4}	−40	.010	.00010
45	178	3.16×10^{4}	−45	.0056	3.16×10^{-5}
50	316	10^{5}	−50	.0032	10^{-5}
55	562	3.16×10^{5}	−55	.0018	3.16×10^{-6}
60	1000	10^{6}	−60	10^{-3}	10^{-6}
65	1.78×10^{3}	3.16×10^{6}	−65	5.62×10^{-4}	3.16×10^{-7}
70	3.16×10^{3}	10^{7}	−70	3.16×10^{-4}	10^{-7}
75	5.62×10^{3}	3.16×10^{7}	−75	1.78×10^{-4}	3.16×10^{-8}
80	10^{4}	10^{8}	−80	10^{-4}	10^{-8}
85	1.78×10^{4}	3.16×10^{8}	−85	5.62×10^{-5}	3.16×10^{-9}
90	3.16×10^{4}	10^{9}	−90	3.16×10^{-5}	10^{-9}
95	5.62×10^{4}	3.16×10^{9}	−95	1.78×10^{-5}	3.16×10^{-10}
100	10^{5}	10^{10}	−100	10^{-5}	10^{-10}
105	1.78×10^{5}	3.16×10^{10}	−105	5.62×10^{-6}	3.16×10^{-11}
110	3.16×10^{5}	10^{11}	−110	3.16×10^{-6}	10^{-11}
115	5.62×10^{5}	3.16×10^{11}	−115	1.78×10^{-6}	3.16×10^{-12}
120	10^{6}	10^{12}	−120	10^{-6}	10^{-12}
130	3.16×10^{6}	10^{13}	−130	3.16×10^{-7}	10^{-13}
140	10^{7}	10^{14}	−140	10^{-7}	10^{-14}
150	3.16×10^{7}	10^{15}	−150	3.16×10^{-8}	10^{-15}
160	10^{8}	10^{16}	−160	10^{-8}	10^{-16}
170	3.16×10^{8}	10^{17}	−170	3.16×10^{-9}	10^{-17}
180	10^{9}	10^{18}	−180	10^{-9}	10^{-18}
190	3.16×10^{9}	10^{19}	−190	3.16×10^{-10}	10^{-19}
200	10^{10}	10^{20}	−200	10^{-10}	10^{-20}

Table A.3—Electric-field Strength-to-Magnetic Field-to-Power Densities

dBμV/m	V/m	dBμA/m	A/m	dBpT	pT & mΓ	nT & Γ	Gauss	Tesla	dBW/m²	dBm/m²	mW/cm²
280	10^8	228.5	2.65×10^5	230.5	3.33×10^{11}	3.33×10^8	3.33×10^3	0.333	134.3	164.3	2.67×10^{12}
260	10^7	208.5	2.65×10^4	210.5	3.33×10^{10}	3.33×10^7	333	0.033	114.3	144.3	2.67×10^{10}
240	10^6	188.5	2.65×10^3	190.5	3.33×10^9	3.33×10^6	33.3	0.003	94.3	124.3	2.67×10^8
220	10^5	168.5	265	170.5	3.33×10^8	3.33×10^5	3.33	3.33×10^{-4}	74.3	104.3	2.67×10^6
200	10^4	148.5	26.5	150.5	3.33×10^7	3.33×10^4	0.333	3.33×10^{-5}	54.3	84.3	2.67×10^4
180	10^3	128.5	2.65	130.5	3.33×10^6	3.33×10^3	0.0333	3.33×10^{-6}	34.3	64.3	267
160	100	108.5	0.265	110.5	3.33×10^5	333	3.33×10^{-3}	3.33×10^{-7}	14.3	44.3	2.67
140	10	88.5	0.0265	90.5	3.33×10^4	33.3	3.33×10^{-4}	3.33×10^{-8}	−5.7	24.3	0.0267
120	1	68.5	2.65×10^{-3}	70.5	3.33×10^3	3.33	3.33×10^{-5}	3.33×10^{-9}	−25.7	4.3	2.67×10^{-4}
100	0.10	48.5	2.65×10^{-4}	50.5	333	0.333	3.33×10^{-6}	3.33×10^{-10}	−45.7	−15.7	2.67×10^{-6}
80	0.01	28.5	2.65×10^{-5}	30.5	33.3	0.0333	3.33×10^{-7}	3.33×10^{-11}	−65.7	−35.7	2.67×10^{-8}
60	10^{-3}	8.5	2.65×10^{-6}	10.5	3.33	3.33×10^{-3}	3.33×10^{-8}	3.33×10^{-12}	−85.7	−55.7	2.67×10^{-10}
40	10^{-4}	−11.5	2.65×10^{-7}	−9.5	0.333	3.33×10^{-4}	3.33×10^{-9}	3.33×10^{-13}	−105.7	−75.7	2.67×10^{-12}
20	10^{-5}	−31.5	2.65×10^{-8}	−29.5	0.0333	3.33×10^{-5}	3.33×10^{-10}	3.33×10^{-14}	−125.7	−95.7	2.67×10^{-14}
19	8.9×10^{-6}	−32.5	2.36×10^{-8}	−30.5	0.0297	2.97×10^{-5}	2.97×10^{-10}	2.97×10^{-14}	−126.7	−96.7	2.12×10^{-14}
18	7.94×10^{-6}	−33.5	2.10×10^{-8}	−31.5	0.0265	2.65×10^{-5}	2.65×10^{-10}	2.65×10^{-14}	−127.7	−97.7	1.68×10^{-14}
17	7.08×10^{-6}	−34.5	1.87×10^{-8}	−32.5	0.0236	2.36×10^{-5}	2.36×10^{-10}	2.36×10^{-14}	−128.7	−98.7	1.34×10^{-14}
16	6.31×10^{-6}	−35.5	1.67×10^{-6}	−33.5	0.0210	2.10×10^{-5}	2.10×10^{-10}	2.10×10^{-14}	−129.7	−99.7	1.06×10^{-14}
15	5.62×10^{-6}	−36.5	1.49×10^{-8}	−34.5	0.0187	1.87×10^{-5}	1.87×10^{-10}	1.87×10^{-14}	−130.7	−100.7	8.43×10^{-15}
14	5.01×10^{-6}	−37.5	1.33×10^{-8}	−35.5	0.0167	1.67×10^{-5}	1.67×10^{-10}	1.67×10^{-14}	−131.7	−101.7	6.70×10^{-15}
13	4.47×10^{-6}	−38.5	1.18×10^{-8}	−36.5	0.0149	1.49×10^{-5}	1.49×10^{-10}	1.49×10^{-14}	−132.7	−102.7	5.32×10^{-15}
12	3.98×10^{-6}	−39.5	1.05×10^{-8}	−37.5	0.0133	1.33×10^{-5}	1.33×10^{-10}	1.33×10^{-14}	−133.7	−103.7	4.23×10^{-15}
11	3.55×10^{-6}	−40.5	9.40×10^{-9}	−38.5	0.0118	1.18×10^{-5}	1.18×10^{-10}	1.18×10^{-14}	−134.7	−104.7	3.36×10^{-15}
10	3.16×10^{-6}	−41.5	8.38×10^{-9}	−39.5	0.0105	1.05×10^{-5}	1.05×10^{-10}	1.05×10^{-14}	−135.7	−105.7	2.67×10^{-15}
9	2.82×10^{-6}	−42.5	7.46×10^{-9}	−40.5	9.40×10^{-3}	9.40×10^{-6}	9.40×10^{-11}	9.40×10^{-15}	−136.7	−106.7	2.12×10^{-15}
8	2.51×10^{-6}	−43.5	6.65×10^{-9}	−41.5	8.38×10^{-3}	8.38×10^{-6}	8.38×10^{-11}	8.38×10^{-15}	−137.7	−107.7	1.68×10^{-15}
7	2.24×10^{-6}	−44.5	5.93×10^{-9}	−42.5	7.46×10^{-3}	7.46×10^{-6}	7.46×10^{-11}	7.46×10^{-15}	−138.7	−108.7	1.34×10^{-15}
6	2×10^{-6}	−45.5	5.28×10^{-9}	−43.5	6.65×10^{-3}	6.65×10^{-6}	6.65×10^{-11}	6.65×10^{-15}	−139.7	−109.7	1.06×10^{-15}
5	1.78×10^{-6}	−46.5	4.71×10^{-9}	−44.5	5.93×10^{-3}	5.93×10^{-6}	5.39×10^{-11}	5.93×10^{-15}	−140.7	−110.7	8.43×10^{-16}
4	1.58×10^{-6}	−47.5	4.20×10^{-9}	−45.5	5.28×10^{-3}	5.28×10^{-6}	5.28×10^{-11}	5.28×10^{-15}	−141.7	−111.7	6.70×10^{-16}
3	1.41×10^{-6}	−48.5	3.74×10^{-9}	−46.5	4.71×10^{-3}	4.71×10^{-6}	4.71×10^{-11}	4.71×10^{-15}	−142.7	−112.7	5.32×10^{-16}
2	1.26×10^{-6}	−49.5	3.33×10^{-9}	−47.5	4.20×10^{-3}	4.20×10^{-6}	4.20×10^{-11}	4.20×10^{-15}	−143.7	−113.7	4.23×10^{-16}
1	1.12×10^{-6}	−50.5	2.97×10^{-9}	−48.5	3.74×10^{-3}	3.74×10^{-6}	3.74×10^{-11}	3.74×10^{-15}	−144.7	−114.7	3.36×10^{-16}
0	10^{-6}	−51.5	2.65×10^{-9}	−49.5	3.33×10^{-3}	3.33×10^{-6}	3.33×10^{-11}	3.33×10^{-15}	−145.7	−115.7	2.67×10^{-16}
−20	10^{-7}	−71.5	2.65×10^{-10}	−69.5	3.33×10^{-4}	3.33×10^{-7}	3.33×10^{-12}	3.33×10^{-16}	−165.7	−135.7	2.67×10^{-18}
−40	10^{-8}	−91.5	2.65×10^{-11}	−89.5	3.33×10^{-5}	3.33×10^{-8}	3.33×10^{-13}	3.33×10^{-17}	−185.7	−155.7	2.67×10^{-20}
−60	10^{-9}	−111.5	2.65×10^{-12}	−109.5	3.33×10^{-6}	3.33×10^{-9}	3.33×10^{-14}	3.33×10^{-18}	−205.7	−175.7	2.67×10^{-22}

Table A.4—Magnetic Flux Density-to-Field Strength-to Power Densities

dBpT	pT/mΓ	nT/Γ	Gauss	Tesla	dBμA/m	A/m	dBμV/m	V/m	dBW/m²	dBm/m²	mW/cm²
240	10^{12}	10^{9}	10^{4}	1	238	8×10^{5}	289.5	3×10^{8}	143.8	173.8	2.4×10^{13}
220	10^{11}	10^{8}	10^{3}	0.1	218	8×10^{4}	269.5	3×10^{7}	123.8	153.8	2.4×10^{11}
200	10^{10}	10^{7}	10^{2}	0.01	198	8×10^{3}	249.5	3×10^{6}	103.8	133.8	2.4×10^{9}
180	10^{9}	10^{6}	10	0.001	178	796	229.5	3×10^{5}	83.8	113.8	2.4×10^{7}
160	10^{8}	10^{5}	1	10^{-4}	158	80	209.5	3×10^{4}	63.8	93.8	2.4×10^{5}
140	10^{7}	10^{4}	0.1	10^{-5}	138	8	189.5	3×10^{3}	43.8	73.8	2.4×10^{3}
120	10^{6}	10^{3}	0.01	10^{-6}	118	0.8	169.5	300	23.8	53.8	23.9
100	10^{5}	10^{2}	0.001	10^{-7}	98	0.08	149.5	30	3.8	33.8	0.24
80	10^{4}	10	10^{-4}	10^{-8}	78	0.008	129.5	3	−16.2	13.8	0.0024
60	10^{3}	1	10^{-5}	10^{-9}	58	8×10^{-4}	109.5	0.3	−36.2	−6.2	2.39×10^{-5}
40	100	0.1	10^{-6}	10^{-10}	38	8×10^{-5}	89.5	0.03	−56.2	−26.2	2.39×10^{-7}
20	10.000	0.01	10^{-7}	10^{-11}	18	7.96×10^{-6}	69.5	0.003	−76.2	−46.2	2.39×10^{-9}
19	8.913	8.91×10^{-3}	8.9×10^{-8}	8.9×10^{-12}	17	7.09×10^{-6}	68.5	2.67×10^{-3}	−77.2	−47.2	1.90×10^{-9}
18	7.943	7.94×10^{-3}	7.94×10^{-8}	7.94×10^{-12}	16	6.32×10^{-6}	67.5	2.38×10^{-3}	−78.2	−48.2	1.51×10^{-9}
17	7.079	7.08×10^{-3}	7.08×10^{-8}	7.08×10^{-12}	15	5.63×10^{-6}	66.5	2.12×10^{-3}	−79.2	−49.2	1.20×10^{-3}
16	6.310	6.31×10^{-3}	6.31×10^{-8}	6.31×10^{-12}	14	5.02×10^{-6}	65.5	1.89×10^{-3}	−80.2	−50.2	9.50×10^{-10}
15	5.623	5.62×10^{-3}	5.62×10^{-8}	5.62×10^{-12}	13	4.47×10^{-6}	64.5	1.69×10^{-3}	−81.2	−51.2	7.55×10^{-10}
14	5.012	5.01×10^{-3}	5.01×10^{-8}	5.01×10^{-12}	12	3.99×10^{-6}	63.5	1.50×10^{-3}	−82.2	−52.2	6.00×10^{-10}
13	4.467	4.47×10^{-3}	4.47×10^{-8}	4.47×10^{-12}	11	3.55×10^{-6}	62.5	1.34×10^{-3}	−83.2	−53.2	4.76×10^{-10}
12	3.981	3.98×10^{-3}	3.98×10^{-8}	3.98×10^{-12}	10	3.17×10^{-6}	61.5	1.19×10^{-3}	−84.2	−54.2	3.78×10^{-10}
11	3.548	3.55×10^{-3}	3.55×10^{-8}	3.55×10^{-12}	9	2.82×10^{-6}	60.5	1.06×10^{-3}	−85.2	−55.2	3.01×10^{-10}
10	3.162	3.16×10^{-3}	3.16×10^{-8}	3.16×10^{-12}	8	2.52×10^{-6}	59.5	9.49×10^{-4}	−86.2	−56.2	2.39×10^{-10}
9	2.818	2.82×10^{-3}	2.82×10^{-8}	2.82×10^{-12}	7	2.24×10^{-6}	58.5	8.46×10^{-4}	−87.2	−57.2	1.90×10^{-10}
8	2.512	2.51×10^{-1}	2.51×10^{-8}	2.51×10^{-12}	6	2.00×10^{-6}	57.5	7.54×10^{-4}	−88.2	−58.2	1.51×10^{-10}
7	2.239	2.24×10^{-3}	2.24×10^{-8}	2.24×10^{-12}	5	1.78×10^{-6}	56.5	6.72×10^{14}	−89.2	−59.2	1.20×10^{-0}
6	1.995	2.00×10^{-3}	2.00×10^{-8}	2.00×10^{-12}	4	1.59×10^{-6}	55.5	5.49×10^{-4}	−90.2	−60.2	9.50×10^{-11}
5	1.778	1.78×10^{-3}	1.78×10^{-8}	1.78×10^{-12}	3	1.42×10^{-6}	54.5	5.33×10^{-4}	−91.2	−61.2	7.55×10^{-11}
4	1.585	1.58×10^{-3}	1.58×10^{-8}	1.58×10^{-12}	2	1.26×10^{-6}	53.5	4.75×10^{-4}	−92.2	−62.2	6.00×10^{-11}
3	1.413	1.41×10^{-3}	1.41×10^{-8}	1.41×10^{-12}	1	1.12×10^{-6}	52.5	4.24×10^{-4}	−93.2	−63.2	4.76×10^{-11}
2	1.259	1.26×10^{-3}	1.26×10^{-8}	1.26×10^{-12}	0	1.00×10^{-6}	51.5	3.78×10^{-4}	−94.2	−64.1	3.78×10^{-11}
1	1.122	1.12×10^{-3}	1.12×10^{-8}	1.12×10^{-12}	−1	8.93×10^{-7}	50.5	3.37×10^{-4}	−95.2	−65.2	3.01×10^{-11}
0	1.000	10^{-3}	10^{-8}	10^{-12}	−2	7.96×10^{-7}	49.5	3.00×10^{-4}	−96.2	−66.2	2.39×10^{-11}
−20	0.1	10^{-4}	10^{-9}	10^{-13}	−22	8×10^{-8}	29.5	3×10^{-5}	−116.2	−86.2	2.4×10^{-11}
−40	0.01	10^{-5}	10^{-10}	10^{-14}	−42	8×10^{-9}	9.5	3×10^{-6}	−136.2	−106.2	2.4×10^{-15}
−60	0.001	10^{-6}	10^{-11}	10^{-15}	−62	8×10^{-10}	−10.5	3×10^{-7}	−156.2	−126.2	2.4×10^{-17}

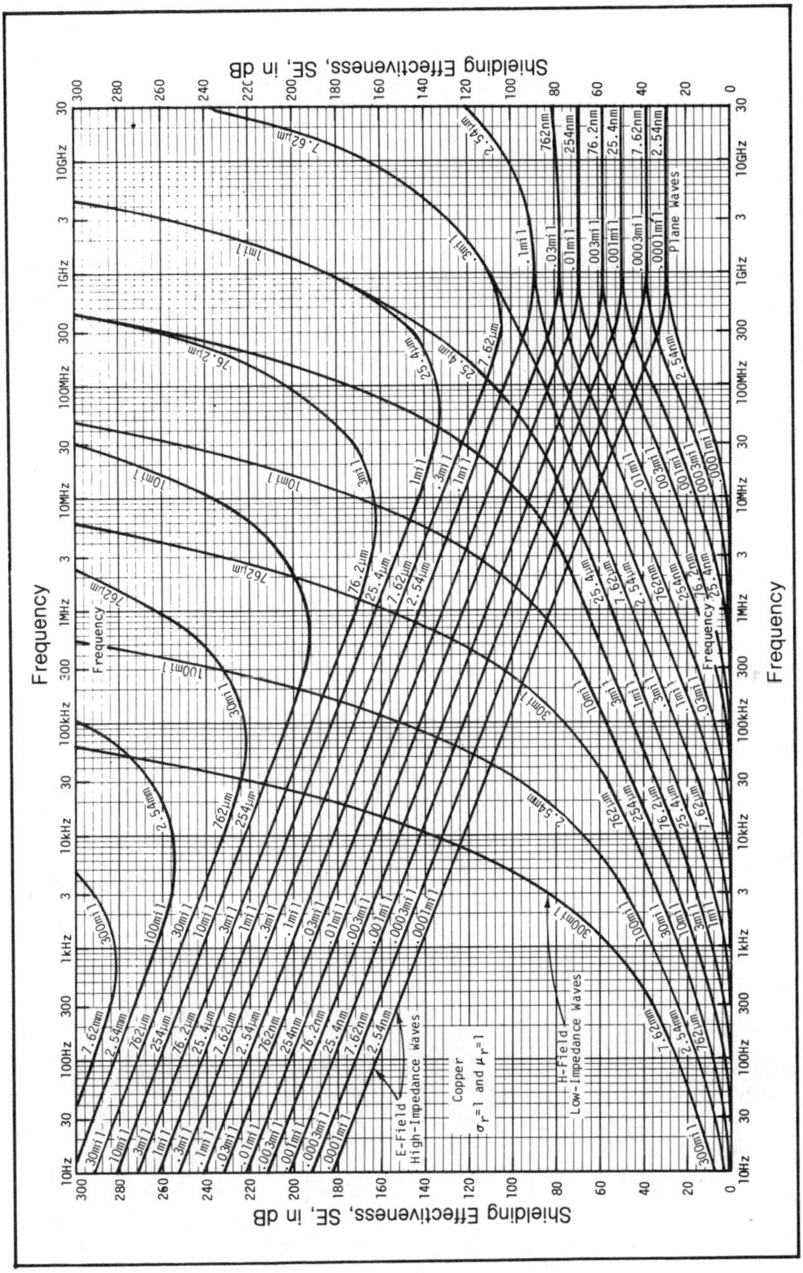

Figure B.1—Shielding Effectiveness of Copper vs. Frequency for Source-to-Metal Distance of 10 cm

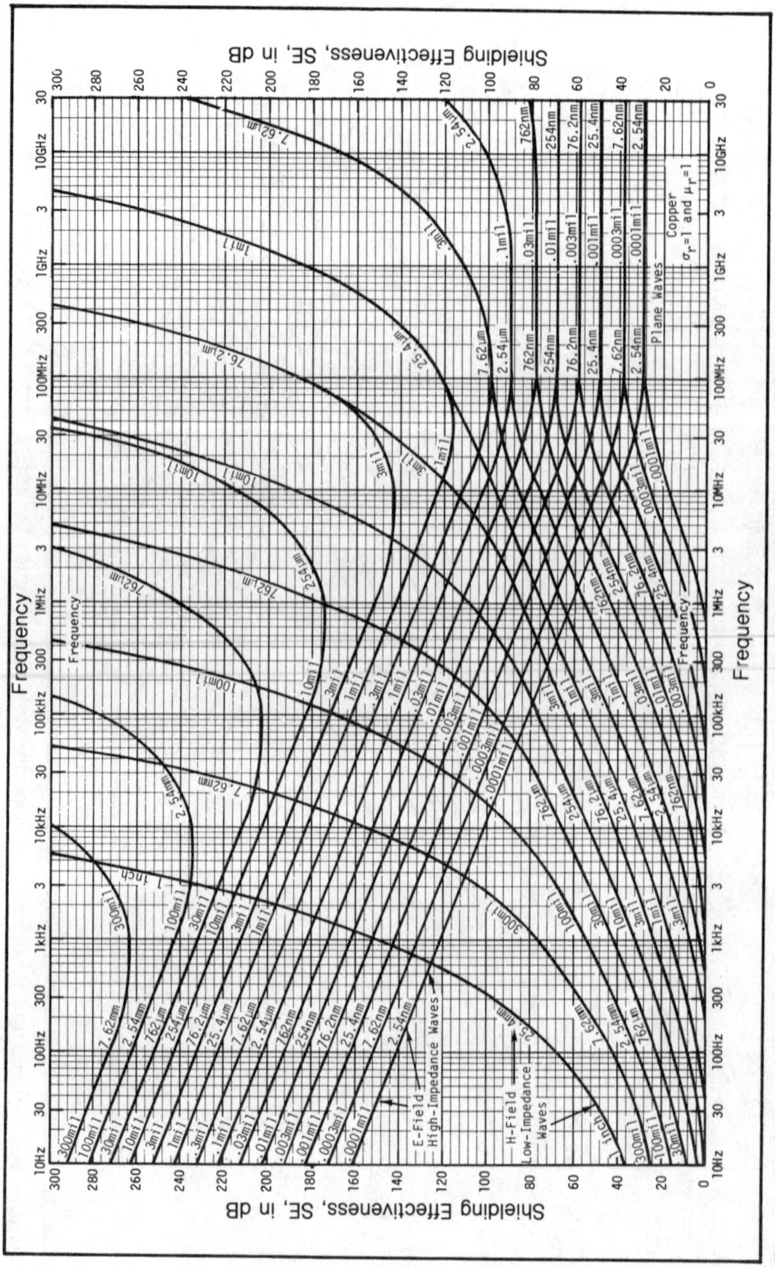

Figure B.2—Shielding Effectiveness of Copper vs. Frequency for Source-to-Metal Distance of 1 m

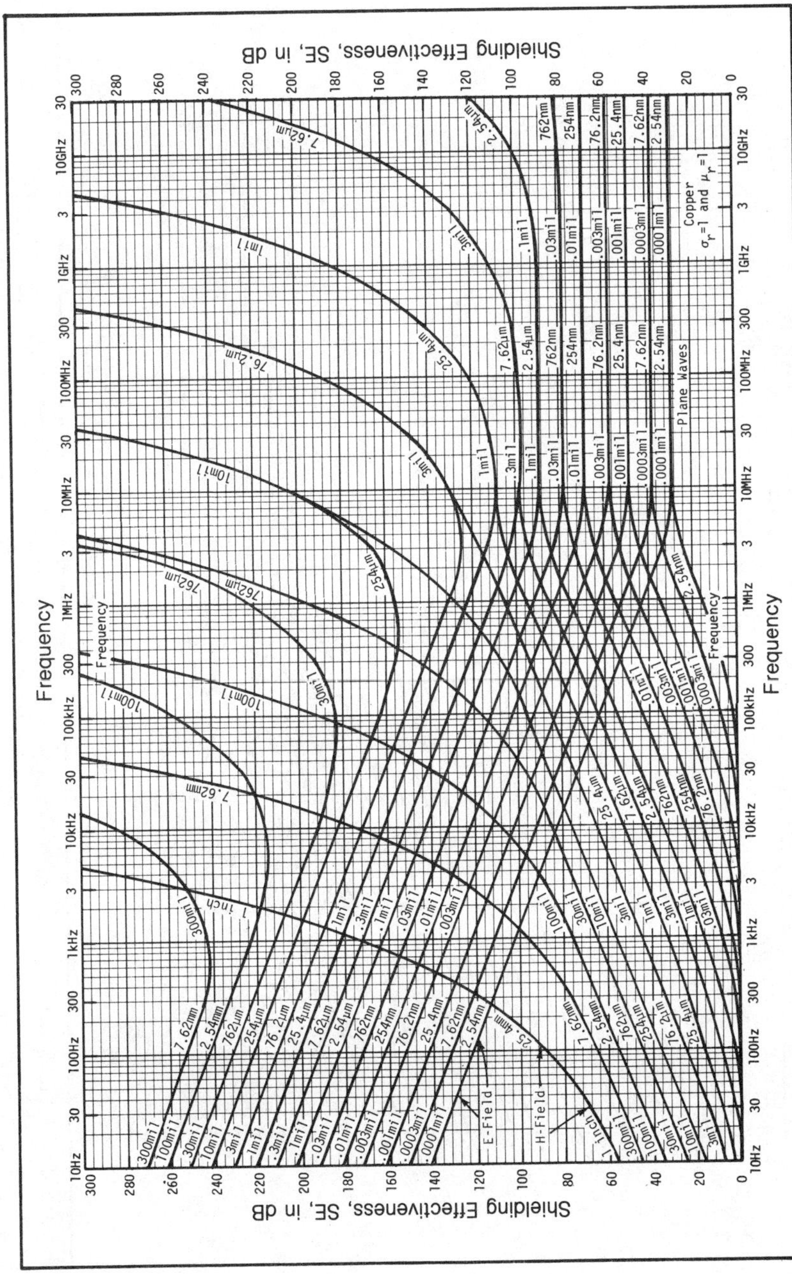

Figure B.3—Shielding Effectiveness of Copper vs. Frequency for Source-to-Metal Distance of 10 m

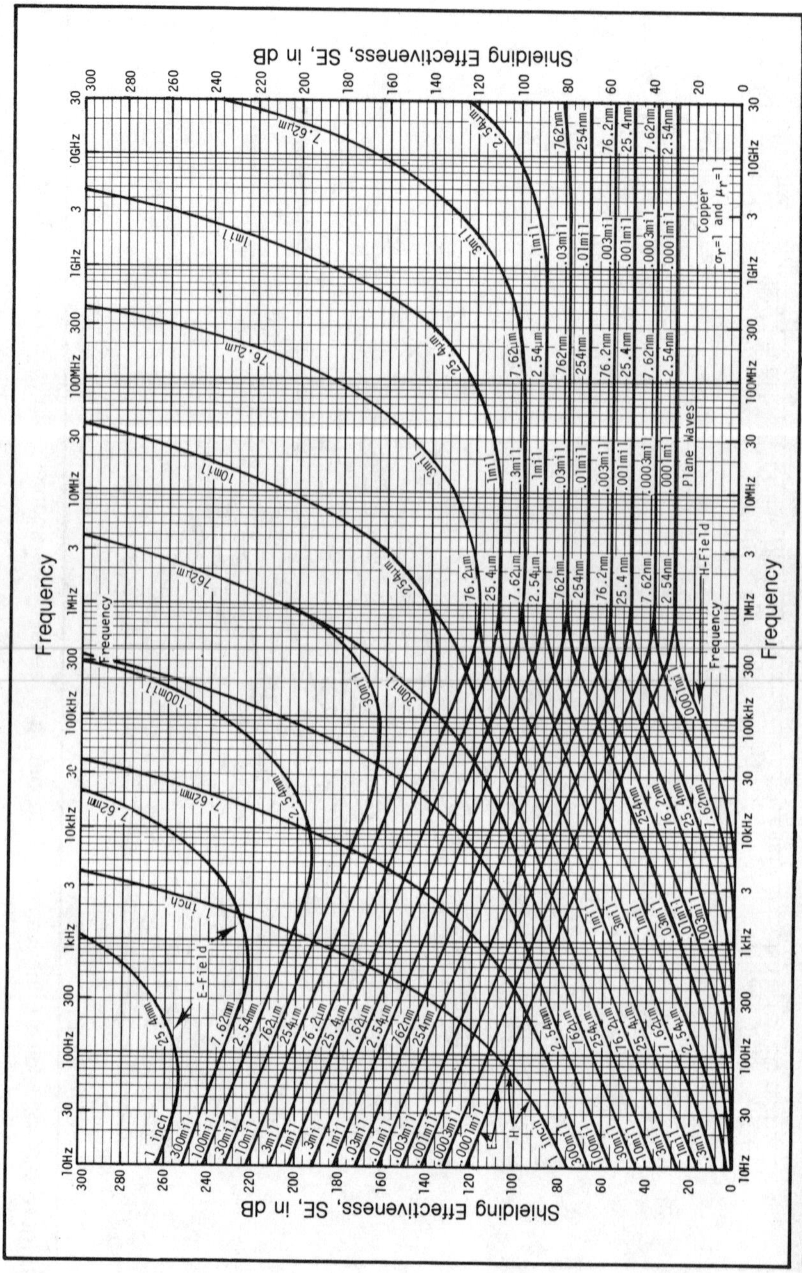

Figure B.4—Shielding Effectiveness of Copper vs. Frequency for Source-to-Metal Distance of 100 m

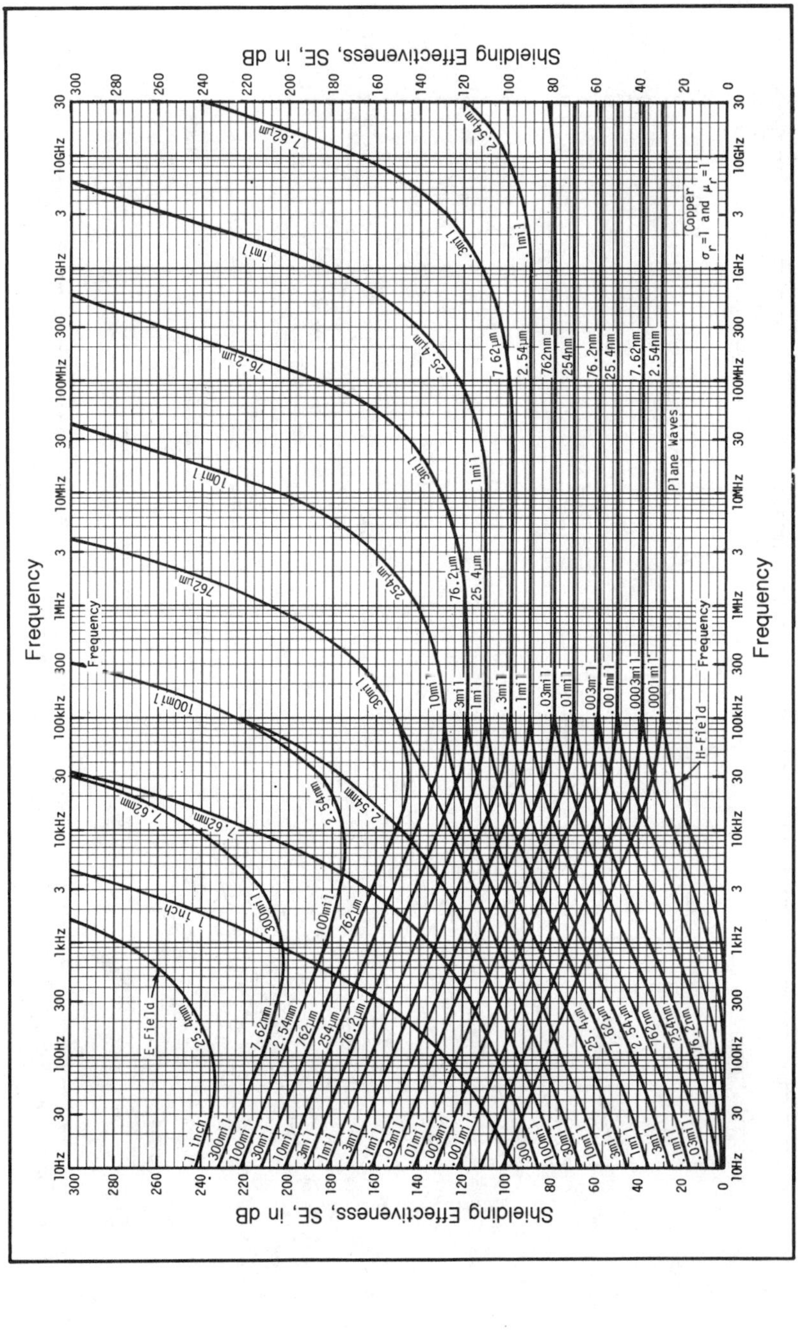

Figure B.5—Shielding Effectiveness of Copper vs. Frequency for Source-to-Metal Distance of 1 km

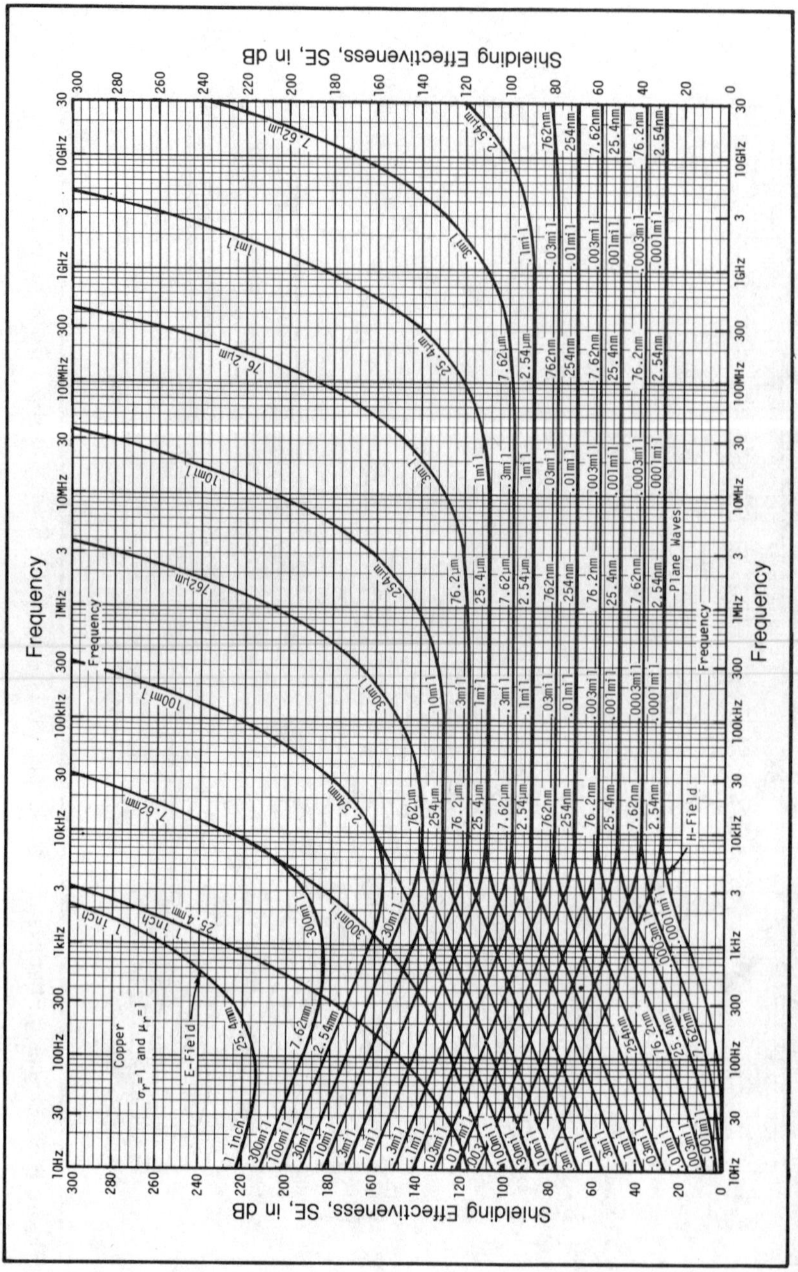

Figure B.6—Shielding Effectiveness of Copper vs. Frequency for Source-to-Metal Distance of 10 km

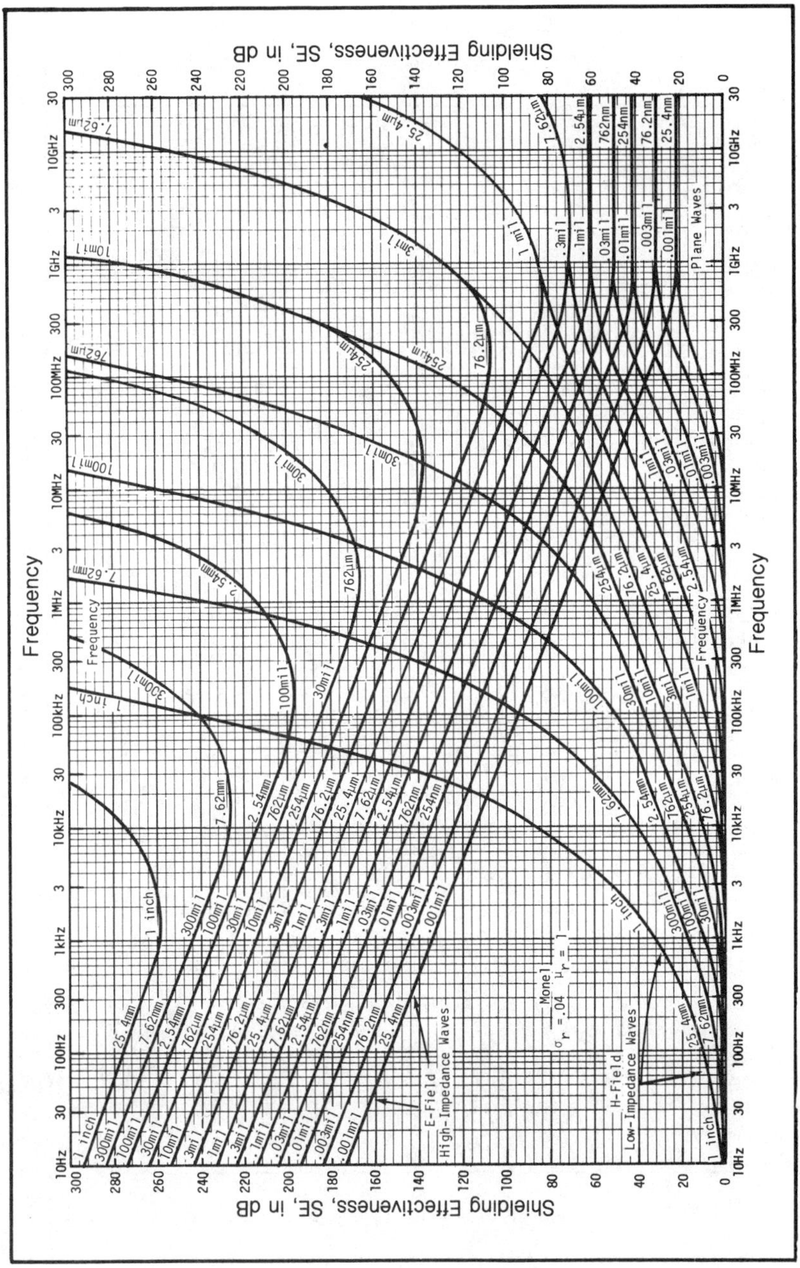

Figure C.1—Shielding Effectiveness of Monel vs. Frequency for Source-to-Metal Distance of 10 cm

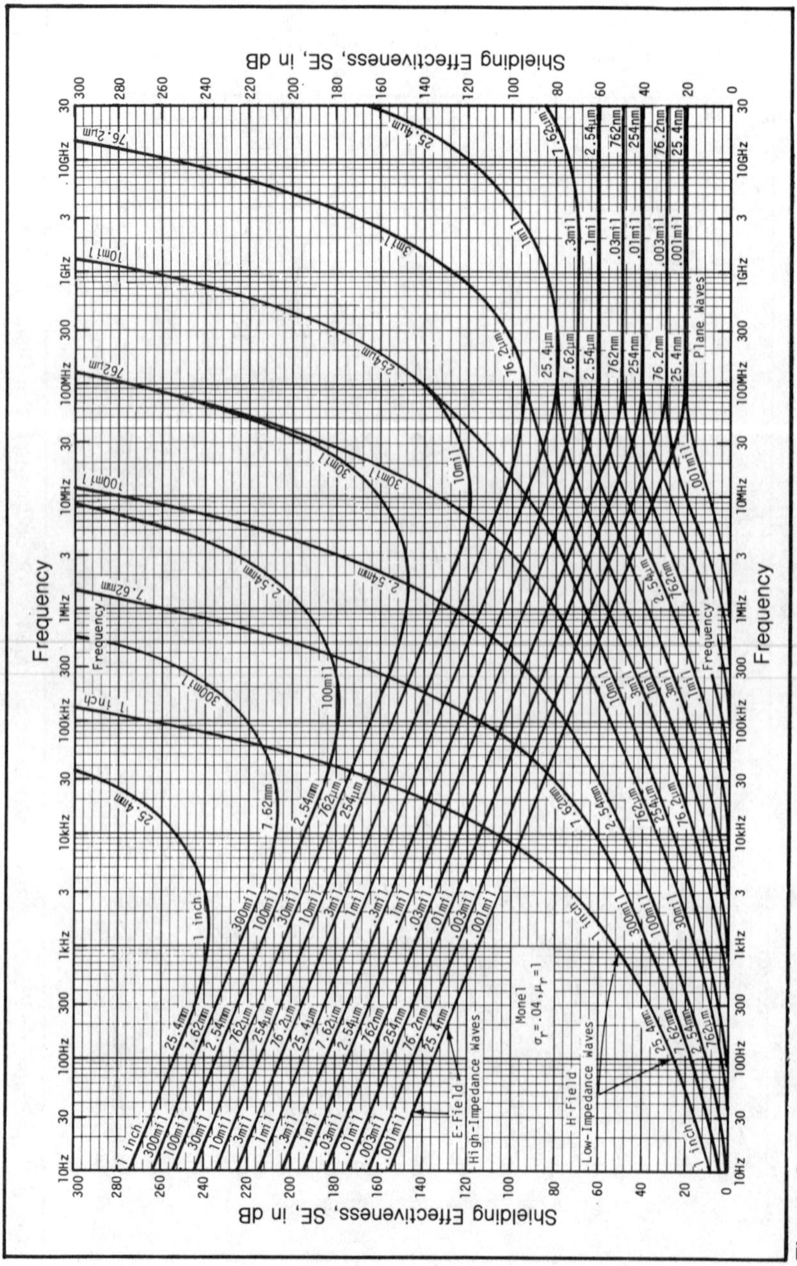

Figure C.2—Shielding Effectiveness of Monel vs. Frequency for Source-to-Metal Distance of 1 m

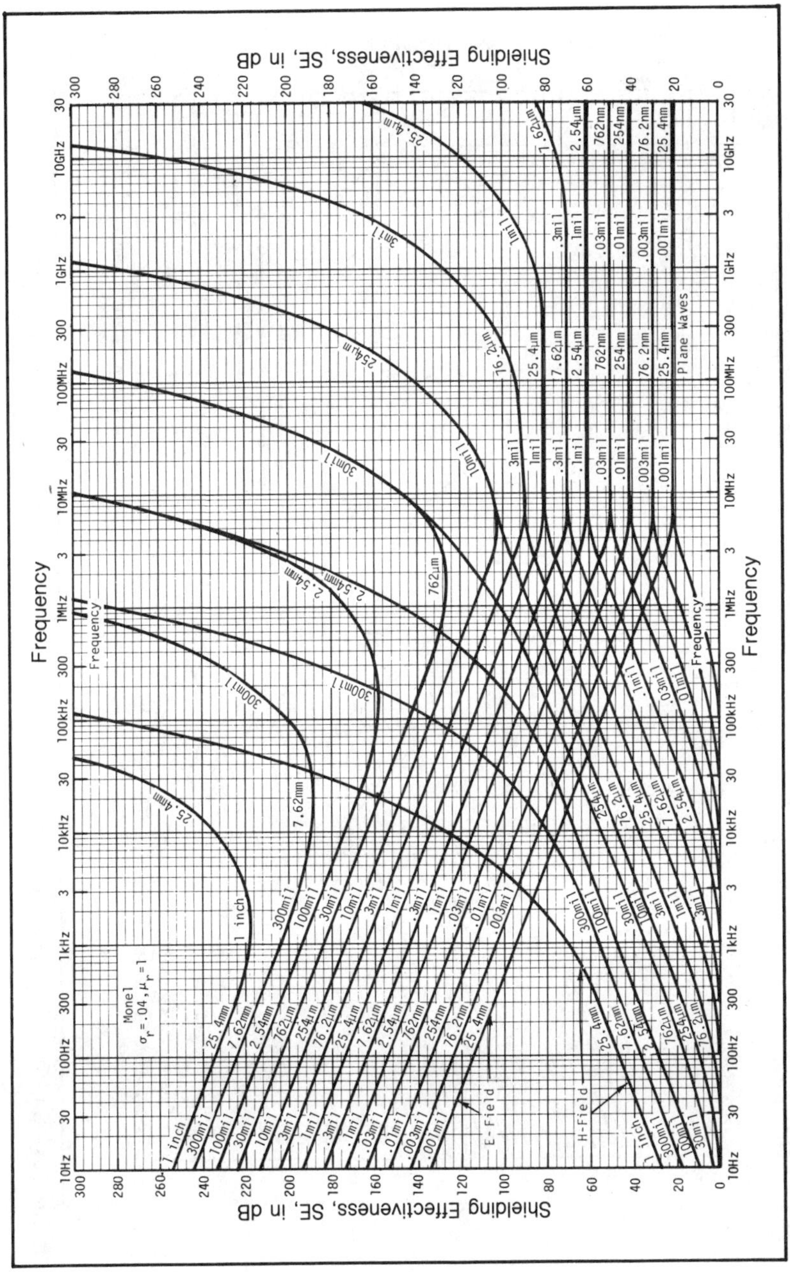

Figure C.3—Shielding Effectiveness of Monel vs. Frequency for Source-to-Metal Distance of 10 m

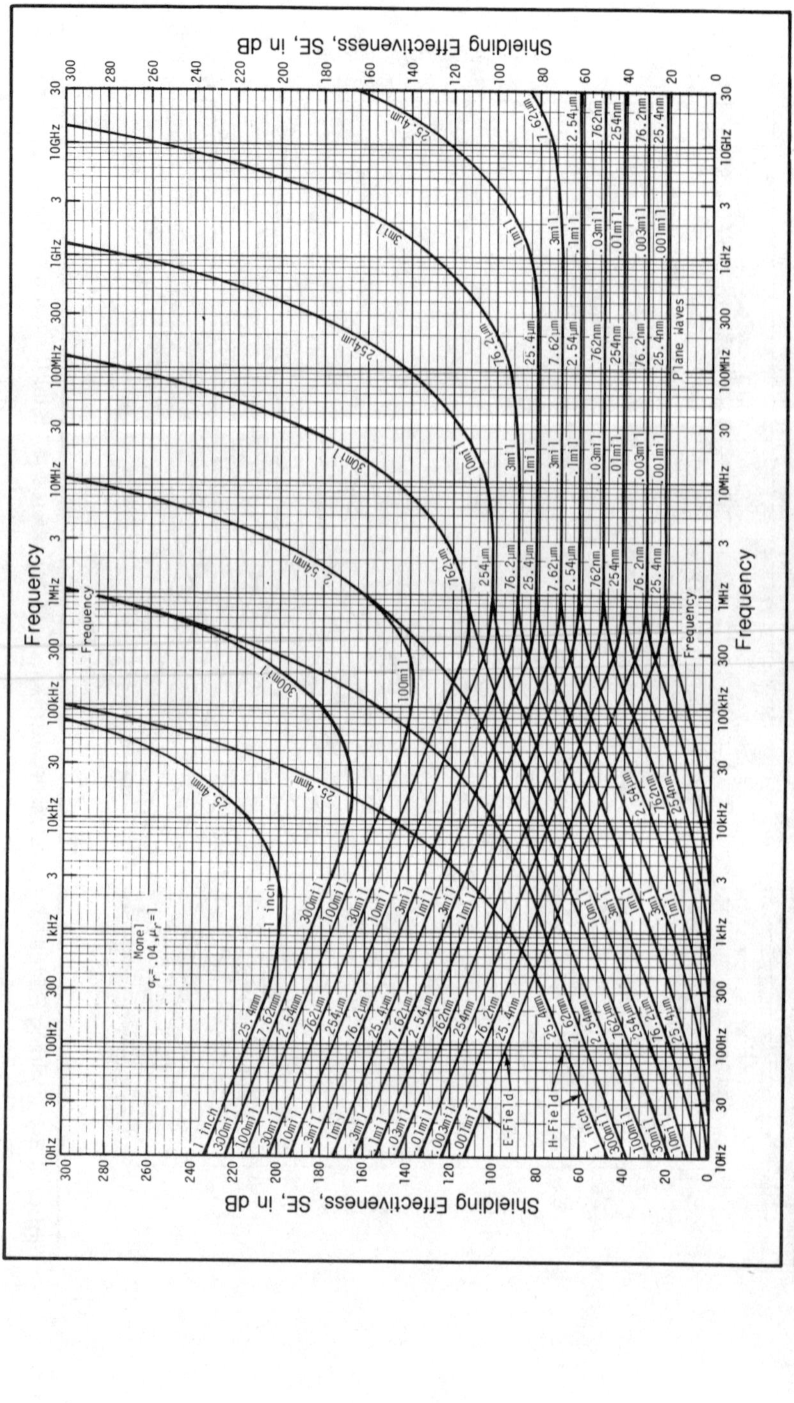

Figure C.4—Shielding Effectiveness of Monel vs. Frequency for Source-to-Metal Distance of 100 m

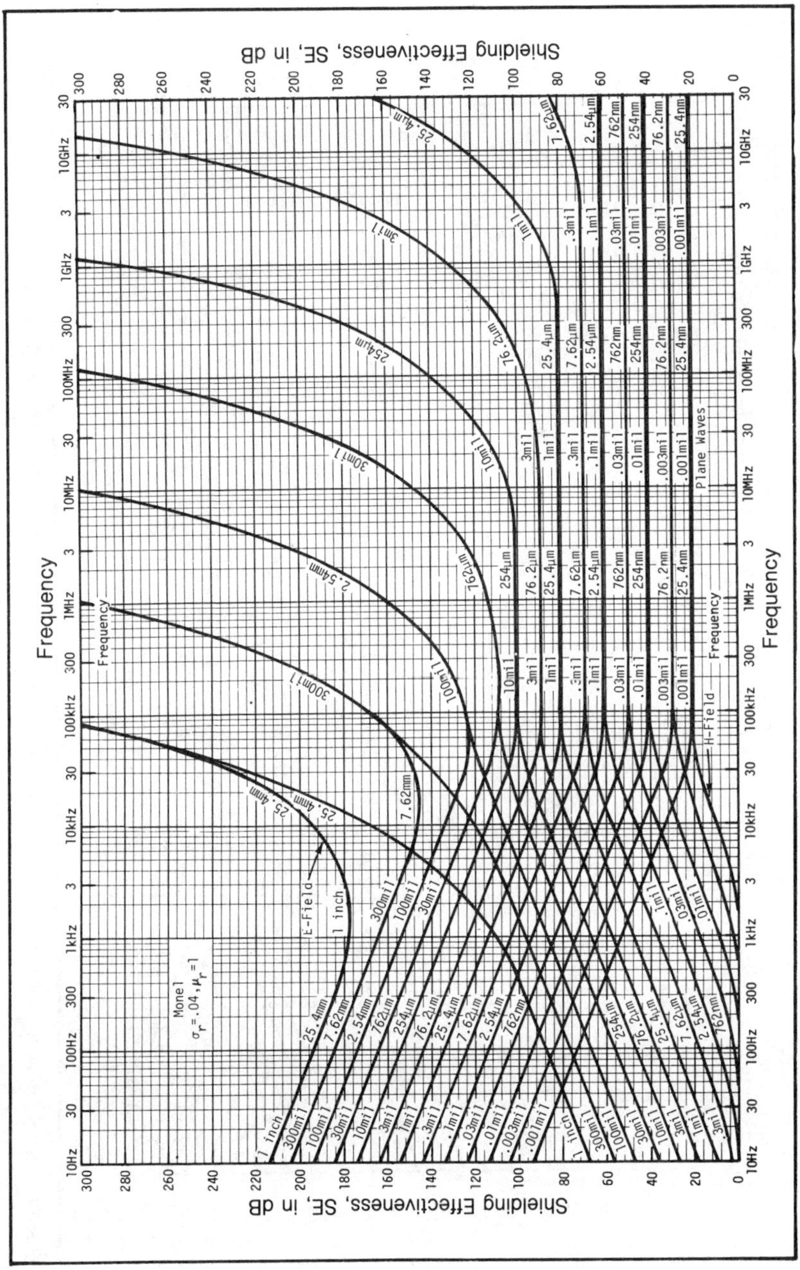

Figure C.5—Shielding Effectiveness of Monel vs. Frequency for Source-to-Metal Distance of 1 km

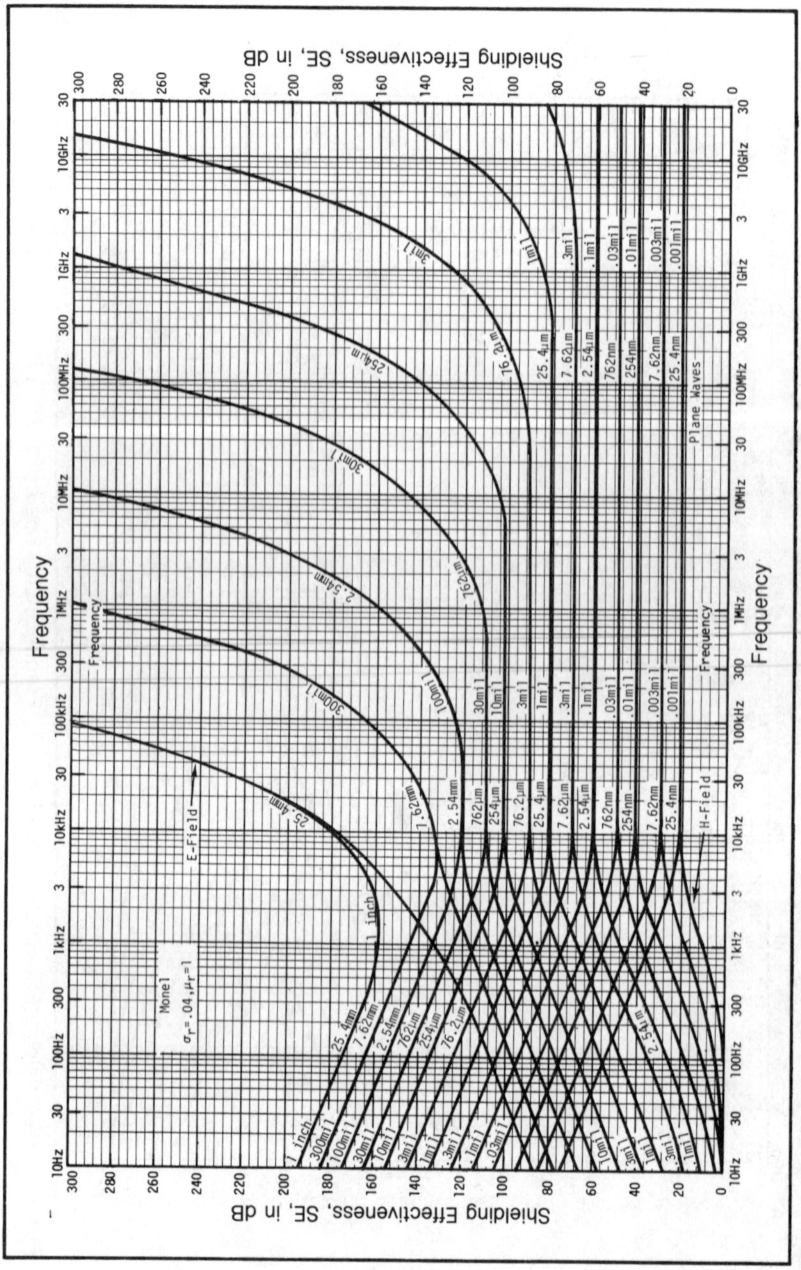

Figure C.6—Shielding Effectiveness of Monel vs. Frequency for Source-to-Metal Distance of 10 km

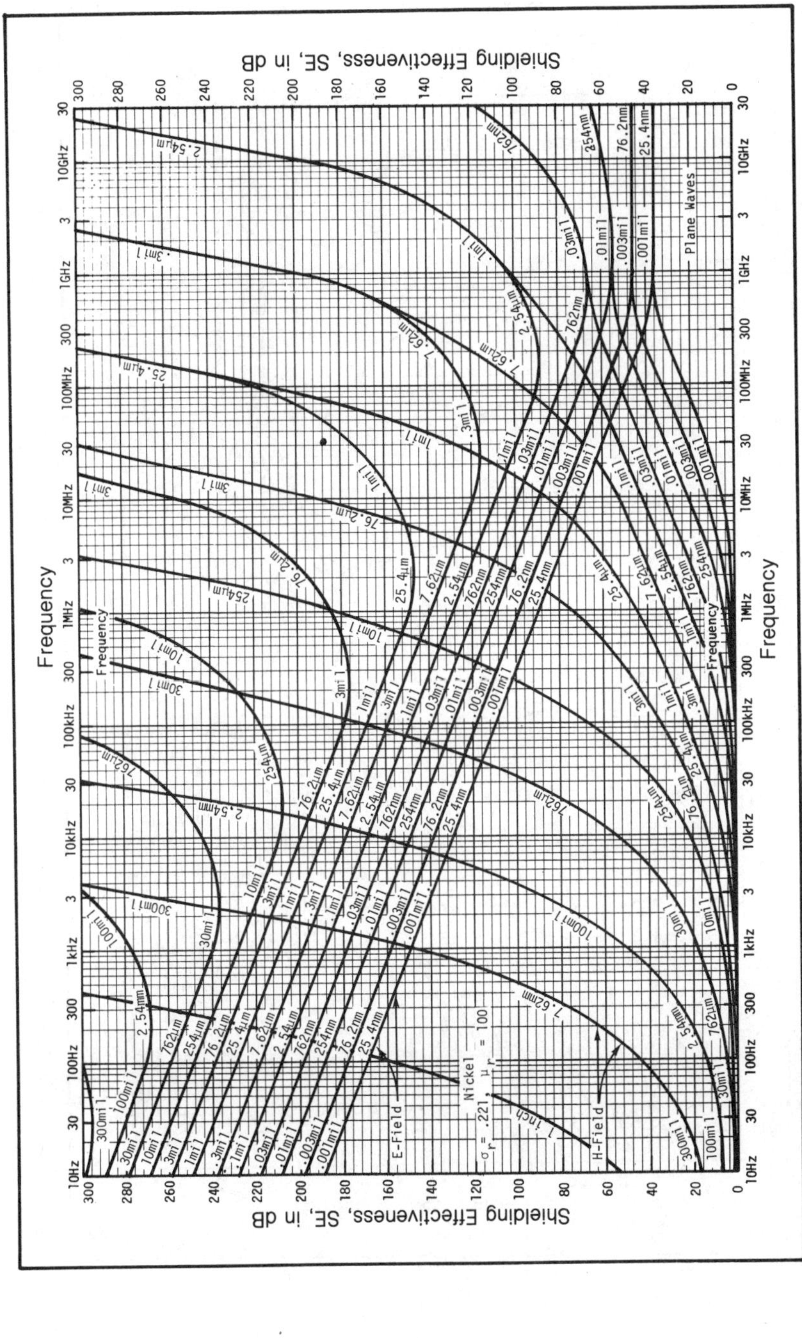

Figure D.1—Shielding Effectiveness of Nickel vs. Frequency for Source-to-Metal Distance of 10 cm

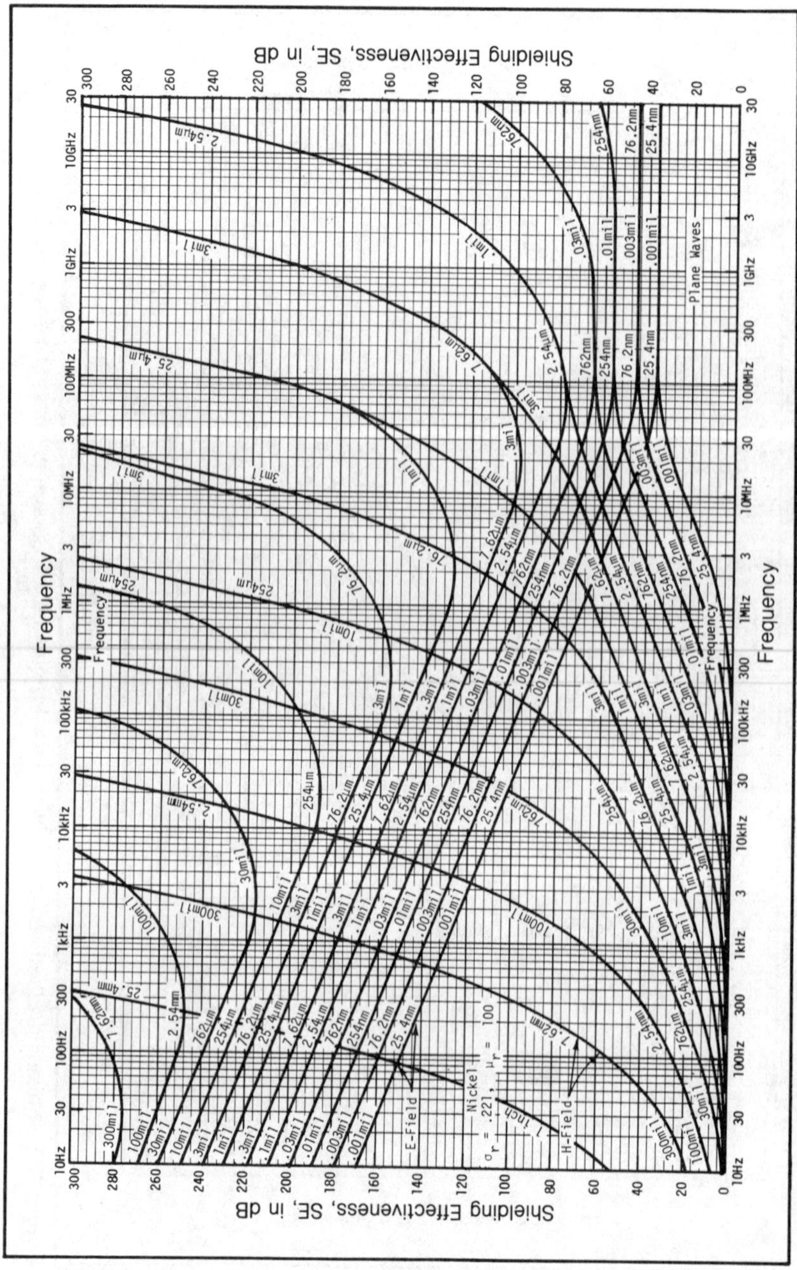

Figure D.2—Shielding Effectiveness of Nickel vs. Frequency for Source-to-Metal Distance of 1 m

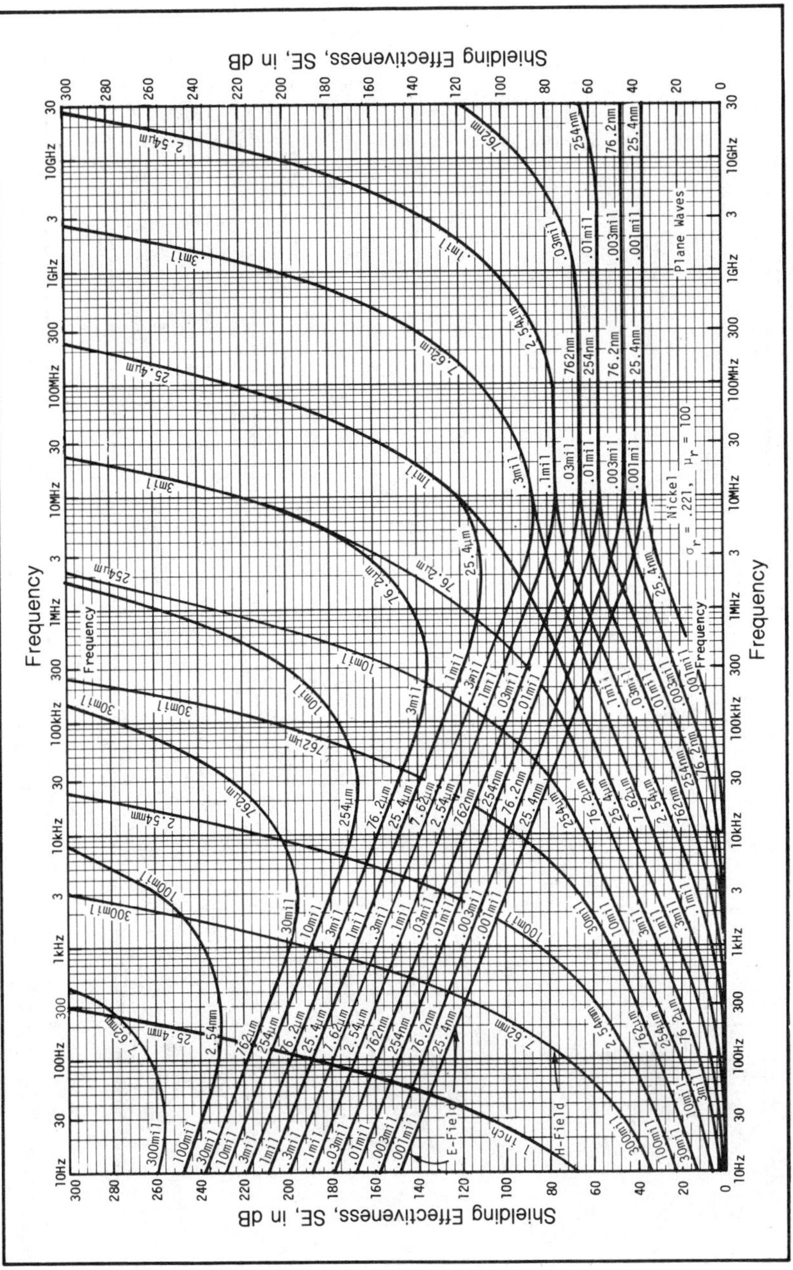

Figure D.3—Shielding Effectiveness of Nickel vs. Frequency for Source-to-Metal Distance of 10 m

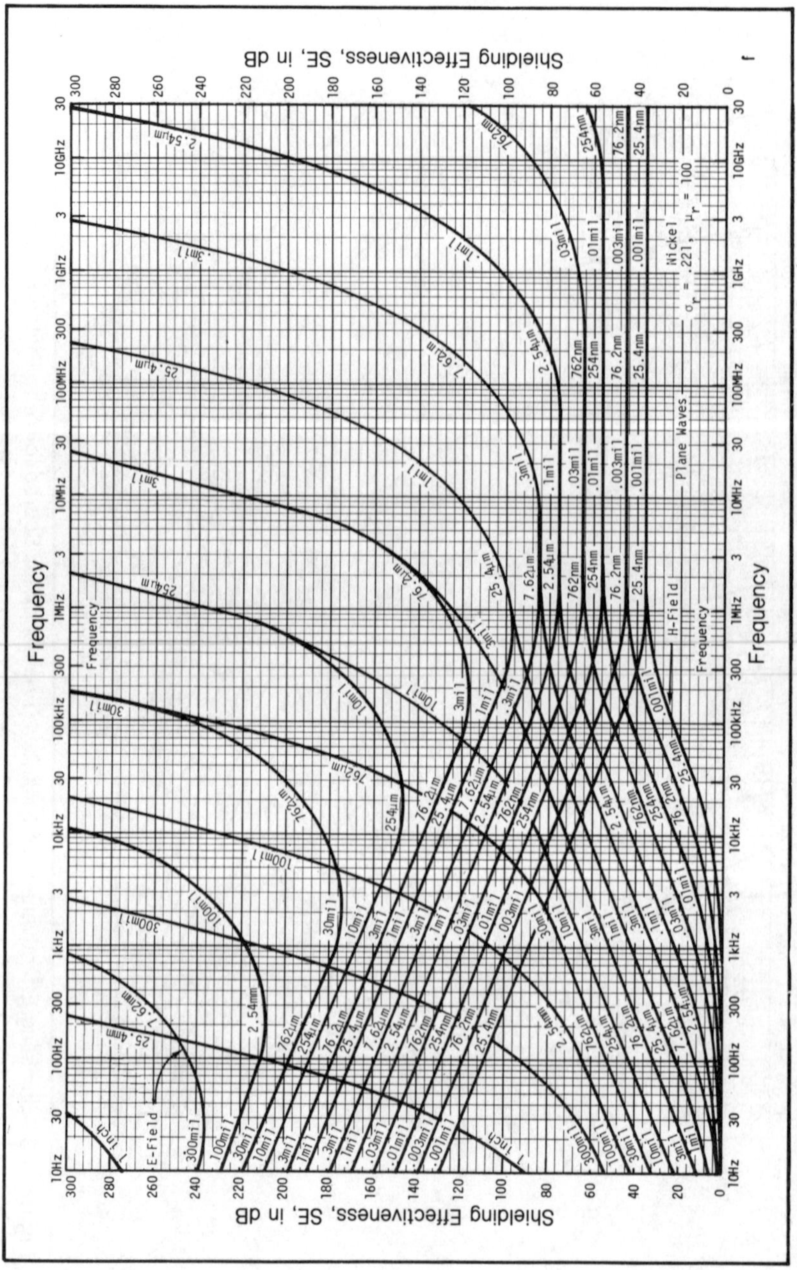

Figure D.4—Shielding Effectiveness of Nickel vs. Frequency for Source-to-Metal Distance of 100 m

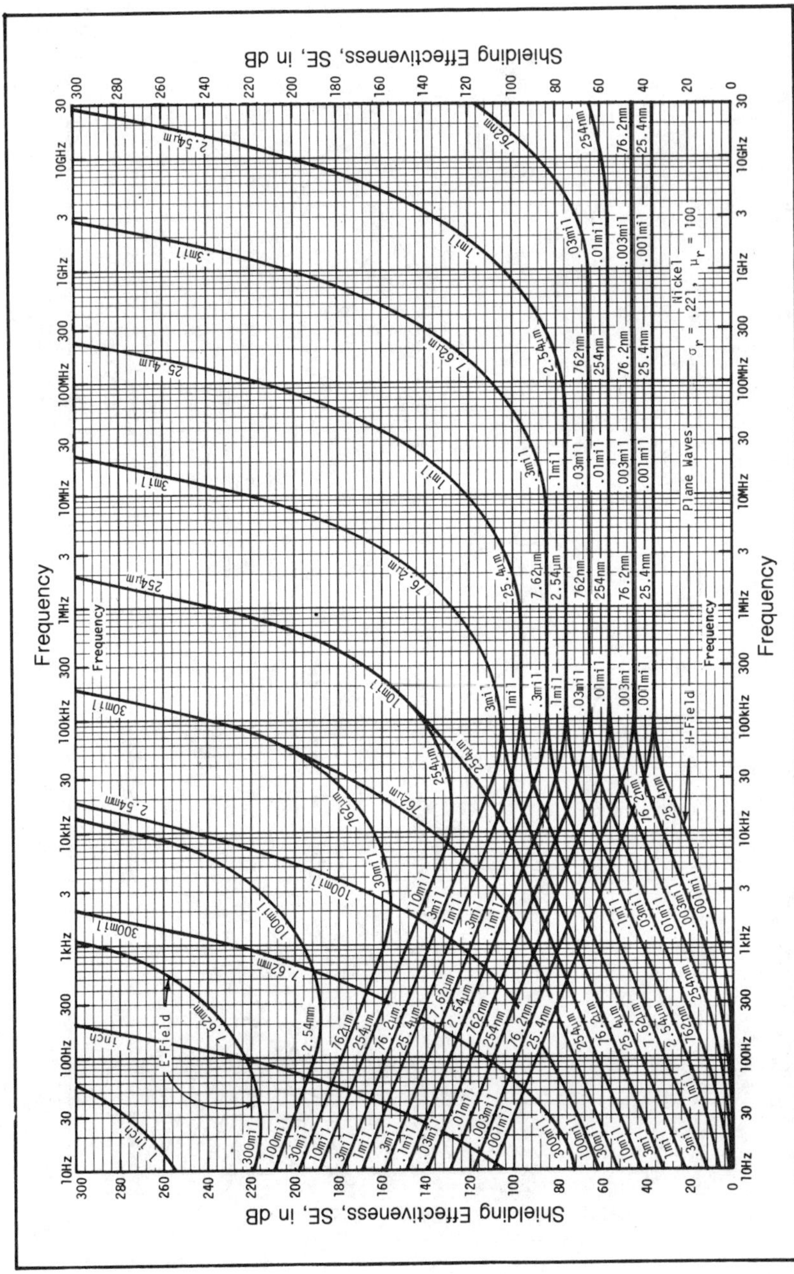

Figure D.5—Shielding Effectiveness of Nickel vs. Frequency for Source-to-Metal Distance of 1 km

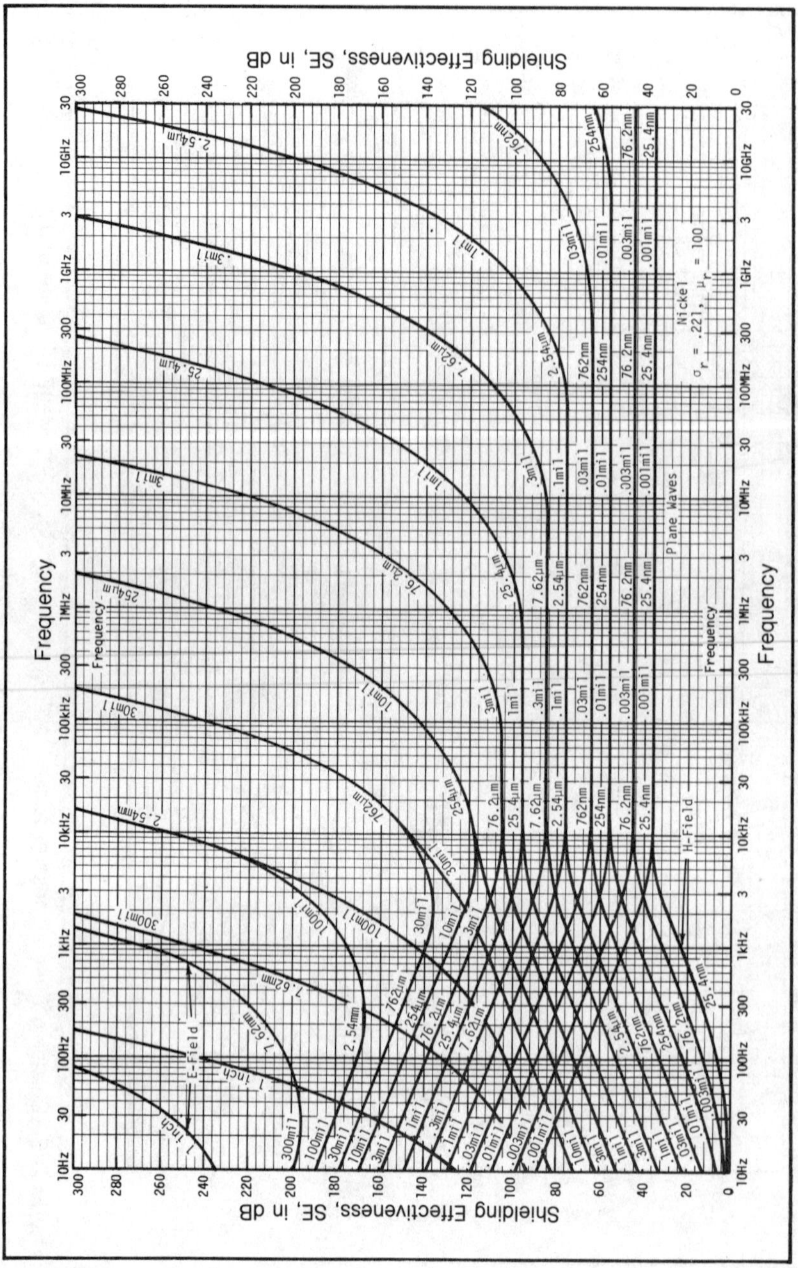

Figure D.6—Shielding Effectiveness of Nickel vs. Frequency for Source-to-Metal Distance of 10 km

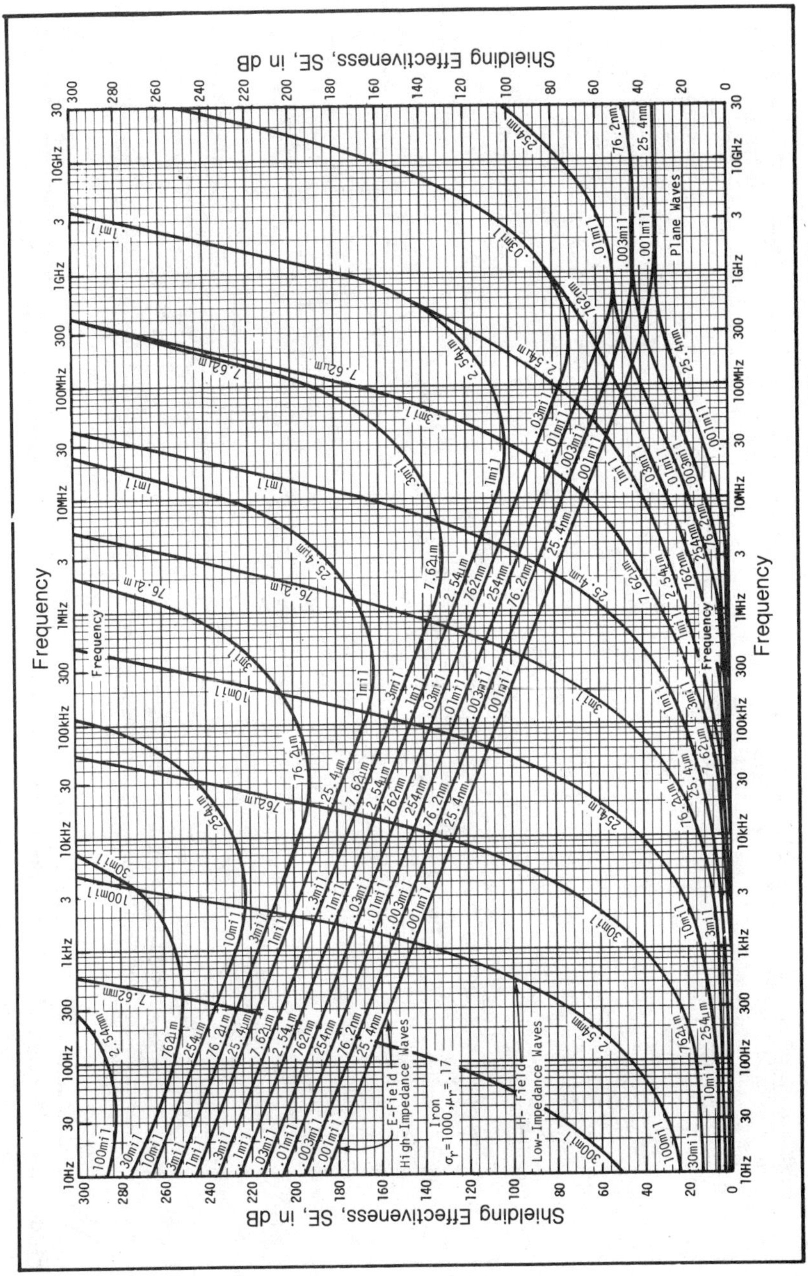

Figure E.1—Shielding Effectiveness of Iron vs. Frequency for Source-to-Metal Distance of 10 cm

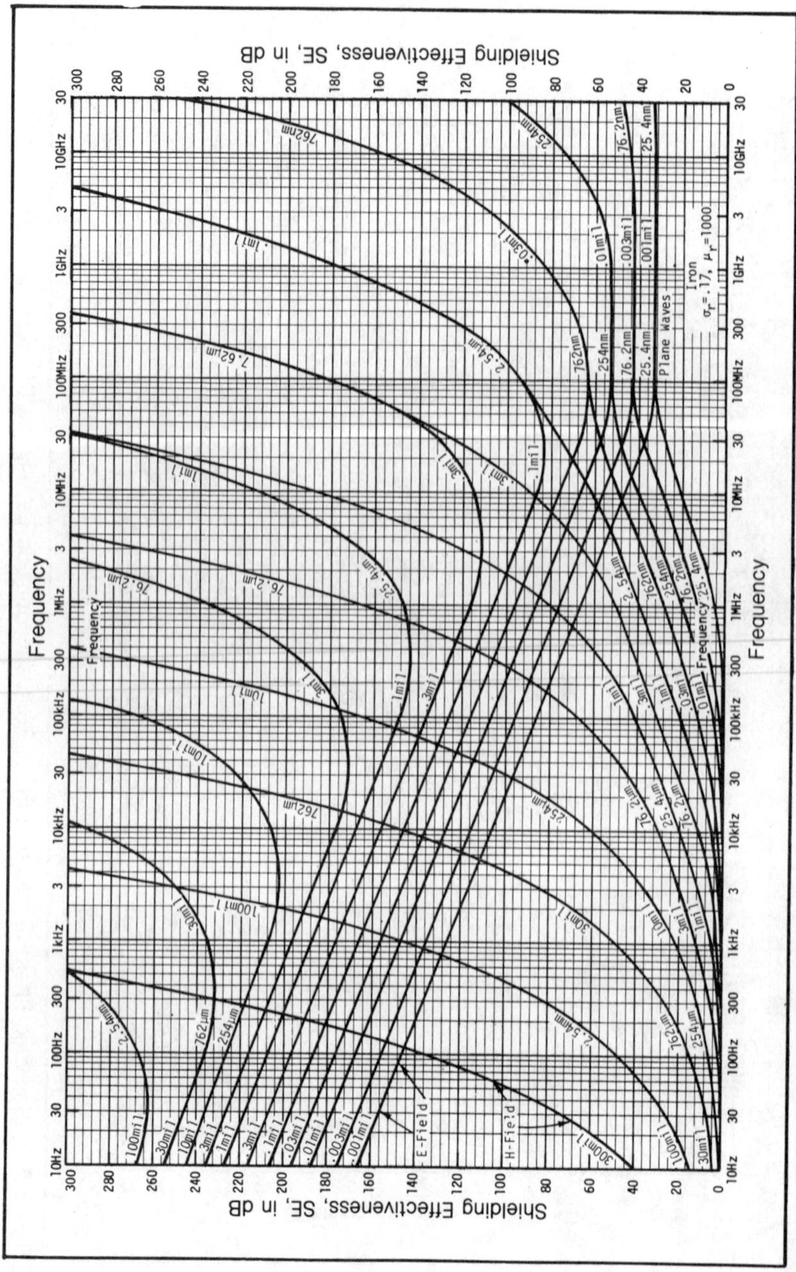

Figure E.2—Shielding Effectiveness of Iron vs. Frequency for Source-to-Metal Distance of 1 m

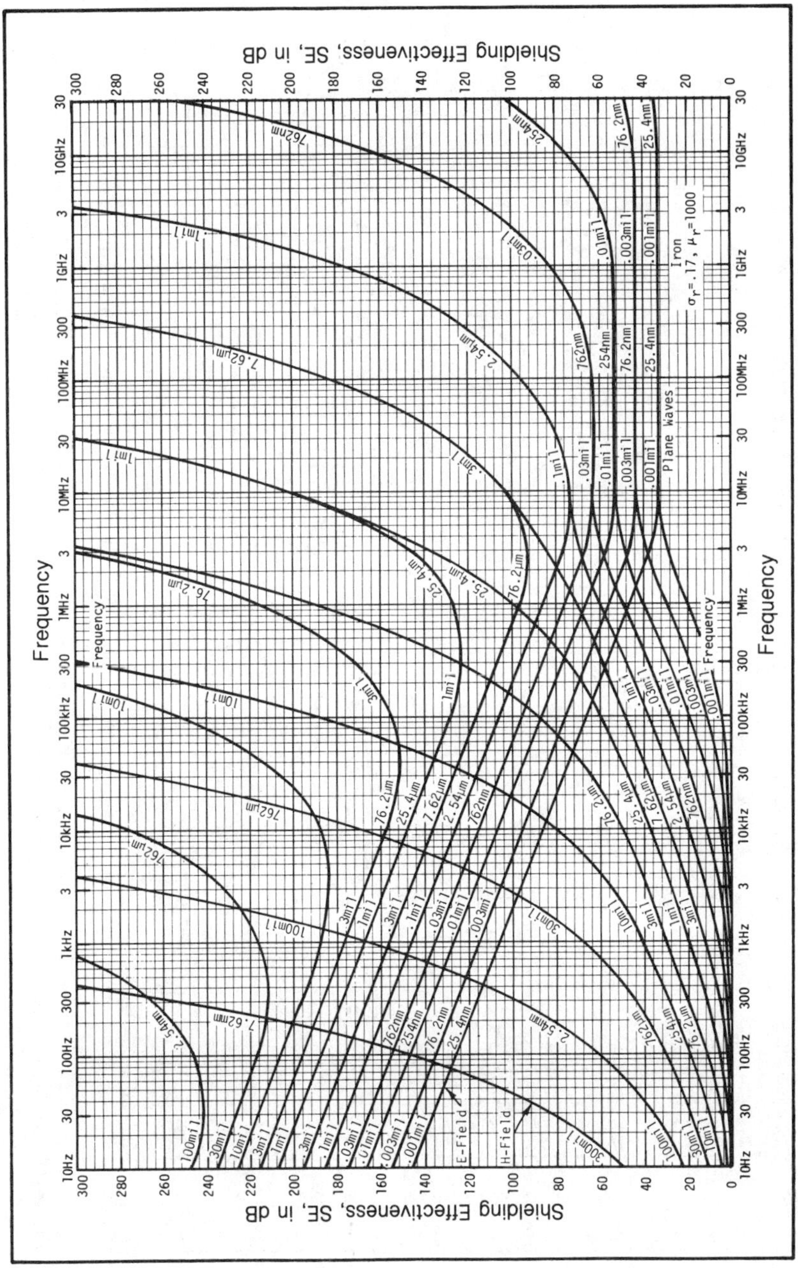

Figure E.3—Shielding Effectiveness of Iron vs. Frequency for Source-to-Metal Distance of 10 m

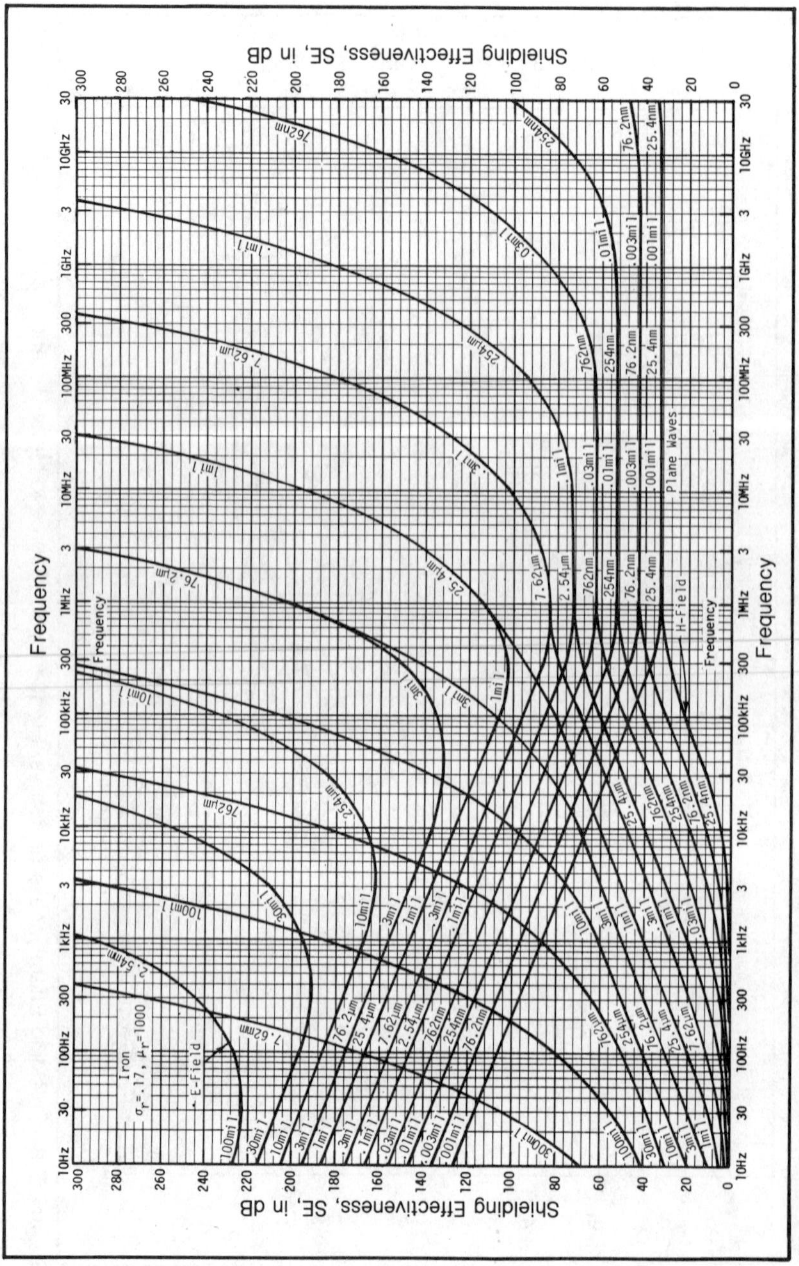

Figure E.4—Shielding Effectiveness of Iron vs. Frequency for Source-to-Metal Distance of 100 m

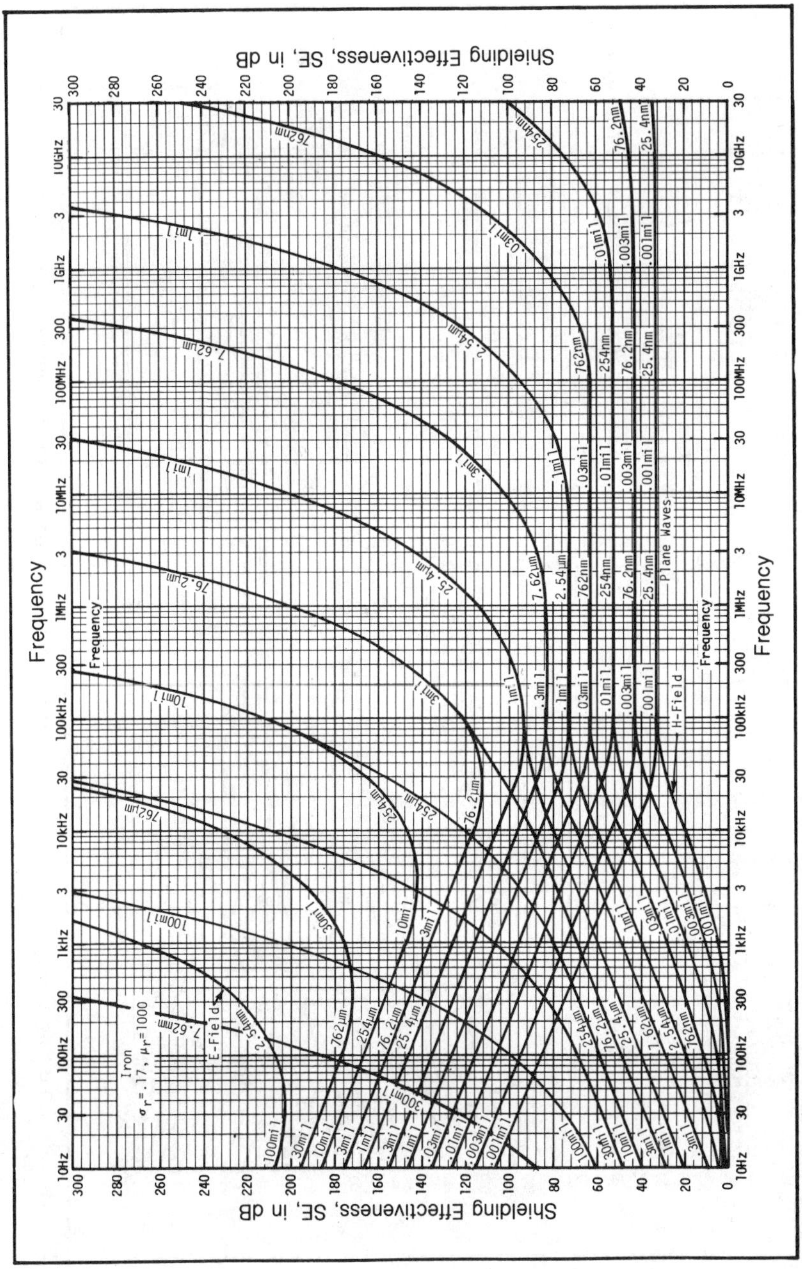

Figure E.5—Shielding Effectiveness of Iron vs. Frequency for Source-to-Metal Distance of 1 km

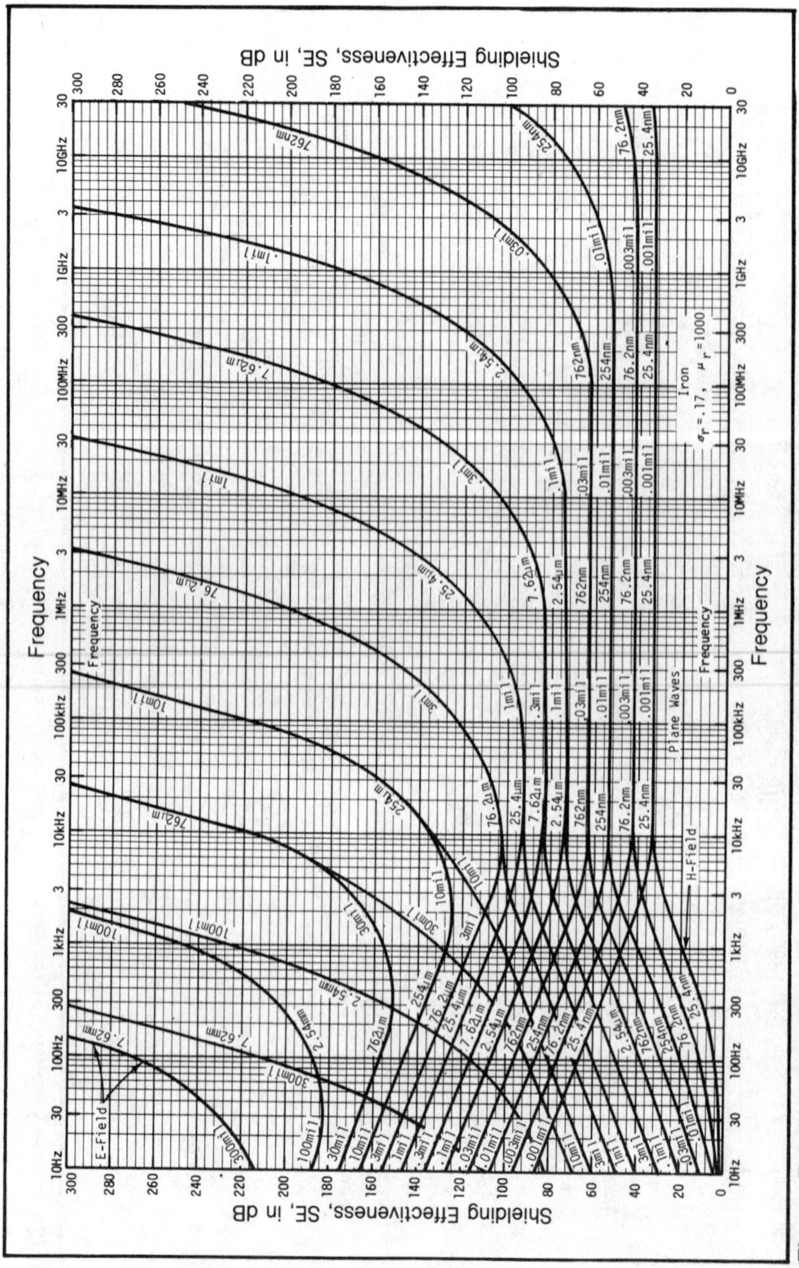

Figure E.6—Shielding Effectiveness of Iron vs. Frequency for Source-to-Metal Distance of 10 km

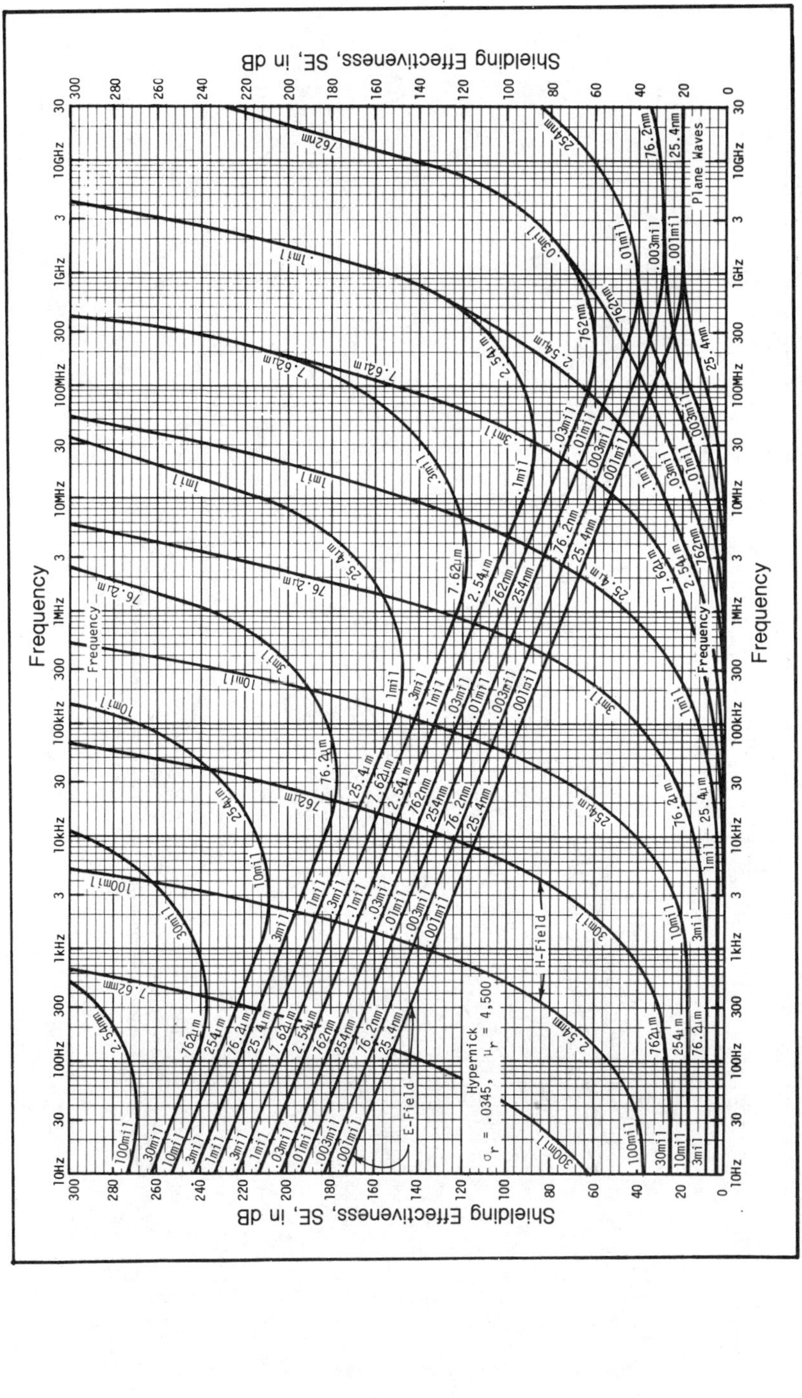

Figure F.1—Shielding Effectiveness of Hypernick vs. Frequency for Source-to-Metal Distance of 10 cm

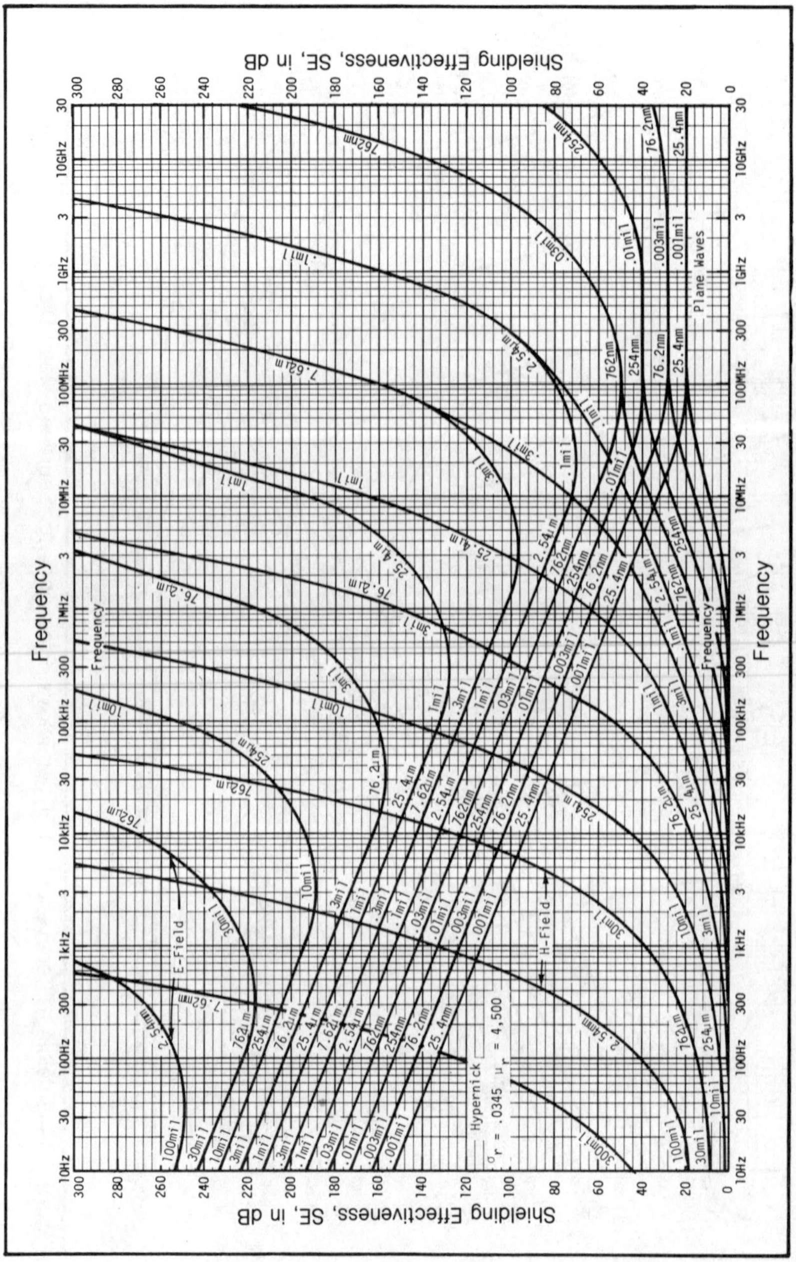

Figure F.2—Shielding Effectiveness of Hypernick vs. Frequency for Source-to-Metal Distance of 1 m

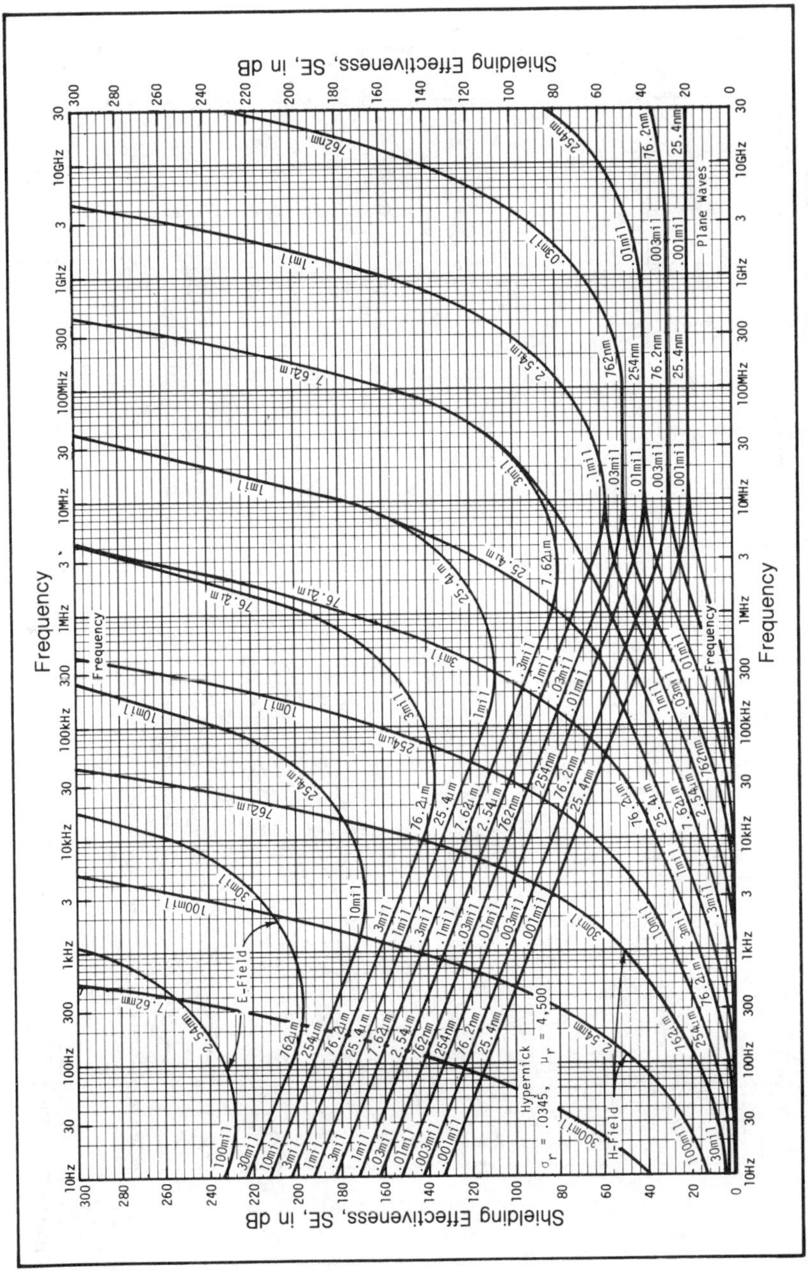

Figure F.3—Shielding Effectiveness of Hypernick vs. Frequency for Source-to-Metal Distance of 10 m

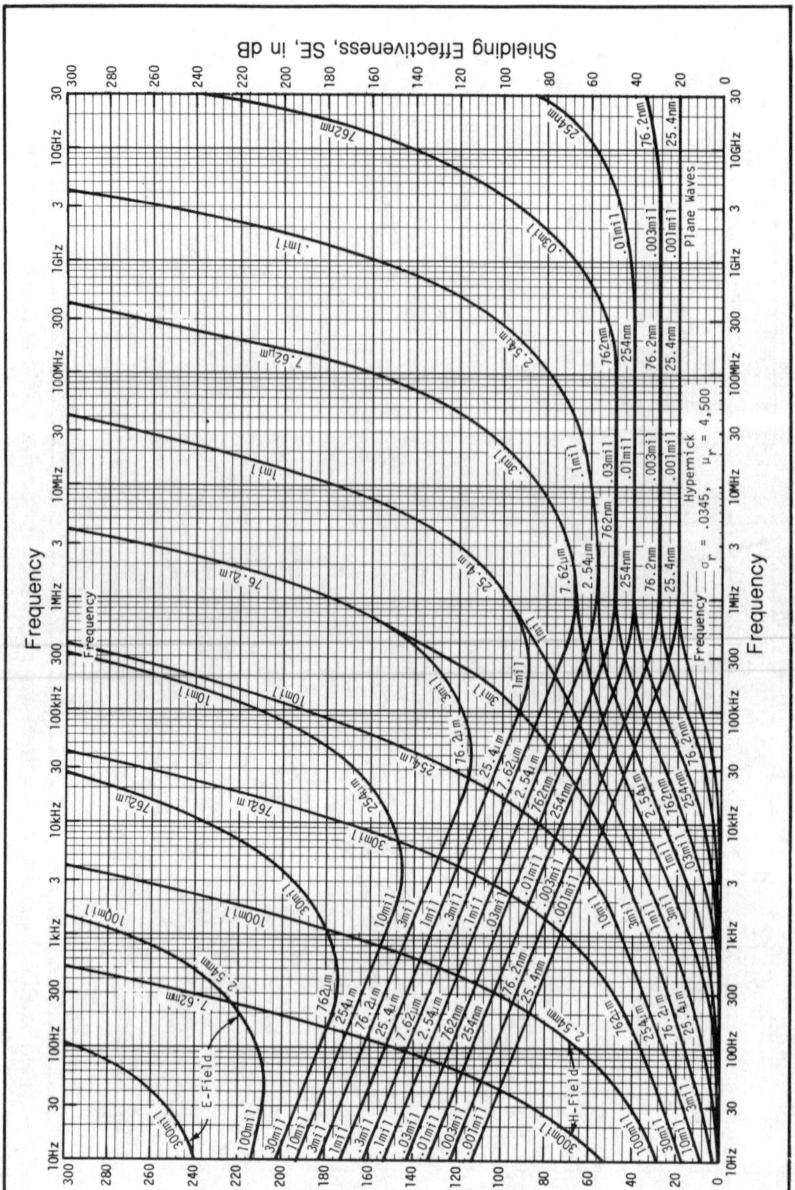

Figure F.4—Shielding Effectiveness of Hypernick vs. Frequency for Source-to-Metal Distance of 100 m

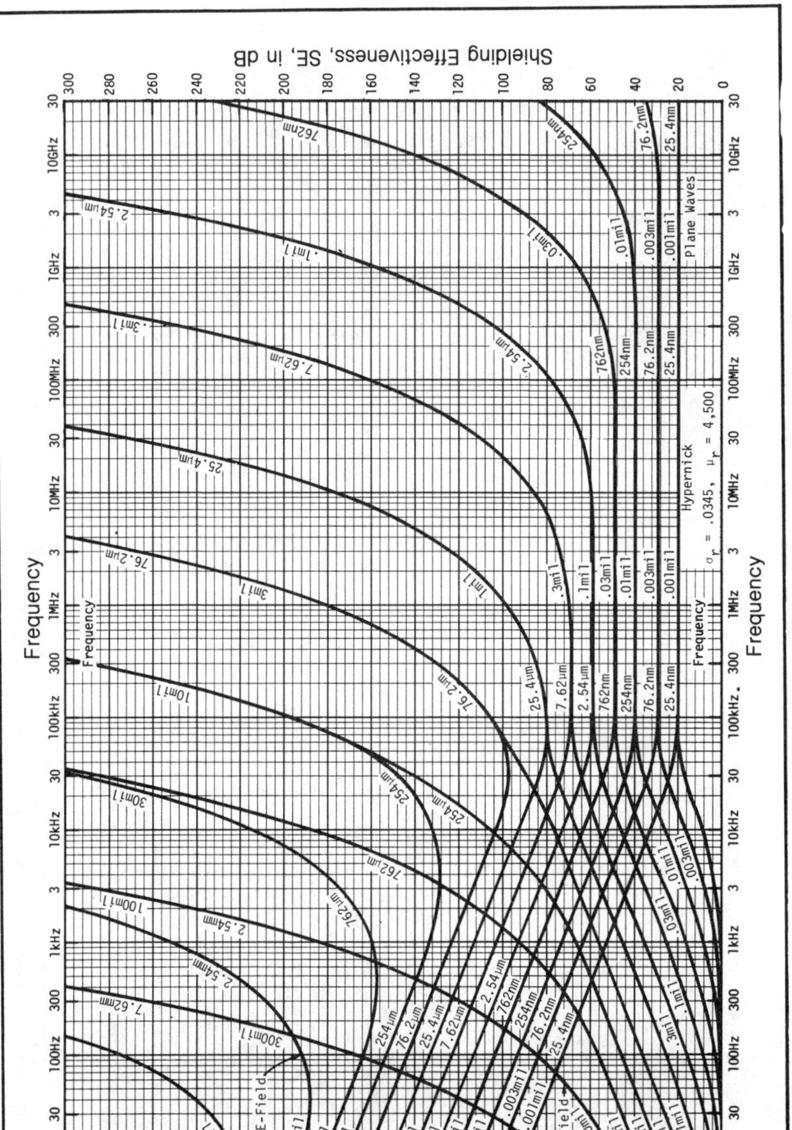

Figure F.5—Shielding Effectiveness of Hypernick vs. Frequency for Source-to-Metal Distance of 1 km

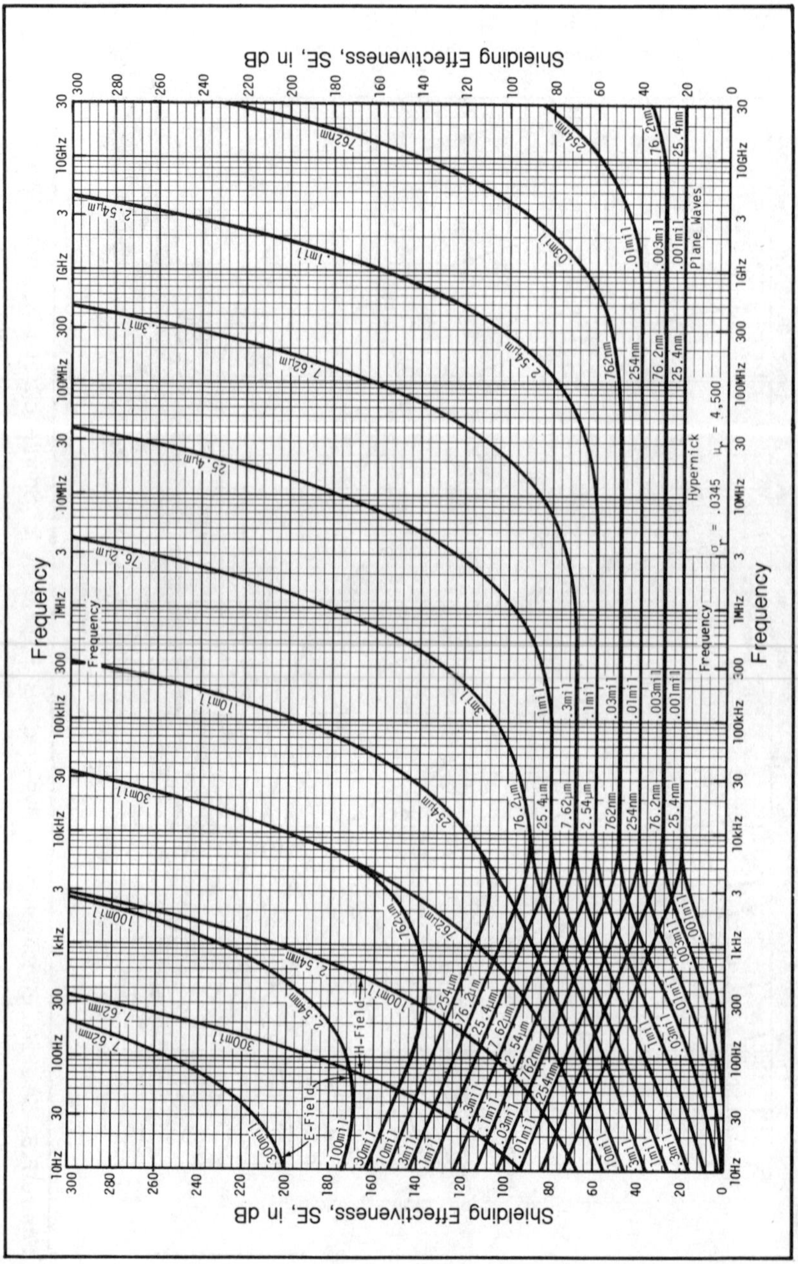

Figure F.6—Shielding Effectiveness of Hypernick vs. Frequency for Source-to-Metal Distance of 10 km

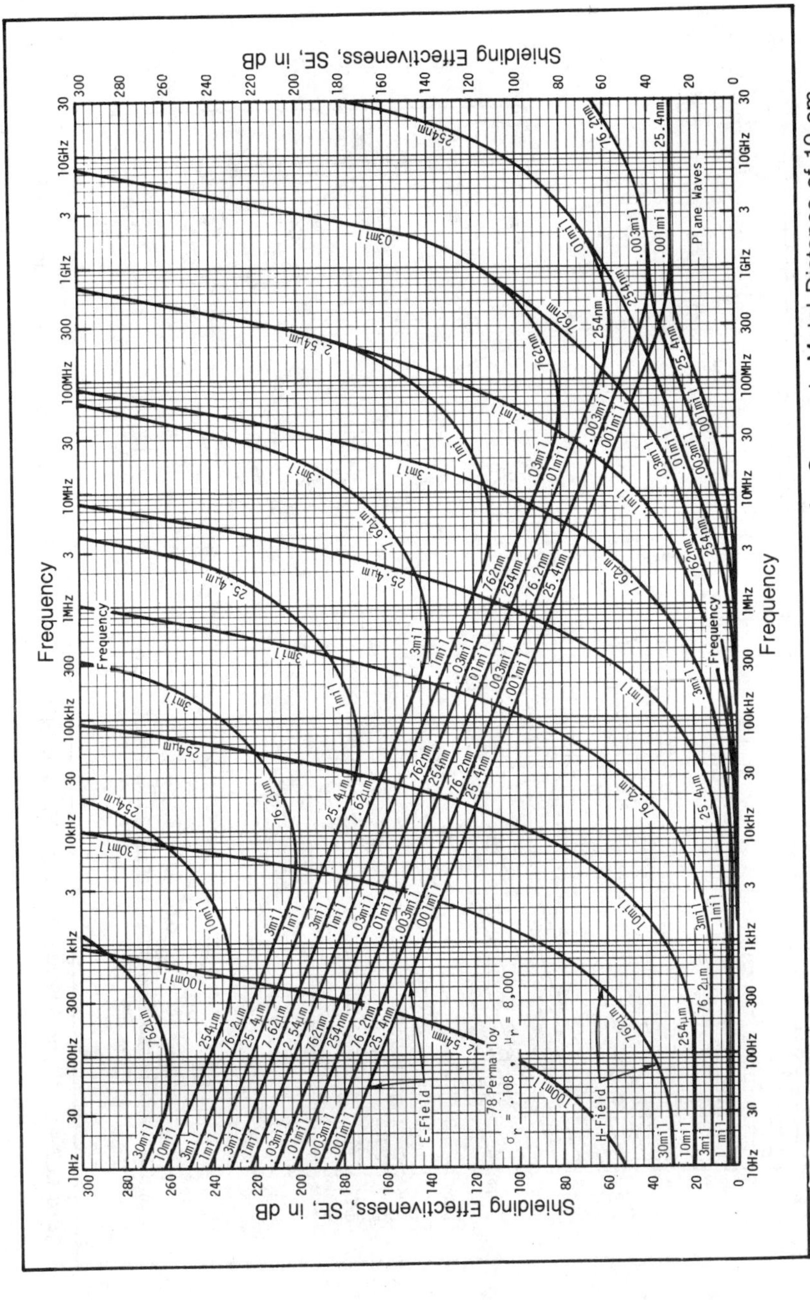

Figure G.1—Shielding Effectiveness of 78 Permalloy vs. Frequency for Source-to-Metal Distance of 10 cm

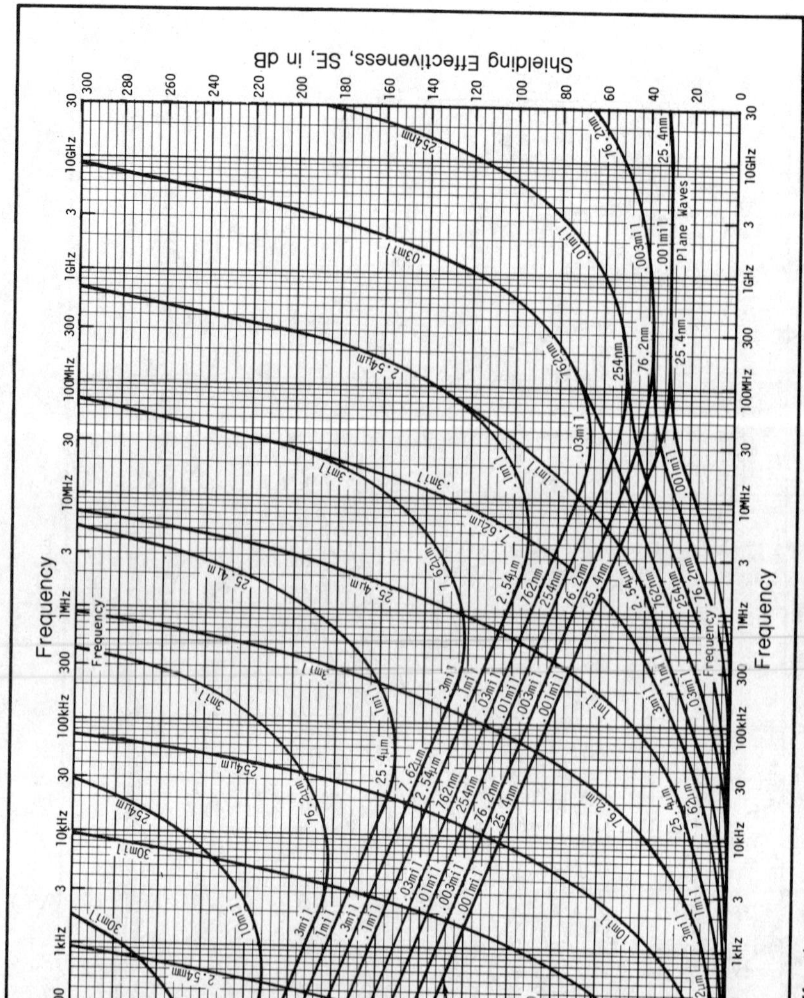

Figure G.2—Shielding Effectiveness of 78 Permalloy vs. Frequency for Source-to-Metal Distance of 1 m

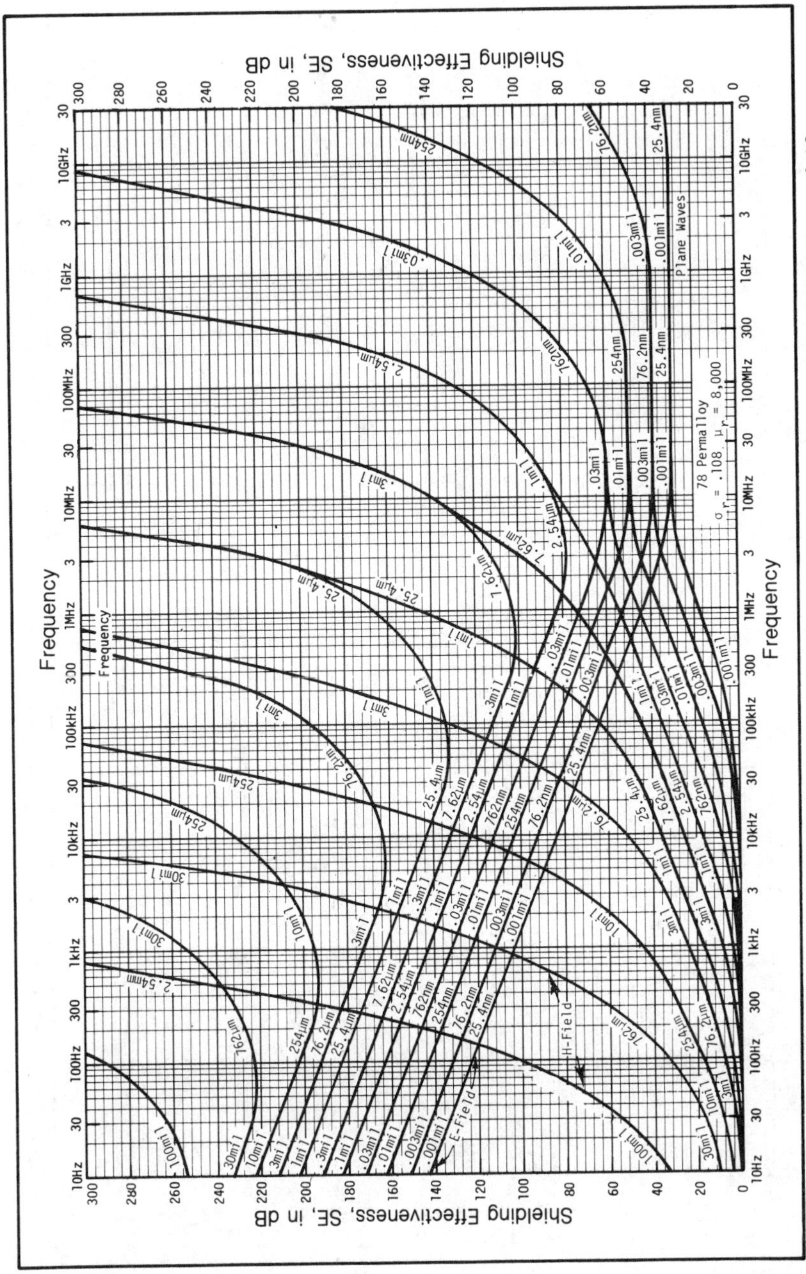

Figure G.3—Shielding Effectiveness of 78 Permalloy vs. Frequency for Source-to-Metal Distance of 10 m

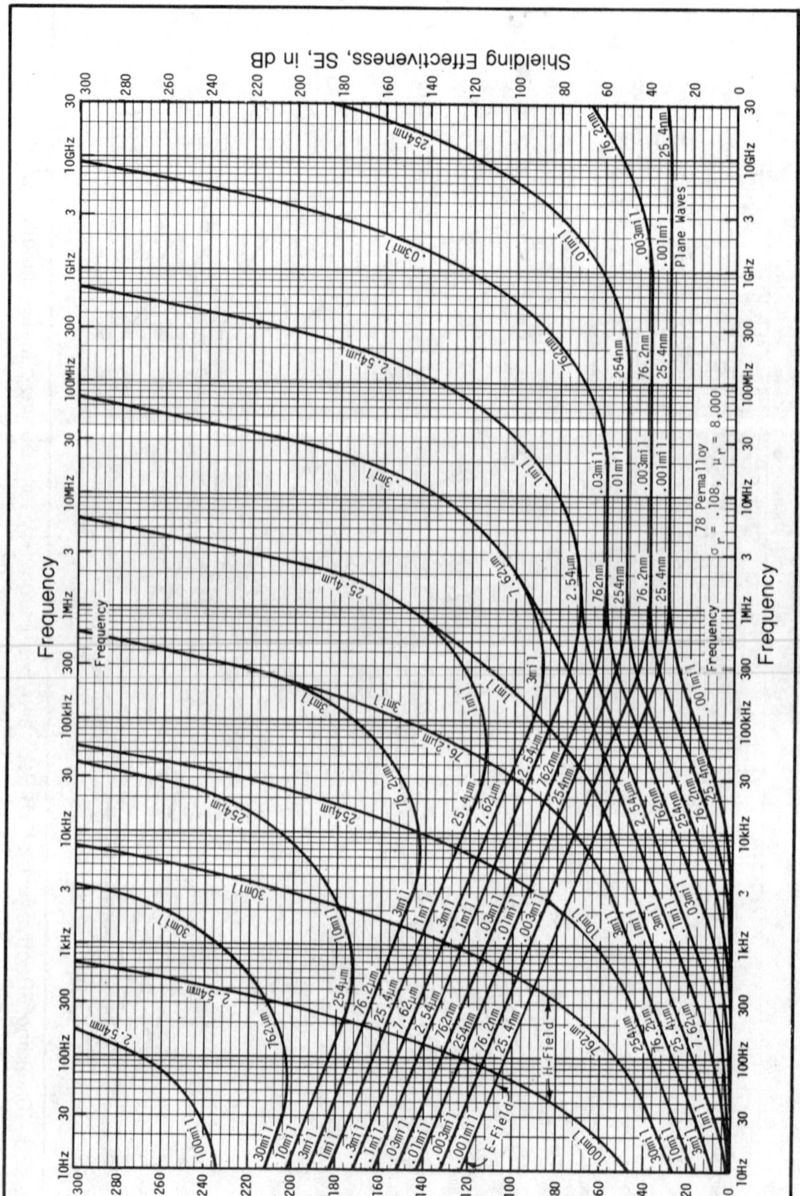

Figure G.4—Shielding Effectiveness of 78 Permalloy vs. Frequency for Source-to-Metal Distance of 100 m

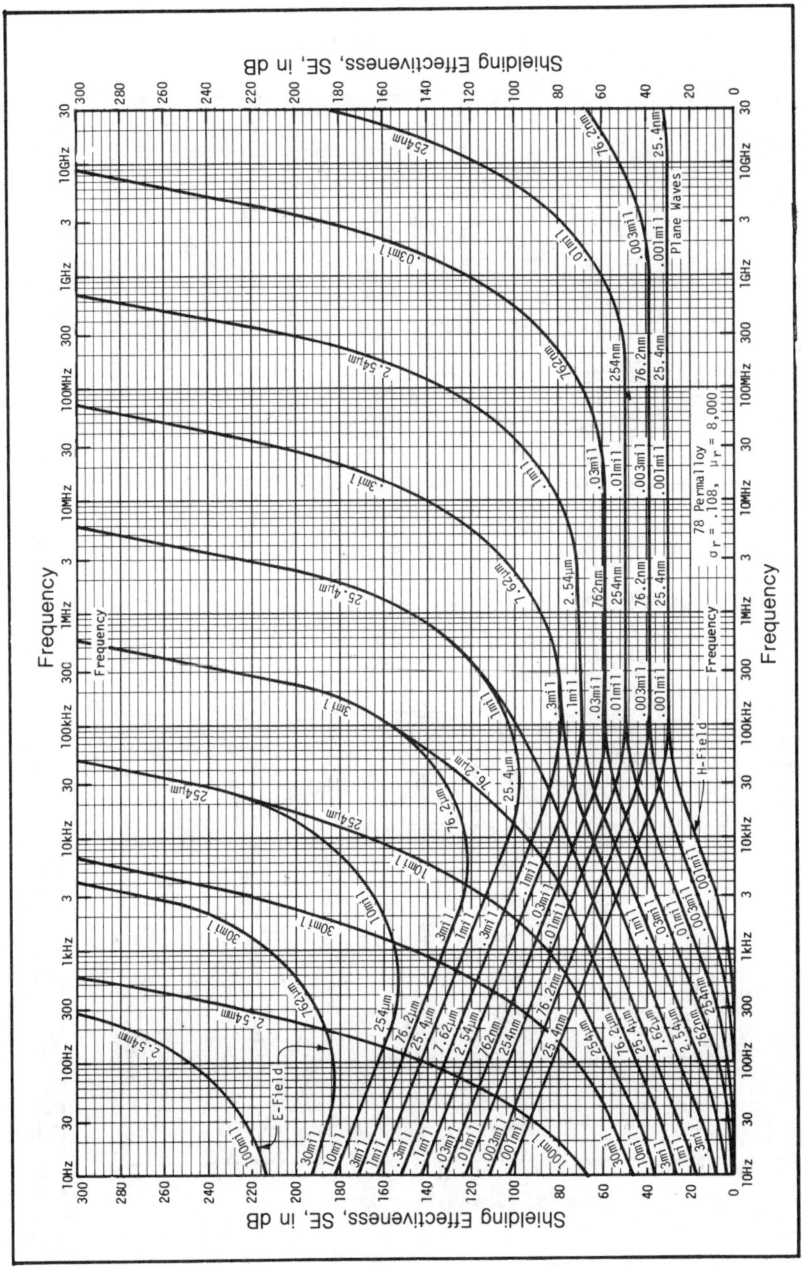

Figure G.5—Shielding Effectiveness of 78 Permalloy vs. Frequency for Source-to-Metal Distance of 1 km

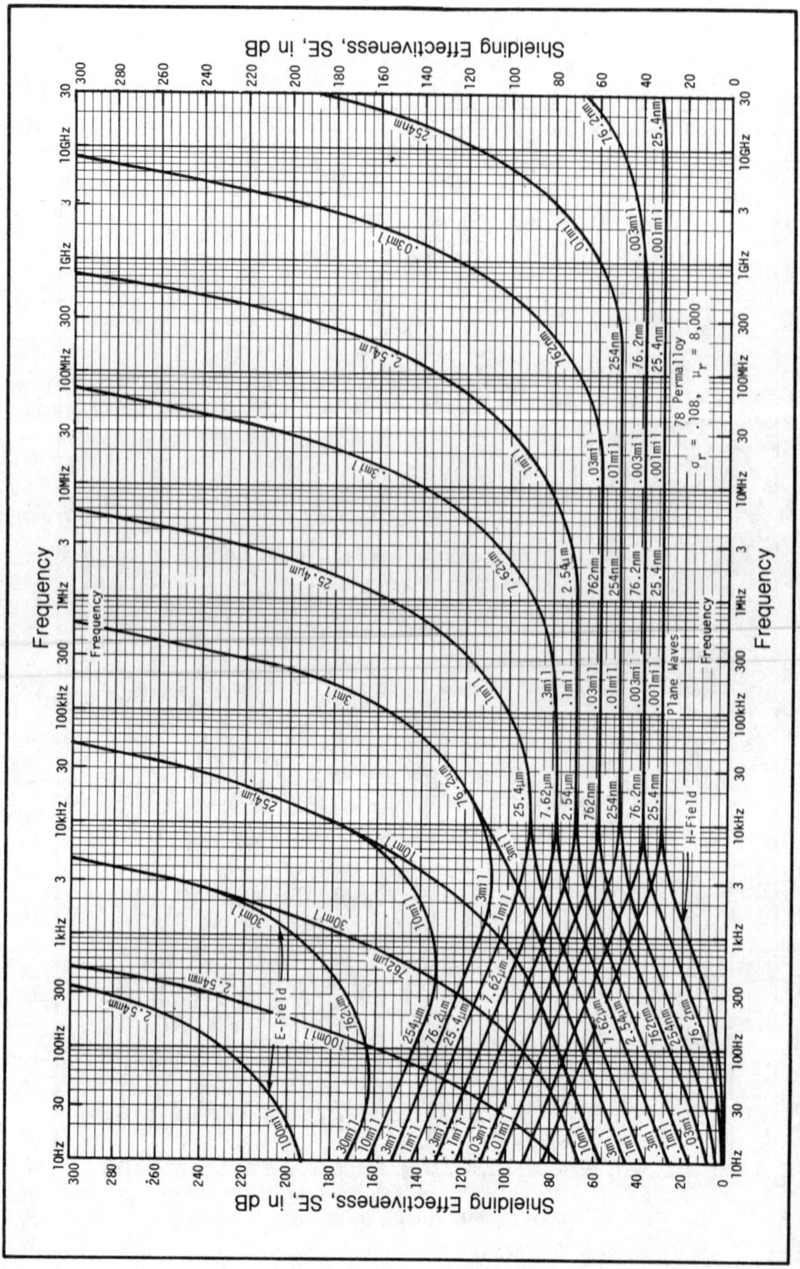

Figure G.6—Shielding Effectiveness of 78 Permalloy vs. Frequency for Source-to-Metal Distance of 10 km

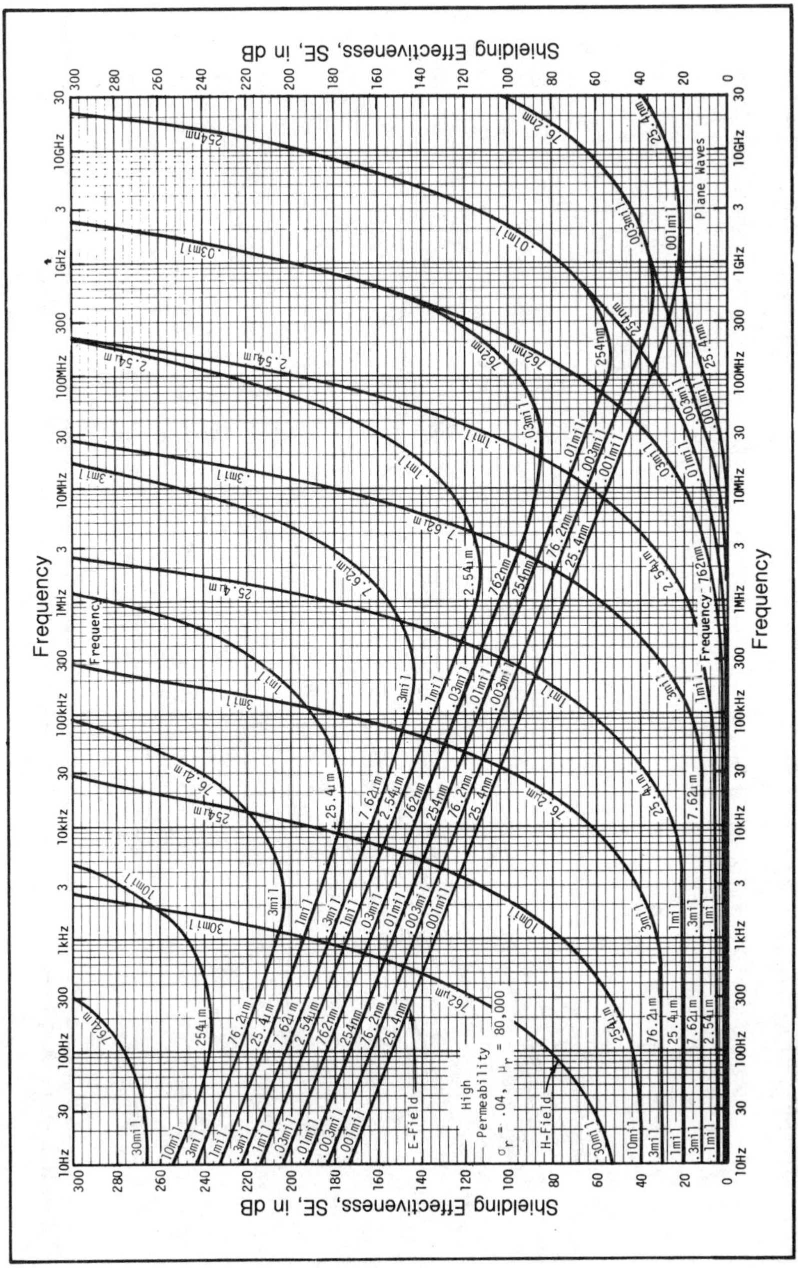

Figure H.1—Shielding Effectiveness of High Permeability vs. Frequency for Source-to-Metal Distance of 10 cm

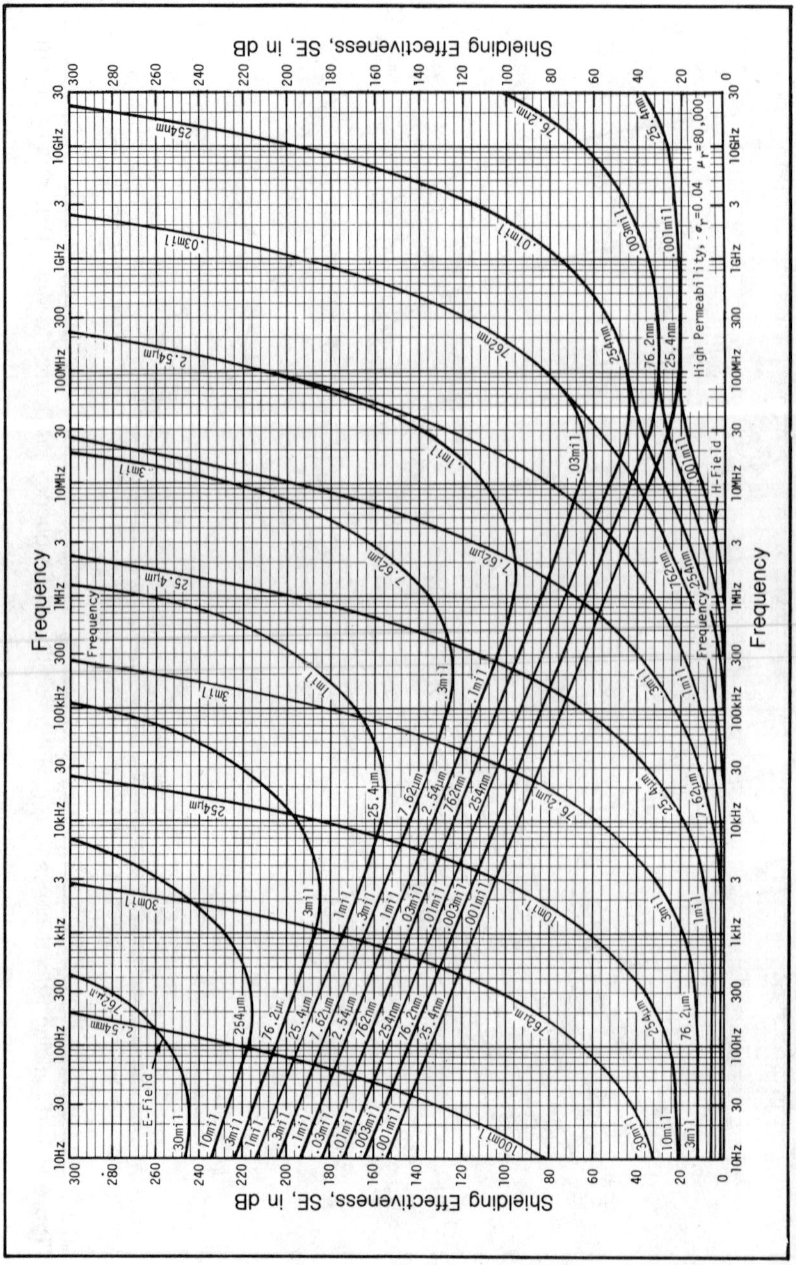

Figure H.2—Shielding Effectiveness of High Permeability vs. Frequency for Source-to-Metal Distance of 1m

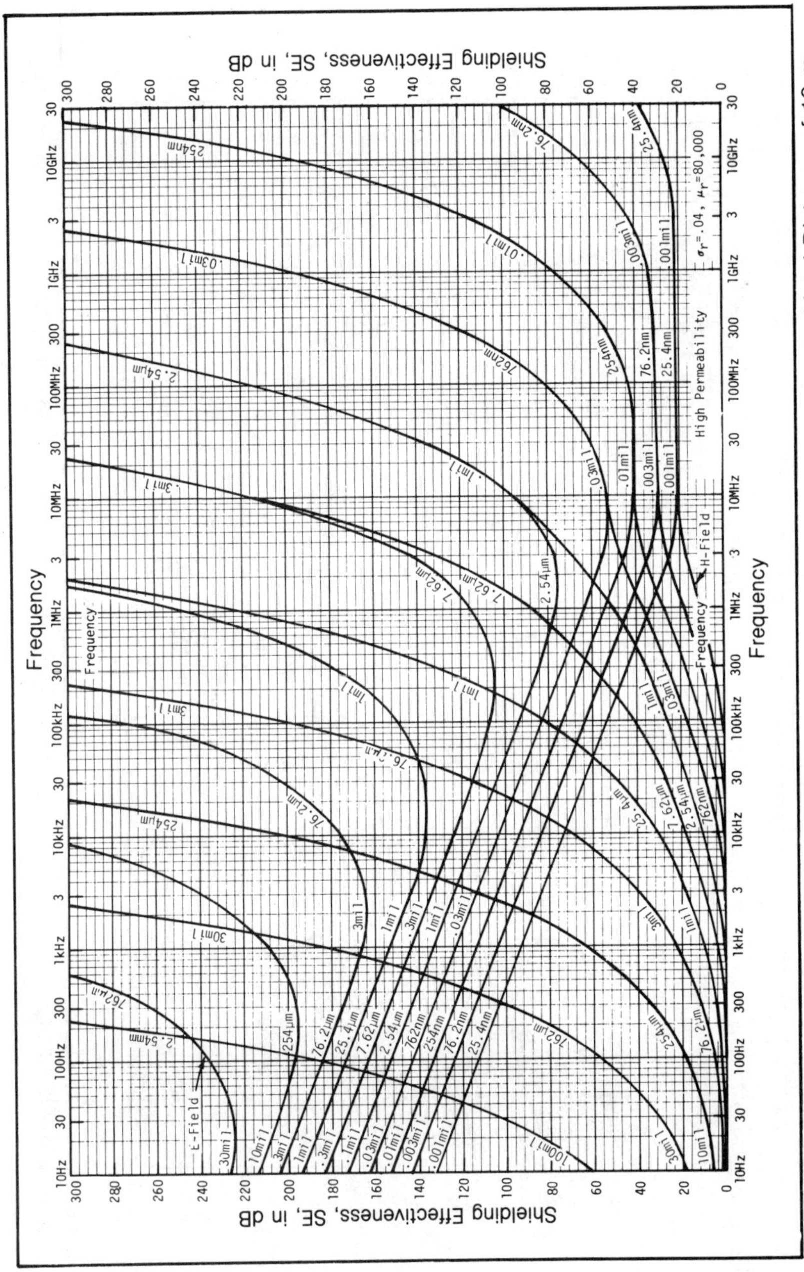

Figure H.3—Shielding Effectiveness of High Permeability vs. Frequency for Source-to-Metal Distance of 10 m

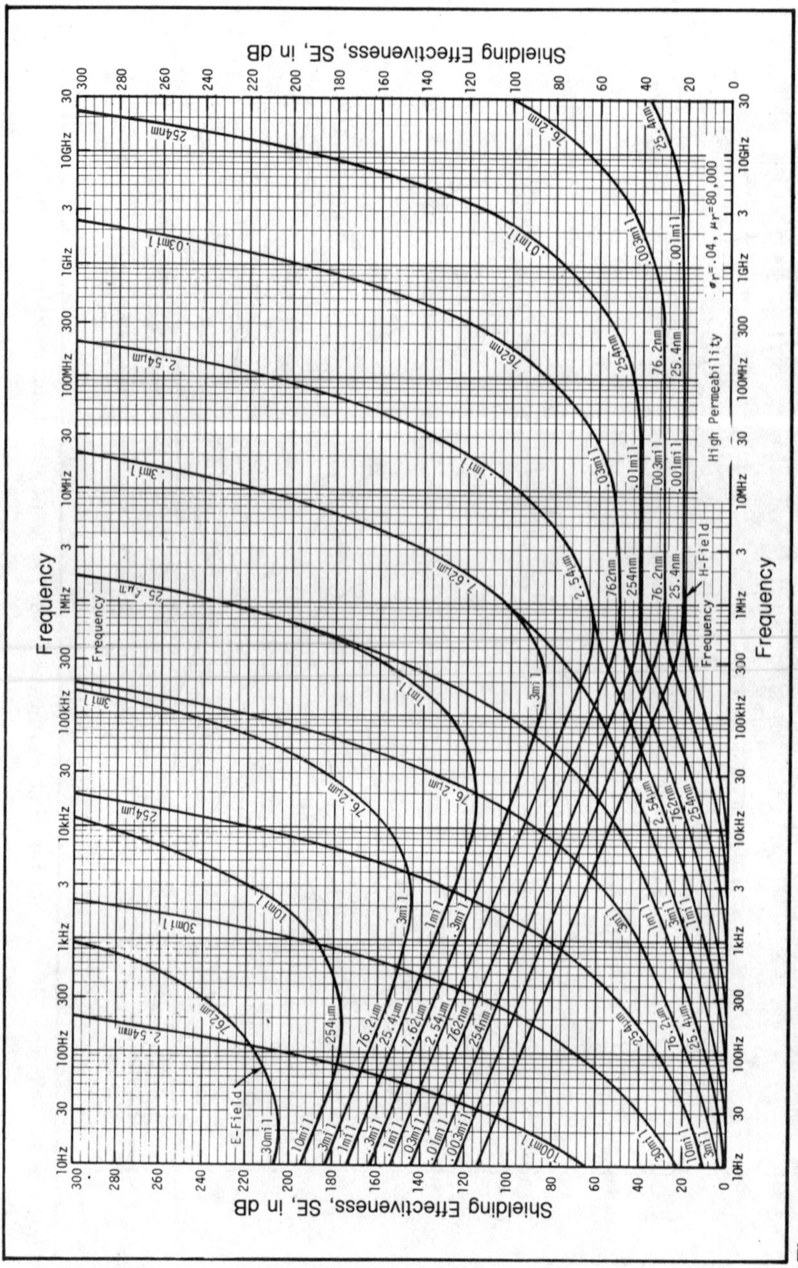

Figure H.4—Shielding Effectiveness of High Permeability vs. Frequency for Source-to-Metal Distance of 100 m

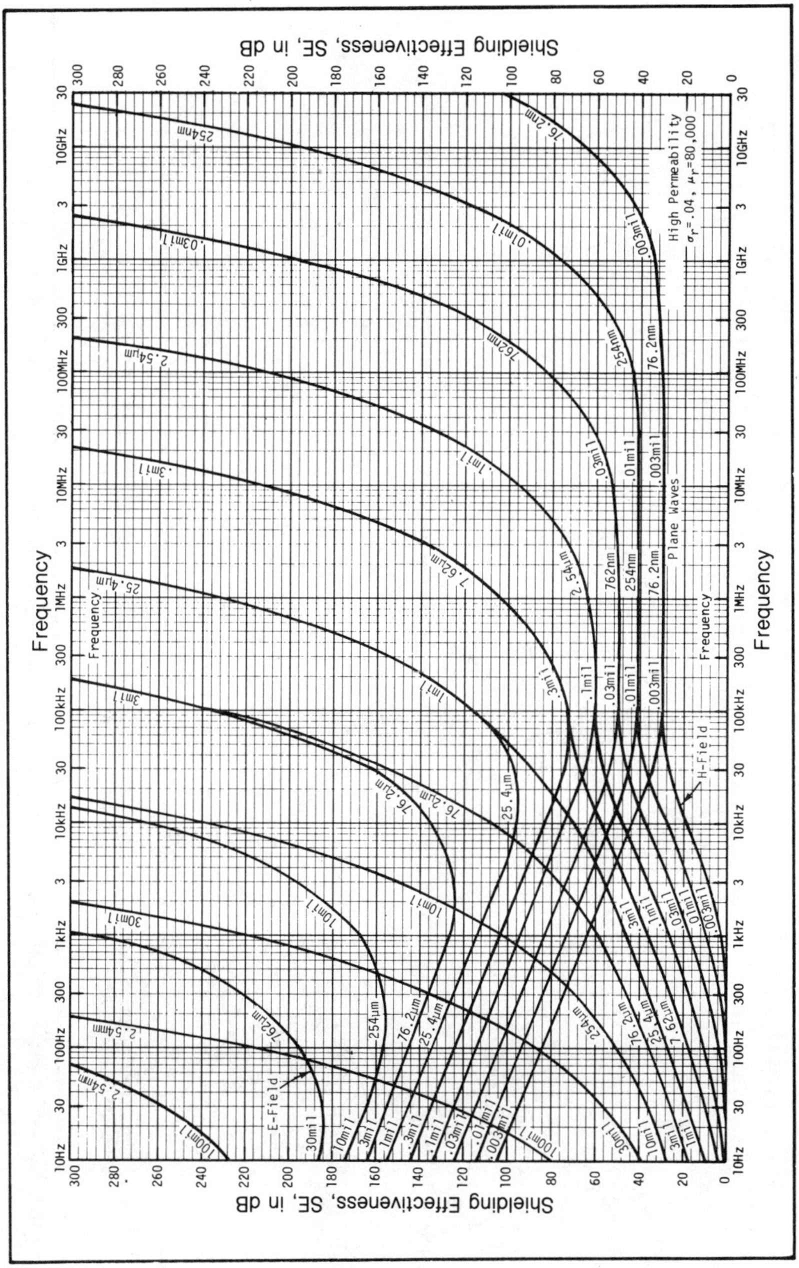

Figure H.5—Shielding Effectiveness of High Permeability vs. Frequency for Source-to-Metal Distance of 1 km

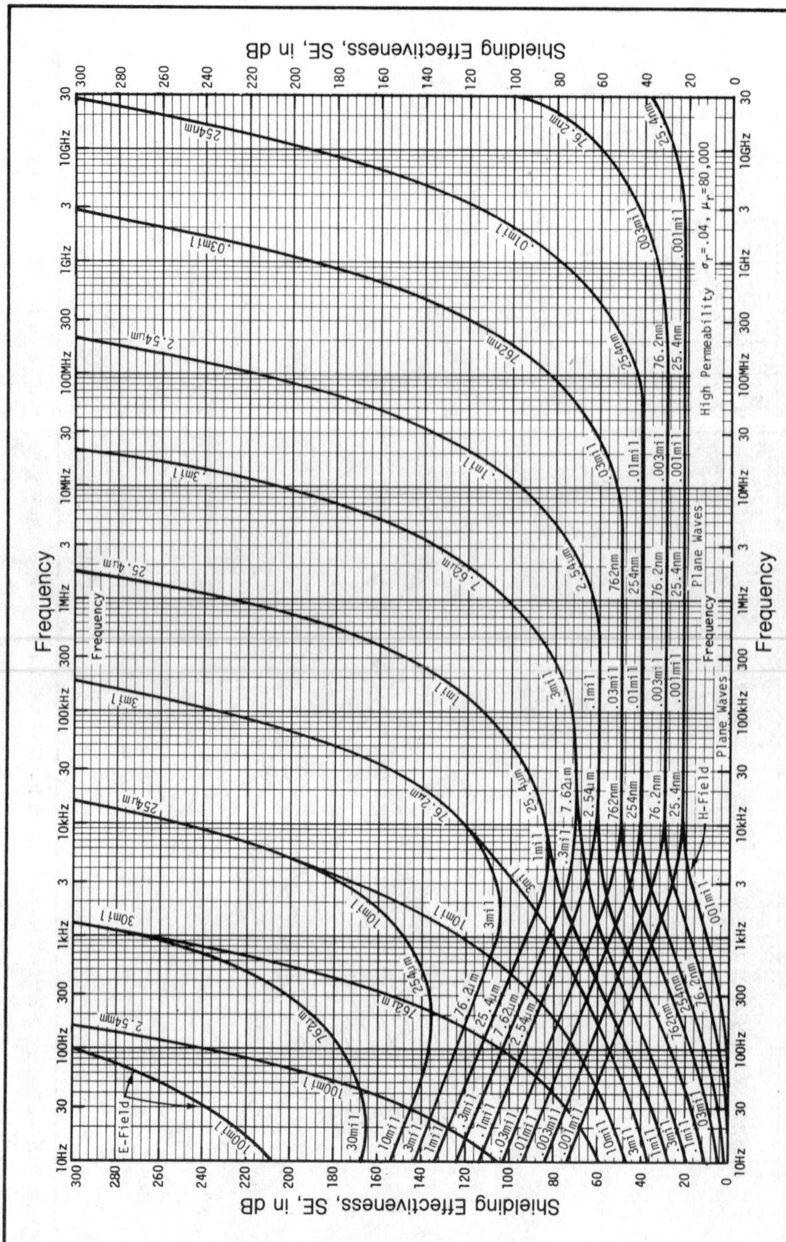

Figure H.6—Shielding Effectiveness of High Permeability vs. Frequency for Source-to-Metal Distance of 10 km

Appendix I

Shielding Effectiveness in the Time Domain

Although not frequently addressed, it can be interesting to apply the SE concept in time domain, specifically when one needs to predict the attenuation of a metal wall against a single pulse.

The typical problem is that of a single-field pulse like EMP or lightning. Knowing that the apertures and leakages have been correctly treated for the high-frequency portion of such a spectrum (i.e., the portion related to rise time), the problem remains to calculate the absorption attenuation for the lowest frequencies, typically the portion within the first arch $1/\tau$ of the Fourier integral. Except for metals with a high μ_r, we can assimilate any metal barrier against a low-impedance field to a low-pass filter since the absorption loss approaches zero at low frequencies (thickness $< \delta$). Figure I.1 shows that, based on absorption only, the total attenuation of the pulsed field through the metallic wall can be approximated grossly by:

$$A \cong \frac{\tau_C}{\tau}, \text{ for } \tau << \tau_C \tag{I.1}$$

where, τ = pulse width in μs (50% amplitude) of the EMI transient

τ_C = time-constant equivalent of the barrier in μs

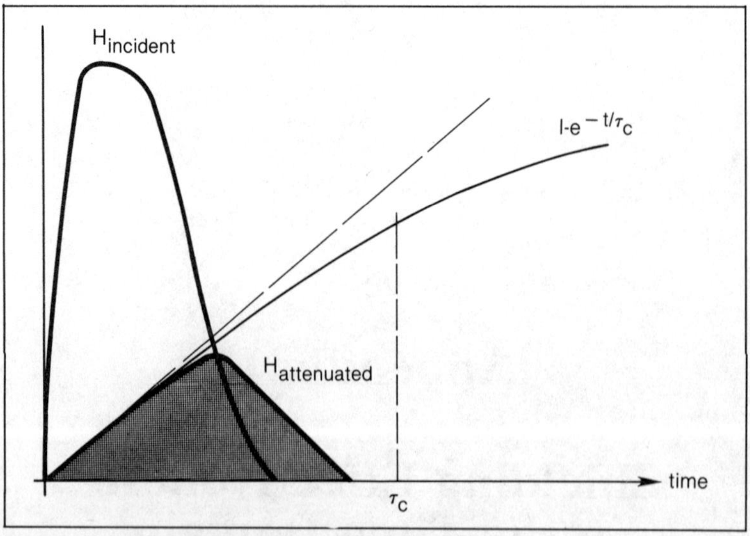

Figure I.1—Total Attenuation of Pulsed Field

$$\text{or } \tau_{c\mu s} = \frac{1}{\pi f_{c\ MHz}}$$

$$f_{c\ MHz} = \text{low cutoff frequency of the shield}$$

$$= \text{frequency at which } t = \delta \text{ (see Chapter 1)}$$

$$= \frac{4.10^{-3}}{t^2_{mm}\ \sigma_r \mu_r}$$

Replacing f and τ_c by their values,

$$A = \frac{79.6\ t^2_{mm}\ \sigma_r \mu_r}{\tau} \qquad \text{(I.2)}$$

The condition $\tau << \tau_c$ allows for assimilating the first part of the exponential to a straight slope.

Example: Aluminum wall with:

$$t = 0.5 \text{ mm}$$

$$\sigma_r = 0.7$$

$$\mu_r = 1$$

EMI pulse:

$$H = 100 \text{ A/m (equivalent to 37 kV/m)},$$

$$\tau = 1 \ \mu\text{s}$$

$$A = \frac{79.6 \times 0.5^2 \times 0.7}{1} = 14 \qquad \text{or } A \text{ dB} = 23 \text{ dB}$$

The peak field inside will be 100 (A/m)/14 = 7.1 A/m, which is still a strong H-field. The field pulse inside will exhibit a certain pulse stretching due to the integrating action through the low pass "filtering" of the shield. This crude method will provide at least an estimate of the minimum amount of shielding, since no reflection nor shape factor of the enclosure has been accounted.

Appendix J

Shielding Products and Services Directory

Find the product/service you need here and scan the list of companies for the one that best meets your needs. Addresses and phone numbers can be found in the directory starting on page J.9.

AIR FILTERS, SHIELDED

Bavaria Elektronik GMBH
Braden, A Div. of AMCA Int'l. Corp.
Brady Co., Inc (The)
Broad Base Industries
DJM Electronics Inc.
EKS-Norland Electro-Kinetic Systems
EMC Technologists
Glitsch Special Products, Inc.
GRU Assoc.
High Country Shielding Corp.
Jacques Dubois
LectroMagnetics, Inc.
Lindgren R. F. Enclosures, Inc.
Micron Air Purification, Inc.
R.F.I. Industries Pty. Ltd.
REMP AG
Scientific Hardware Systems
Spectrum Control, Inc.
Spira Manufacturing Corp.
Technical Environment Control, Inc.
Varinit Corp.

ANECHOIC MATERIALS

Advanced ElectroMagnetics, Inc.
DJM Electronics Inc.
Egis Corp.
Fortifiber Corp.
High Country Shielding Corp.
L.E.A.D.
RANTEC Div. Emerson Electric Co.
Rayproof Shielding System Corp.
Stauffer Japan Limited
SVT Brandschutz Vertriebsgesellschaft
TDK Corp. of America

CABINETS, SHIELDED

3M Static and Electro Magnetic Control
Accu-Tech Cable, Inc.
Alacra Systems
Amco Engineering Company
Antistatic Industries, Inc.
Arvinyl Div. of Arvin Industries, Inc.
Belling Lee Intec, Ltd.
Boekel Industries, Inc.
Calmark Corp.
Capcon, Inc.
Chromium Corp.
Comsearch Applied Technology, Inc.
Devex SA
ECS Composites
EMC Technologists
Emcor Products/Crenlo, Inc.
Enclosure Corp.

Equipto Electronics Corp.
Euroshield OY
Galapogos Corp., The
General Devices Co., Inc.
GRU Assoc.
High Country Shielding Corp.
Ishizaki Trading, Inc.
J. M. Benson Co., Inc.
Magnetic Radiation Laboratories, Inc.
Magnetic Shield Corp/Perfection Mica
Optima Enclosures
Pacific West Electronics
PacTec Div. of LaFrance Corp.
Par-Metal Products, Inc.
Plastic-Plate, Inc.
REMP AG
Shipley Co.
SVT Brandschutz Vertriebsgesellschaft
Wyle Laboratories Electronic Encl

CABLE ASSEMBLIES & HARNESSES—SHIELDED

3M Static and Electro Magnetic Control
A.E.S. Interconnects
Accu-Tech Cable, Inc.
Adams Russell Electronics Co., Inc.
Alpha Wire Corp.
Belden Wire and Cable
Bentley Harris Manufacturing Co.
Breeze Illinois, Inc.
BRIntec Corp., Brand-Rex Cable Systems
Capcon, Inc.
Component Manufacturing Service, Inc.
CW Industries
Daburn Electronics & Cable Corp.
Electro Standards Laboratory, Inc.
G&H Technology, Breech-Lok Div.
GSI Corp.
Huber & Suhner, Inc.
Kyle Technology Corp.
Laminaire Corp.
Magnetic Shield Corp/Perfection Mica
Middco Connectors/Middleburg Corp.
Musorb
Reliable Electric/Utility Products
REMP AG
Rogers Corp/Flexible Interconnections
Sabritec
Santa Fe Textiles
Thomas & Betts Corp.
TRW Connector Div.
Varinit Corp.
Vermillion Enterprises Inc.
Zippertubing Co.

CABLE SHIELD—HEAT SHRINKABLE

Accu-Tech Cable, Inc.
Alpha Wire Corp.
Breeze Illinois, Inc.
Chomerics, Inc.
Daburn Electronics & Cable Corp.
GfO-Gesellschaft fur Oberflächentechnik
Jacques Dubois
Markel Corp.
Schlegel Corp.
Vermillion Enterprises Inc.
Zippertubing Co.

CABLES—FLAT RIBBON

3M Static and Electro Magnetic Control
Accu-Tech Cable, Inc.
Alpha Wire Corp.
Belden Wire and Cable
BRIntec Corp., Brand-Rex Cable Systems
CW Industries
Daburn Electronics & Cable Corp.
Electro Standards Laboratory, Inc.
GfO-Gesellschaft fur Oberflächentechnik
GSI Corp.
Techni-Tool Inc.
Thomas & Betts Corp.

COATINGS, ABSORPTIVE

Capcon, Inc.
INCO Selective Surfaces
Jacques Bubois
Kaltbrunner AG
L.E.A.D.
Musorb
OCLI Optical Coating
  Laboratory, Inc.
Photon Technologies, Inc.
Sierracin Corp. - Intrex Div.
Tokin America Inc.
W.H. Brady Co.

COATINGS, CONDUCTIVE

Alacra Systems
Altoflex
Antistatic Industries, Inc.
Applied Energy Technology, Ltd.
Bavaria Elektronik GMBH
Broad Base Industries
Chomerics, Inc.
Chromium Corp.
Conductive Systems, Inc.
Deep Coat Metallizing, Inc.
Deposition Technology, Inc.
Duracote Corp.
E/M Corp.
EKS-Norland Electro-Kinetic Systems
Enthone, Inc.
Evaporated Coatings Inc.
Furane Products/Isochem Operations
GfO-Gesellschaft fur Oberflächentechnik

Great Union, Inc., Int'l.
INCO Selective Surfaces
Kaltbrunner AG
Master Bond, Inc.
Micro-Circuits Co., Inc.
Miller-Stephenson Chemical Co., Inc.
OCLI Optical Coating
 Laboratory, Inc.
Optical & Conductive Coatings
ORCUS
Photon Technologies, Inc.
Pilkington Glass, Ltd.
Plastic Systems, Inc.
Scientific Materials Corp.
Shipley Co.
Sierracin Corp.—Intrex Div.
Spraylat Corp.
Stauffer Japan Limited
SVT Brandschutz Vertriebsgesellschaft
TBA Industrial Products Ltd.
Tech Spray Inc.
Techni-Tool Inc.
Tecknit, A TWP Co.
W.H. Brady Co.
Walter G. Legge Co., Inc.
WESCORP
Westek
Zymet, Inc.

COMPONENTS, EMI SUPPRESSED

Aerospace & Commercial Components
Altoflex
Capcon, Inc.
Eaton Corporation/Cutler Hammer Prod.
Eichhoff Electronics, Inc.
Eurotest Laboratories, Ltd.
G&H Technology, Breech-Lok Div.
Gowanna Electronics Corp.
GSI Corp.
Middco Connectors/Middleburg Corp.
Middleburg Connectors
Moxness Products, Inc.
MPE Ltd., Liverpool
Plastic-Plate, Inc.
Rockford Engineering Services, Inc.
Sabritec
Sprague Electric Co.
Stauffer Japan Limited
TDK Corp. of America
Tokin America Inc.
Vacuumschmelze
Wyle Laboratories Electronic Encls.

CONDUCTIVE ADHESIVES

3M Static and Electro Magnetic Control
Altoflex
Aremco Products, Inc.
Bavaria Elektronik GMBH
Broad Base Industries
Chomerics, Inc.
Conap, Inc.
Conductive Systems, Inc.
Contact East
EKS-Norland Electro-Kinetic Systems
Formulated Resins Inc.
Furane Products/Isochem Operations
GM Nameplate, Inc.
Haskon Corp.
Jacques Bubois
Lamart Corp.
Master Bond, Inc.
Micro-Circuits Co., Inc.
Miller-Stephenson Chemical Co., Inc.
ORCUS
Scientific Materials Corp.
Stockwell Rubber Co., Inc.
TBA Industrial Products Ltd.
Tecknit, A TWP Co.
W.H. Brady Co.
WESCORP
Zymet, Inc.

CONDUCTIVE FILLERS

Aremco Products, Inc.
Brunswick Corp., Technetics Div.
Chromium Corp.
G&H Technology, Breech-Lok Div.
Haskon Corp.
International Paper
ORCUS
Potters Industries Inc.
Scientific Materials Corp.
Transmet Corp.
Westek

CONDUIT—ELECTRICAL SHIELDED

Breeze Illinois, Inc.
Chomerics, Inc.
G&H Technology, Breech-Lok Div.
GSI Corp.
J. M. Benson Co., Inc.
Santa Fe Textiles
Zippertubing Co.

CONNECTORS, FILTER PIN

Accu-Tech Cable, Inc.
Alpha Wire Corp.
Breeze Illinois, Inc.
Corcom, Inc.
Eurotest Laboratories, Ltd.
Feller AG
G&H Technology, Breech-Lok Div.
ITT Cannon
Middco Connectors/Middleburg Corp.

Middleburg Connectors
MPE Ltd., Liverpool
Sabritec
Spectrum Control, Inc.
Texas Spectrum Electronics, Inc.
Thomas & Betts Corp.
TRW Connector Div.

CONNECTORS, RF

Accu-Tech Cable, Inc.
Alpha Wire Corp.
Breeze Illinois, Inc.
Cemec, Inc.
Dale Electronics, Inc.
Huber & Suhner, Inc.
ITT Cannon
Kyle Technology Corp.
Middco Connectors/Middleburg Corp.
Sabritec
Spectrum Control, Inc.
Tecknit, A TWP Co.
Thomas & Betts Corp.
TRW Connector Div.
Weinschel Engineering Co., Inc.

CONNECTORS, SHIELD ADAPTERS

Accu-Tech Cable, Inc.
Breeze Illinois, Inc.
CW Industries
G&H Technology, Breech-Lok Div.
ITT Cannon
Middco Connectors/Middleburg Corp.
Sabritec
TRW Connector Div.
Zippertubing Co.

CONTACT STRIPS, FINGERS/GASKETS

Agamice
Bavaria Elektronik GMBH
Broad Base Industries
E.C.T. Div. of Tolo, Inc.
Energy Shielding Inc.
Higbee Gaskets & Sealing Prod., Inc.
High Country Shielding Corp.
Instrument Specialties Co., Inc.
International Paper
Irving B. Moore Corp.
Omega Shielding Products
TBA Industrial Products Ltd.
Tech-Etch, Inc.
Vanguard Products Corp.

CONTACT STRIPS, GASKETS

Altoflex
Bavaria Elektronik GMBH
Broad Base Industries
Duracote Corp.
E.C.T. Div. of Tolo, Inc.
GRU Assoc.
Higbee Gaskets & Sealing Prod., Inc.
High Country Shielding Corp.
Instrument Specialties Co., Inc.
International Paper
Irving B. Moore Corp.
Isolation Technology, Inc.
Knitmesh Ltd.
Musorb
Omega Shielding Products
Stockwell Rubber Co., Inc.
TBA Industrial Products Ltd.

DUCTS/WIRING, SHIELDED

GfO-Gesellschaft fur Oberflächentechnik
SVT Brandschutz Vertriebsgesellschaft
Varinit Corp.
Zippertubing Co.

ELASTOMERS, CONDUCTIVE

Altoflex
AME Corp.
Bavaria Elektronik GMBH
Broad Base Industries
Chomerics, Inc.
CHR Industries, Inc.
Conductive Services, Inc.
Duracote Corp.
E.C.T. Div. of Tolo, Inc.
EKS-Norland Electro-Kinetic Systems
Elastomeric Products Co.
Formulated Resins Inc.
Glitsch Special Products, Inc.
GRU Assoc.
Higbee Gaskets & Sealing Products, Inc.
Irving B. Moore Corp.
Isolation Technology, Inc.
Jacques Bubois
Master Bond, Inc.
Moxness Products, Inc.
ORCUS
Parker Seal Group
Pawling Corp.
QRP, Inc.
Scientific Materials Corp.
SFS Industries, Inc.
Stockwell Rubber Co., Inc.
TBA Industrial Products Ltd.
Tech-Etch, Inc.
Tecknit, A TWP Co.
Vanguard Products Corp.

FINGER STOCK

Agamice
Bavaria Elektronik GMBH
High Country Shielding Corp.
Instrument Specialties Co. Inc.
Omega Shielding Products
R.F.I. Industries Pty. Ltd.
TBA Industrial Products Ltd.

FOILS, SHEET

Bavaria Elektronik GMBH
Chromium Corp.
Duracote Corp.
Electronic Instrument Assoc., Inc.
Fortifiber Corp.
GM Nameplate, Inc.
INCO Selective Surfaces
Lam-A-Shield, Inc.
Lamart Corp.
Magnetic Radiation Laboratories, Inc.
Magnetic Shield Corp/Perfection Mica
REMP AG
Set Point Inc.
SpecTape, Inc.
Topflight Corp.
Vacuumschmelze
W.H. Brady Co.

FOILS, SHIELDED TAPE

Agamice
Bavaria Elektronik GMBH
Broad Base Industries
Chomerics, Inc.
Chromium Corp.
Egis Corp.
Fortifiber Corp.
GM Nameplate, Inc.
Great Union, Inc., Int'l.
INCO Selective Surfaces
Insul-Fab of Texas
Lam-A-Shield, Inc.
Lamart Corp.
Magnetic Shield Corp/Perfection Mica
Magnetics Div., Spang & Co.
Schlegel Corp.
Set Point Inc.
SpecTape, Inc.
Stockwell Rubber Co., Inc.
TBA Industrial Products Ltd.
Topflight Corp.
Vacuumschmelze
W.H. Brady Co.

FUSES, EMI SUPPRESSED

FIC Corp.
Genisco Technology Corp.
Rockford Engineering Services, Inc.

GASKETS, MAGNETIC SHIELDING

Abbey Plastic Corp.
Broad Base Industries
Chomerics, Inc.
E.C.T. Div. of Tolo, Inc.
Egis Corp.
EKS-Norland Electro-Kinetic Systems
Haskon Corp.
Higbee Gaskets & Sealing Products
Insul-Fab of Texas
Jacques Dubois
Knitmesh Ltd.
Parker Seal Group
Pawling Corp.
SFS Industries, Inc.
Spaulding Composites Co.
Spira Manufacturing Corp.
TBA Industrial Products Ltd.
TDK Corp. of America
Tecknit, A TWP Co.
Texas Instruments
Topflight Corp.
Vanguard Products Corp.
W.H. Brady Co.

GASKETS, RF SHIELDING

3M Static and Electro Magnetic Control
Abbey Plastic Corp.
Altoflex
AME Corp.
Bavaria Elektronik GMBH
Bentley Harris Manufacturing Co.
Brady Co., Inc. (The)
Broad Base Industries
Capcon, Inc.
Chomerics, Inc.
CHR Industries, Inc.
Conductive Systems, Inc.
E.C.T. Div. of Tolo, Inc.
EEV Inc.
Egis Corp.
EKS-Norland Electro-Kinetic Systems
Elastomeric Products Co.
Glitsch Special Products, Inc.
GRU Assoc.
Haskon Corp.
Higbee Gaskets & Sealing Products, Inc.
High Country Shielding Corp.
Instrument Specialties Co., Inc.
Insul-Fab of Texas
International Paper
Irving B. Moore Corp.
Isolation Technology, Inc.
Knitmesh Ltd.
L.E.A.D.
Musorb
Omega Shielding Products

ORCUS
Parker Seal Group
Pawling Corp.
Plessey Microwave Materials
R.F.I. Industries Pty. Ltd.
REMP AG
Schlegel Corp.
SFS Industries, Inc.
Spaulding Composites Co.
Spectrum Control, Inc.
Spira Manufacturing Corp.
TBA Industrial Products Ltd.
Tech-Etch, Inc.
Tecknit, A TWP Co.
Topflight Corp.
United Foam Plastics Corp.
Vanguard Products Corp.
W.H. Brady Co.

LUBRICANTS—CONDUCTIVE

Altoflex
Aremco Products, Inc.
Micro-Circuits Co., Inc.
Miller-Stephenson Chemical Co., Inc.
Techni-Tool Inc.
Zymet, Inc.

PLASTIC—CONDUCTIVE

3M Static and Electro Magnetic Control
Abbey Plastic Corp.
Advanced Dynamics Corp/Spec Compounds
Agamice
Alacra Systems
Antistatic Industries, Inc.
Broad Base Industries
C. R. Daniels, Inc.
Central Container Corp.
Chromium Corp.
Correct Products
Dan Newman Co. (Techni-Bag)
Duracote Corp.
Great Union, Inc., Int'l.
H&S Industries
HBM Assoc., Inc.
LNP Engineering Plastics
Lourich Co., The
LPS Industries, Inc.
Melmat, Inc.
Molded Fiberglass Tray Co.
NRD Inc.
Optical & Conductive Coatings
Pacific West Electronics
Plastic Systems, Inc.
REMP AG
RTP Co.
Scientific Materials Corp.
Sentinel Foam Products
Set Point Inc.
Sierracin Corp. - Intrex Div.
Spaulding Composites Co.
Stackbin Corp.
TBA Industrial Products Ltd.
Transmet Corp.
Uniroyal Plastics Co.
United Foam Plastics Corp.
W.H. Brady Co.
WESCORP
Westek
Wilson-Fiberfil Int'l.

PLASTICS MOLDERS—EMI SHIELDING

3M Static and Electro Magnetic Control
Chromium Corp.
Crown City Plating Co.
E/M Corp.
ECS Composites
Galapogos Corp., The
GfO-Gesellschaft fur Oberflächentechnik
HBM Assoc., Inc.
Middco Connectors/Middleburg Corp.
Middleburg Connectors
Molded Fiberglass Tray Co.
Musorb
PacTec Div. of LaFrance Corp.
Phillips Plastics Corp.
Plastic Systems, Inc.
Plastic-Plate, Inc.
Set Point Inc.
WEZ Plastic Industries Ltd.

SHIELDED ROOMS

Acoustic Systems
Advanced Measurement Systems
Amuneal Manufacturing Corp.
Aprel Inc.
Ark Electronics Corp.
Atlantic Research Corp., Tempest Div.
Barlows, Inc.
Belling & Lee
Braden, A Div. of AMCA Int'l. Corp.
Brady Co., Inc. (The)
Cemec, Inc.
Comsearch Applied Technology, Inc.
Devex SA
DJM Electronics Inc.
DNC Architects Inc.
E^3 Engineers
Egis Corp.
Electromagnetic Engineering Inc.
Electronic Instrument Assoc., Inc.
EMC Consulting Dr. Rasek
EMC Fribourg SA
EMC Technologists
Energy Shielding Inc.
Euroshield OY
Fortifiber Corp.

GBI Shielding
GRU Assoc.
Hamer Corp.
Hamilton Engineering, Inc.
High Country Shielding Corp.
Industrial Acoustics Co.
Insul-Fab of Texas
Integrated Facilities Inc.
International Paper
Ishizaki Trading, Inc.
J. M. Benson Co., Inc.
Jacques Dubois
LectroMagnetics, Inc.
Lindgren R. F. Enclosures, Inc.
Overly Manufacturing Co.
Pacific West Electronics
Pilkington Glass, Ltd.
R.F.I. Industries Pty. Ltd.
RANTEC Div. Emerson Electric Co.
Rayproof Shielding System Corp.
REMP AG
Rockford Engineering Services, Inc.
Scientific Hardware Systems
Siemens
S.I.D.T.
Stauffer Japan Ltd.
Stetron Int'l., Inc.
SVT Brandschutz Vertriebsgesellschaft
TAFA, Inc.
TDK Corp. of America
Technical Environment Control, Inc.
Trumbull Testing Labs, Inc.
Vacuumschmelze

SHIELDING MATERIALS—MAGNETIC

Abbey Plastic Corp.
ADE, Inc.
Advanced ElectroMagnetics, Inc.
Amuneal Manufacturing Corp.
Baxter Healthcare Corp/Industrial Div.
Belling Lee Intec, Ltd.
Broad Base Industries
Chomerics, Inc.
Chromium Corp.
Deposition Technology, Inc.
E.C.T. Div. of Tolo, Inc.
Egis Corp.
EKS-Norland Electro-Kinetic Systems
Ferronics Inc.
Ferroxcube
GBI Shielding
GM Nameplate, Inc.
Haskon Corp.
HBM Assoc., Inc.
Higbee Gaskets & Sealing Products, Inc
Integrated Facilities Inc.
International Paper
Inter-Technical Group, Inc., The
Irving B. Moore Corp.
J. M. Benson Co., Inc.
Jacques Dubois
Knitmesh Ltd.
Lam-A-Shield, Inc.
Lamart Corp.
Magnetic Radiation Laboratories, Inc.
Magnetic Shield Corp/Perfection Mica
Magnetics Div., Spang & Co.
Master Bond, Inc.
Multi Tool & Manufacturing, Inc.
Mushield Products Div., Bomco Inc.
Neo-Tec Plating, Inc.
Parker Seal Group
Pilkington Glass, Ltd.
Potters Industries Inc.
Rayproof Shielding System Corp.
Richmond Technology Inc.
Santa Fe Textiles
Schlegel Corp.
Scientific Hardware Systems
SFS Industries, Inc.
Spang Specialty Metals
Spaulding Composites Co.
SpecTape, Inc.
Spira Manufacturing Corp.
Spraylat Corp.
TBA Industrial Products Ltd.
TETKO, Inc.
Texas Instruments
Tokin America Inc.
Vacuumschmelze
Vermillion Enterprises Inc.
W.H. Brady Co.
Zippertubing Co.

SHIELDS, CRT

Amuneal Manufacturing Corp.
Bavaria Elektronik GMBH
Broad Base Industries
Chomerics, Inc.
Chromium Corp.
Egis Corp.
EKS-Norland Electro-Kinetic Systems
Inter-Technical Group, Inc., The
Jacques Dubois
Kantek, Inc.
Lamart Corp.
Magnetic Radiation Laboratories, Inc.
Magnetic Shield Corp/Perfection Mica
Mushield Products Div., Bomco Inc.
OCLI Optical Coating Laboratory, Inc.
Pilkington Glass, Ltd.
Set Point Inc.
Spira Manufacturing Corp.
TETKO, Inc.
Tokin America Inc.
Vacuumschmelze
W.H. Brady Co.

TAPES—PRESSURE SENSITIVE

3M Static and Electro Magnetic Control
Antistatic Industries, Inc.
Assoc. Sales & Bag Co.
Broad Base Industries
CHR Industries, Inc.
Contact East
Egis Corp.
GM Nameplate, Inc.
Great Union, Inc., Int'l.
Haskon Corp.
H&S Industries
Insul-Fab of Texas
International Paper
Lamart Corp.
LPS Industries, Inc.
Permacel
Set Point Inc.
SpecTape, Inc.
Stockwell Rubber Co., Inc.
Topflight Corp.
W.H. Brady Co.
Zippertubing Co.

WINDOWS, SHIELDED

Agamice
Altoflex
Applied Energy Technology, Ltd.
Bavaria Elektronik GMBH
Braden, A Div. of AMCA Int'l. Corp.
Brady Co., Inc. (The)
Broad Base Industries
Chomerics, Inc.
Conductive Systems, Inc.
Deposition Technology, Inc.
Dontech Inc.
EKS-Norland Electro-Kinetic Systems
Equipto Electronics Corp.
Euroshield OY
GfO-Gesellschaft fur Oberflächentechnik
Hamer Corp.
J. M. Benson Co., Inc.
Jacques Dubois
Kantek, Inc.
Lindgren R. F. Enclosures, Inc.
OCLI Optical Coating Laboratory, Inc.
Optima Enclosures
Photon Technologies, Inc.
Pilkington Glass, Ltd.
REMP AG
Scientific Hardware Systems
Sierracin Corp.—Intrex Div.
SVT Brandschutz Vertriebsgesellschaft
TBA Industrial Products Ltd.
Tech-Etch, Inc.
Topflight Corp.
WESCORP

Appendix J

Directory of Manufacturers

3M Static and Electromagnetic Control
2111 W. Braker Lane
Bldg. 501-1W-30
Austin, Texas 55789
(512) 834-3118

A.E.S. Interconnects
P.O. Box 278
Plainfield, Indiana 46168
(317) 272-2551

Abbey Plastic Corporation
Eleven Brent Drive
Hudson, Massachusetts 01749
(617) 562-6971

Accu-Tech Cable, Inc.
645 Hembree Parkway
Roswell, Georgia 30076
(404) 751-9473

Acoustic Systems
P.O. Box 3610
Austin, Texas 78764
(800) 531-5412

Adams Russell Electronics Co. Inc.
1370 Main Street
Waltham, Massachusetts 02154

Advanced Dynamics Corporation Speciality Compounds Division
2000 E. Lamar Boulevard MD600
Arlington, Texas 76006

Advanced ElectroMagnetics Inc.
P.O. Box W
Santcc, California 92071 0618
(619) 449-9492

Advanced Measurement Systems
1115 W. Broad Street
Falls Church, Virginia 22046
(703) 534-4214

Aerospace and Commercial Components
Division of Eaton Corporation
2250 Whitfield Avenue
Sarasota, Florida 34243
(813) 758-7726

Agamice
4 rue Philidor
Paris, France 75020
(1) 43 72 10 05

Alacra Systems
Race & Ridge Streets
Dept 18
Ambler, Pennsylvania 19002
(215) 542-9208

Alpha Wire Corporation
711 Lidgerwood Avenue
Elizabeth, New Jersey 07207
(201) 925-8000

Altoflex
36 rue des Maraichers
Paris, France 75020
(1) 43 72 81 88

Amco Engineering Co.
3801 Rose Street
Schiller Park, Illinois 60176
(312) 671-6670

AME Corporation
33 Jacksonville Road
P.O. Box 195
Towaco, New Jersey 07082-0195
(201) 263-1700

Amuneal Manufacturing Corporation
4737 Darrah Street
Philadelphia, Pennsylvania 19124
(215) 535-3000

Antistatic Industries, Inc.
151 Westfield Avenue
Clark, New Jersey 07066-2429
(201) 396-9600

Applied Energy Technology, Ltd.
Highway 965 South
T.P.O. Box 243
Swisher, Iowa 52338
(319) 857-4121

Aprel, Inc.
Aprel Court
38 Antares Drive
Nepean, Ontario
Canada K2E 7V2
(613) 727-0334

Aremco Products, Inc.
P.O. Box 429
Ossining, New York 10562
(914) 762-0685

Ark Electronics Corporation
1325 Industrial Highway
Southampton, Pennsylvania 18966
(215) 322-6510

Arvinyl Division of Arvin Industries, Inc.
1531 13th Street
Columbus, Indiana 47201
(812) 379-3000

Associated Sales and Bag Co.
160 South 2nd Street
Milwaukee, Wisconsin 53204
(414) 272-2380

Atlantic Research Corporation Tempest Division
5390 Cherokee Avenue
Alexandria, Virginia 22312
(703) 642-2000

Barlows Inc.
1115 West Broad Street
Falls Church, Virginia 22046
(703) 549-2310

Bavaria Elektronik GmbH
Tegernseestrasse 7
8200 Rosenheim
West Germany
08031/12089

Belden Wire and Cable
P.O. Box 1980
Richmond, Indiana 47375
(317) 983-5200

Belling Lee Intec Ltd.T
Intec House
540 Great Cambridge Road
Enfield, England EN13 QW
01 367 0080

Bentley Harris Mfg. Co.
241 Welsh Pool Road
Lionville, Pennsylvania 19353
(215) 363-2600

Boekel Industries, Inc.
509-19 Vine Street
Philadelphia, Pennsylvania 19106
(215) 627-1611

Braden Company, Inc. (The)
Div. of AMCA International Corporation
5199 N. Mingo Road
Tulsa, Oklahoma 74117
(918) 272-5371

Brady Co., Inc. (The)
208 N. Jackson Avenue
North Plainfield, New Jersey 07060
(201) 755-7900

Breeze-Illinois, Inc.
8823 N. Industrial Road
Peoria, Illinois 61615-1584
(309) 693-3322
Telex: 510 390-3691

BRIntec Corporation/Brand-Rex Cable Systems Division
1600 West Main Street
Willimantic, Connecticut 06226-1128
(203) 456-8000

Broad Base Industries
5528 Everglades Street
Suite E
Ventura, California 93003
(805) 658-2772

Brunswick Corporation
Technetics Division
2000 Brunswick Lane
Deland, Florida 32720
(904) 736-1700

C.R. Daniels, Inc.
3451 Ellicott Center Drive
Ellicott City, Maryland 21043
(301) 461-2100
Telex: 87-825

Calmark Corporation
4915 Walnut Grove Avenue
San Gabriel, California 91776
(818) 287-0451

Capcon Inc.
147 West 25th Street
New York, New York 10001
(212) 243-6275

Cemec Inc.
1518 Centre Circle
Downers Grove, Illinois 60515
(312) 495-9696

Central Container Corporation
4041 Hiawatha Avenue
Minneapolis, Minnesota 55406
(612) 721-6224

Chomerics, Inc.
77 Dragon Court
Woburn, Massachusetts 01888
(617) 935-4850

Chromium Corporation
Conductive Coating Division
1221 W. Campbell Road
Suite 245
Richardson, Texas 75080
(214) 231-1518

Component Manufacturing Service Inc.
1 Component Park
West Bridgewater, Massachusetts 02379
(617) 580-0111

Comsearch Applied Technology, Inc.
11720 Sunrise Valley Drive
Reston, Virginia 22091
(703) 620-6300

Conap Inc.
1405 Buffalo Street
Olean, New York 14760
(716) 372-9650

Conductive Systems, Inc. (CSI)
Dorine Ind. Park B26
Merry Lane
Hanover, NJ 07936
(201) 887-3055

Contact East
335 Willow Street South
P.O. Box 786
North Andover, Massachusetts 01845
(617) 682-2000

Corcom, Inc.
1600 Winchester Road
Libertyville, Illinois 60048
(312) 680-7400

Correct Products
1701 N. Greenville Avenue
Suite 400
Richardson, Texas 75091
(214) 690-6645

Crown City Plating Co.
4350 Temple City Boulevard
El Monte, California 91731
(818) 444-9291

CW Industries
130 James Way
Southampton, Pennsylvania 18966
(215) 355-7080

Daburn Electronics and Cable Corporation
70 Oak Street
Norwood, New Jersey 07648
(201) 768-5400

Dale Electronics, Inc.
2064 12th Avenue
Columbus, Nebraska 68601
(402) 564-3131

Dale Electronics Ltd.
Dale House
Frimley Green
Camberley, Surrey
England GU16 6LF
0252 835094

Dan Newman Co. (Techni-Bag)
1051 Bloomfield Avenue
Clifton, New Jersey 07012
(201) 778-6677

Deep Coat Metallizing, Inc.
21 West New Avenue
Lemont, Illinois 60439
(312) 257-3322

Deposition Technology, Inc.
4540 Viewridge Avenue
San Diego, California 92123
(619) 576-0200

Devex SA
P.O. Box 43
Chatel - St - Denis
Switzerland 1618
(021) 948 7808

DJM Electronics Inc.
P.O. Box 55146
Valencia, California 91355
(805) 254-2677

DNC Architects Inc.
1370 Piccard Drive
Rockville, Maryland 20850
(301) 840-1100

Duracote Corporation
350 North Diamond Street
P.O. Box 512
Ravenna, Ohio 44266-0512
(800) 321-2252

E/M Corporation
2801 Kent Avenue
P.O. Box 2400
West Lafayette, Indiana 47906
(317) 497-6348

E.C.T. Div. of Tolo, Inc.
2401 S. Pullman Street
Santa Ana, California 92799-5056
(714) 261-9328

E^3 Engineers
P.O. Box 425
Valley Forge, Pennsylvania 19481
(215) 644-3332

Eaton Corporation/Cutler Hammer Products
Aerospace ¿ Commercial Controls Div.
2250 Whitfield Avenue
Sarasota, Florida 34243-9703
(813) 758-7726

ECS Composites
3560 Rogue River Highway
P.O. Box 188
Grants Pass, Oregon 97526
(503) 476-8871

EEV Inc.
4 Westchester Plaza
Elmsford, New York 10523
(914) 592-6050

Egis Corporation
P.O. Box 751
Barrington, Illinois 60011-0751
(312) 639-4448

Eichhoff Electronics, Inc.
205 Hallene Road
Warwick, Rhode Island 02886
(401) 738-1440

EKS-NORLAND
Electro-Kinetic Systems, Inc.
701 Chestnut Street
Trainer, Pennsylvania 19013
(800) 821-9775

Elastomeric Products Co.
2850 Vicksburg Lane
Plymouth, Minnesota 55447
(612) 559-0201

Electro Standards Laboratory, Inc.
P.O. Box 9144
Providence, Rhode Island 02940
(401) 943-1164

Electromagnetic Engineering Inc.
P.O. Box 1888
Poulsbo, Washington 98370-0269
(206) 779-7069

Electronic Instrument Associates, Inc.
1400 Renaissance Drive
Park Ridge, Illinois 60068
(312) 298-2290

EMC Consulting (Dr. Rasek)
Moggast 72-73
D 8553 Ebermannstadt
West Germany D 8553
011-49-9194-9016

EMC Fribourg SA
Rte de la Fonderie 8
1700 Fribourg
Suisse
037 82-11-31

EMC Technologists
620 Valley Road
Suite 310
Brielle, New Jersey 08730
(201) 223-6699

Emcor Products/Crenlo, Inc.
1600 4th Avenue, N.W.
Rochester, Minnesota 55901
(507) 287-3503

Enclosure Corporation
2900 Wharton Road
Bristol, Pennsylvania 19007
(215) 783-2900

Energy Shielding Inc.
838 Main Avenue
Passaic, New Jersey 07055
(201) 778-6200

Enthone, Inc.
P.O. Box 1900
New Haven, Connecticut 06508
(203) 934-8611

Equipto Electronics Corporation
351 Woodlawn Avenue
Aurora, Illinois 60506-9988
(312) 897-4691

Euroshield Oy
SF-27510 Eura
Finland
+358-38-50631

Eurotest Laboratories, Ltd
Matthews Drive
P.O. Box 262
East Haddam, Connecticut 06423
(203) 873-1453

Evaporated Coatings Inc.
2365 Maryland Road
Willow Grove, Pennsylvania 19090
(215) 659-3080

Feller AG
Bergstrasse
CH-8810 Horgen
Switzerland
725 6565

Ferronics Inc.
52 North Main Street
Fairport, New York 14450
(716) 388-1020

Ferroxcube
5083 Kings Highway
Saugerties, New York 12477
(914) 246-2811

FIC Corporation
12216 Parklawn Drive
Rockville, Maryland 20852
(301) 881-8124

Formulated Resins, Inc.
P.O. Box 508
Greenville, Rhode Island 02828
(401) 949-2060

Fortifiber Corporation
55 Starkey Avenue
Attleboro, Massachusetts 02703
(800) 343-3972 or (617) 222-3500

Furane Products/Isochem Operations
99 Cook Street
Lincoln, Rhode Island 02865
(401) 723-2100

G & H Technology
Breech-Lok Div.
750 W. Ventura Boulevard
Camarillo, California 93010
(805) 484-0543

Galapogos Corporation (The)
P.O. Box 1462
Littleton, Massachusetts 01460
(617) 263-6009

GBI Shielding
14420 S. Bloomfield Avenue
Santa Fe Springs, California 90670
(213) 802-0082

General Devices Co., Inc.
P.O. Box 39100
Indianapolis, Indiana 46239
(317) 897-7000

Genisco Technology Corporation
18435 Susana Road
Rancho Dominguez, California 90221
(213) 537-4750T

GfO - Gesellschaft fur Oberflachentechnik mbH
Klarenbergstrasse 79
D-7070 Schwabisch Gmund
West Germany
07171/607-285

Glitsch Special Products, Inc.
4900 Singleton Boulevard
Suite #2
Dallas, Texas 75212
(214) 631-0120

GM Nameplate, Inc.
2040 15th Avenue West
Seattle, Washington 98119
(206) 284-2200

Gowanna Electronics Corporation
1 Industrial Place
Gowanna, New York 14070
(716) 532-2234

Great Union, Inc. International
13321 Alondra Boulevard
Suite G
Santa Fe Springs, California 90670
(213) 802-1008

GRU Associates
1105 E. Commonwealth Avenue
Fullerton, California 92631
(714) 738-0903

GSI Corporation
1917 Greenspring Drive
Timonium, Maryland 21093
(301) 252-1768

H&S Industries
466 W. Arrow Highway
Unit E
San Dimas, California 91773
(714) 599-6090

Hamer Corporation
202 W. Main Street
P.O. Box 138
Hamersville, Ohio 45130
(513) 379-1313

Hamilton Engineering, Inc.
2108 S.W. 152nd Street
Seattle, Washington 98166
(206) 244-0952

Haskon Corporation
336 Weir Street
Taunton, Massachusetts 02780
(617) 824-4011

HBM Associates, Inc.
7100 E. Bellview Avenue
Suite 201
Englewood, Colorado 80111
(303) 796-7506

HBM Associates, Inc.
7100 E. Bellview Avenue
Suite 201
Englewood, Colorado 80111
(303) 796-7506

Higbee Gaskets and Sealing Products, Inc.
4083 New Court Avenue
P.O. Box 6449
Syracuse, New York 13217
(315) 432-8021 or
(800) 255-4800 (outside NY)

High Country Shielding Corporation
4515 Anaheim Avenue N.E.
Albuquerque, New Mexico 87113
(505) 828-2558

High Country Shielding Corporation
P.O. Box 144
Warren, Vermont 05674
(802) 496-2747

Huber & Suhner, Inc.
500 West Cummings Park
Woburn, Massachusetts 01801
(617) 938-1335

INCO Selective Surfaces
Wiggin Street
Ladywood, Birmingham
B16 OAJ England
021 454 4871

Industrial Acoustics Co.
1160 Commerce Avenue
Bronx, New York 10462
(212) 931-8000

Instrument Specialties Company, Inc.
Delaware Water Gap, Pennsylvania 18327
(717) 424-8510

Insul-Fab of Texas
9101 Chancellor Row
Dallas, Texas 75247
(214) 951-7441

Integrated Facilities, Inc.
5250 W. Century Boulevard
#402
Los Angeles, California 90045
(213) 216-6625

Inter-Technical Group (The), Inc.
One Bridge St.
P.O. Box 23
Irvington, New York 10533
(914) 591-8822

International Paper
Longmeadow Road
Tuxedo, New York 10987
(914) 577-7440

Irving B. Moore Corporation
30 Cambridgepark Drive
Cambridge, Massachusetts 02140
(617) 491-0100

Ishizaki Trading, Inc.
20-4
Nakaitabashi
Itabashi-Ku
Tokyo 173 Japan
(03) 816-6582

Isolation Technology, Inc.
P.O. Box 22808
Santa Barbara, California 93121
(805) 683-4445

ITT Cannon
666 E. Dyer Road
Santa Ana, California 92702
(714) 557-4700

J. M. Benson Company, Inc.
6601 Ammendale Road
Beltsville, Maryland 20705
(301) 595-5220

Jacques Dubois
82 Rue Auguste BADIN
76360 BARENTIN
France
T35 912555

Kaltbrunner AG
Schlachthausstr 58
2540 Grenchen
Switzerland 2540
065-52-11-12

Kantek, Inc.
45 White Street
P.O. Box 2092
New York, New York 10013-2092
(212) 925-7850

Knitmesh, Ltd.
Sanderstead Station Approach
South Croydon, Surrey
England CR2 OYY
01-657-0107

Kyle Technology Corporation
3500 N.W. Stewart Parkway
Roseburg, Oregon 97470
(503) 672-5953

L.E.A.D.
92-98 Boulevard Victor Hugo
Clichy
France 92115
(1) 47-39-24-44

LAM-A-SHIELD, Inc.
1700 London Road
Dept. 5143
Cleveland, Ohio 44112
(216) 481-2733

Lamart Corporation
16 Richmond Street
Clifton, New Jersey 07015
(201) 772-6262

Laminaire Corporation
960 E. Hazelwood Avenue
Rahway, New Jersey 07065
(800) 777-4500

LectroMagnetics, Inc.
6056 W. Jefferson Boulevard
Los Angeles, California 90016
(213) 870-9383

Lindgren R.F. Enclosures, Inc.
1228 Capitol DriveT
Addison, Illinois 60101
(312) 628-9100

LNP Engineering Plastics
412 King Street
Malvern, Pennsylvania 19355
(215) 644-5200

Lourich Company (The)
P.O. Box 43
Sussex, New Jersey 07461
(201) 875-7749

LPS Industries, Inc.
113 N. 13th Street
Newark, New Jersey 07107
(201) 485-4400

Magnetic Radiation Laboratories, Inc.
92A N. Lively Boulevard
Elk Grove Village, Illinois 60007
(312) 437-5200

Magnetic Shield Corporation
Perfection Mica Co.
740 N. Thomas Drive
Bensenville, Illinois 60106 USA
(312) 766-7800

Magnetics Division
Spang & Co.
P.O. Box 391
Butler, Pennsylvania 16003
(412) 282-8282

Markel Corporation
School Lane
P.O. Box 752
Norristown, Pennsylvania 19404
(215) 272-8960

Master Bond, Inc.
154 Hobart Street
Hackensack, New Jersey 07601
(201) 343-8983

Melmat, Inc.
1400 W. 240th Street
Harbor City, California 90710
(213) 325-1625

Micro-Circuits Co., Inc.
10792 Maudlin Road
Route 1T
P.O. Box 518
New Buffalo, Michigan 49117
(616) 469-2727

Micron Air Purification, Inc.
57 Cooley Drive
Longmeadow, Massachusetts 01106-1397
(413) 567-0372

Middco Connectors/Middleburg Corporation
95 Albany Turnpike
Canton, Connecticut 06019
(203) 693-8361

Middleburg Connectors
(see Middco Connectors)

Miller-Stephenson Chemical Co., Inc.
George Washington Highway
Danbury, Connecticut 06810
(203) 743-4447

Molded Fiberglass Tray Co.
E. Erie Street
Linesville, Pennsylvania 16424
(814) 683-4500

Moxness Products, Inc.
1914 Indiana Street
Racine, Wisconsin 53405
(414) 554-5050

MPE Ltd.
Liverpool
P.O. Box 11
Hammond Road, Kirkby Industrial Estate
Liverpool, Merseyside
L33 7UL United Kingdom
(051) 547-2600

Mushield Products Division
Bomco Inc.
Route 128
Blackburn Circle
Gloucester, Massachusetts 01930-2294
(617) 283-9000

Musorb
92/98 Boulevard Victor Hugo
92110 Clichy
France
(1) 47 37 17 73

Neo-Tec Plating, Inc.
939 Industrial Parkway
Plainwell, Michigan 49080
(616) 685-9543
James T. Plemmons.

NRD, Inc.
2937 Alt Boulevard North
Grand Island, New York 14072
(716) 773-7634

OCLI Optical Coating Laboratory, Inc.
2789 Northpoint Parkway
Santa Rosa, California 95407-7397
(707) 525-7436

Omega Shielding Products
1384 Pompton Avenue
Cedar Grove, New Jersey 07009
(201) 890-7455

Optical & Conductive Coatings
428 N. Buchanan Circle, #8
Pacheco, California 94553
(415) 798-6066

Optima Enclosures
2166 Mountain Industrial Boulevard
Tucker, Georgia 30084-5088
(404) 496-4025

ORCUS
Box 16022
Kansas City, Missouri 64112
(913) 782-9781

Overly Manufacturing Co.
574 West Otterman Street
Greensburg, Pennsylvania 15601
(412) 834-7300

Pacific West Electronics
1370 Logan Avenue
Suite A/B
Costa Mesa, California 92626
(714) 957-3158

PacTec Division of LaFrance Corporation
Enterprise & Executive Avenues
Philadelphia, Pennsylvania 19153
(215) 365-8400

PAR-METAL Products, Inc.
Par Industrial Complex
1238 Atlantic Avenue
Dept EMC
Brooklyn, New York 11216
(718) 774-5900

Parker Seal Group
P.O. Box 11751
Lexington, Kentucky 40512
(606) 269-6911

Pawling Corporation
157 Charles Colman Boulevard
Pawling, New York 12564
(914) 855-1000

Permacel
U.S. Highway 1
P.O. Box 671
New Brunswick, New Jersey 08903
(201) 418-2752

Phillips Plastics Corporation
One Plastics Drive
P.O. Box 168
Phillips, Wisconsin 54555
(715) 339-2161

Photon Technologies Inc.
437-A Aldo Avenue
Santa Clara, California 95054-2302
(408) 727-0754

Pilkington Glass, Ltd.
New Technology Business Unit
Cowley Hill Works
St. Helens
England WA10 3TT
0744 69 5677

Plastic-Plate, Inc.
4251 Brockton Court, S.E.
Grand Rapids, Michigan 49508
(616) 698-2030

Plastic Systems, Inc.
50 Brigham Street
Marlboro, Massachusetts 01752
(617) 485-7390

Plessey Microwave Materials
9630 Ridgehaven Court
Suite B
San Diego, California 92123
(619) 571-7715

Potters Industries, Inc.T
377 Route 17
Hasbrouck Heights, New Jersey 07604
(201) 288-4700

QRP, Inc.
P.O. Box 1814
Santa Cruz, California 95061
(408) 425-1667

R.F.I. Industries Pty, Ltd.
54 Holloway Drive
Bayswater, Vic
Australia 3153
3-762-6733

Rantec Division
Emerson Electric Company
7659 Alabama Avenue
Canoga Park, California 91304
(818) 704-6944

Rantec Division Emerson Electric Co.
Anechoic/EMI Systems
7659 Alabama Avenue
Canoga Park, California 91304
(818) 704-6944

Ray Proof Shielding System Corporation
50 Keeler Avenue
Norwalk, Connecticut 06856-5060
(203) 838-4555

Reliable Electric/Utility Products
A unit of Reliance Comm/Tec
11333 Addison Street
Franklin Park, Illinois 60131
(312) 455-8010

REMP AG
Jungfrauweg 2
Laupen
Switzerland CH-3177
41 31 948833

Rockford Engineering Services, Inc.
9959 Calaveras Blvd.
P.O. Box 543
Sunol, California 94586-0543
(415) 862-9012

Rogers Corporation
One Technology Drive
Rogers, Connecticut 06263
(203) 774-9605

RTP Company
580 E. Front Street
Winona, Minnesota 55987
(507) 454-6900

Sabritec
1733 Kaiser Avenue
Irvine, California 92714
(714) 660-1970

Santa Fe Textiles
3151 W. Adams
Santa Ana, California 92704
(714) 751-2733

Schlegel Corporation
P.O. Box 23197
Rochester, New York 14692
(800) 828-6237
(800) 462-1727 (in NY State)

Scientific Hardware Systems
160B Constitution Drive
Menlo Park, California 94025
(415) 322-0393

Sentinel Foam Products
130 North Street
Hyannis, Massachusetts 02601
(617) 775-5220

Set Point, Inc.
69 Elm Street
Foxboro, Massachusetts 02035
(617) 543-3800

SFS Industries, Inc.
13767 Freeway Drive
Santa Fe Springs, California 90670
(213) 926-6556

Shipley Company
2300 Washington Street
Newton, Massachusetts 02162
(617) 969-5500

Sierracin Corporation
Intrex Division
12780 San Fernando Road
Sylmar, California 91342
(818) 364-1511

SpecTape, Inc.
2771 Circleport Drive
Erlanger, Kentucky 41018
(606) 283-2044

Spectrum Control, Inc.
2185 W. 8th Street
Erie, Pennsylvania 16505
(814) 455-0966

Spira Manufacturing Corporation
12721 Saticoy St. South
Unit B
North Hollywood, California 91605
(818) 764-8222

Sprague Electric Co.
Marketing Communications
41 Hampden Road
P.O. Box 9102
Mansfield, Massachusetts 02048-9012
(617) 339-8900

Spraylat Corporation
716 S. Columbus Avenue
Mt. Vernon, New York 10550
(914) 699-3030

Stackbin Corporation
1000 Albion Road
P.O. Box 397
Lincoln, Rhode Island 02865
(800) 346-1018 or (401) 333-1600

Stauffer Japan Limited
Godo Kaikan Bldg. 4F
3-27 Kioicho Chiyoda-Ku
Tokyo, Japan 102
81-3-261-8967

Stetron International, Inc.
90 Broadway
Buffalo, New York 14203
(716) 854-3443

Stockwell Rubber Co., Inc.
4749 Tolbut Street
Philadelphia, Pennsylvania 19136
(215) 335-3005

svt Brandschutz Vertriebsgesellschaft mbH International
Maschener Schutzenstrasse 45
Seevetal 3, West Germany 2105
04105/830 15

TAFA, Inc.
P.O. Box 1157
Concord, New Hampshire 03301
(603) 224-9585

TBA Industrial Products Ltd.
P.O. Box 40
Rochdale, Lancashire
England OL12 7EQ
0706-56500

TDK Corporation of America
1600 Feehanville Drive
Mount Prospect, Illinois 60056
(312) 803-6100

Tech-Etch, Inc.
45 Aldrin Road
Plymouth, Massachusetts 02360
(617) 747-0300

Tech Spray Inc.
P.O. Box 949
Amarillo, Texas 79105
(806) 372-8523

Techni-Tool Inc.
5 Apollo Road
Plymouth Meeting, Pennsylvania 19462
(215) 825-4990

Technical Environmental Control, Inc.
7956 Cessna Avenue
Gaithersburg, Maryland 20879
(301) 948-5911

Tecknit
A TWP Company
129 Dermody Street
Cranford, New Jersey 07016
(201) 272-5500

Texas Spectrum Electronics, Inc.
9110 Autobahn Drive
Dallas, Texas 75237
(214) 296-3699

Thomas & Betts Corporation
1001 Frontier Road
Brdigewater, NJ 08807
(201) 685-1600

Tokin America Inc.
155 Nicholson Lane
San Jose, California 95134
(408) 432-8020

Topflight Corporation
Box 2847
York, Pennsylvania 17405-2847
(800) 233-9386

Transmet Corporation
4290 Perimeter Drive
Columbus, Ohio 43228
(614) 276-5522

Trumbull Testing Labs, Inc.
8300 Corona Loop, N.E.
Albuquerque, New Mexico 87113
(505) 823-1818

TRW Connector Division
1501 Morse Avenue
Elk Grove Village, Illinois 60007
(312) 981-6000

Uniroyal Plastics Co.
Thermoplastic Sheet Division
312 N. Hill Street
P.O. Box 2000
Mishawaka, Indiana 46544
(219) 256-8649

United Foam Plastics Corporation
170-D E. Main Street
Georgetown, Massachusetts 01833
(617) 352-2200

Vacuumschmelze
186 Wood Avenue South
Iselin, New Jersey 08830
(201) 494-3530

Vanguard Products Corporation
144 Old Brookfield Road
Danbury, Connecticut 06811
(203) 744-7265

Varinit Corporation
Box 6602
Greenville, South Carolina 29606-6602
(803) 277-7884

Vermillion Enterprises Inc.
4225 West 31st Street South
P.O. Box 12147
Wichita, Kansas 67277
(316) 942-8238

W.H. Brady Co.
Thin Film Products
P.O. Box 571
8225 W.Parkland Court
Milwaukee, Wisconsin 53201
(414) 355-8300

Walter G. Legge Co., Inc.
444 Central Avenue
Peekskill, New York 10566
(800) 345-3443

Weinschel Engineering Co., Inc.
One Weinschel Lane
Gaithersburg, Maryland 20878
(301) 948-3434

WESCORP
144 S. Whisman Road
Mountain View, California 94041
(415) 969-7717

Westek
400 Rolyn Place
Arcadia, California 91006
(818) 446-4444

WEZ Plastic Industries Ltd.
Oberentfelden
5036 Switzerland
(064) 45 11 66

Wilson-Fiberfil International
Akzo Group Member
2267 West Mill Road
P.O. Box 3333
Evansville, Indiana 47732
(812) 424-3831

Wyle Laboratories Electronic Enclosures
225 S. Aviation Boulevard
El Segundo, California 90245
(213) 643-8181

Zippertubing Co.
P.O. Box 61129
Los Angeles, California 90061
(213) 321-3901

Zymet, Inc.
7 Great Meadow Lane
E. Hanover, New Jersey 07936
(201) 428-5245

Index

Other Books Published by ICT

1. Carstensen, Russell V., *EMI Control in Boats and Ships,* 1979.
2. Denny, Hugh W., *Grounding for Control of EMI,* 1983.
3. Duff, Dr. William G., *A Handbook on Mobile Communications,* 1980.
4. Duff, Dr. William G. and White, Donald R.J., Volume 5, *Electromagnetic Interference Prediction & Analysis Techniques,* 1972.
5. Feher, Dr. Kamilo, *Digital Modulation Techniques in an Interference Environment,* 1977.
6. Gabrielson, Bruce C., *The Aerospace Engineer's Handbook of Lightning Protection,* 1987.
7. Gard, Michael F., *Electromagnetic Interference Control in Medical Electronics,* 1979.
8. Georgopoulos, Dr. Chris J., *Fiber Optics and Optical Isolators,* 1982.
9. Georgopoulos, Dr. Chris J., *Interference Control in Cable and Device Interfaces,* 1987.
10. Ghose, Rabindra N., *EMP Environment and System Hardness Design,* 1983.
11. Hart, William C. and Malone, Edgar W., *Lightning and Lightning Protection,* 1979.
12. Herman, John R., *Electromagnetic Ambients and Man-Made Noise,* 1979.
13. Hill, James S. and White, Donald R.J., Volume 6, *Electromagnetic Interference Specifications, Standards & Regulations,* 1975.
14. Jansky, Donald M., *Spectrum Management Techniques,* 1977.
15. Mardiguian, Michel, *Interference Control in Computers and Microprocessor-Based Equipment,* 1984.
16. Mardiguian, Michel, *Electrostatic Discharge—Understand, Simulate and Fix ESD Problems,* 1985.
17. Mardiguian, Michel, *How to Control Electrical Noise,* 1983.
18. Smith, Albert A., *Coupling of External Electromagnetic Fields to Transmission Lines,* 1986.
19. White, Donald R.J., *A Handbook on Electromagnetic Shielding Materials and Performance,* 1980.
20. White, Donald R.J., *Electrical Filters—Synthesis, Design & Applications,* 1980.
21. White, Donald R.J., *EMI Control in the Design of Printed Circuit Boards and Backplanes,* 1982. (Also available in French.)
22. White, Donald R.J. and Mardiguian, Michel, *EMI Control Methodology & Procedures,* 1985.
23. White, Donald R.J., Volume 1, *Electrical Noise and EMI Specifications,* 1971.
24. White, Donald R.J., Volume 2, *Electromagnetic Interference Test Methods and Procedures,* 1980.
25. White, Donald, R.J., Volume 3, *Electromagnetic Interference Control Methods & Techniques,* 1973.
26. White, Donald R.J., Volume 4, *Electromagnetic Interference Test Instrumentation Systems,* 1980.
27. Duff, William G., and White, Donald R.J., Volume 5, *Prediction and Analysis Techniques,* 1970.
28. White, Donald R.J., Volume 6, *EMI Specifications, Standards and Regulations,* 1973.
29. White, Donald R.J., *Shielding Design Methodology and Procedures,* 1986.
30. *EMC Technology 1982 Anthology*
31. *EMC EXPO Records 1986, 1987, 1988*

EMC Books

HANDBOOKS

ICT provides over 30 technical handbooks on EMI/EMC and related disciplines. Written by practicing experts in their field each book is designed to go beyond the tutorial by providing the reader with practical applications, illustrative examples as well as tested and proven methodologies. Each book is packed with illustrations, graphs, tables and math models, and is writen in a clear, consise format to assure the reader understanding and immediate use of the material.

GENERAL TITLES AND APPLICATIONS

Electromagnetic Shielding
Electrical Filters
ESD - Understand, Simulate and Fix
EMC Pocket Primer
EMP Environment and System Hardness Design
EXPO Symposium Records
How to Contol Electrical Noise
Grounding for the Control of EMI
Lightning and Lightning Protection

SPECIFIC TITLES AND APPLICATIONS

Aerospace Engineers Handbook of Lightning Protection, Coupling of External Electromagnetic Fields to Transmition Lines, EMI Control in Computers and Microprocessor-Based Equipment, EMI Control in the Design of PCBs and Backplanes, Fiber Optics and Optical Isolators.

THE MUST EMI/EMC LIBRARY

(1) EMI Control Methodology and Procedures *and* (2) Shielding Design Methodology and Procedures.

THE EMC SOURCE
(A 12-Volume EMI/EMC Series)

Vol 1 Fundamentals of EMC
Vol 2 Grounding and Bonding
Vol 3 Electromagnetic Shielding
Vol 4 Filters & Power Conditioning
Vol 5 EMC in Components & Devices
Vol 6 EMI Test Methods/Procedures
Vol 7 EMC in Telecommunications
Vol 8 EMI Control Methodology
Vol 9 USA Commercial Standards
Vol 10 Int'l Commercial Standards
Vol 11 USA MIL-STDs Part 1
Vol 12 USA MIL-STDs Part 2

for more information ...

Interference Control Technologies
PO Box D
Gainesville, Va 22065
703-347-0030

ICT

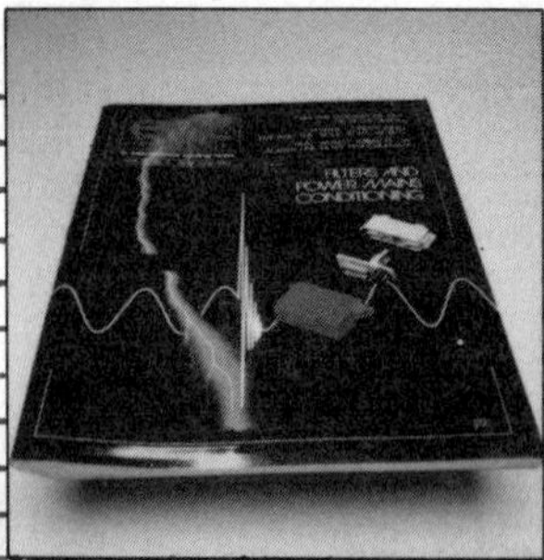
FILTERS AND POWER MAINS CONDITIONING